ADVERSE EFFECTS OF FOODS

ADVERSE EFFECTS OF FOODS

Edited by
E. F. Patrice Jelliffe
and
Derrick B. Jelliffe
University of California
Los Angeles, California

PLENUM PRESS · NEW YORK AND LONDON

Library of Congress Cataloging in Publication Data

Main entry under title:

Adverse effects of foods.

Includes bibliographical references and index.
1. Nutrition disorders. 2. Food poisoning. 3. Food contamination. I. Jelliffe, E. F. Patrice. II. Jelliffe, Derrick Brian.

RC620.5.A38 615.9′54 82-566
ISBN 0-306-40870-8 AACR2

A Division of Plenum Publishing Corporation
233 Spring Street, New York, N.Y. 10013

Printed in the United States of America

CONTRIBUTORS

LILIAN AFTERGOOD, *Nutritional Sciences and Environmental Health Division, School of Public Health, University of California, Los Angeles, California 90024*

KAMAL AHMAD, *Institute of Nutrition and Food Science, University of Dacca, Dacca 2, Bangladesh*

ROSLYN ALFIN-SLATER, *Nutritional Sciences and Environmental Health Division, School of Public Health, University of California, Los Angeles, California 90024*

LAMAN AMIN-ZAKI, *Ministry of Health, Central Hospital, P. O. Box 233, Abu Dhabi, United Arab Emirates*

FRANK APEAGYEI, *Human Nutrition Unit, Biochemistry Department and Commonwealth, Institute of Health, Sydney University, New South Wales 2006, Australia*

ROY BROWN, *Mount Sinai School of Medicine, New York, New York 10029*

HILDE BRUCH, *Department of Psychiatry, Baylor College of Medicine, Texas Medical Center, Houston, Texas 77030*

DENIS P. BURKITT, *Unit of Geographical Pathology, St. Thomas's Hospital, London SE1 United Kingdom*

SYLVIA J. DARKE, *Departments of Health and Social Security, London, United Kingdom*

JOHANNA DWYER, *Frances Stern Nutrition Center, New England Medical Center Hospital, Department of Medicine and Community Health, and Tufts Medical School, Boston, Massachusetts 02111*

SHIRLEY L. FANNIN, *Acute Communicable Disease Control, Public Health Services, Los Angeles, California 90012*

R. J. GILBERT, *Food Hygiene Laboratory, Central Public Health Laboratory, London NW9 5HT, United Kingdom*

MARTIN GORE, *University College Hospital, London, United Kingdom*

LOUIS E. GRIVETTI, *Department of Nutrition, University of California, Davis, California 95616*

J. MICHAEL GURNEY, *Caribbean Food and Nutrition Institute, P.O. Box 140, Kingston 7, Jamaica*

LEIF HAMBRAEUS, *Institute of Nutrition, Faculty of Medicine, University of Uppsala, P.O. Box 551, S-751 22 Uppsala, Sweden*

MASAZUMI HARADA, *Department of Neuropsychiatry, Institute of Constitutional Medicine, Kumamoto University, Kumamoto City, Kuhonji 4-chome, Japan*

TERUO HONDA, *Department of Pediatrics, Tokyo Medical College, Tokyo, Japan*

R. W. HORNABROOK, *Department of Medicine, Wellington Clinical School of Medicine, Wellington, New Zealand*

MALATI A. JADHAV, *Department of Child Health, Medical School, Christian Medical College and Hospital, Vellore 632004, Tamil Nadu, India*

DERRICK B. JELLIFFE, *Population, Family and International Health Division, School of Public Health, University of California, Los Angeles, California 90024*

E. F. PATRICE JELLIFFE, *Population, Family and International Health Division, School of Public Health, University of California, Los Angeles, California 90024*

MELVIN LEE, *Division of Human Nutrition, School of Home Economics, University of British Columbia, Vancouver, British Columbia, V6T 1W5 Canada*

JOSEPH LEICHTER, *Division of Human Nutrition, School of Home Economics, University of British Columbia, Vancouver, British Columbia, V6T 1W5 Canada*

D. G. LINDSAY, *Food Science Division, Ministry of Agriculture, Fisheries, and Food, London SW1P 2AE, United Kingdom*

MORRIS A. LIPTON, *Department of Psychiatry and Biological Science Research Center, University of North Carolina School of Medicine, Chapel Hill, North Carolina 27514*

TATSUO MATSUMURA, *Department of Pediatrics, School of Medicine, Gunma University, Maebashi, Gunma, Japan*

LOIS D. MCBEAN, *National Dairy Council, Rosemont, Illinois 60018*

DONALD S. MCLAREN, *Department of Medicine, Royal Infirmary, University of Edinburgh, EH9 3YW Scotland*

PANATA MIGASENA, *Department of Tropical Nutrition and Food Sciences, Faculty of Tropical Medicine, Mahidol University, Bangkok 4, Thailand*

YONGYOT MONSEREENUSORN, *Department of Biology, Faculty of Science, Ramakamhaeng University, Banggapi, Bangkok 24, Thailand*

ANNE B. MURRAY, *Department of Medicine, University of Minnesota Medical School, Minneapolis, Minnesota 55455*

CHRISTOPHER J. MURRAY, *Department of Medicine, University of Minnesota Medical School, Minneapolis, Minnesota 55455*

MEGAN B. MURRAY, *Department of Medicine, University of Minnesota Medical School, Minneapolis, Minnesota 55455*

M. JOHN MURRAY, *Department of Medicine, University of Minnesota Medical School, Minneapolis, Minnesota 55455*

NIGEL J. MURRAY, *Department of Medicine, University of Minnesota Medical School, Minneapolis, Minnesota 55455*

MOHAMMAD G. MUSTAFA, *Nutritional Sciences and Environmental Health Division, School of Public Health, University of California, Los Angeles, California 90024*

CHARLOTTE G. NEUMANN, *Population, Family and International Health Division, School of Public Health, University of California, Los Angeles, California 90024*

KHIN KYI NYUNT, *Nutrition Research Division, Department of Medical Research, Rangoon, Burma*

MEINHARD ROBINOW, *Department of Pediatrics, School of Medicine, Wright State University, Dayton, Ohio 45404*

JOHN R. K. ROBSON, *Department of Family Practice, Medical University of South Carolina, Charleston, South Carolina 29403*

DAPHNE A. ROE, *Division of Nutritional Sciences, Cornell University, Ithaca, New York 14853*

HOSSAIN A. RONAGHY, *Formerly Department of Community Medicine, Medical School, Shiraz University, Shiraz, Iran. Present address: Trauma Medical Clinic, San Diego, California; and School of Public Health, University of California, Los Angeles, California 90024*

SHANE ROY, III, *Department of Pediatrics, University of Tennessee Center for the Health Sciences, Memphis, Tennessee 38163*

CHARLES L. SENN, *Senn Environmental Consultant Associates, Inc.,Los Angeles, California 90032*

J. C. SHERLOCK, *Food Science Division, Ministry of Agriculture, Fisheries, and Food, London SW1P 2AE, United Kingdom*

ELWOOD W. SPECKMANN, *National Dairy Council, Rosemont, Illinois 60018*

S. G. SRIKANTIA, *National Institute of Nutrition, Indian Council of Medical Research, Hyderabad, 500007, India*

K. L. STUART, *Commonwealth Secretariat, Marlborough House, London SW1Y 5HX, United Kingdom*

A. STEWART TRUSWELL, *Human Nutrition Unit, Biochemistry Department and Commonwealth Institute of Health, Sydney University, New South Wales 2006, Australia*

P. C. B. TURNBULL, *Food Hygiene Laboratory, Central Public Health Laboratory, London NW9 5HT, United Kingdom*

MARK L. WAHLQVIST, *Department of Human Nutrition, Deakin University, Victoria, Australia*

BRIAN A. WHARTON, *Sorrento and Birmingham Maternity Hospitals, Birmingham, United Kingdom*

JEANINE C. WHELESS, *Biological Science Research Center, University of North Carolina School of Medicine, Chapel Hill, North Carolina 27514*

JOHN R. WHITAKER, *Department of Food Science and Technology, University of California, Davis, California 95616*

CHRISTINE S. WILSON, *Department of Epidemiology and International Health, University of California, San Francisco, California 94143*

FAITH W. WINTER, *Mount Sinai School of Medicine, New York, New York 10029*

KUNG-LAI ZHANG, *Beijing Medical College, Beijing, China.*

CONTENTS

III. DIETARY ADDITIVES

IV. SOCIAL TOXICANTS

V. INFECTIONS

VI. GENETICS

VII. ALLERGY

VIII. FOOD ACCIDENTS

IX. EXCESSIVE INTAKES

X. DEFICIENT INTAKES: NATURALLY OCCURRING

XI. DEFICIENT INTAKES: CULTURALLY INDUCED

XII. NUTRITIONAL IMBALANCE

XIII. CONSEQUENCES OF INFANT FEEDING

XIV. AVOIDANCE OF ADVERSE DIETS

OVERVIEW

Derrick B. Jelliffe and E. F. Patrice Jelliffe

As with all living organisms, human beings cannot survive without the chemicals needed to support their basic metabolism and their daily activity. Growing children and pregnant women obviously have higher nutrient requirements, as do breast feeding mothers.

With the exceptions of oxygen, essentially a key nutrient breathed into the system, and vitamin D, synthesized by the skin's exposure to the sun's ultraviolet rays, basic nutrients are obtained naturally from foods ingested, digested, absorbed, and metabolized—or stored or excreted, if taken in amounts larger than needed.

Good health or ill health can, of course, be the direct results of the adequacy or otherwise of the nutrients taken in in the diet. Likewise, the individual's ability to deal with the inevitable infections resulting from exposure to the myriad microflora that continually surround all species, including human beings, depends in considerable measure on the symbiotic relationship of his nutritional status and his ability to resist such infections.

Since the emergence of early man—probably between one to three million years ago—most of the life of our species has been spent as gatherer–hunters. The development of pastoralism and agriculture were very recent in terms of evolutionary time—probably some 10,000 years ago. The rise of large-scale cities, with their special dependency on an efficient system of agricultural food production, transport, and marketing, is still more recent. Even the most ancient of cities, such as Babylon, date back only three thousand years. Finally, huge, industrialized cities are chronologically contemporary, having arisen as a result of the Industrial Revolution, which gained its initial massive momentum in Britain in the 19th century. The accompanying urbanization began in Europe and North America, but after World War II has become a dominant demographic feature throughout almost all parts of the world, including less developed Third World countries. Indeed, in the past few decades untram-

Derrick B. Jelliffe and E. F. Patrice Jelliffe • Population, Family and International Health Division, School of Public Health, University of California, Los Angeles, California 90024.

melled migration to the "septic fringes" of urban centers has been and continues to be overwhelming, accompanied by enormous problems with food supplies, employment, housing, sanitation and family structure (including infant feeding).

Initially these changes were gradual, but they have attained runaway momentum and speed in the past twenty years, with tremendous effects on human dietary habits. Large populations have moved from predominantly food-growing, often subsistence ways of life to a cash economy, which depends, of course, on the availability of food in urban areas and ultimately on the effectiveness of large-scale agricultural production, storage and distribution.

Without lapsing into a sentimental vision of the past, the diet of our gatherer–hunter ancestors appears to have been inherently balanced, usually based on a very wide range of such foods as roots, berries, seeds, nuts, and leaves, together with smaller, or occasionally larger, amounts of meat when hunting was successful. The basic omnivorous habit of humankind seems to be indicated by the types of teeth present—suitable both for grinding and tearing—and on the range of intestinal enzymes found. Mobility was the essential fact of life for *Homo erectus,* with, as a corollary, the availability of very large areas of land over which to wander and hunt. Weather changes obviously could affect the food supply, but the mobility of these small bands would enable migration to less affected areas. The basic diet was, it would seem, most usually *mixed,* consisting of *vegetable products* which had been gathered, and lesser amounts of *animal products.*

By contrast, agricultural communities automatically narrowed the range of foods available to them by their lack of mobility—and usually concentrated their efforts on the most locally productive staple. This increased the amount of available food provided the weather was propitious, especially as regards rainfall.

Most animals, including humankind, come to recognize that some foods have adverse effects—that is, that they are poisonous and to be avoided. However, this does not seem to be absolute, as exemplified in the occurrence of plant poisoning in animals that are range-fed. This may perhaps be more common in domestic animals that have not had the benefit of the maternal training and guidance found in the wild.

Certainly, for early man and for communities still living as gatherer–hunters (Jelliffe *et al.,* 1962), a knowledge of certain foods to avoid and seasons at which others should be eaten or not eaten is a widespread part of the folklore and practical training for life. Nevertheless, even under these circumstances things can go wrong. For example, risks may be taken in eating the wrong foods during famine seasons, while children may accidentally poison themselves with toxic foods, such as bright-colored berries.

In other words, naturally occurring toxicants are always present in dietaries. Well-known examples include solanine in potatoes and hydrocyanic acid in cassava (manioc). These are usually eaten in very small quantities, and therefore are of little significance, or alternatively, they have been recognized as problems by a particular community, and removed or neutralized through ingenious, time-tested techniques. However, sometimes such toxicants are literally unavoidable. The classic case is with the mineral, fluoride; in parts of India and other areas in the world, the only year-round, naturally occurring water supply contains extremely high levels of fluoride, and an excessive intake is the consequence of normal water and foodstuff consumption.

Exotic-seeming examples of naturally occurring toxicants are to be found in many parts of the world. For example, the *ackee (Blighia sapidia)* is one of the ingredients of the national dish of Jamaica, but is recognized to be poisonous if eaten unripe. In Burma, the *djenkol* bean is a prized, but highly acquired taste, whose minor and occasional major ill effects are well recognized by the population.

As a consequence of settling into villages and cities, the acquisition of nonmobile goods rather than the minimum light-weight equipment feasible for the gatherer–hunter, became part of the way of life. This meant, for example, that the implements used in the preparation, processing, storage, and cooking of food themselves became possible sources for environmental toxicants. The classic example in this regard is the lead-glazed pottery of ancient Rome, which continues to be a problem in village-level pottery making in the eastern Mediterranean (Accra *et al.*, 1981).

While toxicants of various sorts have been potential hazards ever since humankind formed settlements, it is only in recent years that *Homo technologicus* has moved by a quantum jump into a way of life in which innumerable environmental chemicals have been introduced into a complex new system of life in which food is produced by agribusiness and processed, preserved, and marketed by the food industry. The benefits of this new technological way of life were appreciated early on—that is, an increased, more secure, more abundant, and more constant supply of foods, especially for cities in industrialized countries. Only comparatively recently—that is, in the last 10–20 years—has it become recognized that this whole process has been accompanied by the widespread use of large numbers of substances in agriculture and in food processing about which too little is known. In fact, probably one of the most unfortunate manifestations of modern man's technological *hubris* was that this was not appreciated early on, i.e., immediately following World War II. Nowadays, the number of such substances is very great and the problem of assessing their adverse effects is infinitely complicated by their number and by their possible interaction one with another. Also, in many cases a chronic situation may have arisen from nonfood industries with other environmental pollutants, such as the PCBs, which are now widespread in the environment of certain industrial countries (Webb, 1975).

Although often rightly thought of as issues of particular concern in industrialized countries, it must be noted that in many so-called less technically developed countries the urge to "catch up" and the lack of stringently applied laws have made pollution *more* of a problem than in industrialized countries.

For example, Mexico City has been stated to be among the worst cities in the world in terms of air pollution, and the highest levels of DDT in humans were recorded in Guatemalan agricultural laborers some years back, as a result of crop dusting. Furthermore, in what is now the most powerful industrial country in the world, Japan, a forceful, but very rapid, transition into industrialization has had some specially striking ill consequences from environmental toxicants, notably from mercury and cadmium.

All over the world, the availability of potent chemicals, used in many aspects of modern life, have the potential for acute disasters, as well as the much more widespread and cumulative chronic consequences. Classic and tragic examples include the outbreak of poisoning with mercury-contaminated grain in Iraq and the well-known "Firemaster" cattle poisoning episode in Michigan. A recent instance is the outbreak of polyneuropathy in Sri Lanka from contamination of *gingili* cooking oil with TCP (tri-cresyl phosphates) (Senanayake and Jeyaratnam, 1981).

As would be expected, much controversy surrounds the significance (or otherwise) of various substances employed in modern food production and preservation, and their benefits and real or potential hazards. Two instances that have generated particular controversy are the possible role of additives in the "hyperkinesis syndrome" of young children, and the significance of monosodium glutamate in the so-called "Chinese Restaurant Syndrome."

"Social toxicants" comprise a quite separate issue. In all cultures, it would seem, humans have discovered foods or drugs that can act as stimulants to the brain to enable them to think better, increase their physical endurance, and blunt the often difficult and disagreeable facts of life. The ill consequences of such social toxicants as alcohol and coffee have been much in the forefront of attention recently.

In particular, the recent rise in alcoholism throughout the world, both in industrialized and less technically developed countries, and the increasing involvement of women and adolescents as well as adult males, has drawn more and more attention to this traditional escape mechanism. Much concern is increasingly generated concerning the effect of alcohol abuse on the body in general, but particularly on the liver. Recent work clearly shows that this abuse occurs as much in the well-nourished individual as in the "skid row" societal reject. Also examined in recent times, the particular ill effect of alcohol on the pregnant woman's fetus has vividly brought home the vulnerability of the young human organism *in utero*. In addition, early reports on the toxic effects of excessive ethyl alcohol on the brain have been supported by studies from several parts of the world detailing severe cerebral syndromes occurring in heavy drinkers.

The intestinal canal is obviously a major portal for bodily infections. This is perhaps emphasized by the fact that the human body essentially surrounds a long and complex alimentary tube of over 30 ft in length, which is normally filled with billions of potentially harmful microorganisms, themselves playing an important role in digestion and nutrition.

Obviously, foods eaten can introduce harmful organisms to the alimentary canal with resulting local disease, often manifested as diarrhea and vomiting, or can introduce dangerous organisms into the body itself. For instance, disseminated tuberculosis can occur from the drinking of cow's milk from animals infected with bovine disease—and this indeed was a major problem prior to the almost universal pasteurization of milk in Western countries. There has been a considerable amount of recent work concerning the ill effects of newly recognized bacteria and viruses, whose presence has become known through innovative laboratory techniques. These have enabled, for example, the identification of various enteroviruses that are known to cause diarrhea very commonly in young children.

The fact that food habits can influence infections is obvious, but it is brought home forcefully to the urbanized Westerner by the occurrence of bizarre-seeming infections in distant places, such as those reported from New Guinea—*kuru* and *pigpel.* At the same time, the existence of "improbable infections" continue to come to light both in industrialized and Third World countries. Recent examples include infant botulism in California and elsewhere and adult botulism after the consumption of fresh white ants in Kenya (Nightingale and Ayim, 1980).

In considering the adverse effects of food, the question of the "seed" and the "soil" needs to be borne in mind. In other words, some individuals or, indeed, some groups of humankind may find it difficult to deal with certain types of food because of inherited errors of metabolism. Sometimes these may be relatively rare, but in some areas they are very widespread and have practical nutritional and public health significance. For example, "favism," or the inability of many individuals in eastern Mediterranean and other countries to eat the broad bean *(Vicia faba)* can be of importance, as this is a major local source of plant protein.

Overlapping the genetic inability to metabolize foods is the phenomenon of allergy, in which individuals may be hypersensitive or allergic to certain items in foods, often protein constituents. A major issue in this regard concerns the relative commonness of these allergies. Some evidence suggests that they may be increasingly more common in the industrialized countries than in less technically developed areas, although comparative statistics are very difficult to obtain or evaluate. Basically, the question is whether the clinical features diagnosed as being allergic are, in fact, really so, as the term *allergy* can be used as a convenient label.

As with any other aspect of life, including breathing, physical accidents may occur in relation to eating. Principally these fall into three categories: those where food is inhaled for various reasons, those where large quantities of indigestible material (sometimes plant or fruit skins) block the intestinal canal, and those where sharp foods, mostly bones, penetrate and puncture the alimentary canal.

Foods can, in addition, also have adverse effects if they are taken in excessive quantities, deficient amounts, or in an imbalanced way. Thus, nontoxic foods taken in overlarge quantities can have adverse effects—the most notable being, of course, calorie-overdosage obesity. In addition, too high a consumption of foods containing certain nutrients can cause ill consequences. For example, carotenemia (aurantiasis) can result from an excessive intake of beta-carotene containing foods, such as carrots or tangerines, while a diet overfortified with vitamin D was shown to have pathological results in infants in postwar Britain.

It is stressing the tragically obvious that deficiency of foods has adverse effects, and a vast literature has arisen concerning the serious question of malnutrition in the present-day world. The issue relates to poverty, lack of available foods, and inequitable distribution of income and foodstuffs, all of which becomes more obvious in times of disaster, such as famines and wars. It is not the intention of the present publication to dwell on these dominant international public health and nutrition issues. Nevertheless, two types of nutrition deficiency will be mentioned which lead to adverse effects. The first of these occurs when diets themselves are naturally deficient. The classic example is the zinc-deficient diet in Iran and some other Middle Eastern countries, leading to dwarfism, among other things. Another such example is the possibility that a low intake of selenium may be responsible in part for *ke-shan* disease in parts of China.

The second type of deficiency occurs where foodstuffs are actually available, but results from cultural practices. In western urbanized countries a classic example is anorexia nervosa, which is undoubtedly an emotionally triggered, gross restriction of food intake, related in part to a desirable slim-body image. Also, in India, the so-called "syndrome of tremors" occurs in vitamin B12-deficient, breast-fed babies of mothers whose diet is restricted in foods containing vitamin B12 by both poverty and cultural restriction. Lastly, in the western world in recent years many new food practices have arisen due in considerable measure to a reaction against an overmechanized way of life, and the processed foods that form part of this lifestyle. Often such moves toward more "natural" foods have been a healthy trend toward a diet that is more time-tested and less uncertain than one based on industrially processed foods. However, this can be carried to harmful extremes, and ill effects have been described from more restrictive diets of this type, in particular the Zen macrobiotic diet.

Until quite recently, the value of the diet was largely considered in relation to its nutritional composition as analyzed *in vitro*—that is, outside the body prior to ingestion. It is now appreciated that matters are much more complicated than this and that the bioavailability of foods depends on their interaction and absorption when mixed together *within* the intestinal canal. Likewise, the question of balance between different nutrients and the physical composition of diets has come to the forefront more recently, especially regarding the question of the highly processed foods of the modern world compared with the more fiber-rich diets eaten by our ancestors for hundreds of thousands of years. The amount of dietary fiber desirable is a major concern at the present time. Also, the question of the correct amount of lipids in the diet is highly controversial and important in numerous ways, including their possible roles in the causation of coronary heart disease and forms of cancer.

Again, only recently have the consequences of the diet in early childhood and the subsequent pattern of health or disease in adult life come to be examined seriously. This will, it is quite certain, continue to be a major, and difficult, area for investigation. It is possible that there are very great differences between bottle-feeding babies with cow's milk-based formulas and breast feeding with human milk. Immediate concerns such as nutrition, prevention of infection and allergy, economics, and child spacing, have become increasingly more evident in this regard. In addition, the long-term consequences of such differences are very much at the forefront of present-day consideration. The emphasis with regard to the "degenerative diseases" of later life is now focused on their prevention by means of change in the way of life of individuals, rather than on radical technological interventions. It seems increasingly possible that the seeds of many adult conditions are, in fact, laid in infancy and related in part to the form of infant feeding. However this will, of course, be most difficult to prove or disprove in view of the many other variables that occur in every human being's "life flight," and the relatively long life span of the subjects investigated.

The avoidance of any adverse effects of food obviously needs to vary greatly in different parts of the world with different societies, ecologies, levels of sophistication, and ways of life. For example, the ill effects of foods in a nomadic group in Sahelian Africa would probably largely be connected with food shortages and infections. By contrast, in an industrialized urban population in Western Europe or North America, problems of environmental pollutants would loom to the forefront.

The selection of an optimal diet today must be based, before anything else, on the existence of adequate supplies of culturally acceptable and nutritionally balanced basic foods (e.g., in many Third World circumstances, the staple foods are cereal grain, a legume, and human milk), *and* methods by which these can be distributed equitably throughout the population.

Basically, it is still important that national policy ensures that adequate amounts of the required nutrients can be obtained by the whole population, and are bioavailable (Tracy, 1980). This can be achieved, for example, by a better understanding of the interaction of foods one with another in the intestinal canal, as, for example, the complementary significance of cereal–legume mixtures. At the same time, in industrialized countries the newer knowledge of the avoidance of *excesses* of such calorie-dense items as sucrose and fats needs much more consideration, as does greater emphasis on cereal grains, vegetables, and fiber-containing foodstuffs in general.

On the whole, the inherent or natural toxicity of some foods is better understood and usually better dealt with than the present-day maze in which humankind finds itself regarding the multitude of environmental toxicants that abound and increase in numbers. All would agree that a primary goal is to develop appropriate techniques for screening, detecting, and eliminating harmful chemicals absorbed into the body as a result of agricultural practices or food processing, and yet to ensure high agricultural yields and effective, widespread distribution and marketing of foods. Such testing is, of course, much easier to write about than to do, because animal studies are prolonged, costly, and difficult to evaluate.

Humankind currently finds itself at many crossroads and crises. One of these is to produce sufficient food for the world as a whole, especially in less technically developed countries. Where these countries are concerned, local production and relative self-sufficiency with the use of time-tested, fuel-economizing technology seems more appropriate and also more clearly required. By contrast, in industrialized countries a widespread and abundant supply of foodstuffs is still available, although with rapidly increasing costs, and with many built-in uncertainties concerning possible unrecognized, long-term consequences of the technological processes involved in agribusiness and food processing and marketing. The key question is how to ensure that such food supplies are geared both to modern knowledge regarding nutritionally desirable food policy and to profitability. At the same time, screening methods must be devised for potentially harmful substances that may enter the modern food chain, and prudence and care exerted before introducing novel processed foods derived from previously unused sources such as algae or single cell protein (SCP).

The need for conservatism is rightly emphasized by Synge (1977) in a recent symposium on "Safety in Man's Food":

> ... in dealing with human beings, the effects of a new dietary constituent may not be manifested as disease until fifty years or more after beginning to eat it. The danger may also completely elude detection in the course of animal experiments or in the practical feeding of our relatively short-lived farm livestock.

This is a vast, uncertain, and rapidly expanding field. The present volume attempts to draw on experienced individuals from numerous parts of the world to give their views on certain aspects of this subject. A totally encyclopedic coverage has not been possible, and would require many volumes. Conversely, overlap between one section and another has occurred and, indeed, seems desirable, as quite often unanimity of opinion does not exist.

All aspects of living can be dangerous to one's health (Lock and Smith, 1976) and certainly this applies to foods and eating. In the present-day world, many new and ancient basic questions exist as to how supplies of appropriate, nutritious foods can be made available to all the world's citizens, while minimizing actual or potential adverse effects, particularly those related to modern agriculture and food processing.

REFERENCES

Accra, A., Degani, R., Raffoul, Z., and Karahagopian, Y., 1981, Lead-glazed pottery: A potential health hazard in the Middle East, *Lancet* **1**:433.

Anonymous, 1981, Chronic fluorosis, *Br. Med. J.* **1**:253.

Gloag, D., 1981, Contamination of food: Mycotoxins and metals, *Br. Med. J.* **1**:879.

Jelliffe, D. B., Woodburn, J., Bennett, F. J., and Jelliffe, E. F. P., 1962, The children of the Hadza hunters, *J. Pediat.* **60**:607.

Lock, S., and Smith, T., 1976, *The Medical Risks of Life*, Penguin, Hammondsworth, U.K.

Nightingale, K. W., and Ayim, E. N., 1980, Outbreak of botulism in Kenya after ingestion of white ants, *Lancet* **2**:1682.

Senanayake, N., and Jeyaratnam, J., 1981, Toxic polyneuropathy due to gingili oil contaminated with tricresyl phosphate affecting adolescent girls in Sri Lanka, *Lancet* **1**:88.

Synge, R. L. M., 1977, The problem of assessing the safety of novel protein-rich foods, *Proc. Nutr. Soc.* **36**:107.

Tracy, M., 1980, Nutrition and food policy—an emerging issue for agricultural economists? *J. Agric. Econ.* **31**:369.

Webb, M. (ed.), 1975, Chemicals in the food and environment, *Br. Med. Bull.* **31**:3.

I

NATURALLY OCCURRING TOXICANTS

1

NATURALLY OCCURRING TOXICANTS IN FOOD

Leif Hambraeus

INTRODUCTION

It is quite obvious that foods contain large numbers of constituents, some of which are of little nutritive value for the consumer. Some components, however, not only lack nutritive value, but interfere with other constituents and reduce their nutritive value. They may, therefore, be termed *antinutritive factors.* The effects of such antinutrients can be manifold. In one situation, the substances interfere with the uptake or metabolism of certain nutrients, which may result in disturbances in the normal physiological function in man. The occurrence of certain oligosaccharides in some legumes may thus result in problems such as flatulence. Thyroid deficiency induced by glucosinolates occurring in plants of the genus *Brassica,* which interfere with the thyroid hormone production, is another example, where the antimetabolic effect is more pronounced. Toxicants produced by various microorganisms which induce intensive diarrhea, and the carcinogenic effect of certain mycotoxins are examples of more potent factors occurring in natural foods that have been contaminated in one or another way. The most potent toxicants of all are reported to occur as normal constituents in some mushrooms.

It is known that the toxicity of a substance will also depend on the specific metabolism in individuals. Thus, a genetic disturbance can increase the susceptibility for the toxicants. This is especially known to occur in favism. In other cases, some normal components in food which are not regarded as toxic for normal persons, are harmful in individuals with abnormal metabolisms or tolerances. Such examples are *lactose,* a normal constituent of dairy products, which in individuals with reduced lactase activity in the intestines may lead to intolerance problems that can be severe; *phenylalanine* in the case of phenylketonuria, and *lactose* and *galactose* in the case of galactosemia, where

Leif Hambraeus • Institute of Nutrition, Faculty of Medicine, University of Uppsala, P.O. Box 551, S-751 22 Uppsala, Sweden.

the deficient metabolic capacities in individuals for such normal constituents of food lead to severe metabolic disturbances that may even be fatal.

Natural foods in everyday diets may on the other hand contain a great number of toxic substances. However, this does not necessarily mean that the food is hazardous to the human being. A substance that is said to be toxic has a more or less pronounced capacity to induce deleterious effects on the organism when tested by itself in certain doses. It does not mean, however, that this always happens under usual conditions on a normal diet. Every healthy individual consumes a multitude of toxic substances in his normal diet every day. Yet he or she survives without any signs of intoxication. This is probably due to the fact that natural toxicants usually exert their effect only when they are consumed under special conditions or when there are other potentiating substances available. Also, the concentration of toxicants occurring in the food is often so low that the item must be consumed in usually not realistic amounts every day for an extended period of time if intoxication is to occur. It should furthermore be remembered that the organism is suited to handle small amounts of various toxicants and most of the toxic effects of various chemicals that are potentially hazardous do not have an additive effect. There also seems to occur antagonistic interactions that make some ingredients interfere and reduce the toxic effect of other components.

Most of the natural toxins are of vegetable origin and are concentrated along the food chain in the animal body, and very few occur *per se* in the animal. Raw fresh meat, where no preservatives or additives are used, could be said to be the least dangerous food item as it is derived from an animal that has used the special detoxifying capacity in its liver. Consequently the toxic problems of food components of animal origin generally occur when the animals have consumed toxic products in their feed, which have not been detoxified in the liver.

There are also a number of man-made toxicants that occur as a result of modern food production and food technology. These include such components as residues from the use of biocides, herbicides, pesticides and fertilizers during food production. Other examples are those chemical substances that are produced during the processing of food or various food additives and preservatives used in food technology, as well as components that might occur in the food items from the packing material. If on the other hand no preservatives are used, the microbial action on foodstuffs, no matter whether it is of animal or vegetable origin, may give rise to the formation of toxic substances. To this situation could furthermore be added the components that result from the pollution of the environment and those that result from mistakes during the handling of the foodstuff or accidents. Although neither of these could be described as natural food toxicants in the strict sense, they may nevertheless occur for various reasons as toxicants in what is thought to be natural food items.

Foods may consequently contain substances of various origins that may be hazardous even when moderate amounts are consumed of what is considered to be natural food products. The compounds can therefore be separated into two groups: those which may occur naturally as a result of the intrinsic metabolism of the animal or plant *(natural or inherent toxins)* and those that are formed as a result of microbial growth, accumulated from the environment, or unintentionally introduced during handling and storage *(acquired toxins).*

In this review, the various naturally occurring toxicants are discussed according to the following outline:

1. *Natural components of natural food products* (inherent toxins)
    a. *simple components,* i.e., carbohydrates, fatty acids, proteins and amino acids, vitamins
    b. *complex components* that have specific antimetabolic effects, i.e., antivitamins, goitrogens, chelates, vasoactive psychoactive factors, carcinogens, other potential antinutrients
2. *Natural contaminants of natural food products* (acquired toxins)
    a. toxicants of *microbiological* origin
    b. *nonmicrobiological* toxicants, i.e., minerals, toxicants consumed by animals used as food sources

NATURAL COMPONENTS OF NATURAL FOOD PRODUCTS (INHERENT TOXINS)

Carbohydrates

Obviously the carbohydrates are natural constituents of all human diets, and play an essential role as the main energy source in the diet. The carbohydrates do not usually contain any substances which can be considered as very toxic in healthy individuals. Nevertheless, the following comments can be made.

There are some inborn errors of carbohydrate metabolism where the consumption of various hexoses and disaccharides leads to very hazardous and sometimes even fatal consequences (Stanbury *et al.,* 1978). Such examples are *galactosemia* where even traces of galactose can lead to severe clinical symptoms and signs, and *fructose* intolerance, which in extreme cases can be fatal. There are also inherited disorders regarding the metabolism of the disaccharides, i.e., congenital lactose and sucrose intolerances, although they are rare. The role of saccharose in the development of caries as well as its disputed role as a trigger for cardiovascular disease (Yudkin and Morland, 1967), obesity (Connor and Connor, 1976), and diabetes (Nikkilä, 1974) is worth consideration because of excessive sugar consumption in industrialized countries.

Milk intolerance is most often due to a *lactose intolerance,* resulting from a deficient lactase activity in the intestinal mucosa (Johnson *et al.,* 1974). Intake of lactose through consumption of dairy products leads to the occurrence of undigested lactose in the intestinal lumen, causing a series of problems. First, it disturbs the osmolar balance as the undigested lactose attracts water from the tissue. In addition, lactose is a good substrate for the intestinal flora, whose enzymes ferment it to lactic acid which further raises the osmolarity. Finally, fermentation also leads to gas formation and abdominal discomfort due to flatulence and diarrhea. *Primary low lactose activity* is one of the most commonly encountered intestinal enzymatical dysfunctions but differs widely in healthy subjects of various racial groups who live under the same conditions. *Secondary low lactase activity* is associated with certain diseases of the small intestine, i.e., infectious disorders, celiac disease, tropical malabsorption including sprue, and protein–calorie malnutrition.

Oligosaccharides such as *raffinose* and *stachyose* have a more definite antinutritive effect. They pass the intestines without splitting as the intestinal tract does not produce the splitting enzyme α-galactosidase, and are then digested by some anaerobic microorganisms, giving rise to gas formation and resulting in flatulence, dyspepsia, and diarrhea (Steggerda, 1968). The content of oligosaccharides is thought to be largely responsible for the often reported problems of flatulence after consumption of diets containing beans and other legumes.

Lipids and Fatty Acids

The fat in the diet is usually derived from food constituents that have long been used as energy sources in the diet (Vergroesen, 1975). Consequently, most of them have been selected throughout time and have very few untoward effects. During the last decades, however, quite a few unconventional agricultural products have been introduced in our diets as sources of fat energy. This means an increased potential risk for adverse effects. The problems of the antinutritive effects of dietary fats could essentially be focused on the following problems: (1) the occurrence of certain toxic fatty acids, (2) the potentially hazardous effect of an increased fat intake, and (3) the effect of an increased intake of polyunsaturated fatty acids (PUFA) *per se* or of the vitamin E requirement.

During the last decades an increasing interest has been devoted to the risks of the ingestion of certain fatty acids in the human diet. Of greatest interest has been *erucic acid* (C 22:1 ω 9), which may account for as much as 50% of the fatty acid content of the oil obtained from some *Brassica* species, i.e., rapeseed and mustard. In 1960 Roine and his collaborators reported myocardial effects in rats that had been fed 50–70% of their calories as rapeseed oil. This inspired an intensive study on the toxicity of erucic acid (Engfeldt, 1975; Ziemlanski, 1977). It has been stated that even an intake of 4 energy per cent in the form of erucic acid can lead to morphological effects. The worldwide production and increasing use of rapeseed and mustard seed oils in margarine, and the important role of rapeseed as an agricultural crop in certain countries such as Canada, Northern Europe, India, and Pakistan, have called for intensive studies. Plant breeding activities have now resulted in low erucic acid-containing rapeseed varieties to be developed and used commercially. The mechanism for the development of the abnormalities shown in experimental animals consuming erucic acid is probably blockage of the β-oxidation of the fatty acids in the mitochondria (Engfeldt, 1975). Erucic acid may even occur in minor amounts in fish (Christopherson *et al.*, 1968). However, the content of the very long chain fatty acids containing 20 or 22 carbons and 1–6 double bonds (Stansby, 1967) in some marine oils is probably worth more intensive studies with regard to possible toxic effects (Ziemlanski, 1977).

It should however be noted that other fatty acids may be potentially harmful. Trans-isomers of fatty acids which occur in the depot fat of animals have been much discussed with regard to their role in the organism (Vergroesen, 1975). There are also a number of uncommon fatty acids with marked physiological effects even in very low concentrations, e.g., cyclopropene fatty acids, which are found in cottonseed oil (Bailey *et al.*, 1966). Epoxy oils occur in soybean and sunflower oil (Sen Gupta, 1972; Mikolajzac *et al.*, 1968) and furanoid fatty acids have been reported to occur in some species of fish (Glass *et al.*, 1975).

As in the case of carbohydrates, there are certain inborn errors of fat metabolism that call for special attention. One example is the disturbed metabolism of branched-chain fatty acids in Refsum's disease (Refsum, 1946), a serious neurologic anomaly that results from a gross infiltration of the tissues by lipids that contain large amounts of phytanic acid.

The fat content in the diet has been much discussed during the last few decades as there has been observed a pronounced increase in the fat energy per cent in the diet of the population of affluent societies (Blix, 1965; Blythe, 1976). At the same time the potential hazardous effect of saturated fatty acids for the development of atherosclerotic disease has been debated (Truswell, 1978). This has led to the recommendation of an increased intake of polyunsaturated fatty acids (see for example Select Committee on Nutrition and Human Needs, 1977; Wretlind *et al.*, 1976). Whether an increased intake of polyunsaturated fatty acids is advantageous for the outcome of atherosclerosis and cardiovascular disease or not, an increased intake of polyunsaturated fatty acids also leads to some potential toxic problems (Truswell, 1978). First, an increased intake of unsaturated fatty acids calls for an increased intake of vitamin E (Horwith, 1962). The vegetable oils that are usually recommended as sources for polyunsaturated fat are fortunately also the best vitamin E sources in the diet. It should however be noticed that no human population has customarily consumed diets high in polyunsaturated acids (Keys, 1970). The storage of fats rich in polyunsaturated fatty acids can result in the formation of hydroperoxides in the presence of oxygen even in the absence of heat. The unpleasant flavor and taste makes this fat organoleptically unacceptable and consequently reduces the risk for its toxic effects as it will probably not be consumed. Of greater interest is the autooxidation and formation of cyclic monomers during heat treatment of polyunsaturated fats in deep fat frying, as these compounds are toxic when consumed alone (Lea, 1962; Billek, 1973). The possible role of dietary fat as an etiologic factor in colon carcinogenesis has also been discussed (Reddy *et al.*, 1976).

Proteins and Amino Acids

Essential organic nitrogen compounds such as proteins and amino acids also impose some adverse and potentially harmful effects on the organism. *Amino acid toxicity, amino acid imbalance* and *amino acid antagonism* represent a few examples of such problems, in addition to the fact that some proteins are toxic as such, or function as enzyme inhibitors.

Amino Acids as Potential Natural Toxicants

It is of course disputable whether amino acids that occur in food proteins considered as essential nutrients should really be classified as naturally occurring toxicants. Nevertheless, a few proteins do contain disproportionate amounts of some amino acids and furthermore studies have shown that increased intake of some amino acids rarely may lead to some disturbances (Harper *et al.*, 1970). In addition to the age of the subject and the adequacy of the diet with respect to the intake of protein and energy as well as certain vitamins, the relative proportions of amino acids in the diet influence the susceptibility of the individual to the amino acid load. Interestingly, the indispensable amino acids seem to be less well tolerated in abundance than the dispensable ones, methionine being the least well tolerated of the nutritionally important amino acids (Harper, 1973).

Amino acid toxicity refers to the fact that some amino acids have a deleterious effect on the organism when they occur in overabundance. The condition can only be counteracted by the reduction of the dietary intake of the toxic amino acid. Amino acid toxicity is to some extent equivalent to the toxic effects on the cells of an inherited amino acidopathy with an accumulation of the amino acids, e.g., phenylketonuria. Usually, the toxic effect is related to the special chemical and metabolic function and features of the amino acid(s) in question. Interestingly, it seems that the toxic effects are enhanced when the subject is given a low-protein diet. So far, toxic effects in humans have been described for methionine and the sulphur amino acids (Klavins and Peacocke, 1964), and for phenylalanine and tyrosine (Boctor and Harper, 1968). It has been postulated that the amino acids that are closely related metabolically seem to be more toxic than those that have only a few metabolic reactions in common (Salmon, 1958).

It is obvious that the human animal can tolerate large intakes of protein and amino acids in a well-balanced pattern. Nevertheless, such large intakes usually lead to liver and kidney hypertrophy, indicating that such a metabolic load may increase the demand for metabolic capacity and regulatory mechanisms in these organs. In relation to food perhaps one of the most interesting amino acids is glutamic acid. Monosodium glutamate has long been used to enhance the flavor of food and may be connected with the *Chinese Restaurant Syndrome* (Kwok, 1968), which is described as a feeling of weakness, palpitation, and numbness on the back of the neck and back and may last for 2 hours, but does not give rise to any aftereffects. According to some authorities the syndrome is associated with a high glutamate intake and occurs when the blood concentration of monosodium glutamate exceeds a certain level.

The term *amino acid imbalance* was introduced as a concept by Harper in 1964 to describe the change in an amino acid requirement as a consequence of an excess of another amino acid. The administration of one amino acid may then lead to a relative deficiency of another amino acid, especially when the latter is the limiting amino acid. The mechanism behind the development of an amino acid imbalance has still not been explained although the concept has been discussed (Harper, 1958, 1964; Harper *et al.*, 1970). Studies in experimental animals, especially rats, have shown that after only a few hours on an imbalanced diet, they will reduce their food intake and the plasma concentration of the limiting amino acid decreases. A metabolic imbalance will appear with a subsequent loss of muscular mass and a rise in the liver content of protein, glycogen, and fat (Sidransky and Baba, 1960). In some cases, amino acid imbalances are remarkably specific in that only a fraction of a percent of an amino acid added to a diet may cause imbalance in an experimental animal (Harper and Rogers, 1965). The imbalance between isoleucine and leucine as an etiological factor in the pathogenesis of endemic pellagra in populations consuming maize and jowar (sorghum) has been postulated (Gopalan and Rao, 1975). This seems to be the first report of an important nutritional defect in humans caused by an amino acid imbalance. Leucine has furthermore been of specific interest, as it is linked to gluconeogenesis, and in infants a special leucine-induced hypoglucemic syndrome is described (DiGeorge and Auerbach, 1960).

Closely related to amino acid imbalance is the problem of *amino acid antagonism*. This concept refers to a state where there is competition between amino acids that are structurally related and share a common step in metabolism or transport. The problem can be solved by the addition of an amino acid. However, it is not necessarily the limiting amino acid that should be supplemented but the amino acid(s) that is chemically related to the amino acid in excess. Such competition has been described among the branched chain amino acids (where the effect of leucine has been most discussed) and between lysine and arginine (Harper *et al.*, 1970).

Amino acid antagonism is of special interest in disorders involving transport defects, i.e., cystinuria and Hartnup disease, as well as in a number of inborn errors of amino acid metabolism. Clinical symptoms of these metabolic problems, resulting from the intake of certain amino acids, often include mental retardation and neurological disorders. The best known example is phenylketonuria. It has been suggested that one of the mechanisms responsible for the impaired development of the central nervous system in phenylketonurics might be the restricted passage of other amino acids into the brain due to the elevated plasma phenylalanine levels (Neame, 1961; Linneweh *et al.*, 1963).

When single doses of individual amino acids are given orally or intravenously exceeding the requirement by tenfold, mild gastric distress symptoms and even nausea, febrile reactions, and headache have been reported in normal, healthy individuals. With the exception of the high leucine–low isoleucine content in maize however, there are few indications that individual amino acids are available in foodstuffs in amounts large enough to cause adverse effects due to amino acid imbalance.

Protein Intolerance

It is a well-known fact that the metabolic capacity of the immature liver during the neonatal period is limited, and even a two- to threefold increase in the protein intake might lead to the metabolic disturbance, late metabolic acidosis (Kildeberg, 1964), or to a pathologic plasma amino acid pattern, *iatrogen aminoacidopathia.*

There are also diseases involving the urea cycle that lead to a very pronounced and reduced protein tolerance (Colombo, 1971). This gives rise to ammonia intoxication when a protein intake above the metabolic capacity of protein synthesis is reached (Hambraeus *et al.*, 1974).

In addition to these toxic effects of protein intake on metabolism there are a number of reports of intolerances against various food proteins on an allergic or immunological basis.

Protein Inhibitors of Enzyme Systems

An enzyme inhibitor can affect the binding and transformation of substances or make the substrate unavailable for the enzyme. It can also interfere with the biosynthesis of the enzyme or affect the turnover of the enzyme, or it can influence the regulator system of the enzyme. It seems to be most common, however, that enzyme inhibitors exert their effects as substrate or cofactor analogues. They also form strong bindings to the active sites of the enzyme, giving rise to enzymatically inactive compounds (Whitaker and Feeney, 1973).

In the normal individual, there is also a delicate balance between the rate of biosynthesis and the rate of degradation of an enzyme. Any substance that affects this balance, e.g., giving rise to an increased turnover or interfering with the hormonal control of the enzyme activity, will appear as an enzyme inhibitor. Human beings differ in their susceptibility to constituents of food because of differences in the levels of their enzymes. Thus, lactose is toxic to many noncaucasian individuals having a low lactase activity in the intestine and phenylalanine is toxic to individuals suffering from phenylketonuria, although both lactose and phenylalanine are normal components of the diet.

As many enzyme inhibitors are proteins, their activity may be lost during denaturation. Osborne and Mendel (1917) have been said to be the first to observe and describe that heat treatment of soybeans improved their nutritive value. It has since been shown that the nutritive value of many legumes could be improved after heating. This is at least to some extent due to the inactivation of the proteolytic enzyme inhibitor during denaturation.

Trypsin inhibitors are probably the most widely distributed among the inhibitors of proteolytic enzymes. Thus many inhibitors have been shown to have activity against α-chymotrypsin as well as against trypsin. Interestingly, trypsin and chymotrypsin inhibitors are said to be somewhat resistant to proteolysis and they are consequently unaffected by commonly used processing and food preparation techniques. However, inhibition of bovine trypsin is by no means an index of activity against human trypsin, but proteolytic enzyme inhibitors often react against a variety of enzymes.

There are also other proteolytic enzymes that play an essential physiological role although not necessarily within the gastrointestinal system. Other examples of proteolytic enzymes are *elastase inhibitor* which has been isolated from kidney beans (Pusztai, 1968), *plasmin inhibitors* from blood serum, kidney bean, lima bean and soybean (Feeney *et al.*, 1969), and *papain inhibitor* from primarily the legumes (Learmonth, 1958) but also from wheat and egg white (Fossum and Whitaker, 1968; Hites *et al.*, 1951).

Protein inhibitors of various amylases have been reported to occur in wheat and beans (Bowman, 1945). Some of them seem to be inactivated by denaturation at cooking temperature, while others are more resistant. As amylase inhibitors have a strong action *in vitro* against pancreatic amylase, it could be assumed that these inhibitors may have antinutritional effects when occurring during a restricted food intake.

Solanine is a nonprotein inhibitor of cholinesterase and may occur in abundance in potatoes under certain circumstances. Other significant sources of solanine are tomatoes, eggplants, sugar beets, and apples (Orgell, 1963). Moderate and even severe poisoning including gastrointestinal disturbances and neurological disorders have been reported as a result of the intake of solanine-containing potatoes (Rühl, 1951).

Specific Antinutrients and Antimetabolites

The *goitrogens* comprise some of the most common toxicants in human food. They have long been recognized and include the glucosinolates which have a limited goitrogenic effect when they are intact. However, they are split in plants by the action of myrosinases (Greer, 1962) and also in the gastrointestinal tract, giving rise to isothiocyanates and oxazolidines. These goitrogenous compounds are present in a variety of plants, e.g., cabbage, rapeseed, mustard seed, brussels sprouts, and cauliflower, and may produce thyroid enlargements (Josefsson, 1967). It is assumed that they act primarily by competitive inhibition of the iodization of thyroxine. There is however little evidence that endemic goiter could be related to the consumption of goitrogenous food in man (Anonymous, 1968). The transmission of goitrogenous compounds through the milk of dairy cattle fed considerable amounts of goitrogens in their feed has not been considered a public health problem (Virtanen *et al.,* 1963). Interestingly, glucosinolates are particularly abundant in green leaves, and the use of various cabbage species for the production of leaf protein concentrates may therefore give rise to unexpected goitrogenous effects.

Hemagglutinins are compounds of glucoprotein origin that affect the red blood corpuscles ultimately resulting in agglutination. Today they are more commonly known as *lectins* owing to their very pronounced specificity of action (Boyd and Shapleigh, 1954). They also have a very selective impact on the intestines and their absorption of various nutrients, which has been postulated to result from a specific combination with the cells of the intestinal wall. Lectins have been found mostly in plants but have also been isolated in animal products (Jaffé, 1953). It has been assumed that the lectins are responsible for a number of the toxic effects from the consumption of raw legumes.

Among the specific antinutrients such components as *oxalic acid* and *phytates* should also be mentioned. These represent compounds that interfere with mineral metabolism as they form chelates with calcium, iron, and zinc, to mention a few examples. Oxalic acid is found primarily in spinach and rhubarb, but also is present in mushrooms and chocolate (Oké, 1969). The urinary excretion of oxalic acid is increased when huge doses of vitamin C are consumed (Lamden, 1971). In addition to its negative influence on the availability of various minerals, an increased urinary excretion of oxalic acid may lead to renal calculus formation and secondary renal failure.

Phytate plays a more pronounced role as a mineral chelator, as it is a strong acid and binds a broad spectrum of minerals (Oberleas, 1973). Thus, nutritional deficiencies of zinc, copper, and iron have been linked to excessive intake of phytic acid. The role of phytic acid has been of special concern during the last few years because of increased interest in the use of dietary fiber such as wheat bran in the human diet. Thus, the increased content of zinc in wholemeal is counteracted by the increased content of phytic acid, the latter leading to chelate formation and a reduction in the availability of zinc (Sandström, 1980). The increased interest in dietary fiber in the human diet as well as the use of meat analogues based on soybean products, which also contain phytic acid, call for further work on the potential nutritional hazards of these compounds in the diet.

Vitamins

There is no doubt that vitamins represent essential nutrients in the diet and it could consequently be disputed whether they might be regarded as natural toxicants as well. However, as in the case of amino acids, there are descriptions of potential hazards when vitamins are consumed in doses far above the recommended allowances. During the last decades the belief in megadoses of vitamins has somewhat increased as a component of the increase in food faddism. This makes the potential risk of vitamin intoxication a reality. Furthermore, there is a tendency to use fortification and enrichment of various food items with vitamin and mineral supplements to a greater extent in fabricated food products. This may secondarily lead to a risk of overconsumption of some vitamins, and calls for some guidelines (Darby and Hambraeus, 1978). As the water-soluble vitamins are usually easily excreted in the urine, the risk of their overdosage is much less than with respect to the fat-soluble vitamins that are stored within the body tissues and excreted to a rather limited extent. Consequently most reports of intoxication are related to the fat-soluble vitamins.

With regard to some of the vitamins it is relevant to discuss three dosages: the *minimal requirement* of a vitamin that is needed to counteract any nutrition deficiency syndromes, a *pharmacological level* that has been used for certain therapeutic purposes, and a *toxicological level* above which certain toxic effects have been reported (Table I).

Vitamin A is present in food in two forms: as *carotenoid pigments,* usually β-carotene in plants, and as *preformed vitamin A* stored in the liver and fat of animals. Although there are reports of vitamin A intoxication it is rarely a result from the intake of vitamin A-rich natural foodstuffs but usually due to prolonged consumption of vitamin supplements. Nevertheless vitamin A intoxications have been reported from the intake of the livers of polar bears and other Arctic mammals, and from livers of large fishes with a capability of storing considerable vitamin A concentrations (Nater and Doeglas, 1970). Interestingly, there seem to be no reports of vitamin A intoxication as a result of high β-carotene intake, except for apparently harmless carotinemia. The symptoms and signs of vitamin A intoxication, both in its acute and chronic form, are alarming and often misinterpreted as signs of brain tumor. However, there are still no recorded deaths ascribed to vitamin A intoxication. Furthermore there is an antagonism between vitamins A and D so far that they decrease the toxic effects of each other. An increased intake of another fat-soluble vitamin, vitamin E, also seems to counteract the toxic effect of vitamin A.

Vitamin D is one of the most potent toxic vitamins where the range between therapeutic, pharmacologic, and toxicologic levels is least. Although it is not a true vitamin—it is synthesized in the body as a result of exposure of the skin to ultraviolet light—there are reports of vitamin D toxicity not from overexposure to sunlight, but rather from the consumption of vitamin D from cod-liver oil or from an excessive intake of vitamin D-enriched products, as noted in infantile hypercalcemia. The toxicity from food products, however, seems to vary, probably due to the fact that the vitamin A content may counteract the toxicity of the vitamin D.

Vitamin E is a vitamin that is still said to be searching for its function. Although very few cases of vitamin E deficiency have so far been described in man, there is a widespread use of preparations containing the vitamin sometimes even in rather large doses. No certain toxic effects have however been observed, although there are reports of headache, nausea, fatigue, and blurred vision on high dosage vitamin E administration (King, 1949).

TABLE I
Physiologic, Proposed Pharmacologic, and Toxic Doses of Some Vitamins[a]

Vitamin	Physiologic dose	Pharmacologic dose[b]	Toxic dose	Manifestations of toxicity
Vitamin A	5,000 i.u.	50,000–100,000 i.u. (acne)	100,000–500,000 i.u.	Headache, nausea, vomiting, pseudotumor cerebri
Vitamin B_6	2.0 mg	50 mg (against symptoms of oral contraceptives)	10,000 mg	Abnormal hepatic enzymes
Niacin	20 mg	2,000–5,000 mg (hypercholesterolemi)	100 mg	Cutaneous flush
			1,000 mg	Carbohydrate intolerance
			2,000–6,000 mg	Gastritis
β-tocopherol	15 mg	300–1,200 mg (cardiovascular disorders)	1,000 mg/kg (animals)	Increased deposition of cholesterol in aorta, decreased tolerance to ethanol
Vitamin D	400 i.u.	50,000–100,000 i.u. (hypophosphatemic rickets)	1,000–3,000 i.u. (children), 150,000 (adults)	Hypercalcemia, renal failure
Vitamin C	45 mg	100–2,500 mg (colds)	2,000–4,000 mg	Induces nephrolithiasis, induction of vitamin C-dependent syndrome, inactivates vitamin B_{12}, reverses effects of anticoagulants, interferes with test for glucosuria, reproductive failure

[a] Adapted from Hodges, 1976.
[b] Not always accepted as proper treatment.

Vitamin K is needed for prothrombin synthesis and is usually supplied through the intestinal microflora. Nevertheless, the vitamin is often administered parenterally to pregnant mothers in order to counteract the hemorrhagic disease of the newborn that was earlier found as a result of vitamin K deficiency. On the other hand, there are descriptions of vitamin K intoxication that have included the deaths of premature infants (Allison, 1963). However, vitamin K intoxication through food intake does not seem to represent any public health problem.

With respect to the water-soluble vitamins, greatest interest has been devoted to *vitamin C* and *niacin.* Ascorbic acid is known to be excreted rather efficiently in the urine even when high doses are consumed. However, this can result in a more than doubled excretion of oxalic acid and may secondarily lead urinary calculus formation (Lamden, 1971). Furthermore, there are reports that ascorbic acid can block the anticoagulant effects of heparin and dicoumarol although it is not completely understood how this happens (Rosenthal, 1971). Finally, an increased consumption of ascorbic acid has been said to also increase the requirement of the vitamin, resulting in the risk of the development of scurvy (Hodges, 1976).

Niacin, but not nicotinamide, in high doses can cause capillary vasodilation and increased intracranial blood flow. This has been observed when the vitamin has been used in large doses for the lowering of serum cholesterol (Anonymous, 1961). It was also accidentally observed when some cured meat and provision products were treated with niacin for the preservation of their red color.

Antivitamins

A number of substances have been reported to act on the availability of vitamins, and are commonly referred to as *antivitamins.* Their effect can be due to one or several of the following factors: they may have similar chemical structures, they compete with the vitamin in various metabolic reactions, they might react with the vitamin *per se* and make it unavailable, or finally they might give rise to symptoms resembling vitamin deficiencies.

Somogyi (1978) has proposed dividing the antivitamins into two major groups: (a) structurally similar antivitamins that compete with the vitamins due to resemblance of chemical structure, and (b) structure-modifying antivitamins that destroy or decrease the effect of a vitamin by modifying the molecule or forming a complex molecule.

Most of the antivitamins occurring in the food seem to belong to the latter group. The first antivitamin to be described seems to be thiaminase, which was reported in the early 1930s (Green, 1937) and occurs in viscera of various fishes. But a thiamin inactivity effect has also been described in some fruits and vegetables, e.g., blueberries, black currants, red beets, brussels sprouts, and red cabbage. Antivitamins directed toward other vitamins have also been reported, for example, the niacin inhibitors that occur in some cereals (Christensson *et al.*, 1968). A pyridoxine antagonist, linatine, was described a few decades ago in flax seed (Kratzer *et al.*, 1954), and a pantothenic acid inhibitor has been isolated from pea seed lingers (Smaskevskii, 1966). The biotin antagonist, avidin, occurs in raw egg white and represents a classical example of an antivitamin (Eakin *et al.*, 1940).

Toxins with Specific Biological Action

Vasoactive Substances

Quite a few of the compounds occurring in natural foods are chemically classified as amines and it is well known that biogen amines usually are pharmacologically active compounds. Tyramine, histamine, and tryptamine, as well as noradrenalin and dihydroxyphenylamine, are examples of amines that are vasoactive. They are often found in greatest amounts in some fruits, i.e., bananas, avocado, and pineapple, but also in plantains and tomatoes (Lovenberg, 1973). As bacteria often contain decarboxylases that attack amino acids, fermented products such as cheeses, or aged food items often contain various amines. Thus, tyramine occurs in cheddar cheese and may induce acute attacks of hypertension (Horwitz *et al.*, 1964), and even fatal attacks are reported in individuals receiving inhibitors of monoamine oxidase for some reason. It has also been assumed that amines may induce attacks of migraine in susceptible individuals (Blackwell *et al.*, 1967).

Psychoactive Substances

Among these could be mentioned such stimulants as caffeine and other xanthines. There are however also compounds that are thought of as depressants, the most important one of course being ethyl alcohol, which is consumed in alcoholic beverages.

Carcinogens

During the last few years an increased interest has been devoted to the occurrence of various carcinogens in food. Some scientists have even claimed that not only an increasing number of cancer cases are induced by the presence of carcinogens in fabricated foods, but also that carcinogenic components in the food are directly or indirectly responsible for more than 50% of all cancer in man (Gustafsson, 1980).

The fact that nitrite can react with secondary amines to form potentially carcinogenic nitrosamines *in vivo* has led to an increased interest for studies on the use of nitrite and nitrate in food processing and technology. Nitrite has long been used in the manufacture of cured meat and provisions. From the beginning, it was used mainly to induce the characteristic red color of the products. However, of greater importance is that nitrite is a potent inhibitor of pathogenic microorganisms, especially the toxin-forming *Clostridium botulinum.*

Regarding the possible role of nitrite and nitrate in food in the etiology of human cancer (Lijinsky and Epstein, 1968; Smith, 1980), it is of interest that these compounds occur as normal constituents in several vegetables, e.g., spinach, beets, and lettuce (Keeney, 1970), and that nitrite and nitrate can be formed by bacteria in the intestine (Hawksworth and Hill, 1971) or by endogenous synthesis in the human intestine (Tannenbaum *et al.,* 1978).

Among the naturally occurring carcinogenic compounds in food, however, the most discussed and perhaps also the most potent are those derived from fungi; the aflatoxins, which will be commented on later, are the most studied of this group. Another example of carcinogenic compounds in plants is cycasine, a toxic glucoside that is reported to have a carcinogenic effect similar to that of dimethylnitrosamine (Shank and Magee, 1967).

Organspecific Toxins, i.e., Hepatotoxins and Lung Toxins

Certain alkaloids are reported to cause liver damage and necrosis or liver tumors and sometimes lung toxicity in animals. Pyrrolizidine alkaloids occur in a wide range of plant species and can induce acute and chronic poisoning especially in livestock, but they may also function as potent liver and lung toxins (Bull *et al.,* 1968). A specific hepatic veno-occlusive disease has been reported especially in Jamaican children as a result of the drinking of certain bush teas, probably due to their content of pyrrolizidone alkaloids (Hill, 1960). Of special interest is the report that bees can accumulate significant amounts of toxic alkaloids in their honey, making honey from certain districts a potentially harmful product (Deinzer *et al.,* 1977). Significant amounts of glucosides occur in the seeds of a number of fruits, i.e., apple, apricot, peach, cherry, and plum. Cyanogenic glucosides occur in a number of plants and can give rise to hydrogen cyanide, a very potent cytochrome oxidase inhibitor. However, only a few of those plants are really used as human food, cassava and almond being the most classical examples (Wilson, 1979).

NATURAL CONTAMINANTS OF NATURAL FOOD PRODUCTS (ACQUIRED TOXINS)

Microbial Toxins

Many toxicants are produced by microorganisms and are consumed by man through a concentration during the food chain. It should of course be remembered that infection of the gastrointestinal tract by means of gram-negative organisms from such genera as *Salmonella, Yersinia* and *Vibrio* are also examples of food-borne "intoxications." Infections due to animal parasites such as *Entamoeba histolytica* and *Giardia lamblia* are further examples of this type of intoxication. In addition, a number of intestinal virus infections are commonly induced by contaminated food products, e.g., virus-induced hepatitis. These are however examples of food-borne infections primarily related to problems with food hygiene, rather than examples of natural food toxins (Hobbs and Gilbert, 1978).

Toxins are usually produced prior to ingestion in contaminated foods where the microorganisms have proliferated under suitable conditions. They are usually divided into two different types, the *endotoxins* and the *exotoxins*. Many types of exotoxins have been described as originating from *Staphylococcus pyogenes*. They can induce extreme nausea, vomiting, diarrhea, and dehydration, but although severe there is a low mortality. This intoxication is a typical example of contaminated food, as this bacteria is normally found on the skin and in the upper respiratory tract.

Clostridium perfringens, the causative object of gas gangrene, may induce food poisoning leading to severe diarrhea. Some spores can survive normal cooking for even an hour or more at 100°C. The spores are then known to be able to germinate and produce numbers of spores in the intestines but not in foodstuffs. The toxin released in the intestines during sporulation cause the attacks of diarrhea and cramps.

Botulism is a neurological disease which may be fatal if not properly treated in time. It is caused by the heat-labile immunogenic protein botulism neurotoxins produced by *Clostridium botulinum,* and occurs through ingestion of anaerobically preformed toxins in foods. Ingested spores of the bacteria that may have emanated from honey have also been reported to result in toxin production in the intestines (Arnon *et al.,* 1978).

A number of intoxications have been reported from various seafood products. They are often produced by microorganisms such as marine algae. *Ciguatera poisoning* occurs mainly in the Pacific and is caused by a microscopic plant. *Gambierdiscus toxicus,* which lives on the surface of coral algae, passes its toxin up the food chain to larger fish, such as red snapper and barracuda. Characteristic symptoms of the disease are tingling of the lips and tongue, abdominal problems, muscle pains and general weakness. Ciguatera poisoning is sporadic. If cases develop, animal testing is indicated, and the public should be warned of the situation. There is a great need for better methods of detecting poisonous fish and perhaps detoxifying them before eating them. They are difficult to control as they occur sporadically and their poisons are often quite stable against heat and cooking. *Saxitoxin* is one of the most potent low molecular weight poisons known. It is concentrated in the gills and hepatopancreas of the shellfish (Schantz *et al.,* 1975) and survives conventional cooking. Intoxication leads to gastrointestinal and neurological symptoms and may even be fatal in a small percentage of those affected.

Mycotoxins

During the last few decades a special interest has been devoted to diseases caused by compounds produced by molds and fungi, i.e., mycotoxins. They result from the contamination of various food stuffs at all stages of production and storage. As toxicogenic molds are widespread in the environment, contamination of foods occurs easily.

Ergotism seems to be the first example of toxicosis due to a mycotoxin. It is caused by a complex mixture of various pharmacologically active alkaloids, ergot, which are produced by a parasitic fungus and associated with the consumption of flour prepared from cereals, most often rye, that are infected with *Claviceps purpurea.* The ergot alkaloids cause vasoconstriction, which leads to abortion, and gangrene of extremities when relatively large amounts are consumed for long periods. The disease was earlier known as "St Anthony's Fire" (Barger, 1931). The ergot alkaloids have also been used therapeutically for the treatment of migraine as well as for induction of parturition. However the alkaloids seem to have a limited lifespan and are often destroyed during storage. The consumption of fresh, newly harvested rye infected with *C. purpurea* therefore may induce ergotism to a greater extent than consumption of rye that has been stored for one year or more. The last outbreak of ergotism in Sweden occurred in 1853, while the last notable local epidemic outbreak in man is reported as late as in 1951 (Wilson and Hayes, 1973). The centralized milling of flour, in addition to strict control of the quality of cereals, seems to be the most effective way to counteract the risks of ergotism.

As moisture is essential for the growth of mold, mycotoxins occur commonly in moist tropical areas. However, optimal growth can today also be obtained in temperate zones of the globe and consequently problems of mycotoxins are often occurring in cooler industrialized countries. Renwick (1972) reported a correlation in geographical space between the average severity of winter-stored blighted potatoes infected by *Phytophtora infestans* and the incidence of anencephaly and spina bifida in man. This led to an intensive discussion regarding the potential harmful effects of potato consumption in pregnant women in the early 1970s (Poswillo *et al.,* 1972). A great deal more work, however, seems to be needed before this hypothesis based on epidemiological evidence can be accepted.

So far the greatest attention has been devoted to the *aflatoxins,* which are derived from the common mold of *Aspergillus* strains. Some of them represent the most potent carcinogens so far known. Aflatoxin occurs naturally as a contaminant of many foods, peanuts and corn perhaps being the most well known examples. A correlation between aflatoxin intake in food and the incidence of hepatoma in certain populations has been postulated (Wilson, 1978). Aflatoxin consequently is an example of a natural toxin that has been known to not only affect domestic animals but also humans. It should be mentioned that aflatoxin represents another digniti of problem than other toxicants discussed, e.g., goitrogens, as the latter may be reduced by plant breeding. Strict control of storage and import products seem to be the only valid approaches to reduce the risk of intoxication from aflatoxins.

Various strains of *Penicillium* also seem to produce mycotoxins, some of them neurotoxins (also known as tremorgens), and all of them having an indole structure. This has been shown to affect animals but reports from

Nigeria and India support the hypothesis that some neurological signs observed in certain communities could also be caused in humans by a tremorgen-contaminated food (Wilson and Hayes, 1973). The mechanism behind the neurological action is not known.

The *Fusarium* strains also comprise fungi that may contaminate food frequently and may lead to intoxications. Some of them are reported as dermotoxins, while others act as depressors on the bone marrow or as carcinogens (Wilson *et al.*, 1972). Hepatotoxins found in injured sweet potatoes are examples of more organ-specific toxins, another example being the lung toxins described to emanate from infections with fusarium (Doster *et al.*, 1978).

Mushrooms

Another group of mycotoxins are those derived from various mushrooms that are accidentally or normally used in the human diet. One and the same species of mushroom may show a variability in its toxicity. This might be due to the extent to which the mushroom has been processed, the effect of the environmental conditions, or to an increased sensitivity of the consumer.

The toxic substances in mushrooms vary in structure as well as with respect to the toxic symptoms they induce. Some of them are polypeptides, which makes it possible to produce antiserum against them. Others are alkaloids, the best known perhaps being *muscarine,* which is a derivative from choline. It gives rise to symptoms from the gastrointestinal tract and also has hallucinatory properties. It has been used for years in Siberia as a hallucinatory agent.

Some of the toxins found in mushrooms are very potent and may cause fatal intoxications. The most potent toxicants are reported to occur in *Amanita phalloides.*

There are also some toxicants derived from mushrooms which are only toxic when they interact with other compounds. Thus the ingestion of *Coprinus atramentarius* (Inky cap) followed by the consumption of alcoholic beverages leads to production of a compound related in its activity to "antabuse," which has been used in the treatment of alcoholism (Kingsbury, 1964).

Contamination of Heavy Metals

At the end of the 1960s, organic metal mercury derivatives were found to occur in seafood. This increased the interest in the presence of various heavy metals such as mercury, lead, cadmium, and copper, as contaminants of food. This is of special importance in the industrialized world, where air pollution might be deleterious. The use of lead tetraethyl in petrol for motor vehicles, in combination with the enormous expansion of the automobile industry, has led to increased pollution problems; high contents of lead have been shown in food crops grown in the neighborhood of motorways.

The daily intake of various trace heavy minerals such as lead, mercury, and cadmium may be low. However, as there is a very limited capacity in the body to excrete them, even low intakes can result in the accumulation of toxic levels of these constituents in the body. The occurrence of methyl mercury in fish and shellfish as a result of environmental pollution has been of great concern during the last decade. The World Health Organization (WHO) has dealt with the problem in a special technical report (WHO, 1972). However, as the occurrence of heavy metals in natural products such as fish and shellfish is the result of manmade environmental pollution, they are not considered as naturally occurring toxicants, and will therefore not be dealt with in this survey.

Favism and Lathyrism

Favism occurs in special susceptible individuals who consume the legume *Vicia faba,* more commonly known as broad bean. It is also thought possible to induce favism by the inhalation of *Vicia faba* pollen. Favism includes hemolysis with high fever, icterus, liver enlargement, and splenomagaly. *Vicia faba* is cultivated in many areas of the world; however, favism occurs mainly in the Mediterranean countries (although cases are not rare in other districts as well, i.e., China, Iran, and Iraq).

The disease is caused by an inherited genetic enzyme defect, a deficiency of glucose-6-phosphate-dehydrogenase (Patwardhan and White, 1973). Extracts of *V. faba* that contain substances such as divicine and irouramil, oxidize glutathione to its disulphide and this cannot be sufficiently reduced to glutathione. Although the exact mechanism is still unknown this in some way has a destructive effect on the cell membranes of the red blood corpuscles.

Lathyrism is another type of disease caused by some toxic factors occurring in Lathyrus species. Two clinical varieties are described (Somogyi, 1978): *osteolathyrism,* which seems to occur exclusively in animals, and *neurolathyrism,* which also is described to occur in humans. The latter is the classical form that was known to Hippocrates. The symptoms in neurolathyrism comprise paresthesis, spastic paralysis, and may in extreme cases lead to death. Outbreaks of this disease are usually associated with famine when large quantities of lathyrus meal are consumed (Bell, 1973). There is still insufficient knowledge regarding the exact mechanism of the action of the toxins in neurolathyrism, although a number of substances have been described as exerting a neurolathyrogenic effect.

CONCLUSION

It is obvious from this short review that there are a number of toxicants that occur naturally in food. It is furthermore evident that most of them occur in plants, while those naturally occurring in the animal products usually are derived directly or indirectly from vegetable sources.

It is often said that the greatest risks of intoxication today are through the contamination of the food chain by various chemicals and environmental pollutants that are ultimately consumed by man. It should however never be forgotten that by far the greatest problem of toxicity of today among environmental carcinogens are the polycyclic hydrocarbons present in tobacco smoke.

In the present day debate, it is sometimes said that modern food technology and industrial preservation techniques lead to increased risks for the occurrence of toxic components in food, and that consumers demand food "without any chemicals." This is of course not correct as all foods are composed of chemicals.

It is furthermore often said that a food is unacceptable if it contains any harmful substances. This is another unrealistic statement as most if not all substances are toxic to some degree, and consequently also potentially harmful; it is always a question of dose–response. Thus, oxygen which is an essential component for life is harmful when increased 3- to 5-fold, similarly a 2- to 3-fold increase in protein intake might be deleterious for the neonate due to the immature metabolic capacity of its liver. *Consequently, it cannot be stated that any component in our environment or in our food could never be harmful.* More than 150 distinct chemical substances have been identified in the potato, among which certain antinutritional components as solanin, arsenic, alkaloids, tannin, and nitrate could be mentioned. There are also a number of substances that have no significant or recognized nutritional importance for man. Nevertheless, the potato is regarded as a natural food and a most valuable, cheap, and nutritious component in many basic human diets. This is furthermore illustrated by the fact that already in 1913 Hindhede had reported that the consumption of 2–4 kg of potatoes as the single source of nutrition every day for 300 days did not lead to any nutritional disturbances or deficiencies, neither to any toxical problems in man.

This review has tried to illustrate that there are a number of toxicants more or less harmful that occur naturally in food. Obviously, man has a capacity to handle these toxic components to a certain degree through the detoxifying capacity of the liver. Exposure to toxic components in the food is by no means a problem confined to the last decades or century. The use of certain additives and preservatives in food technology is often based on the assumption that they will maintain, prolong or even enhance the nutritional quality of the product throughout storage, as well as reduce microbial contamination or growth of microorganisms during the handling and storage. It still seems to be a valid statement that more people to date have died because of naturally occurring toxicants of chemical and microbiological origin in food than due to the use of preservatives in food technology. Nevertheless, specific legislation regarding the use of additives and preservatives in food is extremely more vast than legislation dealing with the problem of occurrence of natural toxicants in food. People still die of botulinum intoxication, but do we have any evidence that anybody has died as a result of the use of nitrite for the preservation of cured meat and provisions? The expanded use of organic or "natural foods" and marketing of health food stores, as well as the increased use of ancient recipes for homemade preservation of food items as an ingredient of food faddism during recent years, calls for intensified studies on the role of naturally occurring toxicants, and increased information to the public regarding their potential harm.

REFERENCES

Allison, A. C., 1963, Danger of vitamin K to newborn, *Lancet* **1:**669.

Anonymous, 1961, Treatment of hypercholesterolemia with nicotinic acid, *Nutr. Rev.* **19:**325.

Anonymous, 1968, Present knowledge of naturally occurring toxicants in food, *Nutr. Rev.* **26:**129.

Arnon, S. S., Midura, T. E., Damus, K., Wood, R. M., and Chin, J., 1978, Intestinal infection and toxin production by *Clostridium botulinum* as one cause of sudden infant death syndrome, *Lancet* **1:**1273.

Bailey, A. V., Harris, J. A., Skau, E. L., and Kerr, T., 1966, Cyclopropenoid fatty acid content and fatty acid composition of crude oils from twentyfive varieties of cottonseed, *J. Am. Oil Chem. Soc.* **43:**107.

Bell, E. A., 1973, Aminonitriles and aminoacids not derived from proteins, in: *Toxicants Occurring Naturally in Foods,* 2nd ed., p. 153, National Academy of Sciences, Washington, D.C.

Barger, G., 1931, *Ergot and Ergotism,* Guerney and Jackson, London.

Billek, G., 1973, Veränderungen der Fette unter Fritierbedingungen und deren analytische Erfassung. Chemische Veränderungen der Fette beim Fritieren, *Fette Seifen Anstrichm.* **75:**582.

Blackwell, B., Marley, M., Price, J., and Taylor, D. 1967, Hypertensive interaction between monoamine oxidase inhibitors and foodstuffs, *Br. J. Psychiatry* **113:**349.

Blix, G., 1965, A study on the relation between total calories and single nutrients in Swedish food, *Acta. Soc. Med. Ups.* **70:**117.

Blythe, C., 1976, Problems of diet and affluence, *Food Policy* **1:**91.

Boctor, A. M., and Harper, E. E., 1968, Tyrosine toxicity in the rat: The effect of high intake of *p*-hydroxyphenylpyruvic acid and force-feeding high tyrosine diet, *J. Nutr.* **95:**535.

Bowman, D. E., 1945, Amylase inhibitor of navy beans, *Science* **102:**358.

Boyd, W., and Shapleigh, E., 1954, Specific precipitating activity of plants agglutinins (lectins), *Science* **119:**419.

Bull, L. B., Culvenor, C. C. J., and Dick, A. T., 1968, *The Pyrrolizidine Alkaloids,* North-Holland Pub., Amsterdam.

Christensson, D. D., Wall, J. S., Dimler, R. J., and Booth, A. N., 1968, Nutritionally unavailable niacin in corn. Isolation and biological activity, *J. Agric. Food Chem.* **16:**100.

Christopherson, B. O., Svaar, H., Langmark, F. T., Gumpen, S. A., and Norum, K. R., 1976, Rapeseed oil and hydrogenated marine oils in nutrition, *Ambio* **5:**169.

Colombo, J. P., 1971, Congenital disorders of the urea cycle and ammonia detoxication, *Monographs on Pediatrics,* Vol. 1, S. Karger, Basel.

Connor, W. E., and Connor, S. L., 1976, Sucrose and Carbohydrate, in: *Present Knowledge in Nutrition,* 4th ed., p. 33, Nutrition Foundation, New York.

Darby, W. J., and Hambraeus, L., 1978, Proposed nutritional guidelines for utilization of industrially produced nutrients, *Nutr. Rev.* **36:**65.

Deinzer, M. L., Thomson, P. A., Burgett, D. M., and Isaacson, D. L., 1977, Pyrrolizidine alkaloids: Their occurrence in honey from tansy ragworth (*Senecio jacobaea* L.), *Science* **195:**497.

DiGeorge, A. M., and Auerbach, V. H., 1960, Leucine-induced hypoglycemia. A review and speculations, *Am. J. Med. Sci.* **240:**792.

Doster, A. R., Mitchell, F. E., Farrell, R. L., and Wilson, B. J., 1978, The effect of 4-ipomeanol, a product from mold-damaged sweet potatoes, on the bovine lung, *Vet. Pathol.* **15:**367.

Eakin, R. E., Snell, E. E., and Williams, R. J., 1940, A constituent of raw egg white capable of inactivating biotin *in vitro, J. Biol. Chem.* **136:**801.

Engfeldt, B. (ed.), 1975, Morphological and biochemical effects of orally administered rapeseed oil on rat myocardium, *Acta. Med. Scand. Suppl.* **585.**

Feeney, R. E., Means, G. E., and Bigler, J. C., 1969, Inhibition of human trypsin, plasmin, and thrombin by naturally occurring inhibitors of proteolytic enzymes, *J. Biol. Chem.* **244:**1957.

Fossum, K., and Whitaker, J. R., 1968, Ficin and papain inhibitor from chicken egg white, *Arch. Biochem. Biophys.* **125:**536.

Glass, R. L., Krick, T. P., Sand, D. M., Rahn, C. R., and Schlenk, H., 1975, Furanoid fatty acids from fish lipids, *Lipids* **10:**695.

Gopalan, C., and Rao, K. S. J., 1975, Pellagra and amino acid imbalance, *Vitam. Horm.* **33:**505.

Green, R. G., 1937, Chastek paralysis, *Minn. Wildl. Dis. Inv.* **3:**83.

Greer, M. A., 1962, The natural occurrence of goitrogenic agents, *Rec. Progr. Horm. Res.* **18:**197.

Gustafsson, J.-Å., 1980, personal communication.

Hambraeus, L., Hardell, L. I., Westphal, O., Lorentsson, R., and Hjorth, G., 1974, Argininosuccinic aciduria. Report of three cases and the effect of high and reduced protein intake on the clinical state, *Acta Paediatr. Scand.* **63:**525.

Harper, A. E., 1958, Balance and imbalance of amino acids, *Ann. N. Y. Acad. Sci.* **69:**1025.

Harper, A. E., 1964, Amino acid toxicities and imbalances, in: *Mammalian Protein Metabolism* (H. N. Munro and J. B. Allison, eds.), *Vol. II,* p. 87, Academic Press, New York.

Harper, A. E., 1973, Amino acids of nutritional importance, in: *Toxicants Occurring Naturally in Foods,* 2nd ed., p. 130, National Academy of Sciences, Washington, D.C.

Harper, A. E., and Rogers, O. E., 1965, Amino acid imbalance, *Proc. Nutr. Soc.* **24:**173.

Harper, A. E., Benevenga, N. J., and Wohlhueter, R. M., 1970, Effects of ingestion of disproportionate amounts of amino acids, *Physiol. Rev.* **50:**428.

Hawksworth, G., and Hill, M. J., 1971, The formation of nitrosamines by human intestinal bacteria, *Biochem. J. Proc.* **122:**28.

Hill, K. R., 1960, Discussion on seneciosis in man and animals, *Proc. R. Soc. Med.* **53:**281.

Herting, D. C., 1966, Perspective on vitamin E, *Am. J. Clin. Nutr.* **19:**210.

Hindhede, M., 1913, Studien über Eiweissminimum, *Skand. Arch. Physiol.* **30:**97.

Hites, B. D., Sandstedt, R. M., and Schaumburg, L., 1951, Study of proteolytic activity in wheat flour doughs and suspensions. II. A papain inhibitor in flour, *Cereal Chem.* **28:**1.

Hobbs, B. C., and Gilbert, R. J., 1978, *Food Poisoning and Food Hygiene,* 4th ed., E. Arnold, London.

Hodges, R. E., 1976, Ascorbic acid, in: *Present Knowledge in Nutrition,* 4th ed. p. 119, Nutrition Foundation, New York.

Horwith, M. K., 1962, Interrelations between vitamin E and polyunsaturated fatty acids in adult man, *Vitam. Horm.* **20:**541.

Horwitz, D., Lovenberg, W., Engelman, K., and Sjoerdsma, A., 1964, Monoamine oxidase inhibitors, tyramine and cheese, *J. Am. Med. Assoc.* **188:**1108.

Jaffé, W. G., 1973, Toxic proteins and peptides, in: *Toxicants Occurring Naturally in Foods,* 2nd ed., p. 106, National Academy of Sciences, Washington, D.C.

Johnson, J. D., Kertchner, N., and Simoons, F. J., 1974, Lactose malabsorption: Its biology and history, *Adv. Pediatr.* **1:**197.

Josefsson, E., 1967, Distribution of thioglucosides in different parts of Brassica plants, *Phytochemistry* **6:**1617.

Keeney, D. R., 1970, Nitrates in plants and waters, *J. Milk Food Technol.* **33:**425.

Keys, A. (ed.), 1970, Coronary heart disease in seven countries, *Circulation Suppl.* **1:**61.

Kildeberg, P., 1964, Disturbances of hydrogen ion balance occurring in premature infants. II. Late metabolic acidoses, *Acta Paediat. Scand.* **53:**517.

King, R. A., 1949, Vitamin E therapy in Dupuytren's contracture, *J. Bone Jt. Surg.* **31B:**443.

Kingsbury, J. M., 1964, *Poisonous Plants of the United States and Canada,* Prentice-Hall, Englewood Cliffs, New Jersey.

Klavins, J. V., and Peacocke, I. L., 1964, Pathology of amino acid excess. III. Effects of administration of excessive amounts of sulphur-containing amino acids: Methionine with equimolar amounts of glycine and arginine, *Br. J. Exp. Pathol.* **45:**533.

Kratzer, F. H., Williams, D. E., Marshall, B., and Davis, P. N., 1954, Some properties of the chick growth inhibitor in linseed oil meal, *J. Nutr.* **52**:555.

Kwok, R. H., 1968, Chinese restaurant syndrome, *N. Engl. J. Med.* **278**:796.

Lamden, M. P., 1971, Dangers of massive vitamin C intake, *N. Engl. J. Med.* **284**:336.

Lea, C. H., 1962, The oxidative deterioriation of food lipids, in: *Symposium on Foods: Lipids and their Oxidation.* (H. W. Schultz, ed.), p. 3, Avi Pub., Westport, Connecticut.

Learmonth, E. M., 1958, The influence of soya flour on bread doughs. III. The distribution of the papain-inhibiting factor in soya beans, *J. Sci. Food Agric.* **9**:269.

Lijinsky, W., and Epstein, S. S., 1968, Nitrosamines as environmental carcinogens, *Nature* (London) **225**:21.

Linneweh, F., Ehrlich, M., Graul, E. H., and Hundeshagen, H., 1963, Über den Aminosäurentransport bei Phenylketonuricher Oligophrenia, *Klin. Wochenschr.* **41**:253.

Lovenberg, W., 1973, Some vaso- and psychoactive substances in food: amines, stimulants, depressants, and hallucinogens, in: *Toxicants Occurring Naturally in Foods,* 2nd ed., p. 170, National Academy of Sciences, Washington, D. C.

Mikolajzac, K. I., Freidinger, R. M., Smith, C. R., Jr., and Wolff, I. A., 1968, Oxygenated fatty acids of oil from sunflower seeds after prolonged storage, *Lipids* **3**:489.

Nater, J. P., and Doeglas, H. M. G., 1970, Halibut liver poisoning in 11 fishermen, *Acta Dermatol.* **50**:109.

Neame, K. D., 1961, Phenylalanine as inhibitor of transport of amino acids in brain, *Nature (London)* **192**:173.

Nikkilä, E. A., 1974, Influence of dietary fructose and sucrose on serum triglycerides in hypertriglyceridaemia and diabetes, in: *Sugars in Nutrition* (H. L. Sipple and K. W. McNutt, eds.), p. 439, Academic Press, New York.

Oberleas, D., 1973, Phytates, in: *Toxicants Occurring Naturally in Foods,* 2nd ed., p. 363, National Academy of Sciences, Washington, D.C.

Oké, O. L., 1969, Oxalic acid in plants and in nutrition, *World Rev. Nutr. Diet* **10**:262.

Orgell, W. H., 1963, Inhibition of human plasma cholinesterase *in vitro* by alkaloids, glycosides, and other natural substances, *Lloydia* **26**:36.

Osborne, T. B., and Mendel, L. B., 1917, The use of soybean as food, *J. Biol. Chem.* **32**: 369.

Patwardhan, V. N., and White, J. W., Jr., 1973, Problems associated with particular foods, in: *Toxicants Occurring Naturally in Foods,* 2nd ed., p. 477, National Academy of Sciences, Washington, D.C.

Poswillo, D. E., Sopher, D., and Mitchell, S., 1972, Experimental induction of foetal malnutrition with "blighted" potato: A preliminary report, *Nature* (London) **239**:462.

Pusztai, A., 1968, General properties of a protein inhibitor from the seeds of kidney bean, *Eur. J. Biochem.* **5**:252.

Reddy, B. S., Narisawa, T., Vukurich, D., Weisburger, J. H., and Wynder, E. L., 1976, Effect of quality and quantity of dietary fat and dimethylhydrazine in colon carcinogenesis in rats, *Proc. Soc. Exp. Biol. Med.* **151**:237.

Refsum, S., 1946, Heredopathia atactica polyneuritiformis, *Acta Psychiatr. Scand. Suppl.* **38**:9.

Renwick, J. H., 1972, Hypothesis Anencephaly and spina bifida are usually preventable by avoidance of a specific but unidentified substance present in certain potato tubes, *Br. J. Prev. Soc. Med.* **26**:67.

Roine, P., Uksila, E., Teir, H., and Rapola, J., 1960, Histopathological changes in rats and pigs fed rapeseed oil, *Z. Ernährungswiss.* **1**:118.

Rosenthal, G., 1971, Interaction of ascorbic acid and warfarin, *J. Am. Med. Assoc.* **215**:1671.

Rühl, R., 1951, Beitrag zur Pathologie und Toxikologie des Solanins, *Arch. Pharm. (Paris)* **284**:67.

Salmon, W. D., 1958, The significance of amino acid imbalance in nutrition, *Am. J. Clin. Nutr.* **6**:487.

Sandström, B.-M., 1980, Zinc absorbtion from composite meal, *Näringsforskn.* **24**:110.

Schantz, E. J., Ghazarossian, V. E., Schnoes, H. K., Strong, F. M., Springer, J. P., Pezzanite, J. O., and Clardy, J., 1975, The structure of saxitonin, *J. Am. Chem. Soc.* **97**:1238.

Select Committee on Nutrition and Human Needs, U.S. Senate, 1977, *Dietary Goals for the United States,* 2nd ed., Government Printing Office, Washington, D.C.
Sen Gupta, A. K., 1972, Epoxy-triglycerides in soyabean oil, *Chem. Ind.* **March 1972:**257.
Shank, R. C., and Magee, P. N., 1967, Similarities between the biochemical actions of cycasin and dimethylnitrosamine, *Biochem. J.* **105:**521.
Sidransky, H., and Baba, T., 1960, Chemical pathology of acute amino acid deficiencies. III. Morphological and biochemical changes in young rats fed valine or lysine-devoid diet, *J. Nutr.* **70:**463.
Smaskevskii, N. D., 1966, A natural antivitamin of pantothenic acid, *Chem. Abstr.* **65:**2677e.
Smith, R. J., 1980, Nitrites: FDA beats a surprising retreat, *Science* **209:**1100.
Somogyi, J. C., 1978, Natural toxic substances in food, in: *Foreign Substances and Nutrition* (J. C. Somogyi and R. Tarjan, eds.), *Wld. Rev. Nutr. Diet* **29:**42.
Stanbury, J. B., Wyngaarden, J. B., and Fredrickson, D. S., 1978, *The Metabolic Basis of Inherited Disease,* 4th ed., McGraw-Hill Book Co., New York.
Stansby, M. E., 1967, *Fish Oils: Their Chemistry, Technology, Stability, Nutritional Properties and Uses,* Avi Pub., Westport, Connecticut.
Steggerda, F. R., 1968, Gastrointestinal gas following food consumption, *Ann. N. Y. Acad. Sci.* **150:**57.
Tannenbaum, S. R., Fett, D., Young, V. R., Land, P. D., and Bruce, W. R., 1978, Nitrite and nitrate are formed by endogenous synthesis in the human intestine, *Science* **200:**1487.
Truswell, A. S., 1978, Diet and plasma lipids: A reappraisal, *Am. J. Clin. Nutr.* **31:**977.
Vergroesen, A. J. (ed.), 1975, *The Role of Fats in Human Nutrition,* Academic Press, New York.
Virtanen, A. I., Kreula, M., and Kiesvaara, M., 1963, Investigations on the alleged goitrogenic properties of cow's milk, *Z. Ernaehrungswiss. Suppl.* **3:**23.
Whitaker, J. R., and Feeney, R. E., 1973, Enzyme inhibitors in foods, in: *Toxicants Occurring Naturally in Foods,* 2nd ed., p. 276, National Academy of Sciences, Washington, D.C.
Wilson, B. J., 1978, Hazards of mycotoxins to public health, *J. Food Prot.* **41:**375.
Wilson, B. J., 1979, Naturally occurring toxicants in foods, *Nutr. Rev.* **37:**305.
Wilson, B. J., and Hayes, A. W., 1973, Microbial toxins, in: *Toxicants Occurring Naturally in Foods,* 2nd ed., p. 372, National Academy of Sciences, Washington, D.C.
Wilson, B. J., Hoekman, T., and Dettbarn, W. D., 1972, Effects of a fungus tremorgenic toxin (penitrem A) on transmission in rat phrenic nerve-diaphragm preparations, *Brain Res.* **40:**540.
World Health Organization, 1972, Evaluation of certain food additives and the contaminants mercury, lead and cadmium, *W. H. O. Tech. Rep. Ser. No. 505.*
Wretlind, A., Truswell, A. S., Hejda, S., Isaksson, B., Kübler, W., and Vivanco, F., 1976, Round table on comparison of dietary recommendations in different European countries, *Nutr. Metab.* **21:**210.
Yudkin, J., and Morland, J., 1967, Sugar intake and myocardial infarction, *Am. J. Clin. Nutr.* **20:**503.
Ziemlanski, S., 1977, Pathophysiological effects of long-chain fatty acids, *Bibl. Nutr. Dieta* **25:**134.

2

INHIBITORS OF ENZYMES IN BIOLOGICAL MATERIALS USED FOR FOODS

John R. Whitaker

INTRODUCTION

Any compound that reduces the observed activity of an enzyme is an inhibitor, by definition. Compounds inhibitory to enzymes occur naturally in biological materials as well as being added intentionally or inadvertently during production, harvesting, or storage. Naturally occurring inhibitors include polyphenolic materials such as the tannins; chelators of essential metal cofactors of enzymes such as phytic acid and ascorbic acid; the peptide inhibitors, especially those excreted by microorganisms; and specific proteins found in plants, animals, and microorganisms. Most of the pesticides and insecticides used are specific enzyme inhibitors, especially of acetylcholinesterase. Compounds such as ascorbic acid and sodium bisulfite are added to raw materials such as peaches and apples to prevent browning; these compounds are inhibitors of polyphenol oxidase.

With the exception of the phenolic compounds, which react generally with proteins and often cause precipitation, the other compounds are rather specific in enzyme inhibition. In the following discussion major emphasis will be on the naturally occurring enzyme inhibitors.

EXAMPLES OF ENZYME INHIBITORS

Cholinesterase Inhibitors

Many of the compounds, such as the organophosphorus and carbamate insecticides are designed specifically to inhibit the acetylcholinesterase of in-

John R. Whitaker • Department of Food Science and Technology, University of California, Davis, California 95616.

sects, thereby disrupting the nervous system. These compounds, when used properly, are slowly decomposed after application to the growing plant and generally do not appear in significant amounts in the harvested food. When improperly used, they can occur not only in plant raw materials but also in animal products through the food chain. These compounds are not destroyed by heat processing.

Inhibitors of acetylcholinesterase also occur naturally. The best known example is the alkaloid, solanine, found in white potatoes (Whitaker and Feeney, 1973). The solanine level of potatoes increases markedly with prolonged exposure to sunlight ("greening") or through breeding. Potatoes in the U.S. market place should contain less than 200 ppm of solanine.

Polyphenol Oxidase Inhibitors

The addition of ascorbic acid, sodium bisulfite, and chelating agents such as EDTA are permitted in some food materials in order to control browning due to polyphenol oxidase activity on phenolic compounds occurring naturally in plant materials. Ascorbic acid (vitamin C) is effective because it lowers the pH, decreasing the activity of the enzyme, and because it reduces the colored product back to the colorless substrate. Sodium bisulfite acts primarily by reducing the colored product to a colorless compound but it may also directly inhibit the enzyme. Chelating agents, such as EDTA, act by removing the essential metal ion, copper, from the enzyme.

Naturally occurring polyphenol oxidase inhibitors have also been reported, especially in fungi that grow well on rotting and decaying plant materials. These may eventually prove useful in control of the undesirable browning of plant materials.

Peptide Inhibitors

These enzyme inhibitors are low molecular weight compounds containing amino acids (Whitaker, 1981). They are found in some snake and bee venoms, in leeches, and in cuttlefish, and are produced by a number of microorganisms, especially the streptomyces. Baker's yeast contains several peptide inhibitors effective against the proteinases A, B, and C produced by the same organism. The peptide inhibitors range in molecular weight from about 200 to 6000. They are effective in microgram amounts, and, being quite specific for certain types of proteinases, may prove useful in the treatment of certain types of diseases such as muscle dystrophy and emphysema.

Some of the peptide inhibitors have unique structures. For example, the two inhibitors from Brazilian snake *(Bothrops jararaca)* venom have the structures

[Gln-Lys-Try-Ala-Pro (mol. wt. = 608)

[Gln-Try-Pro-Arg-Pro-Gln-Ile-Pro-Pro (mol. wt. = 1100)

Both inhibitors contain pyrrolidone carboxylic acid ([Gln) as the NH_2-terminal amino acid and are particularly rich in proline.

Several of the low molecular weight peptide inhibitors produced by microorganisms contain an aldehyde group (for example, the leupeptins, antipain, chymostatin A, and elastatinal). The basic structure of leupeptins is

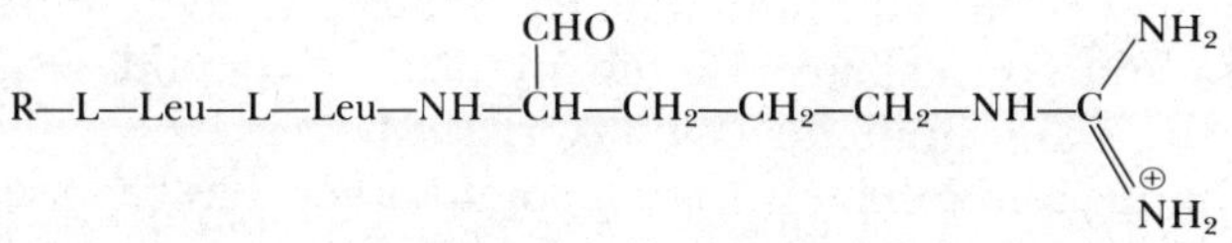

$$\text{R—L—Leu—L—Leu—NH—}\underset{}{\overset{\text{CHO}}{\overset{|}{\text{CH}}}}\text{—CH}_2\text{—CH}_2\text{—CH}_2\text{—NH—C}\begin{matrix}\diagup \text{NH}_2 \\ \\ \diagdown\!\!\diagdown \overset{\oplus}{\text{NH}_2}\end{matrix}$$

The R group is either CH_3CO- or CH_3-CH_2CO-. The peptide is composed of the regular amino acids L-leucine and L-arginine. However, it is unique in that the COOH-terminal carboxyl group (of arginine) has been reduced to an aldehyde group. The leupeptins inhibit plasmin, trypsin, papain, and cathepsin B at the 10–50 μg/ml level. They have been shown to decrease protein degradation in normal and diseased muscles (Libby and Goldberg, 1978).

Phosphoramidon, produced by *Streptomyces tanashiensis,* specifically inhibits thermolysin at the nM level. Its structure is

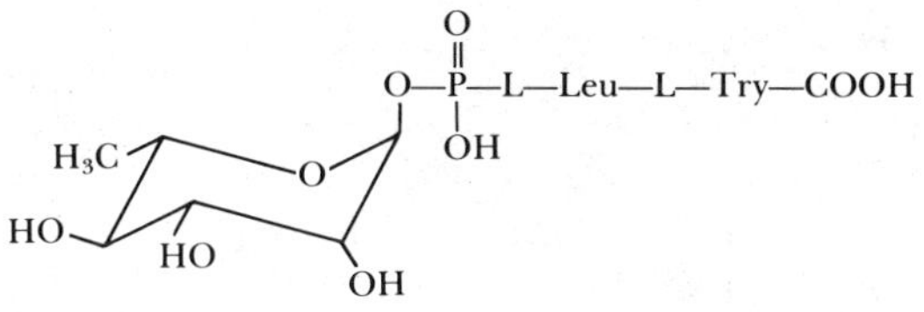

$$\text{(}\alpha\text{-L-rhamnopyranosyl; H}_3\text{C, HO, HO, OH)—O—}\underset{\text{OH}}{\underset{|}{\overset{\text{O}}{\overset{\|}{\text{P}}}}}\text{—L—Leu—L—Try—COOH}$$

It is even more active when the α-L-rhamnopyranosyl group is removed.

Protein Inhibitors

All animals, plants and microorganisms contain specific proteins able to inactivate selected enzymes by formation of stable enzyme-inhibitor complexes (Whitaker, 1981). The most studied of these inhibitors are the trypsin inhibitors of the blood and pancreas, egg whites, and legumes. However, protein inhibitors of other proteinases and of nonproteinases also are found. Examples include the chymotrypsin and elastase inhibitors, the inhibitors of α-amylase, the inhibitors of DNase and RNase and of peroxidase and catalase.

The enormous complexity and diversity of protein inhibitors of enzymes have only recently been appreciated. The complexity and diversity include the many types of inhibitors found in a single tissue, the presence of multiple molecular forms of the same inhibitor, and multiple binding sites on the same inhibitor. A few examples will illustrate this point.

Human plasma contains at least nine different types of inhibitors [α_1-antitrypsin inhibitor, α_1-antichymotrypsin inhibitor, inter-α-trypsin inhibitor, antithrombin III inhibitor, $C\bar{1}$ inactivator, α_2-macroglobulin, a β-lipoprotein-neutralizing thrombin, an inhibitor of plasminogen activation and a thiol protease inhibitor (Heimburger, 1974)]. Potato tubers contain at least six different types of proteinase inhibitors [chymotrypsin inhibitor I, proteinase inhibitors IIa and IIb, kallikrein inhibitors, carboxypeptidases A and B inhibitors, and a papain inhibitor (Whitaker, 1981)].

The chymotrypsin inhibitor I from potato tubers has been separated into at least ten isoinhibitors. There are at least seven forms of the acidic sulfhydryl protease inhibitor from pineapple and four to six forms of the lima bean trypsin inhibitor.

Some of the inhibitors contain multiple binding sites for enzymes. Chicken (egg white) ovomucoid complexes with one molecule of trypsin only, turkey ovomucoid complexes with one molecule of trypsin and one molecule of chymotrypsin simultaneously and independently, while duck ovomucoid complexes with two molecules of trypsin and one molecule of chymotrypsin simultaneously and independently. The ovoinhibitors of chicken and Japanese quail egg whites have six tandem domains of remarkable homology, each potentially able to complex with proteinases (Laskowski *et al.*, 1978).

IMPORTANCE OF INHIBITORS OF ENZYMES

Nutritional Importance

Initial interest in the protein inhibitors of enzymes came from nutritionists concerned with the effect of these inhibitors in feeds and foods on animal and human nutrition. Raw soybean flour inhibits growth in rats, chicks, and other monogastric animals, and raw garden bean flour (*Phaseolus vulgaris* L.) can cause death of animals. Perhaps 40% of the growth retardation is caused by the proteinase inhibitors (Kakade *et al.*, 1973). Soybean trypsin inhibitor enhances the formation of a humoral pancreozymic-like substance that markedly stimulates external secretion by the pancreas. A net result of the enhanced secretion is enlargement of the pancreas due to hyperplasia of some of the cells and an increased stress caused by loss of limiting amino acids, especially the sulfur-containing amino acids. Assessment of the contributions of the enzyme inhibitors to growth retardation is complicated by the presence of other antinutritive factors in beans such as the hemagglutinins (lectins), goitrogenic substances, phytic acid, estrogenic substances, and phenolic compounds.

The nutritional significance of α-amylase inhibitors is not clear. Feeding a single dose of purified wheat amylase inhibitor, along with a high starch diet, has resulted in a reduction in the rate of appearance of glucose in the blood of dogs, rats, and humans. *Ad libitum* feeding of massive doses of purified red kidney bean amylase inhibitor, free of all detectable trypsin inhibitor and hemagglutinin activity, had no effect on growth of rats, but there may be an effect in mice. Undigested starch was excreted in the feces of rats fed raw white kidney beans and purified wheat inhibitors.

Medical Importance

Much of the stimulus for work on the protein and peptide inhibitors is due to potential or actual medical importance. The α_1-antitrypsin inhibitor appears to play a key role in controlling proteolytic activity in lung tissue, due to proteinases of the granulocytes and macrophages elicited as a result of constituents of smog, etc. Lowered levels of α_1-antitrypsin inhibitor may also be associated with respiratory distress syndromes, adult liver disease and hepatic cirrhosis in infancy (Evans *et al.,* 1970).

α_2-Macroglobulin is present in an inactive form in patients with various joint diseases, and low levels of $C\bar{1}$ inactivator in plasma have been correlated with hereditary angioneurotic edema. Decreased levels of inhibitors of the blood clotting system lead to abnormal control of intravascular clotting and fibrinolysis.

The low molecular weight peptide inhibitors from microorganisms have been suggested as therapeutic agents for hypertension, stomach ulcers, cartilage disorders such as osteoarthritis, pancreatitis, burns, contraceptive agents, and muscular dystrophy (Wingender, 1974).

Physiological Significance

In a few cases, the physiological role of the inhibitors is known and either involves a control mechanism or a protective mechanism. The pancreatic proteinase inhibitors protect against the premature activation of trypsinogen, chymotrypsinogen, and procarboxypeptidases A and B. The plasma enzyme inhibitors protect against premature activation of the proteinase zymogens of the blood clotting cascade, and serve as a regulatory mechanism between coagulation and fibrinolysis. This may also protect against liberation of pancreatic proteinases in such diseases as pancreatitis. The proteinase inhibitors of the respiratory tract appear to serve as a protective mechanism against the proteinases of granulocytes and macrophages that accumulate as a result of irritation and/or diseased conditions or through inhalation of microorganisms.

Inhibitors in plants are thought to be a protection against insects or microorganisms or both, although this needs further study. Infestation of potato leaves with Colorado potato beetle larva results in increased levels of protease inhibitor I in the leaves. All insect α-amylases tested are inhibited by red kidney bean amylase inhibitor although the inhibitor is not active against microbial or higher plant α-amylases. It has been suggested that wheat α-amylase inhibitors may protect against attack of wheat by insects during storage. The proteinase inhibitors may also be physiologically important in plant metabolism.

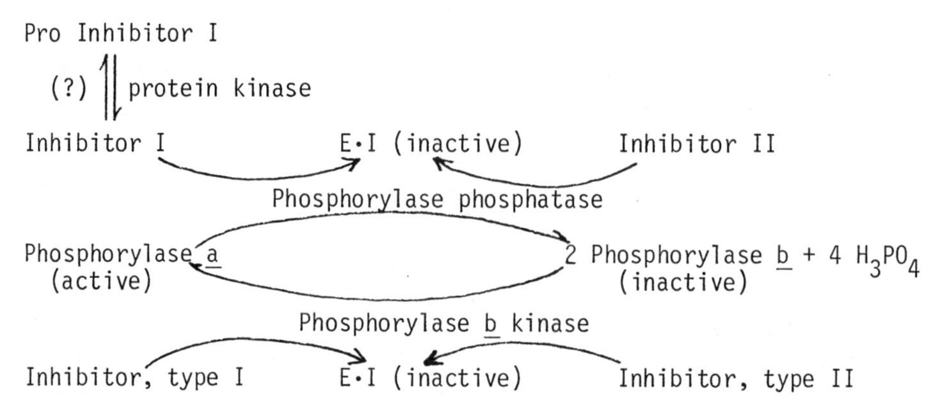

FIGURE 1. Regulation of glycogen breakdown and synthesis by four enzyme inhibitors.

Protein inhibitors often play a regulatory role *in vivo*. For example, the two key enzymes in glycogen breakdown and synthesis, phosphorylase and glycogen synthase, are regulated by at least four inhibitors, as shown in Fig. 1. Phosphorylase phosphatase is regulated by two inhibitors, protein phosphatase Inhibitor I (which must be activated by phosphorylation by a protein kinase) and protein phosphatase Inhibitor II.(Cohen *et al.*, 1977). There are also two inhibitors of phosphorylase *b* kinase (Szmigielski *et al.*, 1977). Protein kinase Inhibitor, type I, inhibits only cAMP-dependent protein kinases while protein kinase Inhibitor, type II, has activity toward nucleotide-independent protein kinases.

Food Storage and Processing Significance

Enzymatic darkening of fruits and vegetables such as apples, peaches and potatoes caused by polyphenol oxidase is a major problem. Not only is there loss in color, there are detrimental changes in taste and nutritional quality. Therefore, much attention has been given to the use of sodium bisulfite, ascorbic acid, and chelating agents, for example, as inhibitors of polyphenol oxidase, with only partial success.

While much is known about some of these inhibitors, such as the trypsin inhibitors, much still remains to be done in determining the physiological role, medical uses, and the mechanism of action of these inhibitors.

REFERENCES

Cohen, P., Nimmo, G. A., and Antoniw, J. F., 1977, Specificity of a protein phosphatase inhibitor from rabbit skeletal muscle, *Biochem. J.* **162:**435.

Evans, H. E., Levi, M., and Mandl, I., 1970, Serum enzyme inhibitor concentrations in respiratory distress syndrome, *Am. Rev. Respir. Dis.* **101:**359.

Heimburger, N., 1974, Biochemistry of proteinase inhibitors from human plasma: A review of recent developments, in: *Proteinase Inhibitors* (H. Fritz, H. Tschesche, L. J. Greene, and E. Truscheit, eds.), Proc. Int. Res. Conf., 2nd (Bayer Symp. V), pp. 14–22, Springer-Verlag, Berlin and New York.

Kakade, M. L., Hoffa, D. E., and Liener, I. E., 1973, Contribution of trypsin inhibitors to the deleterious effects of unheated soybeans fed to rats, *J. Nutri.* **103:**1772.

Laskowski, M., Jr., Kato, I., and Kohr, W. J., 1978, Protein inhibitors of serine proteinases; convergent evolution, multiple domains and hypervariability of reactive sites, in: *Versatility of Proteins* (C. H. Li, ed.), pp. 307–318, Academic Press, New York.

Libby P., and Goldberg, A. L., 1978, Leupeptin, a protease inhibitor, decreases protein degradation in normal and diseased muscles, *Science* **199:**534.

Szmigielski, A., Guidotti, A., and Costa, E., 1977, Endogenous protein kinase inhibitors. Purification, characterization and distribution in different tissues, *J. Biol. Chem.* **252:**3848.

Whitaker, J. R., 1981, Naturally occurring peptide and protein inhibitors of enzymes, in: *Impact of Toxicology on Food Processing* (J. C. Ayres and J. C. Kirschman, eds.), pp. 57–104, Avi Pub., Westport, Connecticut.

Whitaker, J. R., and Feeney, R. E., 1973, Enzyme inhibitors in foods, in: *Toxicants Occurring Naturally in Foods* 2nd ed. (F. M. Strong, ed.), pp. 276–298, National Academy of Sciences, Washington, D.C.

Wingender, W., 1974, Proteinase inhibitors of microbial origin. A review, in: *Proteinase Inhibitors* (H. Fritz, H. Tschesche, L. J. Greene, and E. Truscheit, eds.), Proc. Int. Res. Conf., 2nd (Bayer Symp. V), pp. 548–559, Springer-Verlag, Berlin and New York.

3

AN OUTBREAK OF AFLATOXICOSIS IN MAN

S. G. Srikantia

INTRODUCTION

There has been in recent years growing concern about the possible health hazards of mycotoxins. Under unsatisfactory conditions of food storage, particularly when environmental temperature and moisture are high, fungal contamination can easily occur. Several food commodities have been known to be attacked by a variety of storage fungi, and several of these fungi produce toxic metabolites. Among the more important of these is *Aspergillus flavus*—a fungus that produces a group of toxins collectively known as aflatoxins. Although the results of many studies clearly indicated that aflatoxin is hepatotoxic to farm animals and a number of experimental animals, its possible toxicity to man has largely been speculative.

Toward the end of 1974 there was an epidemic of an unusual liver disease in circumscribed rural areas of two northwestern states of India—Rajasthan and Gujarat. This outbreak was investigated separately by teams of scientists from the National Institute of Nutrition, Hyderabad, (Krishnamachari *et al.*, 1975 a,b; Krishnamachari and Bhat, 1977) and the All India Institute of Medical Sciences, New Delhi (Tandon *et al.*, 1977; Tandon *et al.*, 1978). The results have strongly suggested that the disease outbreak was the result of consumption of aflatoxin contaminated maize.

CLINICAL FEATURES

The illness had a subacute onset and was characterized by low-grade fever and general uneasiness as early manifestations. Rapidly deepening jaundice

S. G. Srikantia • National Institute of Nutrition, Indian Council of Medical Research, Hyderabad 500007, India.

soon made its appearance, which was followed by ascites. Collateral vessels on the abdomen were prominent in a few. Edema of the lower extremities was present in a substantial proportion of cases. Gastrointestinal manifestations were not marked, although vomiting did occur in some subjects early in the course of the disease. Anorexia was not a dominant symptom. Enlargement of the liver was more common in patients (80%) than was enlargement of the spleen (30%). While 60–80% of patients recovered within 2–8 weeks, the rest died generally 6–8 weeks after the onset of the disease. Massive gastrointestinal bleeding was the cause of death and hepatic coma was rare. The clinical picture was thus one of acute liver injury.

BIOCHEMICAL AND PATHOLOGICAL CHANGES

Most subjects had moderate leucocytosis, due mainly to an increase in neutrophils. Levels of serum bilirubin were considerably raised. Alkaline phosphatase and transaminase activity in serum were elevated in 60% of subjects. Hypoproteinemia and hypoalbuminemia were also seen (Tandon *et al.*, 1977).

Examination of biopsies of the liver showed that the normal lobular pattern was absent. Characteristic features were extensive centrizonal scattering, hepatic venous occlusion, marked ductular proliferation, periductal fibrosis, and multinucleated giant cells. There was considerable bile stasis in some of the proliferated bile ducts, dilated biliary canaliculi, and in the cytoplasm of both parenchymal and Kupffer cells (Krishnamachari *et al.*, 1975a,b; Tandon *et al.*, 1978). The histological picture was unlike that seen either in viral hepatitis or in Budd–Chiari syndrome. Nor was it like that in veno-occlusive disease. There was no evidence of recent or past thrombosis. In fact, the histological features of the liver did not resemble those described in any human disease, but did somewhat resemble changes seen in experimentally induced aflatoxin damage in animals. In two out of seven serum samples examined, aflatoxin B_1 could be detected. None of the urine samples, however, was positive for the toxin. Thin layer chromatographic analysis of extracts of the liver showed three prominent spots with green, light blue, and bright blue-violet fluorescence under ultraviolet light (Krishnamachari *et al.*, 1975). None of these spots was aflatoxin, and the real nature of these spots was uncertain.

EPIDEMIOLOGY

The disease broke out in drought stricken and chronic scarcity areas and the epidemic was strictly confined to rural areas. The affected villages were inhabited by tribal populations of very poor socioeconomic status. The habitual food intake was inadequate and the general nutrition of the subjects was unsatisfactory. The disease had an explosive onset and broke out in several widely scattered villages simultaneously. The involvement of more than one member in a household was a characteristic feature and dogs in the affected households also suffered from the same disease. The disease was confined to households in which maize was the staple food. During October 1974 there were unseasonal rains which drenched the standing maize crop, and the outbreak commenced within a few weeks after the harvesting and consumption of this rain-affected and badly stored maize. The maize had been stored as cobs, with the outer coverings intact. No new cases were seen after the locally grown maize had been exhausted. Males appeared to suffer twice as frequently as did females. A great majority of patients were aged between 5 and 35 years, with a peak between 25 and 35 years. There was not a single case below the age of one year and very few children in the group 1–5 years were affected. The exact number of subjects who suffered from the disease is not known, but it is clear that several hundred were involved with a high mortality rate (Krishnamachari *et al.,* 1975, Tandon, *et al.,* 1977).

This outbreak of hepatitis had several characteristics distinctive of an outbreak of mycotoxicosis, viz., nontransmissibility, nonresponsiveness to drugs, seasonal character of the outbreak, and involvement of a specific food item. Detailed studies were, therefore, undertaken to determine whether there were fungal toxins in the food consumed. Analysis of randomly selected samples of maize, sorghum, wheat, and millet, which were commonly consumed in the area, showed that only in maize samples was aflatoxin present. In samples of maize obtained from households in which members were affected, levels of aflatoxin ranged between 6 and 16 ppm (Krishnamachari *et al.,* 1975a,b). The affected maize grains were shriveled and in many instances black in color. Many households were aware of the spoiled nature of the maize, but it was still eaten. In fact, in most houses, the relatively better cobs were preserved for seed purposes and for later use, and visibly spoiled maize was consumed quickly over a period of a few weeks. On an average, an adult had consumed about 350 g of maize daily. Several subjects would, thus, have consumed between 2 and 6 mg of the toxin daily for several weeks.

These findings strongly suggest that the outbreak of hepatitis was due to aflatoxicosis. Unseasonal rains and unsatisfactory storage methods of maize appeared to have been responsible for the heavy contamination of cobs and grains with fungus, and subsequent elaboration of the toxin. Poverty, lack of alternate food, and ignorance about the potential damage of consuming moldy grains were major causes for the outbreak of the disease.

It needs to be pointed out that although there is considerable evidence that aflatoxin is concerned in the etiology of the disease, there are certain epidemiological observations that require explanation. For instance, the difference in the frequency of consumption of moldy corn among the affected and nonaffected families was not marked. Also, aflatoxin was isolated with equal frequency from foods collected from affected and unaffected families (Tandon *et al.,* 1978). These do not, at first sight, support a straight-forward aflatoxin etiology. Possible differences in aflatoxin concentration in different corn samples, amount of contaminated corn consumed, duration over which such corn was eaten, the nutritional status of individual subjects, and coexistent parasitic infections may, however, explain these observations (Tandon *et al.*, 1978). Biological variations in response to aflatoxin ingestion, which have been demonstrated in rhesus monkeys, may also have operated in the human situation. The presence of detectable levels of aflatoxin in plasma and the histopathological changes in the liver, taken in conjunction with the epidemiological features of the disease, strongly support a causative role for aflatoxin. While there are, therefore, strong reasons to believe that aflatoxin was etiologically related, the question as to whether or not other factors were also involved cannot readily be answered.

In experimental animals, aflatoxin injury is known to lead to both hepatic cirrhosis and carcinoma. Although there have been reports that the incidence of hepatoma in some parts of the world bears a relationship to the consumption of foods known to be contaminated with aflatoxin, whether they are mere associations or are cause and effect is not known. It would be of considerable importance to determine in the population that suffered from the disease, if over time, the prevalence of either cirrhosis or liver cancer will be unusually high.

The results of a follow-up study carried out in parts of Gujarat three years after the outbreak of the epidemic showed that a small number of people had died within a year following the onset after having gone through a subacute course of the disease (Krishnamachari and Bhat, 1977). There were virtually no new cases of the type seen during the epidemic. Very few food samples showed the presence of aflatoxin and even when found, levels never exceeded 1 ppb.

CONCLUSION

This outbreak of aflatoxicosis in parts of India has brought out several important points. It has shown that a staple can get contaminated with aflatoxin to an extent that it can pose acute and serious health problems, that unseasonal rains and improper drying and storage are important factors in aflatoxin contamination of foods, that poverty and ignorance lead to the consumption of grains that are visibly and severely damaged, and that man responds to the consumption of aflatoxin with acute, severe liver injury. It is imperative that appropriate measures are taken to prevent the recurrence of such large-scale outbreaks.

REFERENCES

Krishnamachari K. A. V. R., Bhat, R. V., Nagarajan, V., and Tilak, T. B. G., 1975a. Hepatitis due to aflatoxicosin—an outbreak in western India, *Lancet* **1**:1061.

Krishnamachari, K. A. V. R., Bhat, R. V., Nagarajan, V., and Tilak, T. B. G., 1975b. Investigations into an outbreak of hepatitis in part of Western India, *Ind. J. Med. Res.* **63:**1036.

Krishnamachari, K. A. V. R., and Bhat, R. V., 1977. Follow-up study of aflatoxic hepatitis in part of Western India, *Ind. J. Med. Res.* **66:**55.

Tandon, B. N., Krishnamurthy, L., Koshy, A., Tandon, H. D., Ramalingaswami, V., Bhandari, J. R., Mathur, M. M., and Mathur, P. D., 1977, Study of jaundice, presumably due to toxic hepatitis in Northern Western India, *Gastroenterology* **72:**488.

Tandon, H. D., Tandon, B. N., and Ramalingaswami, V., 1978, Epidemic of toxic hepatitis in India of possible mycotoxic origin, *Arch. Path. Lab. Med.* **102:**372.

4

COTURNISM
Poisoning by European Migratory Quail

Louis E. Grivetti

Old World migratory quail *(Coturnix)* are classified by territorial range into three subspecies. The European *C. coturnix coturnix,* common to Europe, North Africa, and western Asia, has been reported toxic while both Asian *C. coturnix japonica* and African *C. coturnix africana* subspecies have not been, heretofore, identified as poisonous. This essay examines the antiquity of coturnism or human food poisoning by migratory quail, its symptoms, contemporary distribution, and possible etiologies, and concludes with suggestions for research on this ancient, curious dietary problem.

EUROPEAN MIGRATORY QUAIL: DISTRIBUTION, HABITAT, AND MIGRATION

The European migratory quail (Fig. 1) is a Palaearctic species distributed from Arctic Europe southward to equatorial Africa. It possesses a west–east range from the Atlantic islands near Africa through Pakistan, northern India, and the central Soviet Union (Bannerman, 1953; Mackworth-Praed and Grant, 1957; Dement'ev *et al.,* 1967; Etchécopar and Hüe, 1967). Within this broad area quail maintain habitat from sea level to above 2500 m, preferring agricultural fields, grasslands, and other protective vegetation offering a diet of mixed seeds, ground-dwelling insects, and small invertebrates (Voous, 1960; Reese and Reese, 1962).

In preparation for migration quail gain nearly 50% body weight. Such fat stores result in large, plump quail that are sharply differentiated from representatives at other times. Quail migration northward from equatorial Africa begins in late winter and lasts through spring. Quail breed in European grain fields from late spring through summer, whereupon they return to sub-

---

Louis E. Grivetti • Department of Nutrition, University of California, Davis, California 95616.

FIGURE 1. European migratory quail *(Coturnix coturnix coturnix).*

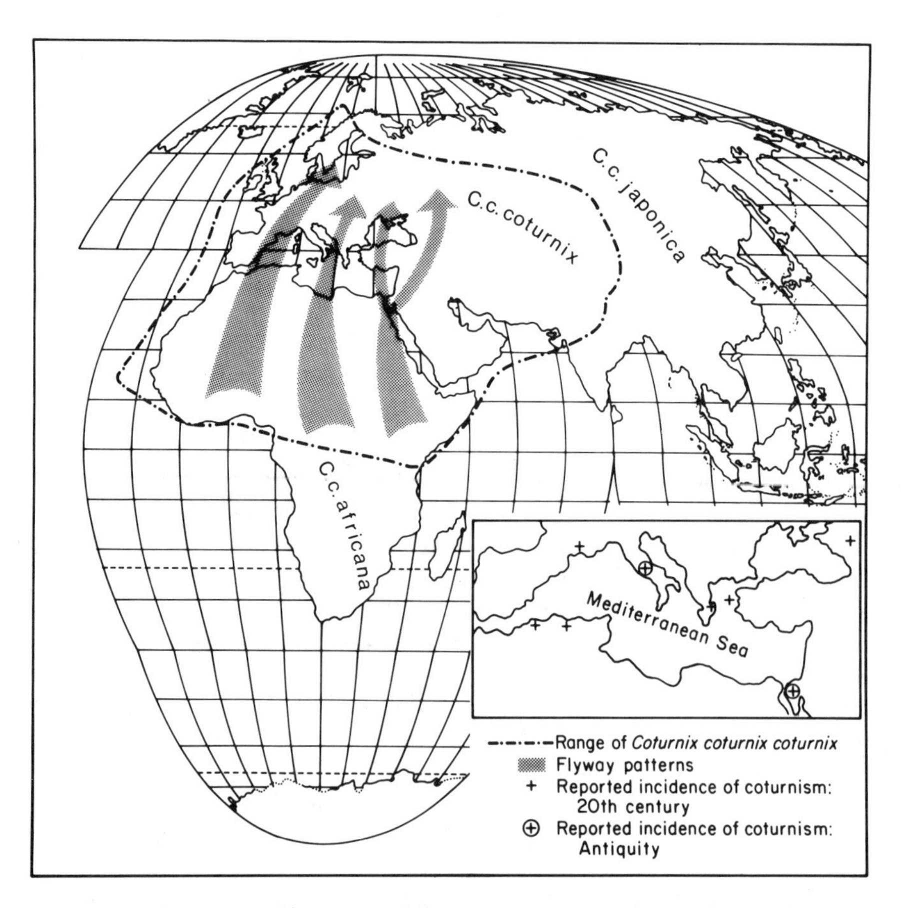

FIGURE 2. Geographic range and flyway patterns *(Coturnix coturnix coturnix).*

Saharan Africa between August and October (Lynes, 1909; Moreau, 1951; Bannerman, 1953). While their migration pattern is complicated and not well understood, quail generally follow one of three major north–south flyways (Fig. 2). The western flyway extends northward across West Africa, the western Sahara, northwest Africa, and western Europe. This route is characterized by a difficult trans-Saharan crossing, but relatively easy passage over the narrow western Mediterranean Sea. The central flyway includes a difficult trans-Saharan and Mediterranean crossing. Beginning in equatorial central Africa the central route extends northward to western Libya and Tunisia, ultimately reaching west coastal Italy and west-central Europe. The eastern flyway begins in the Great Lake region of East Africa and is characterized by a relatively easy northward flight along the Nile River basin to Egypt, where the flyway splits into eastern and western components. The eastern branch crosses southwest Asia (Israel, Lebanon, Syria, and Turkey) and terminates in the western Soviet Union, while the western branch crosses the Mediterranean at its broadest point and extends across mainland and insular Greece, ultimately reaching eastern Europe.

MIGRATORY QUAIL AND HUMAN POISONING

Coturnism, or food poisoning by migratory quail, while recent in name has been documented for at least 2500, possibly 3500 years. The early date stems from a Biblical account (*Numbers,* 11:31–34) when Israelite consumption of migratory quail in Sinai during the Exodus resulted in numerous deaths. Whatever the validity of the Biblical account, coturnism certainly was known to ancient Greek and Roman naturalists, physicians, and theologians, among them Aristotle (*On Plants,* 820:6–7), Didymus (*Geoponics,* 14:24), Lucretius (*On the Nature of Things,* 4:639–640), Philo (*The Special Laws,* 4:120–131), Galen (*De Temperamentis,* 3:4), and Sextus Empiricus (*Outlines of Pyrrhonism,* 1:57).

Central to these Greek and Roman reports is the thesis that migratory quail become toxic after consuming seed from hemlock *(Conium maculatum)* or henbane *(Hyoscyamus niger).* The ancient reports state that quail were not affected adversely by the seed but could transmit plant toxins to humans who dined on such birds. Indeed, the Roman Pliny (*Natural History,* 10:33) revealed that migratory quail were banned as human food in Italy during the 1st century A.D., presumably because of potential toxicity.

Medical–toxicological interest in coturnism continued into the Medieval period as evidenced by writings of prominent Islamic and Jewish scientists, among them Ibn Sina (*The Canon,* 2:2:2:5), Moses ben Maimon (*Comment. Epidemiarum,* 6:5), Qazwiny (*Kitab Aga'il,* 2:250), and al-Demiry (*Hayat al-Hayawan,* 1:505). Such Islamic and Jewish accounts focused on the relationships between quail diet, the plants aconite *(Aconitum napellus)* and hellebore (*Veratrum* spp.), and subsequent human poisoning.

Nineteenth and twentieth century writers, among them Cornevin (1887),

Lewin (1898), and Grieve (1931), suggested that quail are insensitive to hemlock. More recent authors, however, have challenged the thesis that coturnism is due to hemlock by suggesting that quail become poisonous after consuming insects with elevated levels of Aristolochic acid (Rotschild, 1970), or after eating seeds from cyanogenic species (Tullis, 1977). Work by Kennedy and the present author (Kennedy, 1980; Kennedy and Grivetti, 1980; and Grivetti, 1980) has contradicted the hemlock thesis by demonstrating that quail are sensitive to hemlock alkaloids and die after consuming only 0.1 g hemlock seed. Thus, the etiological mechanism(s) for coturnism transmission to humans remain unclear.

COTURNISM: DISTRIBUTION AND SYMPTOMOLOGY

Human poisoning by quail has been reported during the twentieth century from Algeria, France, Greece, and Russia; ancient accounts suggest coturnism in Italy and the northern Sinai peninsula (Kennedy and Grivetti, 1980). Reports by scientists and physicians reveal substantial variation in symptoms attributed to coturnism. In Algeria reports include colic, diarrhea, vomiting, fatigue, sensation of heaviness in the lower extremities, impaired speech, and lower limb paralysis (Sergent, 1941). Across the Mediterranean in southern France symptoms associated with quail poisoning include respiratory distress and appearance of red, itchy urticaria (Sergent, 1948). In Russia, in the region between the Black and Caspian Seas, symptoms of coturnism include general weakness in the legs occurring 3–4 hr after consumption, sharp pains extending from the legs upward through the torso into the neck lasting 2–12 hr, followed by generalized pain and weakness continuing for 3–10 days (Khovanskii, 1954, 1957, 1964).

Coturnism is described in detail from the Greek island of Mytelene. There, symptoms have included anuresis, myoglobinuria, sharp muscular pains along the trunk and extremities graduating to paralysis in muscles recently used, oliguric to anuric urinary output, azothemic blood, elevated uric acid levels, elevated SGOT levels (Serum Glutamic Oxaloacetic Transaminase), respiratory distress, and death (Hadjigeorge, 1952; Ouzounellis, 1968a, 1968b, 1970). Demographic evaluation of coturnism cases from Mytelene reveals a greater propensity for poisoning among active, adult males than among females (Ouzounellis, personal communication, 1970). Unpublished accounts document coturnism on the Greek mainland, specifically at Marathon northeast of Athens (Ouzounellis, personal communication, 1970).

COTURNISM: CONTINUED SEARCH FOR THE TRANSMISSION MECHANISM

It has been widely held from antiquity through the twentieth century that poisoning from migratory quail is due to quail diet, specifically an intake pattern based on hemlock or another toxic alkaloid, perhaps aconite, hellebore, or henbane. First scientific experiments into the mechanism for transmission of coturnism to humans were conducted by Sergent (1941). Quail used in his experiments were healthy, evicerated, thoroughly cooked, and free of pesticide residue (a twentieth century precaution!). Sergent reported that quail in Algeria consumed hemlock *(Conium maculatum),* hellebore *(Hyoscyamus niger),* and black morelle *(Solanum nigrum)* and deduced that the birds became toxic after consuming seeds of these plants. He reportedly fed hemlock to quail but could not produce symptoms typical of hemlock poisoning; such dosed quail were fed to dogs and the canines subsequently exhibited hind-leg paralysis, one of the classical symptoms of hemlock poisoning. Sergent concluded that quail were insensitive to hemlock and that quail flesh would transmit hemlock toxins to animals.

Experiments on migratory quail (Asian subspecies available as an experimental animal) by Kennedy and Grivetti (1980) could not reproduce Sergent's findings and demonstrated that hemlock ingested by quail at the quantity of 1.3–0.1 g daily produced classical symptoms of hemlock poisoning and death. Additional experiments demonstrated that quail, if provided a choice, would not select hemlock as food. Other experiments reported that hemlock ingested by quail in quantities of less than 0.1 g daily would produce a significant appetite depressant effect.

Among the questions to be resolved, therefore is, do migratory quail consume hemlock under free-ranging conditions? Despite ancient nineteenth and twentieth century accounts suggesting hemlock as a regular component of migratory quail diet, proof must lie in botanical examination of seeds found in quail crops. While numerous accounts suggest that quail consume, even thrive on hemlock, this species has not been repeatedly identified by botanists/ornithologists as a component of quail nutriture (see Cheesman and Sclater, 1935; Uljanin, 1941; Witherby *et al.,* 1949; Keve *et al.,* 1953; Heim de Balsac and Mayaud, 1962; Dement'ev *et al.,* 1967; von Blotzheim, 1973; and Bateman, 1977). The only twentieth century account reporting hemlock in quail crops is Sergent (1941) and a close reading of his text suggests that the identification was made by hunters, not botanists.

More significant, however, is the finding that ingestion of hemlock by quail produces an appetite depressant effect (Kennedy, 1980). Given the supposition that quail might consume hemlock, to do so during the premigratory stage when body weight must be nearly doubled would be counter-productive. Since quail are toxic during the migratory period and since the difficult trans-Saharan and trans-Mediterranean flights necessitate increased fat stores, one must logically exclude hemlock as the mechanism for transmission of coturnism.

A further argument against hemlock is based on seasonality and timing of the quail migration. In North Africa quail migrate and reach Algeria primarily between April and June, yet hemlock does not drop seeds until late summer and fall. Consequently, before hemlock is available for quail consumption the quail flocks have already crossed the Mediterranean to Europe.

When quail return to Algeria from France during the fall, hemlock seed is available yet the birds are not toxic at this time.

Another major question for consideration is why are not all quail toxic, or phrased a different way, why are there not more reported cases of coturnism? It is apparent that coturnism is only recently receiving epidemiological and medical attention in detail. Ouzounellis is surely correct when he suggests that the incidence of coturnism is underreported. But given underreporting, other nagging questions remain. Why should quail be toxic on the northward flight to Algeria, but not the southern? Why are males more susceptible than females to coturnism? Could such a pattern be due to cultural traditions of serving quail primarily to men as returning hunters, or could coturnism have a genetic basis?

Given the wide range of symptoms identified with quail poisoning, is coturnism a single or multiple disease? Do different symptoms reflect differences in quail diet at the local level? Does the absence or presence of coturnism, after correlation with quail migration direction, support a dietary basis for the disease, or might flight stress in one direction and ease on the return flight suggest production of a metabolite that elicits coturnism in sensitive individuals or populations?

Thus, it becomes important to clearly document each instance of coturnism to permit evaluation of both dietary and metabolite hypotheses. If quail diet is central to understanding coturnism, logic holds that the dietary elements investigated be lipophyllic, produce no effect on quail, but cause poisoning in humans. In this light, the report by Rotschild (1970) that migratory quail are insensitive to aristolochic acid—a compound common to certain insects on which quail dine—should receive experimental attention. If quail physiology and migration stress underlies coturnism, one would expect outbreaks along the flyway stress localities, especially among hunters who capture birds, dispatch them quickly, and consume them within a few hours.

In addition to clearly documenting instances of coturnism, data should be gathered on patient demography, exercise activities prior to onset of symptoms, time and seasonality of the attack, and cultural information on methods used to obtain quail. Penning or cooping wild birds would permit physiological recovery and excretion of potentially toxic metabolites. Finally, data should be collected on cooking and preservation techniques to clearly differentiate between coturnism in the strict sense and poisoning resulting from unsanitary food preparation.

SUMMARY

Coturnism is ancient, traceable certainly to the Greeks and Romans, possibly earlier. Human poisoning by quail has been reported during the twentieth century from Algeria, France, Greece, and Russia; ancient accounts suggest coturnism in Italy and northern Sinai. Multiple symptoms exhibited locally by humans consuming toxic quail suggest that coturnism may have multiple origins founded in local quail diet or in the production of a metabolite during stress of migration. Such hypotheses, however, remain unconfirmed and the mechanism of transmission to humans remains unclear.

Present knowledge of coturnism, its distribution and its history, permits an additional conclusion to be drawn. Based on contemporary data of quail migration direction, flyway patterns, and seasonality of flight, the Biblical account of quail poisoning in Sinai may reflect an actual event. If so, Israelites of the Exodus would have been poisoned by quail during the months of August–October and the Israelite encampment mentioned in *Numbers* (11:31–34) would have been adjacent to the Mediterranean coast. The disease now called coturnism thus provides support for Biblical scholars who have argued against tradition for a northern Exodus route in contrast with the more widely held southern, tortuous passage.

REFERENCES

al-Demiry [Kamal ed-Din Mohammad Ibn Moussa], 1957, *Hayat al-Hayawan al Kubra,* 3rd ed., Mustafa al-Baby al-Halaby, Cairo.

Aristotle, 1955, *On Plants,* translated by W. S. Hett, William Heinemann, London.

Bannerman, D. A., 1953, *The Birds of West and Equatorial Africa,* 2 Vols., Oliver and Boyd, London.

Bateman, J. B., 1977, *Acute Rhabdomyolysis from Eating Quail,* Report No. R-8-77 (Aug. 12, 1977), Office of Naval Research, London.

Cheesman, R. E., and Sclater, W. L., 1935, On a collection of birds from northwestern Abyssinia, *Ibis* **5(series 13):**151–191.

Cornevin, C., 1887, *Des Plantes Reneneuses et des Empoisonnements qui Elles Determinent,* Firmin Didot, Paris.

Dement'ev, G. P., Gladkov, N. A., Isakov, Y. A., Kartashev, N. N., Kirikov, S. V., Mikheev, A. V., and Ptushenko, E. S., 1967, *Birds of the Soviet Union,* translated by A. Birron and Z. S. Cole, Israel Program for Scientific Translations, Jerusalem.

Didymus, 1805–1806, *Agricultural Pursuits (Geoponics),* 2 Vols., translated by T. Owens, W. Spilsbury, London.

Etchécopar, R. D., and Hüe, F., 1967, *The Birds of North Africa from the Canary Islands to the Red Sea,* translated by P. A. D. Hollom, Oliver and Boyd, London.

Galenus [Galen], 1527, *Galenide. Temperamentis Liberites,* Latin translation by Thomas Linacro, no publisher, London.

Grieve, M., 1931, *A Modern Herbal,* Butler and Tanner, London.

Grivetti, L. E., 1980, Poisoning by Coturnix quail: Continued search for a toxic mechanism, Program Abstract No. 4, *Fourth Annual Meeting West Coast Nutritional Anthropologists,* University of California, Davis Campus, May 3, 1980.

Hadjigeorge, E., 1952, Les Cailles empoisinneuses, *Presse Med.* **68:**1469.

Heim de Balsac, H., and Mayaud, N., 1962, *Les Oiseaux du Nord–Ouest de l'Afrique,* Paul Lechevalier, Paris.

Holy Bible, 1953, King James Edition, Collin's Clear Type Press, London.

Ibn Sina [Avicenna], 1930, *A Treatise on the Canon of Medicine of Avicenna Incorporating a Translation of the First Book,* translated by O. C. Gruner, Luzac and Company, London.

Kennedy, B. W., 1980, *Coturnix Quail, Poison Hemlock, and the Exodus: A Toxicological Inquiry of an Historical Problem,* Unpublished M.Sc. Thesis, Department of Avian Science, University of California, Davis.

Kennedy, B. W., and Grivetti, L. E., 1980, Toxic quail: A cultural–ecological investigation of coturnism, *Ecol. Food Nutr.* **9:**15–42.

Keve, A., Zsák, Z., and Kascab, Z, 1953, A Fùrj Gazdasagi Jelentosege [The agricultural significance of the quail], *Budapest Magyar Nemzeti Museum-Annales Historico-Naturales* **4:**197–209.

Khovanskii, D. V., 1954 [Concerning quail meat poisoning], in: [*Papers of the First Stavropol Physicians' Conference*], pp. 65–70, Stavropol.

Khovanskii, D. V., 1957 [On food poisoning caused by quail meat], *Okhota I Okhotniche* **7:**26.

Khovanskii, D. V., 1964 [Quail meat poisoning], in: [*Problems in Geographical Pathology*], pp. 172–174, Moscow.

Lewin, L., 1898, *Lehrbuch der Toxikologie für Arzte, Studirende, und Apotheker,* Urban and Schwarzenberg, Leipzig.

Lucretius, 1910, *On the Nature of Things,* translated by C. Bailey, Clarendon Press, Oxford.

Lynes, H., 1909, Observations on the migration of birds in the Mediterranean, *Br. Birds* **3:**99–104, 133–150.

Mackworth-Praed, C. W., and Grant, C. H. B., 1957, *Birds of Eastern and North Eastern Africa,* African Handbook of Birds, Ser. 1, Vol. 1, 2nd ed., Longmans, Green and Co., London.

Moreau, R. E., 1951, The British status of quail and some problems of its biology, *Br. Birds* **44:**257–276.

Moses ben Maimon [Maimonides], 1970–1971, *The Medical Aphorisms of Moses Maimonides,* Studies in Judaica Series, translated and Edited by F. Rosner and S. Munter, 2 Vols. (combined), Yeshiva Universtiy Press, New York.

Ouzounellis, T. I., 1968a, Myoglobinuries par ingestion de cailles, *Presse Med.* **76:**1863–1864.

Ouzounellis, T. I., 1968b, Myosphairinouria apo Ortaki, *Iatrike* **14:**213–217.

Ouzounellis, T. I., 1970, Some notes on quail poisoning, *J. Am. Med. Assoc.* **211:**1186–1187.

Philo, 1954, *The Special Laws,* translated by F. H. Colson, Vol. 8, William Heinemann, London.

Pliny, 1942, *Natural History,* 10 Vols., translated by H. Rackham and W. H. S. Jones, Harvard University Press, Cambridge, Massachusetts.

Qazwiny [Zakaria ibn Mohammed ibn Mahmoud], 1957, *Kitab Aga'il el-Makhlouquat wa Ghara'ib el-Mawgoudat,* 2 vols., Mustafa al-Halaby al-Baby Press, Cairo.

Reese, E. P. and Reese, T. W., 1962, The quail, *Coturnix coturnix,* as a laboratory animal, *J. Exp. Anal. Behav.* **5:**265–270.

Rotschild, M., 1970, Les papillons qui se déguisent, *Sci. Vie* **127(632).**

Sergent, E., 1941, Les cailles empoisonneuses dans la Bible, et en Algerie de nos jours: apercu historique et recherces experimentales, *Arch. Inst. Pasteur Alger.* **19:**161–192.

Sergent, E., 1948, Les cailles empoisonneuses en France, *Arch. Inst. Pasteur Alger.* **26:**249–252.

Sextus Empiricus, 1933, Outlines of Pyrrhonisn, in: *The Works of Sextus Empiricus,* Vol. 1, translated by R. G. Bury, G. P. Putnam's Sons, New York.

Tullis, J. L., 1977, Annual discourse. Don't eat the quail, *N. Eng. J. Med.* **297:**472–475.

Uljanin, N., 1941, [Materials on ecology of the quail in north Kasahstan], *Zoologescheskii Mugei Sbornik Trudov* [Moscow University] **6:**153–166.

von Blotzheim, N. G., 1973, *Handbuch der Vögel Mitteleuropas,* Vol. 5, *Galliformes und Gruiformes,* Akademische Verlagsgesellschaft, Frankfurt am Main.

Voous, K. H., 1960, *Atlas of European Birds,* Nelson and Sons, Amsterdam.

Witherby, H. F., Jourdain, F. C. R., Ticehurst, N. F., and Tucker, B. W., 1949, *The Handbook of British Birds,* Vol. 5, H. F. and G. Witherby, London.

5

DIETARY TOXINS AND THE EYE

Donald S. McLaren

INTRODUCTION

The ocular manifestations of dietary toxins considered here are confined to those resulting from the ingestion of substances that are not nutrients. Excessive intake of some carbohydrates, lipids, elements, and vitamin A may give rise to ocular disturbances and these are dealt with elsewhere. Some toxins in animal feeds damage the eye. Among these are bracken fern *(Pteris aquilina)*, male fern *(Felix mas)* and various *Leguminosae*. These conditions have been reviewed recently (McLaren, 1980).

ETHANOL POISONING

Acute Alcoholism

Visual acuity and visual fields are almost normal in cases of acute alcoholism, and in the few authenticated cases of total blindness the fundi have been normal and the vision has invariably returned to normal within a few days. The common term "blind drunk" is therefore a misnomer.

Disturbances of higher visual functions are very well known and, apart from the diplopia, are usually denied by the subject. Sensitivity to light is lowered and the brightness difference threshold is increased. Complex judgments such as depth perception and reaction time are considerably impaired. Ocular movements are affected at an early stage (Newman and Fletcher, 1941). Levett and Karras (1977) found in three volunteers that accommodation times were increased by 10–30% over those of controls when blood alcohol was 50–100 mg/100 ml.

Donald S. McLaren • Department of Medicine, The Royal Infirmary, University of Edinburgh, EH9 3YW Scotland.

Chronic Alcoholism

Wernicke–Korsakoff Syndrome

The Wernicke–Korsakoff syndrome (cerebral beriberi, acute superior hemorrhagic polioencephalitis) due to acute and severe deficiency of thiamin is not an infrequent complication of chronic alcoholism (Victor *et al.*, 1971). The syndrome is characterized by mental confusion accompanied by sixth nerve weakness resulting in nystagmus in the early stages and going on to paralysis as the disease progresses. Less frequently observed eye signs include small retinal hemorrhages, bilateral scotomata, papilledema, and ptosis.

The major pathological changes, consisting of vacuolation of tissues, increase in histiocytes and erythrocytes, and occasionally pinpoint hemorrhages, occur symmetrically in the following parts of the brain: paraventricular parts of nuclei of the thalamus, the mammillary bodies, the periaqueductal region, the floor of the fourth ventricle, and the anterior lobe of the cerebellum.

Prompt therapy with 200 mg thiamin hydrochloride, IM or IV, results in rapid improvement in many cases but the fatality rate is high.

Retrobulbar Neuropathy

This is an occasional feature of chronic alcoholism (McLaren, 1980). Blurring of central vision of gradual onset is the main complaint. A central scotoma for red and green vision develops and there may be a slight general contraction of the visual fields. There is a pathological increase in the normal pallor of the temporal side of the optic disc. All of these changes can be explained by the degeneration that has been demonstrated at autopsy in the papillo-macular bundle that connects the macular ganglion cells of the retina with the lateral geniculate ganglia.

It is generally accepted that the disorder is due to nutritional deficiency secondary to some other factor, of which chronic alcoholism is only one instance. Response has been claimed for therapy with several vitamins of the B complex. Carroll (1966) reported good results with thiamin, even when alcohol consumption was not curtailed.

Liver Cirrhosis

Liver cirrhosis secondary to chronic alcoholism is frequently accompanied by ocular disorders, usually disturbances of dark adaptation or color vision. In some instances these are due to deficiency of vitamin A or zinc and respond to appropriate treatment (McLaren, 1980).

Cruz-Coke (1965) claimed that X-linked genes for defective color vision predispose either to alcoholism or cirrhosis. Fialkow *et al.* (1966) were unable to confirm this. In 46 patients with alcoholic cirrhosis one had the very rare blue-yellow defect and 19 had red-green defects. Most of these latter defects cleared up during convalescence. In another study (Rothstein *et al.*, 1973) 14 out of 16 patients with proven alcoholic cirrhosis had significant dyschromatopsia (color vision defects) on the Farnsworth–Munsell 100-hue test.

Fetal Alcohol Syndrome

Jones *et al.* (1973) reported a fetal alcohol syndrome in babies born to mothers who were chronic alcoholics. Microcephaly, mental retardation, short palpebral fissures, and small eyes have been commonly reported and some have had strabismus and asymmetric ptosis.

CASSAVA POISONING

A polyneuropathy has been reported from several parts of Africa associated with the consumption of cassava that has not received the customary prolonged soaking in water to remove all the cyanide (Kneuttgen, 1955; Money, 1958; Osuntokun *et al.*, 1968; Osuntokun and Osuntokun, 1971). A high proportion of these patients had retrobulbar neuropathy similar to that described in those with chronic alcoholism. Raised plasma thiocyanate levels suggest that the eye and other neurological signs in this syndrome may be due to cyanide poisoning.

EPIDEMIC DROPSY

It is necessary to point out that the disease known as epidemic dropsy has no connection with beriberi, in the acute form of which disease dropsy is a prominent sign. Some of the early writers suggested such an association before the etiology of epidemic dropsy had been discovered and this may account for the persistence of this wrong impression in some quarters. Clinical descriptions include bullous eruptions on the skin and glaucoma.

The disease was first described in India in the nineteenth century; since then it has occurred in epidemic form not only in India but also in Indian communities in other parts of the world such as Mauritius, Fiji, and South Africa. It was shown by Lal and Roy (1937) and Lal *et al.* (1940) that the cause was the ingestion of argemone oil. This oil is obtained from the seeds of the poppy *Argemone mexicana* which is a common contaminant of the mustard plant, the oil from which is used for cooking by many Indian communities. Typical signs of the disease have been produced in human volunteers by feeding 40–50 oz of argemone oil over a period of about 30 days (Chopra *et al.*, 1939; Lal *et al.*, 1941).

In neither of these experimental studies in man was glaucoma produced, although in one instance two out of five volunteers complained of dimness of vision. There was also no rise in intraocular tension in the eyes of hens, rabbits, or cats fed argemone oil by Dobbie and Langham (1961), contrary to the results of other workers (Leach and Lloyd, 1956). Maynard (1909) in Calcutta appears to have been the first to report glaucoma in association with an epidemic of the condition, and Kirwan (1934, 1936) and Dutt (1950) have also stated that glaucoma is common in those with epidemic dropsy. As Dobbie and Langham (1961) pointed out there is really no convincing clinical evidence for a causal relationship. One piece of evidence in favor of the existence of such a relationship is, however, that the histological appearance of the eye in epidemic dropsy is quite different from that of ordinary open-angle glaucoma. Kirwan (1935) reported that there was a dilatation of the capillaries of the whole of the uveal tract, especially marked in the ciliary processes and the choroid. It was postulated that this led to an increased permeability of the endothelial walls and a consequently raised output of aqueous humor into the anterior chamber. It is evident, however, that there is now some doubt as to whether or not the eye is really affected in epidemic dropsy.

Sanguinarine has been isolated from nearly 50 species of poppy-fumiraria weeds. Hakim (1957) showed that the ingestion of these plants by cattle, goats, and sheep led to the secretion of sanguinarine in their milk. He suggested that consumption of contaminated milk was responsible for widespread glaucoma, but this has never been substantiated.

ERGOTISM

The widespread consumption of rye contaminated with ergot fungus *(Claviceps purpurea)* led to many devastating epidemics of the disease in Europe during the Middle Ages (Barger, 1931). The two main groups of symptoms, the gangrenous and the nervous, had been described for centuries but it was not until the middle of the nineteenth century that cataract was included in the symptomatology. Meier (1862) found that 23 out of 283 affected people of Siebenburgen, Transylvania, developed cataract months after an epidemic in 1857 due to eating contaminated rye, young people being more affected than old. Several subsequent reports (Bellows, 1944; Duke-Elder, 1954) largely from Russia in the later years of the nineteenth and earlier part of the twentieth century confirmed the occurrence of cataract usually a matter of months or a year or two after recovery from the acute systemic effects. The latent period is reported to be shorter in children than in older people, and in the aged the lens changes are described as being indistinguishable from those of senile cataract.

Retinal changes occur during the acute illness and consist of vasoconstriction and edema with an amblyopia in which there is peripheral contraction of the fields and a central scotoma. These symptoms are all transitory and optic atrophy does not occur (Kaunitz, 1932).

There seems little doubt from their nature that the retinal signs are caused by the alkaloids of ergot. The nature of the lens damage is less certain, although many theories have been devised to explain it, such as spasm of the ciliary vessels, alteration of the composition of the aqueous, damage to the lens epithelium, and so on.

Scott (1962) pointed out that no instance of cataract has been reported following the medical use of ergot or its alkaloids. He suggested that ergotism of the convulsive type might have been confused with tetany, with which cataract is not infrequently associated.

Cataract has never been produced by ergot in experimental animals, but Peters (1902) described degenerative changes in the retinal ganglia cells and vessel walls.

FAVISM

Choremis *et al.* (1960) described severe impairment of vision in two boys aged 3 and 6 years who had eaten broad beans. Retinal hemorrhages in one appeared to be part of a general hemorrhagic state and in the other papilledema and constriction of retinal vessels were present. They reviewed the few case reports in the literature and noted that visual symptoms usually appear 3–7 days after the onset of hemorrhages elsewhere. Complete blindness may ensue within a few hours and the ultimate prognosis is poor; half showing no improvement and only 10% recovering full vision.

OXALOSIS

Toxicity may result from excessive intake, increased intestinal absorption, increased endogenous synthesis, or retention due to renal dysfunction. Rhubarb and spinach are especially rich sources. Deposits may occasionally occur in the retina, optic nerve, uvea, and oculomotor muscles (Timm, 1963). Garner (1974) described a case with oxalate deposits in the outer layers of a detached and degenerate retina.

FINGER CHERRY TREE POISONING

The fruit of the finger cherry tree (Australian horror tree, *Rhodomyrtus macrocarpa*) native to Queensland and New Guinea, may give rise to blindness when eaten (Flecker, 1944). The local people rub off the skin before eating it and are unaffected. The toxin is unknown.

REFERENCES

Barger, G., 1931, *Ergot and Ergotism,* Oxford Medical Pub., London.

Bellows, J. G., 1944, *Cataract and Anomalies of the Lens,* Kimpton, London.

Carroll, F. D., 1966, Nutritional amblyopia, *Arch. Ophthalmol.* **76:**406.

Chopra, R. N., Pasricha, C. L., Goyal, R. K., Lal, S., and Sen, A. K., 1939, Experimental production of syndrome of epidemic dropsy in man, *Indian Med. Gaz.* **74:**193.

Choremis, C., Joannides, T., and Kyriakides, B., 1960, Severe ophthalmic complications following favism, *Br. J. Ophthalmol.* **44:**353.

Cruz-Coke, R., 1965, Colour-blindness and cirrhosis, *Lancet* **1:**1131.

Dobbie, G. C., and Langham, M. E., 1961, Reaction of animal eyes to sanguinarine and argemone oil, *Br. J. Ophthalmol.* **45:**81.

Duke-Elder, W. S., 1954, *Textbook of Ophthalmology*, Vol. 6, Mosby, St. Louis, Missouri.

Dutt, S. C., 1950, *XVI Conc. Ophthalmol. Br. Acta* **2:**872.

Fialkow, P. J., Thuline, H. C., and Fenster, F., 1966, Lack of association between cirrhosis and the common types of color blindness, *N. Engl. J. Med.* **275:**584.

Flecker, H., 1944, Sudden blindness after eating "finger cherries" *(Rhodomyrtus macrocarpa), Med J. Aust.* **2:** 183.

Garner, A., 1974, Retinal oxalosis, *Br. J. Ophthalmol.* **58:**613.

Hakim, S. A. E., 1957, Poppy alkaloids and glaucoma, *J. Physiol. (London)*, **138:**40P.

Jones, K. L., Smith, D. W., Ulleland, C. N., and Streissguth, P., 1973, Pattern of malformation in offspring of chronic alcoholic mothers, *Lancet* **1:**1267.

Kaunitz, J., 1932, Chronic endemic ergotism: Its relation to thromboangiitis obliterans, *Arch. Surg.* **25:**1135.

Kirwan, E. W. O'G., 1934, Primary glaucoma: Symptom complex of epidemic dropsy, *Arch. Ophthalmol.* **12:**1.

Kirwan, E. W. O'G., 1935, Ocular complications of epidemic dropsy, *Indian Med. Gaz.* **70:**485.

Kirwan, E. W. O'G., 1936, Aetiology of chronic primary glaucoma, *Br. J. Ophthalmol.* **20:**321.

Knuettgen, H., 1955, Über ein Ataziesyndrom bei liberianischen Eingeborenen (Strachan-Svelt-Syndrom), *Z. Tropenmed. Parasitol.* **6:**472.

Lal, R. B., and Roy, S. C., 1937, Investigations into epidemiology of epidemic dropsy: Experiments to test validity of infection theory in semi-isolated community, *Indian J. Med. Res.* **25:**233.

Lal, R. B., Mukjerji, S. P., Das Gupta, A. C., and Chatterji, S. R., 1940, Investigations into epidemiology of epidemic dropsy: Quantitative aspects of problem of toxicity of mustard oil, *Indian J. Med Res.* **28:**163.

Lal, R. B., Das Gupta, A. C., Mukjerji, S. P., and Adak, B., 1941, Investigations into epidemiology of epidemic dropsy: Feeding experiments on human subjects to test toxicity of some of the derivatives and modifications of argemone oil, *Indian J. Med. Res.* **29:**839.

Leach, E. H., and Lloyd, J. P., 1956, Experimental ocular hypertension in animals, *Trans. Ophthalmol. Soc. U.K.* **76:**453.

Levett, J., and Karras, L., 1977, Effects of alcohol on human accommodation, *Aviat. Space Environ. Med.* **48:**434.

McLaren, D. S., 1980, *Nutritional Ophthalmology*, Academic Press, London.

Maynard, F. P., 1909, Preliminary note on increased intraocular tension met with in cases of epidemic dropsy, *Indian Med. Gaz.* **44:**373.

Meier, L., 1892, *Albrecht von Graeffes Arch. Klin. Exp. Ophthalmol.* **8(2):**120.

Money, G. L., 1958, Endemic neuropathies in the Epe district of Southern Nigeria, *West Afr. Med. J.* **7:**58.

Newman, H., and Fletcher, E., 1941, The effect of alcohol on vision, *Am. J. Med. Sci.* **202:**723.

Osuntokun, B. O., and Osuntokun, O., 1971, Tropical amblyopia in Nigerians, *Am. J. Ophthalmol.* **72:**708.

Osuntokun, B. O., Durowoju, J. E., McFarlane, H., and Wilson, J., 1968, Plasma amino acids in the Nigerian nutritional ataxic neuropathy, *Br. Med. J.* **3:**647.

Peters, A., 1902, Uber Veränderungen an den Ciliarepithelien bei Naphthalin-und Ergotinin-vergiftung, *Ber. Versamml. Ophthalmol. Ges.* **30:**20.

Rothstein, T. B., Shapiro, M. W., Sacks, J. G., and Weis, M. J., 1973, Dyschromatopsia with hepatic cirrhosis: Relation to serum B_{12} and folic acid, *Am. J. Ophthalmol.* **75:**889.

Scott, J. G., 1962, Does ergot cause cataract? *Med. Proc.* **8:**4.

Timm, G., 1963, Oxalose und Auge, *Ophthalmologica* **146:**1.

Victor, M., Adams, R. D., and Follins, G. H., 1971, The Wernicke–Korsakoff syndrome. A clinical and pathological study of 245 patients, 82 with postmortem examinations, *Contemp. Neurol. Ser.* **7:**1.

6

ACKEE POISONING

K. L. Stuart

BACKGROUND AND ETIOLOGY

This illness, otherwise known as the Vomiting Sickness of Jamaica, has been recognized on this island for many years. Its clinical features have been reviewed by Hill (1952), Patrick *et al.* (1955), and Stuart and co-workers (1955). It characteristically occurs in poorly nourished Jamaicans either sporadically or in small family outbreaks. The highest incidence is during the months from December to March although occasional outbreaks occur at any time of year. The disorder is uncommon in infants. This is thought to be due to the fact that infants are usually breast-fed and thus protected from any noxious dietary agents or severe protein depletion. Children between one and ten years are most commonly affected.

A number of possible etiological agents have been suggested for the condition and for the severe hypoglycemia that characterizes it. Although a background of malnutrition is common its precise role is still undetermined; but it seems likely that a specific toxin plays a dominant role and that such a toxin may more easily precipitate an undernourished than a normal child into severe hypoglycemia. In many of the cases that have been studied a history of 24–48 hr of near starvation was common (Stuart *et al.,* 1955). It was noted also that attacks characteristically occurred in the early hours of the morning when the blood sugar level is likely to be at its lowest. The occurrence of low blood sugar levels and increased insulin sensitivity in other malnourished states, e.g., pellagra, kwashiorkor, and marasums, has also been noted.

Present evidence, clinical and circumstantial, places the ackee *(Blighia sapida)* at the head of the list of suspected agents. Sir Harold Scott, in his original investigation of the disease in 1915, considered that all cases were

K. L. Stuart • Commonwealth Secretariat, Marlborough House, London SW1P 5HX, United Kingdom.

due to eating the unripe ackee and felt that the condition should be known as "ackee poisoning." The current hypothesis of its action is that a specific hypoglycemic agent elaborated by this fruit, together with an unstable carbohydrate metabolism in poorly nourished subjects, may share joint responsibility for the hypoglycemia which is the unvarying biochemical characteristic of the disorder. Further evidence incriminating the ackee is the fact that two polypeptides, hypoglycin A and B, have been isolated from the ackee that show marked hypoglycemic activity in experimental animals; furthermore, the toxicity has been shown to more than double when fasting animals are used (Hassall and Reyle, 1955).

Additional circumstantial support for this hypothesis derives from the fact that ackee poisoning has not been reported from other Caribbean areas where the ackee is not grown. In addition, the explosive type of family outbreak, whereby several members of a household are affected at one time, suggests a specific ingested toxin. This hypothesis is further supported by the almost complete absence of the disorder in very young infants who are still being fed on human or cows' milk.

When these cases were originally studied in the 1950s the hospital ambulance was commonly sent to bring in for admission and study all the siblings and sometimes the parents of affected children even if they had no conspicuous symptoms. This was justified because a sibling left at home apparently well would be found occasionally on arrival at the hospital to be completely comatose, and would almost certainly have died had he not been admitted. Indeed, the rapidity with which an apparently well child could be precipitated into a fatal coma was a conspicuous feature of the disorder.

An important observation reported from the Department of Microbiology during these initial studies was that the ackee is an excellent culture medium for the hemolytic staphylococcus. The toxin from this organism is heat-stable, so that even if the contaminated food is cooked thoroughly the toxin remains active. It is possible that in some cases ackees, particularly if they had been lying on the ground for too long, may have been contaminated with the hemolytic staphylococcus or some other food-poisoning organism, accounting for food poisoning in association with the ingestion of the ackee fruit.

PATHOLOGY

Liver biopsy and autopsy material show fatty changes in the liver with almost complete absence of liver glycogen, both histologically and on quantitative assay (Patrick *et al.*, 1955). Serial liver biopsies also show that the glycogen is rapidly replaced following successful glucose therapy. Other autopsy findings are fatty metamorphosis of the kidneys and other organs, tissue edema, and alimentary lymphoid hyperplasia.

CLINICAL FEATURES

Detailed clinical descriptions of the first series of cases reported were given by Stuart and co-workers in 1955. All grades of severity were seen. Vomiting, although a common feature, was not invariable, and the central symptoms varied from mere drowsiness to full coma. Two main clinical types were recognized:

1. The abrupt onset of vomiting without pain, fever, diarrhea, or previous malaise. This was followed by rapid prostration, sometimes followed by periods of remission. The vomiting, however, commonly recurred followed by drowsiness, twitching of the limbs, convulsions, coma, and death. Dehydration was not a conspicuous feature. If recovery took place it was usually complete within 48–72 hr and there were no after-effects.
2. A fulminating variety in which vomiting did not occur, the patient being precipitated dramatically into the stage of collapse, drowsiness, convulsions, and fatal coma.

Both mild and severe cases are characterized by severe hypoglycemia with blood sugar levels as low as 10 mg/100 ml. The unvarying occurrence of this finding not only established the disorder as a true clinical entity, but suggested an immediate and logical line of emergency treatment. All other laboratory tests are usually negative. There is no ketosis or albuminuria, and no clinical or laboratory evidence of hepatic dysfunction apart from moderate reduction in serum albumin and cholinesterase levels, as is commonly found in malnutrition.

DIAGNOSIS

The features listed above—the frequent occurrence of family outbreaks, their seasonal pattern, the invariable blurring of consciousness—all readily suggest the diagnosis. It has been emphasized that the symptoms associated with all levels of severity of ackee poisoning are themselves nonspecific and consistent with hypoglycemic attacks from other causes (Stuart *et al.*, 1955). The diagnosis may often, therefore, be confused with other causes of hypoglycemia and clouding of consciousness, and with other causes of vomiting. Confirmation is not usually difficult, however, once the diagnosis is suspected. The laboratory findings and the therapeutic response to oral or intravenous glucose can be relied upon to clinch the diagnosis.

After recovery attacks do not recur as in cases of spontaneous hypoglycemia with which it may be confused.

PROGNOSIS

Apart from the severity of the clinical manifestations the prognosis depends on the rapidity with which treatment can be initiated. The mortality has been shown to be highest in young children. Characteristically, young children die; older children are affected but many recover; usually adults are only slightly affected. Mild cases recover without sequelae. Twitching and convulsions occur in more severe cases and are of serious prognostic import.

TREATMENT

Treatment should be commenced immediately after the diagnosis is made or confidently suspected. Oral or intravenous injection of 50% glucose solution should be administered at a recommended initial dose of 1 g/kg body weight. Subsequent dosages of sugar, fluid, and electrolytes, administered either orally or intravenously, will be determined by the clinical response and laboratory findings, if available.

Most cases occur in rural areas often far from immediate medical facilities or assistance. The following practical advice would, therefore, seem appropriate to first aid workers, parents, and teachers: quarter-hourly, concentrated glucose or cane sugar drinks should be given to *all* suspected cases, and affected children in particular should be removed to a hospital for observation as soon as possible.

PREVENTION

The ackee is almost universally eaten in Jamaica and, in season, is often an important source of food. It would, therefore, be both impractical and futile to attempt to ban it from the average Jamaican diet. Public health propaganda, however, emphasizing particularly the possible dangers of eating the unripe fruit, should be intensified. School teachers, village health workers, public health nurses, as well as doctors, should be reminded of the risk of outbreaks occurring particularly during the ackee season, and of the most practical lines of treatment to be followed.

In addition to specific measures aimed at treating hypoglycemia, more general approaches should be considered. Improved child care with the provision of better welfare centers, community education schemes, more public health nurses, sanitary inspectors etc., should contribute significantly to the eventual eradication of the disease. There is no doubt that the prevalence of the outbreaks have a direct relation to the socioeconomic conditions existing in the country as a whole and particularly in areas where the ackee has become a staple of the diet.

Cicely Williams (1954) pointed out that, whatever the immediate cause of the illness, the main underlying factor is widespread neglect of children. The circumstances surrounding most reported outbreaks also support this hypothesis. This approach to prevention, therefore, cannot be overemphasized.

REFERENCES

Hassall, C. H., and Reyle, K., 1955, The toxicity of the ackee *(Blighia sapida), West Indies Med. J.* **4**:83.

Hill, K. R. 1952, Vomiting sickness of Jamaica: A review, *West Indian Med J.* **1**:243.

Patrick, S. J., Jelliffe, D. B., and Stuart, K. L., 1955, The hepatic glycogen content in acute toxic hypoglycemia, *J. Trop. Pediatr.* **1**:88.

Stuart, K. L., Jelliffe, D. B., and Hill, K. R., 1955, Acute toxic hypoglycemia occurring in the vomiting sickness of Jamaica, *J. Trop. Pediatr.* **1**:69.

Williams, C. D., 1954, *Report of Vomiting Sickness,* Medical Department, Kingston, Jamaica.

7

LATHYRISM

Kamal Ahmad

The crippling disease lathyrism has been known in man at least since the days of Hippocrates, who reported that all men and women who ate peas developed extreme weakness in the legs and that this state persisted (Seley, 1957). The Indian publication *Bhavaprakash* (1550), attributed the crippling condition to *Triputa,* a pulse causing irritation of nerves (Chopra, 1938). The Duke of Wurtemberg was reported to have prohibited the use of lathyrus flour in the making of bread as he recognized its paralyzing effect upon the legs (Buchanan, 1927). The condition was, however, known by different names prior to Cantoni (1873), who introduced the word *lathyrism* in 1873 in describing the disease in Italy.

Two types of lathyrism are known: osteolathryism and neurolathyrism. Osteolathyrism (also known as odoratism) is caused in rats and other animals by the ingestion of *Lathyrus odoratus, hirsutus,* and *pusillus,* while neurolathyrism is caused in man by prolonged ingestion of seeds of *L. sativa.*

Osteolathyrism was observed in rats by Geiger *et al.* (1933) on the ingestion of *L. odoratus* (the common flowering sweet peas). The condition was characterized by extensive abnormality of the skeleton including forward bending of the body, increased shaft diameter of the long bone, and generalized osteoporosis. In pregnant rats *L. odoratus* caused death of the fetus (Stamler, 1955).

The toxic principle responsible for odoratism was isolated by Dupuy and Lee (1954) from *L. pussillus* as y-n-glutamyl b-aminoproprionitrile. The same substance was isolated by McKay *et al.* (1954) from *L. odoratus.* Dasler (1954) demonstrated that the active portion of the molecule was b-aminopropionitrile (BAPN). The primary lesion in osteolathyrism is considered to be due to impairment of collagen metabolism (Rao *et al.,* 1969). According to Page and Benditt (1967), the lathyrogenic effect results from the inhibition of the amino group of lysine residues, preventing its participation in forming cross linkages in collagen.

Kamal Ahmad • Institute of Nutrition and Food Science, University of Dacca, Dacca 2, Bangladesh.

Neurolathyrism is a nervous disease of the upper motor neuron (Jacoby, 1946) characterized by the spastic paraplegia of lower limbs after consumption of *L. sativa* (known as *kheshari*) in the Indian subcontinent. The onset of neurolathyrism is sudden. The symptoms start with aching of the waist, and aching and rigidity of calf muscles. In some cases, a sudden stiffness of the legs develops, resulting in a tendency to walk on the toes. Some of the patients show exaggerated knee and ankle jerks and ankle clonus.

Neurolathyrism may occur at any age, but usually between 15 and 30 years in either sex. In the human, the syndrome appears when *L. sativa* is taken as staple food for 3–6 months.

The first detailed description of the epidemic of lathyrism appeared in writings of General Sleeman (1844), who depicted the situation as it occurred in Madhya Pradesh in 1831 following prolonged consumption of *L. sativa* as the principal constituent of diet. Kojeewnikoff (1859) reported cases of lathyrism in South Russia during the famine of 1894 when a large number of people lived on *Schina (L. sativa).* There had been many reports on the incidence of the disease in the Indian subcontinent such as those of Irvin (1859), Young (1927), Dwivedi and Prashad (1964), and Gopalan (1950). Buchanan (1927) estimated that some 7600 people were afflicted in the period 1896–1902 in the central provinces of India. Acton estimated 60,000 lathyrism patients in Northern Rewa state in 1922. More recent reports indicate that there were 32,000 lathyrism victims in Madhya Pradesh in 1962 and 164,000 in Orissa, Bihar, and West Bengal in 1975. The districts of Rajshahi and Kustia in Bangladesh are especially affected; according to one report some 11,000 victims are to be found in Bangladesh.

Acton (1922) classified the progress or severity of the disease by stages. In the latent stage, the patients complain of a sudden agonizing pain in the calf muscles (myospasm) at night. The contracted muscles take the shape of a ball or lump known locally in India as *lodakas.* Similar contractions may be seen in the muscles of the back of the thigh. While the pain is soon gone, some patients develop neurological disorders in 1–2 weeks, with impairment of their ability to walk, especially downhill, or to run. When the condition deteriorates further the patient goes into what is known as first stage. He takes short steps and makes jerky movements. Due to stiffness of the muscles, he tends to walk on the toes with the knees slightly bent or the ankle extended. He is described to have a scissor or crossed gait. This condition is also known as the "non-stick" stage, as he does not need any stick. The patient may remain at this stage for the rest of his life, but may pass to the second stage, termed the "one-stick" stage. In this second stage, all the earlier signs and symptoms, such as muscular stiffness and bent knee joints, become more pronounced. Patients are unable to walk without the help of a stick. In the third stage the patient can only walk with the help of two sticks. In the fourth stage, which comes following continuous spasticity of the hamstring and calf muscles, the patient can only crawl using his palms and knees.

Many investigations were carried out to identify the biochemical cause of lathyrism. Lal (1949) noted vitamin A deficiency in a population with lathyrism. Mellanby (1930) put forward the theory that lathyrism was due to a neurotoxin, the effect of which could be prevented by food rich in vitamin A, even when the toxin-bearing seeds of *L. sativus* were consumed in large amounts. Ganapathy and Dwivedi (1961) reported deficiencies of both vitamins A and C in communities that had the incidence of the disease, but also reported lathyrism in certain families that consumed vitamin A-rich diets. Ganapathy further reported that the administration of vitamin A did not result in any clinical improvement of the condition. Jacoby (1946) reported that treatment with vitamins A, B, and B complex was ineffective. Many animal species such as ducks, chicks, peacocks, pigeons, rabbits, guinea pigs, rats, mice, and monkeys were fed seeds of *L. sativa* in an attempt to induce lathyrism. The results were either confusing or clearly unsuccessful. Stockman (1929) claimed to have produced symptoms of human lathyrism in monkeys but others failed in their attempts with monkeys, as well as with dogs and rats. Similar investigations failed in guinea pigs and in ponies (Stott, 1930). Roy *et al.* (1973) reported that the intraperitoneal injection of a dilute alcoholic extract of *L. sativa* into one-day-old chicks produced symptoms suggestive of neurological involvement. It included retraction of head, twisting and rigidity of neck, and paralysis in some instances resulting in the inability of the bird to stand and walk. The factor responsible for these symptoms was isolated and characterized to be β-oxalylaminoalanine (BOAA) (Murti *et al.*, 1965). A dose of 30 mg/100 g body weight of this compound produced neural features and opisthotonos in young chicks from which they recovered after 2–5 hr. On the basis of these neurotoxic effects on chicks, it has been claimed that BOAA is the compound responsible for the neurolathyrism in man.

In view of the fact that BOAA has been considered to be the toxin of lathyrism and that it is water soluble, it is advised that the seeds be soaked in three changes of water so that the toxin is leached out. The seed may also be soaked for 4–5 hr in hot water (60–70°C) and the rejected "soak water" will presumably contain the toxin. Attempts have also been made to find or evolve a variety of *L. sativus* low in BOAA by genetic screening or breeding (Anonymous, 1962).

In view of the fact that the physiology of chicks is different from man, and, as the signs and symptoms noted in chicks would not be regarded as quite equivalent to those seen in human neurolathyrism, there has remained a great deal of doubt whether BOAA is really the toxin responsible for human lathyrism. However, the full investigation of the problem has presented difficulties for want of a suitable experimental animal model with comparable neurological manifestations and disabilities to those seen in human patients. It is to be recalled that BOAA is nontoxic to adult laboratory animals.

The population that suffers from lathyrism lives in poor economic conditions and has been reported to be deficient in many nutrients, as mentioned earlier, though no specific deficiency could be confirmed as having been responsible. It has appeared to us that the vitamin C deficiency could be a predisposing factor to the condition, not only because most of the sufferers of the disease had various degrees of deficiency of the vitamin, but also because of the fact that most of the experimental animals on which experimental lathyrism was attempted are those that did not need vitamin C. This led us to our own attempt to produce experimental lathyrism in guinea pigs. We did find that at the stage of subclinical deficiency of vitamin C, feeding of *L. sativus* seed produced a condition comparable to that seen in man.

As would be seen (Table I) a group of adult male guinea pigs (group A) was put on a diet of cooked *L. sativus* (*khesari*) supplemented with all vitamins (including vitamins A and D) except vitamin C. A control group (group B) of the animals was given the same diet as group A but including vitamin C (5 mg/animal daily). Then two other sets of the animals—group C and group D—were fed diets in which *L. sativus* was substituted by another legume, namely *Phaseolus radiatus*, but which were otherwise similar to groups A and B, respectively (i.e., without or with vitamin C). These two groups were included in the study in order to preclude or isolate any effect of vitamin C deficiency itself in absence of *L. sativus*.

While not a single animal of group B (which received vitamin C) showed any neural symptoms, 26 of the 35 animals in group A (which did not have vitamin C) developed progressively neurological symptoms—namely, monoplegia, paraplegia, and finally hemiplegia in 11 weeks' time. They showed various conditions such as tremor, ataxic gait, dragging of legs, and rigidity of neck.

TABLE I

Effect of Lathyrus Sativus on Adult Male Guinea Pigs[a]: Protective Action of Vitamin C

Group[b]	Diet	Observation	Remarks
A	Cooked *L. sativus* with supplement of all B vitamins, vitamins A and D, but no vitamin C	26 animals developed neurological symptoms of lathyrism within 11 weeks; the remaining 9 showed no neurological symptom but lost weight	Extremely significant
B	As above but with vitamin C, 5 mg/animal daily	None showed neurological symptoms	P = 0.000005
C	Cooked *P. radiatus* with added B vitamins, and vitamins A and D but no vitamin C	Normal in all respects; none showed neurological symptoms	No neurological symptoms appeared in the experimental animals just due to the absence of vitamin C in the diet when *L. sativus* was replaced by *P. radiatus*
D	As above but with added vitamin C, 5 mg/animal daily	Same as group C	

[a] 280–360 g/animal.
[b] Group A comprised 35 animals; groups B, C, and D comprised 20 animals each.

At the stage of monoplegia, an affected animal can be cured by including vitamin C (0.1%) in the daily diet. The reversal of the clinical features took place rapidly, and complete or near-complete recovery could be seen in 3–7 days. None of the animals which developed hemiplegia could be saved.

CONCLUSION

Neurolathyrism is a public health problem of significant dimensions. It afflicts large numbers of poor people in the Indian subcontinent. Even though the disease has been known since the days of Hippocrates, and many researchers have devoted their time and attention to the condition, little progress had been made because of the failure to produce experimental lathyrism in laboratory animals.

The recent findings of Ahmad and Jahan (1980) that subclinical deficiency of vitamin C in guinea pigs renders the animal susceptible to the toxin present in *L. sativus,* has made possible the induction of experimental lathyrism in guinea pigs. Undoubtedly, this finding will be useful in the isolation of a new toxin of lathyrism, or the confirmation that BOAA is the real toxin. The biochemistry of the condition, including the mechanism as to how vitamin C exerts its protective action against neurolathyrism, may also unfold in the near future.

REFERENCES

Acton, H. W., 1922, An investigation into the causation of lathyrism in man, *Indian Med. Gaz.* **57:**241 247.

Ahmad, K., and Jahan, K., 1980, *Third Asian Congress of Nutrition, Abstracts and Programmes,* p. 190, Djakarta.

Anonymous, 1962, *Lathyrism—A Preventable Paralysis,* Indian Council of Medical Research, New Delhi.

Buchanan, A., 1927, *Report of Lathyrism in Central Provinces in 1896–1902,* pp. 1–71, Government Press, Nagpur.

Cantoni, A., 1873, Lathyrisimo (Lathyrism) illustrata de tre casi clinic, *Morgagni* **15:**745.

Chopra, R. N., 1938, *British Encyclopedia of Medical Practice, VII,* pp. 615–657, Butterworths, London.

Dasler, W., 1954, Isolation of toxic crystals from sweet peas, *Science* **120:**307–308.

Dupuy, H. P., and Lee, J. G., 1954, The isolation of a material capable of producing experimental lathyrism, *J. Am. Pharm. Assoc.* **43:**61–62.

Dwivedi, M. P., and Prasad, B. G., 1964, An epidemiological study of lathyrism in the district of Rewa, Madhya, Pradesh, *Indian J. Med. Res.* **1:**81.

Ganapathy, K. T., and Dwivedi, M.P., 1961, *Report of Indian Council of Medical Research,* Gandhi Memorial Hospital, Rewa.

Geiger, B. J., Steenhock, H., and Parson, H. T., 1933, Lathyrism in the rat, *J. Nutr.* **6:**427.

Gopalan, C., 1950, Lathyrism syndrome, *Trans. R. Soc. Trop. Med. Hyg.* **44:**333.

Irving, J., 1859, Notice of a form of paralysis of the lower extremities extensively prevailing in part of the district of Alahabad produced by use of *Lathyrus sativus* as an article of food, *Ann. Ind. Med. Sci.* **6:**424.

Jacoby, H., 1946, Curative treatment of lathyrism, disease of the nervous system, *Indian Med. Gaz.* **81:** 246–247.
Kojeewnikoff, K., 1894, Lathyrism in Russia, *Vestn. Psych.* **10:**2–15.
Lal, S. B., 1949, Lathyrism in Bihar, *Indian Med. Gaz.* **84:**468.
McKay, G. F., Lalich, J. J., Schilling, E. D., and Strong, F. M., 1954, A crystalline "Lathyrus Factor" form *Lathyrus odoratus, Arch. Biochem. Biophys.* **52:**313–322.
Mellanby, E., 1930, The relation of diet to health and disease, *Br. Med. J.* **1:**677.
Murti, V. V. S., Sherardi, T. R., and Venkitasubramaniam, T. A., 1965, Neurotoxic compounds of the seeds of *Lathyrus sativus, Photochemistry* **3:**73.
Page, R. C., and Benditt, E. P., 1967, Molecular diseases of connective and vascular tissue II. Amine oxidase inhibition by the lathyrogen b-aminopropionitrile, *Biochemistry* **6:**1142–1148.
Rao, S. L. N., Malathi, K., and Sarma, P. S., 1969, *World Review of Nutrition and Dietics, Vol. 10,* pp. 214–238, Karger, Basel.
Seley, H., 1957, Lathyrism, *Rev. Can. Biol.* **16:**1–82.
Stamler, F. W., 1955, Reproduction in rats fed Lathyrus peas or aminonitriles, *Proc. Soc. Exp. Biol. Med.* **90:**294.
Sleeman, W. H., 1844, Rambles and Recollections of an Indian Official, Constable, London.
Stockman, R. 1929, Lathyrism, *J. Pharmacol. Exp. Ther.* **37:**43.
Stott, H., 1930, On the distribution of lathyrism in the United Provinces and on its cause, with a description of a 4-3/4 feeding experiment on Tonga Ponies with botanically pure cultures of *Lathyrus sativus* and *Vicia sativa, Indian Med. Res.,* **18:**51.
Young, T. C. M., 1927, A field survey of lathyrism, *Indian J. Med. Res.* **15:**453–479.

8

THE DJENKOL BEAN

Khin Kyi Nyunt

The botanical name for this starchy legume is *Pithecollobium lobatum.* The English common name is the djenkol bean. However, nutritionists know it by the more descriptive name of Ape's Earring, *djonkol,* or stink-bean. The Burmese name is *da-nyin-thee.* It grows on a tall tree of the *Mimosa* family. Its fruit pods are horseshoe shaped or spirally twisted. It flowers from January to April and fruits appear from August to October. The djenkol bean is popular in most countries of southeast Asia, wherever it flourishes—in Burma, Thailand, Malaysia, and Indonesia.

This terse botanical description cannot give an idea of the graceful proportions of the shape of the crown of the tree with its spreading branches, affording dense shade, and its great trunk; it is a magnificent tree when viewed from a distance, similar to a giant oak. The lax-spiral twisted fruit pod is truly a work of abstract sculpture. When the glossy dark brown pod is opened, each creamy colored bean is nestled in its own section like an oyster in its shell. The color of the inner skin of the bean becomes a deeper shade of reddish brown when exposed to the air. The tree is often planted deliberately by people near their homes. The mature bean germinates easily and begins to bear fruit in about five years. The beans are retailed in numerical lots, in multiples of five, and wholesaled by the hundreds. They are not sold by weight. The price varies according to the grade in size. The boiled beans are sold as half cotyledons, five on one skewer, ready to eat.

The role of *da-nyin-thee* in the Burmese diet is as a vegetable relish, to dunk in a salty fish sauce called *ngapi-ye,* to eat with the bland staple food of boiled rice. Normally, one or two beans may be consumed by a person at a meal, but the number may vary with the size of the bean. If one is fortunate enough to have the tree in one's own garden, early in the season the immature leathery, glossy brown pod is levered open with the point of a knife, and the small three-quarter-inch diameter bean with the creamy skin is peeled to

Khin Kyi Nyunt • Nutrition Research Division, Department of Medical Research, Rangoon, Burma

reveal a pale, jade green color. Eaten raw, it has a crisp texture and a bitter–sweet delicate flavor, contrasting with the taste of salty fish sauce, and bland boiled rice staple. As many as ten of these small djenkol beans may be eaten by a person in one meal.

As it matures in the next month, the bean becomes more starchy and develops a stronger smell. The medium-sized beans are about an inch in diameter, and, still in their thin reddish brown skins, are often preserved in brine. Their cost was about 6¢/hundred (U.S. rates) in July, 1980, at Toungoo. The preservation process requires a prior soaking in water for three days, changing the water daily. The beans swell and the texture becomes crisper. The beans are then placed in earthen or glass jars, a cooled strong brine is poured in, and crossed sticks are inserted to keep the floating beans immersed for about ten days. These are served in the same way as before, as a vegetable relish with salty fish sauce and boiled rice. The brine has the distinctive smell and is known to be used in the treatment of diabetes.

Later in the season, the mature djenkol beans are at their greatest size, and are prized for their waxy texture and slightly bitter-sweet rich taste after having been boiled an hour or two. After boiling, the brown skins are discarded, and the beans are again boiled in fresh water, with a little peanut oil added, until the water has evaporated. This method gives them a glossy, attractive brown color outside and a pale, olive color within. These are also served the same way as before—that is, as a vegetable relish. The boiled beans are sometimes pounded flat and added to a pork curry to serve with rice. In the Tenasserim region, in the town of Tavoy, the boiled beans are cooked with pounded, dried shrimp, pounded onions, garlic, shrimp paste *(kapi),* peanut oil, and a little water, and served with rice. Burmese girls pound the boiled djenkol beans flat and add it to *la-pet,* a pickled tea leaf salad, which is a national favorite. The typical way the Mon ethnic group of people use this bean is to slice the boiled beans into a soup of fried Dolichos beans that has germinated for a couple of days and hulled. To this soup is added a variety of thorny and leafy vegetables called *su-boke-ywet,* which is also strong smelling. The soup is flavored with dried shrimps, onions, garlic, and seasoned with shrimp paste and salt. This is served with boiled rice. A favorite dish that requires more elaborate, time-consuming preparation is *danyin-thee-ohn-no-san.* One usually prepares a batch of at least 25–50 mature djenkol beans of the largest size from 1.5–2 inch diameter. This is sufficient to eat with the sauce from a large mature coconut or two smaller coconuts. The beans are boiled the usual way, with a spoon of peanut oil added to the water, for an hour or two until the water is evaporated. The beans are then split and each half is pounded carefully on an oiled board with an oiled pestle into a very thin, flat, round discs. They then look like mottled tortoise shells in color, or a very fragile lace about 3 inches or more in diameter. This is served with a teaspoon of creamy coconut-milk sauce seasoned with salt and a little sugar. The sauce is very simple; the coconut is grated, the coconut milk is extracted by squeezing it in a clean napkin or cloth; hot water and salt is added and the grated coconut is squeezed in the cloth to obtain the second extract. The process is repeated for the third extract. All the coconut extract is combined and allowed to evaporate to a thick creamy consistency, and seasoned with salt and a teaspoon or two of sugar. As this dish is a seldom-made delicacy, as well as being rather delicious and out of the ordinary, it is easy to understand why a lot more beans are eaten through sheer self-indulgence than would be normally consumed. At least once in a season, the family is given this treat. It is also served to guests, and frequently sent to friends, neighbors, and relatives. This delicacy is also sold in the market where a busy housewife may eat it on the spot, and take some home in a plastic bag for the family. It is a snack for tea-time, or a savory after lunch or dinner. It is not served with rice.

Da-nyin-thee-mee-poke means baked or grilled djenkol beans. The traditional preparation is to bake the mature beans slowly, buried under hot ashes with small, glowing embers for half an hour or until a fragrant aroma develops and the beans are tender. The brown skin is discarded, and the baked djenkol beans are served the usual way with salty fish sauce and boiled rice.

Da-nyin-gwet means germinated djenkol beans. These fetch a higher price in the market. The mature beans are soaked in a tub of water till swollen, and then left moist to drain in a wicker sieve until they germinate. The color changes to an emerald green, the starch hydrolyzes so that a sweeter taste develops, the texture becomes less waxy and the side effects are less. *Da-nyin-gwet* is also baked slowly as described above or grilled about 3 in from the fire for about 3–5 min. They may be boiled in water until tender, but slow baking is the method of choice to bring out the aroma.

Da-nyin-thee is definitely one of the foods which demands an acquired or cultivated taste, perhaps from early youth. It is well known that the unpleasant smell of *da-nyin-thee* lingers on the breath and is offensive to those who have not shared in the feast. It is probably excreted in the saliva. Its excretion in the urine leaves the toilet room with a strong distinctive odor even after flushing. One may imbibe huge quantities of water to induce diuresis in an effort to rid the smell on one's breath. The traditional way is to chew some raw rice. Adolescent girls and boys refuse the food they love because of this side effect. If the husband objects, the wife has to avoid eating these djenkol beans. The beans are sought during the Buddhist lent, when many people go on a strict vegetarian diet. This food can be considered as a form of "soul food" being served to monks and nuns, and to people who are fasting after noon on the full moon days, on the eighth waning day, on "no moon" day, and on the eighth waxing day, all of which are fasting days in Buddhist Burma.

It is also well known that consuming large quantities of djenkol beans leaves one feeling very drowsy, with the limbs feeling heavy and the neck stiff. These transient syndromes are termed *tet*. There are several other vegetables and fish foods which are believed to cause this syndrome; it is not thought to be specific to djenkol bean consumption. Moreover, these syndromes are not observed in every case of heavy djenkol bean consumption. Some heavy eaters never complain of even the mildest side effects. In the Arakan region, it is not at all unusual for people to eat about 20 or 30 of the small spherical variety of djenkol beans, boiled, nibbling each half flat and dipping it into individual saucers of peanut oil and salt, as a snack. However, it has been verbally reported from medical officers in Myitkyina that after consuming large quantities of djenkol beans, persons have been admitted to hospitals with complaints of severe abdominal pains, oliguria, anuria, edema, and sometimes even hematuria. Unfortunately these persons were not asked whether consumption of large quantities had ever troubled them prior to hospitalization. It may best be considered as a personal food idiosyncracy. Up to now, there is no unequivocal clinical evidence, except for a strong association, to incriminate the djenkol bean in the etiology of acute renal failure.

Some research done on the chemical composition (Sein-Gwan and Chit-Maung 1968) and pharmacological investiagion (Chit-Maung, 1980) of the djenkol bean show a high concentration of manganese chloride, which is said to posses a hypoglycemic effect (Rubenstein *et al.*, 1962). Like other legumes, it is deficient in sulfur amino acids and phenylalanine. Sixteen common acids have been definitely identified in the protein. Of eighteen amino acids found in the free amino acid portion of the fruit, most of them are the common amino acids present in animal proteins. The unusual amino acids are N-methyl α-amino acid, pipecolic acid, γ-amino-butyric acid and djenkolic acid. The last three are present in very high concentrations. Chemical analyses have shown that *Pithecolobine* has been isolated from the bark and seed of the plant. It is a yellowish, amorphous powder on a brown, oily mass with narcotic odor, and is soluble in water, alcohol, chloroform, ether, and petroleum ether. It forms crystalline salts.

Djenkolic acid is an amino acid isolated from the djenkol bean; it is known to occur free in the plant. The common salt solution extract of djenkol beans contains the protein and the free amino acids of the fruit. These are djenkolic acid, aspartic acid, glutamic acid, aspargine, serine, arginine, threonine (trace) alanine, γ-amino-butyric acid, proline, pipecolic acid, valine, leucine, histidine, tyrosine (trace), methioine (trace), *N*-methyl α-amino acid (trace), and glycine. Cysteine, cystine, phenyl-alanine, trytophane and lysine were absent. Djenkolic acid is not metobolized by mammals and is excreted in its free form in the urine. γ-amino-butyric acid is a metabolite present in the brain of mammals. It is said to possess a depressive effect on the brain. Drowsiness induced after taking large quantities of djenkol bean may be due to this substance. The pharmacology of pipecolic acid is unknown. The ether-soluble extract of djenkol bean is found to be a sulfur-containing organic compound. This substance seems to be absorbed from the gastrointestinal tract because the urine voided after consuming the djenkol bean has the same smell. Three extracts of the djenkol beans were prepared to test their antidiabetic effects: common salt solution, water-insoluble but ether-soluble extract, and the ash. Experiments on fasting albino rats, carried out individually using the three extracts, showed no effect on the blood sugar level.

The pharmacological properties of the djenkol bean may depend as much on regional differences in soil, water, and climatic conditions of temperature, humidity, and rainfall in the area in which it grows as on the genetic difference of the plant varieties. The Arakan region in Burma produces a smaller, more spherical and waxy variety of djenkol bean that is popular with most people in the rest of Burma. When consumed in large quantities, the sequelae are the symptoms of acute renal failure, that is to say oliguria and anuria. For treatment of this condition, indigenous medicine practioners grind the hoof of rhinoceros with water, and this extract is given orally to the patient to relieve oliguria or scanty micturition. The unpleasant smell of the breath and urine as a side effect is said to be minimal after consumption of this variety of djenkol bean. The variety of djenkol bean from Myitkyina region in Burma also appears to cause a minimal amount of unpleasant smell in the breath and urine after its consumption. This may be a mixed blessing, for when the social deterrent to eating large quantities is absent, the pathological symptoms are more likely to occur. It is known that eating 18 djenkol beans has been followed by the onset of acute renal failure. Some persons experience nausea and vomiting of a transient nature after djenkol beans.

As Ape's Earring, djenkol is listed as item number 363, under food group five—vegetable and vegetable products—in the Food and Agriculture Food Composition Table for use in East Asia (1972). In 100 g of edible portion there are 93 kcal. food energy, 76.3% moisture, 6.2 g protein. 0.2 fat, 16.9 g carbohydrate, 1.3 g fiber, 0.4 g ash, 23 mg calcium, 38 mg phosphorous, 0.7 mg iron, 505 μg β-carotene equivalent, 0.14 mg thiamine, 0.01 mg riboflavin, 0.4 mg niacin, and 8 mg ascorbic acid.

Djenkol bean is the perfect example of that old axiom that one man's food is another man's poison.

REFERENCES

Food and Agriculture Organization, 1972, *Food Composition for Use in East Asia,* U.S. Department of Health, Education and Welfare, Public Health Service, National Institute of Health, Washington, D.C.

U-San-Hla, 1979, *Handbook of Natural Foods,* Jiva Daya. 536–537. 180. Botstaung Pagoda Road, East Rangoon, Burma.

Sein-Gwan and Chit-Maung, 1968, Chemical composition of Djenkol bean, *Union Burma J. Sci. Technol.* **1**:221.

Chit-Maung, 1980, Personal communication, Senior Research Officer, Pharmacology Research Division, Department of Medical Research, 5. Zafar Shah Road, Rangoon, Burma.

Merck Index of Chemicals and Drugs, 7th ed., 1960, pp. 389, 825, Merck and Co., Rahway, New Jersey.

Rubenstein, A. H., Levin, N. W., and Eliots, G. A., 1962, Hypoglycemia induced by manganese, *Nature* **194:**188.

Van Veen, A. G., and Latuasan, H. E., 1949, The state of djenkolic acid in the plant, *Chron. Nat.* **105:**288.

II

ENVIRONMENTAL TOXICANTS

9

ENVIRONMENTAL CONTAMINANTS

D. G. Lindsay and J. C. Sherlock

INTRODUCTION

Since antiquity there have been descriptions of episodes of acute ill health that have arisen from the consumption of beverages heavily contaminated by a toxic chemical. It has only been in the last 20 years, however, that any detailed consideration has been given to the adverse effects that can arise from the contamination of food by environmental chemicals.

By the mid-1960s there was substantial evidence that the widespread use of certain organochlorine insecticides, such as DDT and dieldrin, had resulted in the gradual buildup of these chemicals in soils, fish and animal tissues, and human adipose tissues. Foodstuffs became contaminated by residues of these insecticides, even when the chemicals had not been directly used on the food. The factors leading to the persistence of these pesticides and their widespread occurrence in the environment are not peculiar to this class of chemical, and other industrial chemicals have since been shown to be widespread contaminants of the environment. For the majority of industrial chemicals, there is generally insufficient toxicological information available with which to assess any potential hazards to human health, if the chemicals appear in food or beverages. Pesticides, on the other hand, which are used because they exhibit biological toxicity, have been subjected to extensive toxicological studies prior to their approval for marketing, and their safety-in-use is evaluated on a continuing basis in the light of the latest advances in toxicology. It is only very recently that preregistration schemes for approval prior to the manufacture and use of nonpesticidal chemicals have been introduced. In some countries such schemes will apply only to those chemicals that are new; for many of the 70,000 chemicals that are currently in production in the U.S. alone, insuffi-

---

D. G. Lindsay and J. C. Sherlock • Food Science Division, Ministry of Agriculture, Fisheries and Food, London SW1P 2AE, United Kingdom.

cient toxicological data will be available to do more than qualitatively assess possible risks to human health.

CHEMICAL NATURE OF ENVIRONMENTAL CONTAMINANTS

Chemical Stability

The factors that determine the presence of an environmental chemical residue in food do not differ from those that determine the persistence of chemicals in the environment as a whole. The properties of chemicals that determine their persistence are:

1. Resistance to degradation by heat leading to the dispersion of the chemical during disposal and incineration of wastes.
2. Stability to electromagnetic radiation such that terrestrial sources of radiation (principally wavelengths in the IR region of the electromagnetic spectrum) are unable to degrade the chemical bonds.
3. Low solubility in water resulting in the strong absorption of the chemical onto sediments or soils and the absence of effective dilution and dispersion.
4. Resistance to chemical degradation both in acid or alkaline media, or both.

Biological Stability

The factors that determine chemical stability are also responsible in part for the inhibition of biological degradation processes through the action of microorganisms, plants, and animals. The catabolic processes that have been developed to utilize energy from food degradation are unable to operate on chemicals with certain large and multiple functional groups. In addition, if these functional groups are not normally encountered in naturally occurring chemical compounds there is the possibility that the degradation pathways, which have been evolved by microorganisms and higher organisms, are ineffective.

Biomagnification of Environmental Contaminants

The resistance of some chemicals to catabolism can result in the bioaccumulation of the substance within a living organism, whereby it is absorbed and stored in tissues; for example, this occurs with inorganic substances such as cadmium and lead. Organic chemicals that are nonpolar may be associated with fat-storage tissues used as energy reserves. This in turn can lead to biomagnification through successive trophic levels in the food chain, which may eventually result in concentrations in human food that are many times higher than those that were present in the original discharge. Thus, the lipophilic character of the chemical will be an important factor in determining the biomagnification potential of the chemical. Certain organometallic chemicals, e.g., methyl mercury compounds, are sufficiently lipophilic to undergo bioaccumulation.

SOURCES OF FOOD CONTAMINATION

Environmental chemical residues in food occur as a result of agricultural activities, mining, energy production, and industrial chemical production and use. Apart from the use of pesticides on certain crops, in no instance is it intended that residues of environmental chemicals be present in food apart from those inorganic elements that are naturally present in foods at a certain level. The chemicals enter the food chain through various routes depending on their use and the mechanisms leading to persistence. These routes are shown in Fig. 1. Chemical residues in food can also arise purely accidentally through misuse of chemicals.

SUSCEPTIBLE FOODS

Accidental contamination of the food supply can affect any foodstuff at all stages in its production up to its eventual consumption. In general, however, the contamination of food by organic environmental chemicals affects mostly those foods that have a high fat content such as meat, dairy products, and fish. Persistent organic compounds are invariably strongly bound to soil colloids and are not readily translocated from the soil into the growing crop, and soil microbiological activity will eventually degrade the compounds through the carbon cycle, except for the most persistent compounds. On the other hand inorganic elements will undergo no degradation—they may be translocated into plants either before or after microbial activity—and for some elements, such as cadmium, plants are the major dietary source.

HEALTH SIGNIFICANCE OF FOOD CONTAMINANTS

The contamination of food and drinking water by industrial chemicals is the major source of exposure of the general population to environmental contaminants for all except the most volatile chemicals. Potential effects from the contamination of food may arise from either of the following:

1. The gradual diffusion of persistent chemicals throughout the environment resulting in a low-level and long-term exposure to a major proportion of the population, or

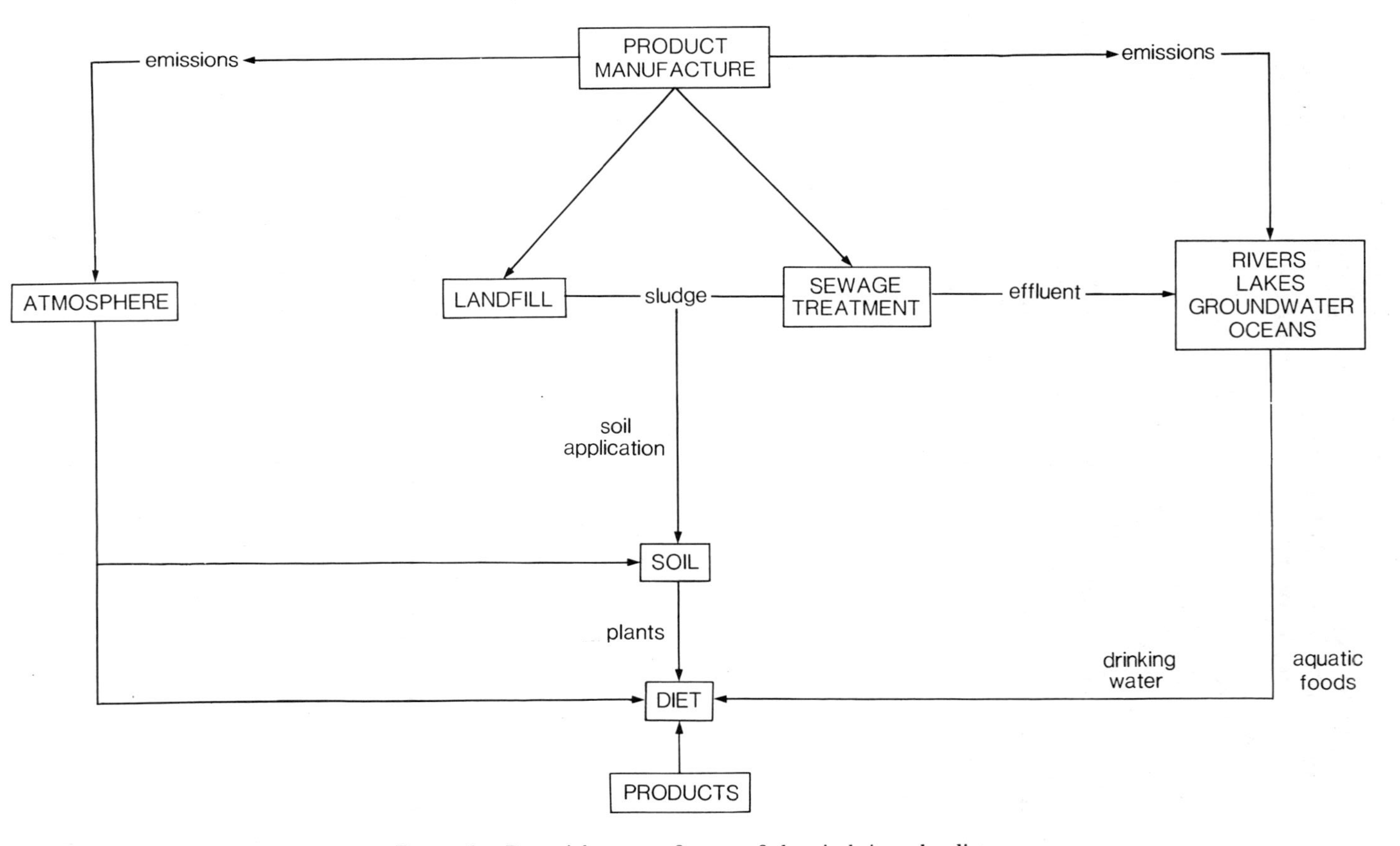

FIGURE 1. Potential routes of entry of chemicals into the diet.

2. The exposure to a higher dose of a contaminant as a result of the localized discharge of chemical, (e.g., accidental discharge) which results in subacute effects in a restricted section of the population consuming the food.

It is only in the latter case that health effects have actually been shown to occur. This is hardly surprising since it would be extremely difficult to demonstrate any long-term effects due to a continuous low-level exposure to a toxic substance in the general population, unless this gave rise to an unusual pattern of disease.

A major public health concern arises from the possibility of the presence of residues of chemicals known or suspected to be human carcinogens in the food chain. The concern arises from the fact that at least for certain types of carcinogenic substances, namely the proximate electrophilic carcinogens such as dimethylnitrosamine and vinyl chloride, effects can be observed in experimental animals even when low doses are administered. The effect of continuous low-level administration of the carcinogen is to delay the time-to-tumor formation. It does not remove the ultimate carcinogenic response. Table I illustrates this point from experimental studies on the administration of aflatoxin B_1 to male Fisher rats. Similar effects are noticed in radiation-induced carcinogenesis (UNSCEAR, 1977).

Apart from the potential health risks associated with environmental chemicals that are carcinogenic, potential chronic health effects may arise from the accumulation of these substances in human tissues. The factors that cause these chemicals to accumulate in animal, fish, and plant tissues are also responsible for their accumulation in man. Organic and inorganic chemicals can accumulate in tissues and in some instances cross usually effective barriers such as the placenta or the blood–brain barrier through transport associated with plasma proteins or bindings to erythrocytes, and may concentrate in excretory and other organs. Thus, potential long-term health effects may include birth defects and reproductive disorders in addition to systemic tox-

TABLE I

Dose–response Data for Aflatoxin B_1 Carcinogenesis in Male Fisher Strain Rats[a]

Dietary aflatoxin level (μg/kg)	Duration of feeding (weeks)	Incidence of liver tumors	Time of appearance of earliest tumor (weeks)
0	74–109	0/18	—
1	78–105	2/22	104
5	65–93	1/22	93
15	60–96	4/21	96
50	71–97	20/25	82
100	54–88	28/28	54

[a] Wogan *et al.*, 1974.

icity. Under conditions of stress or unusual metabolic activity resulting in the mobilization of fat stores, high levels of the substance may be liberated into the blood stream and into breast milk, resulting in levels of exposure to the neonate that are greater than those which would normally be encountered in infant foods.

Many of the food contamination episodes that have led to human disease have produced the severest effects on infants. The neonate is particularly susceptible to the effects of toxic chemicals because of the following reasons:

1. It is subject to a higher dose than an adult through its greater food consumption on a body-weight basis.
2. It invariably consumes a much more restricted diet.
3. It is incapable of metabolizing substances as effectively as an adult because the adult complement of detoxifying enzymes has not yet been developed.
4. Actively differentiating and developing cells are more sensitive to toxicity.

Limitations of Epidemiology

In the following section instances are described whereby epidemiology has been used successfully to associate disease with a particular chemical contaminant of the diet. The common features of these episodes are the following:

1. A limited number of people were exposed.
2. Their diets consisted in large measure of both highly contaminated and locally produced and consumed food.
3. Livestock and domestic animals were invariably affected.
4. The characteristics of the disease were invariably unusual.

The studies have not been very successful in accurately defining the level of exposure of individuals, and it is therefore difficult to determine dose–response relationships and enable levels of exposure showing no effects to be derived.

The estimation of the exposure of a population to an environmental contaminant in food is fraught with difficulties, particularly since dietary patterns are neither constant nor homogenous within a community. In consequence it is almost impossible to obtain accurate information about food consumption habits, through dietary recall over periods of months or even years. In addition it is difficult to find a suitable control population because of the variability in human susceptibility, personal behavior patterns, and emotional responses to stress.

Exposure to low doses of an environmental contaminant over many years can give rise to ultimate pathological effects, although for neoplastic disease this can take some 40 years. Even if an environmental contaminant in the diet were to drastically increase the cancer rate from 1 in 10^6 in the normal population to 1 in 10^4, a sample population of at least 3000 people would need to be studied to provide convincing statistical evidence that such an increase had in fact occurred. Even then, if such a change were detected, this would not prove causality. Further studies would be necessary in which exposure to the agent was eliminated from the statistical design before this could be established. Because of such limitations, epidemiological studies are likely to be of limited use in defining the role played by the environmental contamination of food in disease induction. The need to exert legislative controls on contaminants will invariably be based on the use of *in vitro* experiments or on *in vivo* techniques utilizing experimental animals in the laboratory for such risk assessments.

Risk Assessment

The detection and identification of a residue of a toxic chemical in the diet or in beverages usually requires some assessment of the potential risk to the population. This is necessary not only because it would be socially unacceptable to wait until health effects were observed, but also because of the difficulties previously described in actually demonstrating that a causal relationship exists.

The information required to make any assessment of risk can include its correct identification together with information on the systemic effects of the substance in a number of animal species, and information about its potential genetic toxicity and carcinogenicity. An estimate of the likely degree of exposure of the population at risk is also required.

Toxicological Data

In the absence of any information on human health effects, inevitably one must resort to data obtained from tests on experimental animals. The minimum practical information necessary is likely to be a 90-day continuous-exposure or paired-feeding study (Food Safety Council, 1978), as well as short-term tests for mutagenesis before and after metabolic activation (Venitt, 1980). That spectrum of tests that reflect as nearly as possible the effects of human metabolism on the production of a DNA-reactive substrate are clearly the most appropriate. The most potent carcinogens can sometimes be detected in a 90-day test, but where carcinogenicity is suspected, it is almost invariably necessary to carry out chronic toxicity studies in two mammalian species extending over a major proportion of their lifespan to evaluate the potential carcinogenicity of a chemical.

Estimation of Dose

An assessment of the dose ingested by a population requires information about the average levels of contamination in the diet, the amount of food consumed, and the frequency of consumption. Inevitably, in communities where diets consist of a wide range of foods, information on dietary habits will only be approximate and can be obtained by calculation from national food consumption or food purchase surveys. Other techniques include the use of diaries and food weighing to record the food consumed, or the analysis of duplicate samples of the food consumed at chosen intervals (Marr, 1971). It must be accepted that these procedures are rarely an exact reflection of exposure because of the considerable practical difficulties involved. However, in those instances where the population consumes a limited type of food, as typified by the diets of young infants, or where only one component of the diet is contaminated, much more accurate estimations of dose can be obtained from the use of diary studies in conjunction with the analysis of duplicate samples of the food consumed over a period of time.

It is also possible to determine human exposure through biological monitoring. Measurement of mercury levels in blood and hair (Pallotti *et al.*, 1979), blood lead levels (Sartor and Rondia, 1980), and persistent organic contaminants in breast milk (Polishuk *et al.*, 1977) are all examples of data that has been used in the past to reflect time-average exposure of a population to a contaminant. This approach can only be used satisfactorily, however, if the diet is the major route of exposure.

Evaluation of the Risk

Animal toxicity data should enable an estimate to be made of the dose at which no effects are observed—the *no detectable adverse effect level* (NEL). To take account of the uncertainty of translating animal data to humans, the insufficiency of the data available, or the severity and reversibility of toxic effects it is customary to apply an appropriate safety factor to the NEL (which can vary between 10 and 1000 or more to estimate the arbitrary acceptable level of risk), the acceptable daily intake (ADI), or the tolerable weekly intake (TWI). Thus the ADI is the daily dosage of a chemical which, during an entire lifetime, appears to be without appreciable risk on the basis of all the known facts. From a comparison between the estimated intake of the population and the ADI it can be determined whether or not controls over exposure should be considered. Although the process has no scientific basis, it has been shown, in practice, to be a socially acceptable yardstick in the approval for use of chemicals that are deliberately added to food, i.e., food additives and pesticides.

The major difficulty arises in the estimation of risks from environmental chemicals that are carcinogens. Attempts have been made to calculate mathematically the risks of exposure to low doses of carcinogens (Food Safety Council, 1978). However, the choice of mathematical procedures for low-dose extrapolation from animal data has no firm biological basis. There is also no reason to suppose that should such mathematical techniques be applied an "acceptable" or virtually safe dose for some carcinogens would result, which would allow a maximum level of a carcinogenic chemical in food to be above the limit of detection of present day analytical techniques, that is in the range of 1 μg/kg to 1 mg/kg. The present inability to quantify such risks adequately has resulted in action being taken to reduce human exposure to these substances wherever practicable even though they are detectable only in very low concentrations.

Contamination of human milk or cow's milk by an environmental chemical could continuously expose an infant to a toxic substance. Because of the particular sensitivity of the neonate to toxic substances, there are often sufficient grounds to justify controls over the source of the contamination even in the absence of definite evidence of harm to human health.

ENVIRONMENTAL CONTAMINANTS OF FOOD AND THEIR EFFECTS

Although there are many instances of human health effects caused by toxic chemicals, the majority of these have occurred as a result of industrial exposure or of incidents, such as the Seveso accident, where the general population were most likely exposed to the toxic agent through direct contact with the toxin rather than solely through the consumption of food contaminated by 2,3,7,8-tetrachlorodibenzodioxin.

There are a number of well documented accounts of where food has been so heavily contaminated by the industrial discharge of chemicals, or through careless controls over the use of certain chemicals, that food consumption has led to chronic disease. These effects have ranged from nonreversible and debilitating disease, such as occurred in the Minamata tragedy in Japan, to the less severe effects which affected residents of Michigan when they consumed beef contaminated with polybrominated biphenyls (PBBs).

Organic Contaminants

Polychlorinated Biphenyls

Polychlorinated biphenyls (PCBs) are a group of chlorinated compounds (Fig. 2) whose chemical properties and whose use in a variety of applications—primarily as dielectrics in transformers and capacitors and as heat exchangers—have resulted in their broad dispersal into the environment over the past 50 years.

PCBs were detected as low-level environmental pollutants in 1966 when environmental samples were being examined for levels of the persistent organochlorine pesticides. PCBs are almost ubiquitous contaminants of fish and, to a lesser extent, of other lipid-rich foods. This in turn has resulted in the accumulation of these compounds in man and they are to be detected in human breast milk and adipose tissue (Polishuk *et al.*, 1977). Commercial PCB mixtures are a complex series of products since there are on the biphenyl

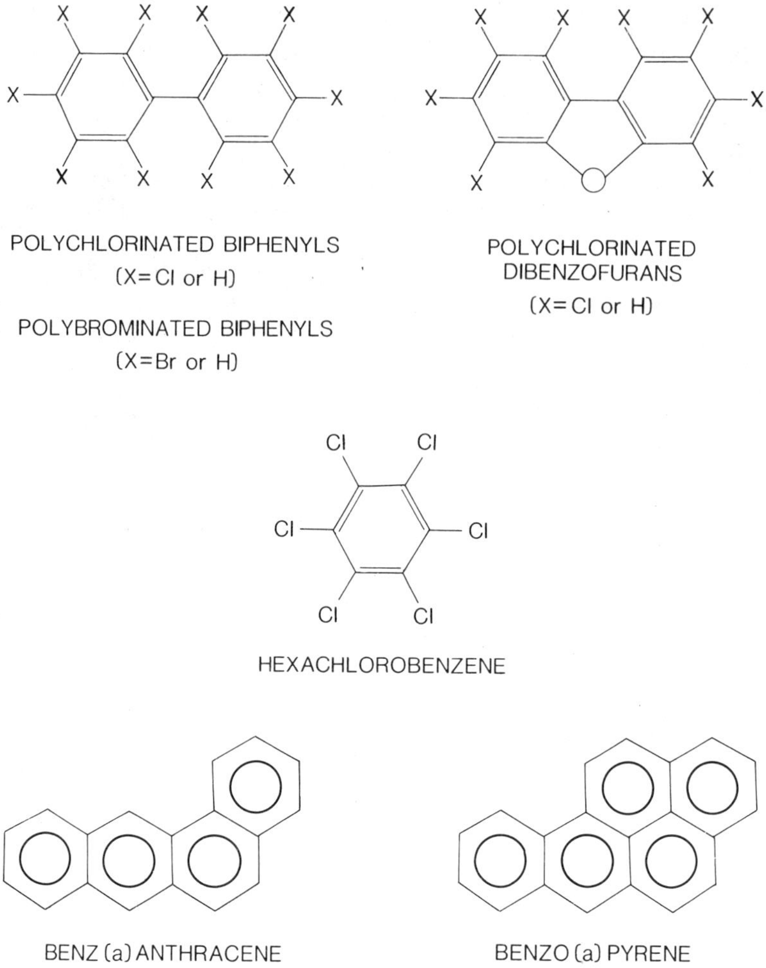

POLYCHLORINATED BIPHENYLS
(X=Cl or H)
POLYBROMINATED BIPHENYLS
(X=Br or H)

POLYCHLORINATED DIBENZOFURANS
(X=Cl or H)

HEXACHLOROBENZENE

BENZ (a) ANTHRACENE

BENZO (a) PYRENE

FIGURE 2. Chemical structures of some known organic environmental contaminants of foods.

molecule ten possible sites for chlorination which could lead to the theoretical formation of 209 individual PCBs. They also contain other persistent compounds as impurities, and the concentration of these impurities increases if PCBs are heated. This is especially so for the polychlorinated dibenzofurans (PCDFs) and the polychlorinated quaterphenyls (PCQs). PCDFs have been detected in ranges varying from 0.8–18 mg/kg in commercial PCBs (Roach and Pomerantz, 1974; Nagayama *et al.*, 1976).

The only appreciable sources of PCBs in the human diet are meat, poultry and related products together with fish (Jelinek and Corneliussen, 1975), although the levels normally encountered in meat and poultry are appreciably lower than those found in fish, unless these foodstuffs have been subject to gross contamination. Fish have been extensively monitored for PCB residues. In general, the higher the fat content of the species, the higher the concentration of PCBs that are found. In fish caught around the U.K. levels of PCBs in muscle tissue range between 0.01 mg/kg and 1.7 mg/kg (Portmann, 1979). However, the highest levels which have been reported are from freshwater fish taken from the Great Lakes and Hudson River. Coho salmon and lake trout taken from the Great Lakes were found with levels ranging as high as 39 mg/kg while Hudson River species ranged as high as 178 mg/kg (Anonymous, 1976). PCBs have been detected in human fat in Europe with levels reported ranging from 0.5 to 20.5 mg/kg (Gatti, 1975). Levels of residues of PCBs found in human breast milk ranged from 0.01 to 9.7 mg/kg in whole milk (Polishuk *et al.*, 1977).

In 1968 an outbreak of chloracne and eye irritation occurred amongst the local population in Fukuoka, Japan which was described as "Yusho" disease. After an intensive investigation into the source of the disease, it was eventually traced to the consumption of a batch of rice oil that had become contaminated by PCBs as a result of a leak from a heat-exchange system used in the production of the oil. Siblings of exposed mothers were smaller than normal and showed hyperpigmentation that disappeared after a few months (Yagamuchi *et al.*, 1971). Diminished growth was also observed in boys who had consumed the oil (Yoshimura, 1971). Associated symptoms such as anorexia, fatigue, together with neurological and respiratory effects continued in the population for many years, and by 1973 the number of patients exhibiting symptoms had reached 1200, and 22 deaths were attributed to the incident (Urabe, 1974). Even in 1980 it is reported that there are many Yusho patients still suffering from acne-type eruptions and fatigue.

On the basis of the epidemiological work undertaken, it was estimated that the smallest amount of contaminated rice oil producing symptoms when ingested over a period of about 120 days contained about 0.5 g of PCBs, equivalent to about 0.07 mg/kg body weight per day (Kuratsune and Masuda, 1972).

In contrast to the effects observed in Japan, Finnish workers who were occupationally exposed to PCBs, at levels corresponding to those causing skin lesions in Yusho patients, were asymptomatic (Karppanen and Kolho, 1973). Yusho patients who died were found to have average PCB levels in blood of 6.7 μg/kg, in liver of 0.1 μg/kg and in adipose tissue of 2.5 mg/kg. These levels are only 2–3 times greater than the average levels found in the general population. If it is accepted that some isomers will be more toxic than others, then even on an individual isomer basis the concentrations found were only sevenfold greater for that isomer. These facts have led to speculation that Yusho disease may be a result of the PCB impurities present in the rice oil rather than the PCBs themselves. PCDFs and PCQs have been detected in the contaminated rice oil and have also been found in the liver, adipose tissue and blood from post-mortem studies (Masuda, 1980).

Attempts to reproduce similar effects in experimental animals are hindered by the fact that few species show effects which either qualitatively or quantitatively approach those which have been observed in man. So far such effects have only been shown in the monkey (McNulty *et al.,* 1980). Female monkeys show reproductive dysfunction when exposed to PCBs in amounts comparable to that of Yusho patients (Allen and Barscotti, 1976; McNulty, 1975). Breast-fed infant monkeys suffered illness when their mothers' milk contained levels of PCBs within the range 7–16 mg/kg on a fat basis (Allen and Barscotti, 1976). Those infant monkeys that survived recovered from the overt toxic effects, although behavioral abnormalities were noted.

Some strains of rats and mice developed hyperplastic liver nodules and hepatomas after exposure to high levels of PCBs (Kimbrough, 1974). The significance of these experiments is unknown, given the impurities present in PCB mixtures, the species variability, and the high doses which were administered to produce effects.

Hexachlorobenzene (HCB)

Hexachlorobenzene is a derivative of benzene in which the six hydrogen atoms are replaced by chlorine (Fig. 2) and it is environmentally persistent. In the 1940s HCB was introduced as a fungicidal seed treatment which proved outstanding in the control of wheat bunt. This use has, however, declined and it is likely that at present the major source of this chemical in the environment results from its formation as a waste product during the synthesis of many important industrial chlorinated chemicals (e.g., tetrachloroethylene), or from the use of pesticides which degrade to HCB or contain small quantities of HCB as an impurity. Like the PCBs, HCB can be detected as a low-level contaminant of animal fats and fish.

As with the PCBs, the most important sources of HCBs in the human diet are meat, poultry, and fish, together with products in which they are ingredients. In dietary surveys undertaken in the U.S. during 1972–1974, HCB was found in about 5% of the food samples examined in ranges ranging from 0.3 to 7.0 μg/kg (Manske and Johnson, 1977). Similar results have been found in the UK (Laboratory of the Government Chemist, 1977).

HCB can be readily detected in animal fats (Dejonckheere *et al.,* 1975) and the most likely sources result either from feeding livestock with HCB-treated grains, or from HCB waste disposal.

Few data are available on the levels of HCB in marine fish. Zitko (1971) reports levels in the range 1–19 μg/kg in muscle tissue. Higher levels have been reported in fresh-water fish, especially from the Mississippi system and the Rhine (Courtney, 1979). HCB residues have been detected in human fat. Levels reported range between 0.01–21.80 mg/kg (Gatti, 1975). HCB levels in breast milk range between 0.5–3.5 mg/kg on a fat basis (Luquet *et al.,* 1975).

A sudden outbreak of porphyria cutanea tarda in the southeastern region of Turkey was first observed in 1955. By 1959 over 3000 cases had been examined. The rarity of the disease, combined with its explosive occurrence, strongly suggested that the possible etiological factor was the ingestion of a toxic substance. This was eventually traced to the ingestion of seed-treated wheat (Cam and Nigogosyan, 1963). HCB-treated wheat seed began to be imported into Turkey in 1954 and was unintentionally diverted to replace edible stores of wheat during a period of temporary scarcity. Estimates of the amount of HCB ingested over a relatively long period were 0.05–2 g/day before the skin effects became apparent.

The essential chemical features of the disease were facial lesions, porphyria, weakness, hirsuitism, hyperpigmentation and arthritic changes (in decreasing order of occurrence). About 80% of the patients were young children between the ages of 6 and 15. In some villages almost all of the children under the age of two years, breast-fed by mothers who had eaten the contaminated wheat, subsequently died of a distinct disease known as "pink sore" characterized by the formation of skin papules which were red in color. A large number of stillbirths were also reported.

Work on experimental animals, designed to determine if the qualitative and quantitative effects of human HCB ingestion could be reproduced, has shown that porphyria can be induced by the oral administration of HCB (Ockner and Schmid, 1961; De Mateeis *et al.*, 1961). In a rat multigeneration reproduction study, HCB produced stillbirths and a low postnatal survival (Grant *et al.*, 1977). In addition, HCB causes dose-related skeletal malformations in the rat (Khera, 1974).

In a study on low-level administration of HCB to rats, subsequent food deprivation enhanced the toxic response, indicative of the mobilization of HCB from adipose tissue stores (Villeneuve *et al.*, 1977).

Polynuclear Aromatic Hydrocarbons (PAHs)

PAHs are organic compounds containing two or more aromatic rings which may have substituted groups attached to one or more of these rings. As a general class of chemicals they can arise in food during food treatment, packaging, heating, and smoking, and as a result of environmental pollution arising from combustion processes. Thus they may be detected in a wide range of foodstuffs at levels varying between parts per 10^9 and in excess of parts per 10^7. Human exposure to PAHs also results from general atmospheric pollution and cigarette smoke.

Individual PAHs generally recognized as carcinogens and which are present in food include benz(a)anthracene; dibenzanthracenes; benzofluranthrenes, cholanthrene, the benzopyrenes and chrysen (International Association for Research on Cancer, 1973). Little is known about the relevance to human health of environmental sources of PAHs to the overall dietary intake.

Epidemiological surveys have been undertaken to determine the significance of exposure to PAHs but there is no firm evidence that these compounds are carcinogenic to man when present in food.

Polybrominated Biphenyls (PBBs)

PBBs, the brominated equivalent of the PCBs, were manufactured in the United States for use as flame retardant additives in fibers and plastics. In 1973 this chemical was accidentally supplied to a feed compounder instead of the chemical nutrient additive that had been requested. Animals fed this contaminated feed stored the very persistent PBBs in their fat and other tissues. These entered the human food chain when contaminated beef, pork, poultry and dairy products were sold and eaten. Although ill effects were observed in farm animals after a relatively short period of feeding, it was almost a year before the cause of these effects was identified.

Epidemiological studies on people exposed to PBBs have shown that the effects were mainly neurological (Lillis *et al.*, 1978). However, it has not been possible to show a relationship between serum PBB levels and the symptoms reported which included fatigue, headache, loss of appetite, abdominal pain and diarrhea, swelling of joints and effects such as acne. Functional abnormalities were detected in lymphocytes from the exposed group, although again no constant relationship was found between these effects and the degree of exposure as estimated by serum PBB levels. The significance of these results is unknown (Bekesi *et al.*, 1978).

Potential Organic Contaminants of Food

Table II summarizes some of the classes of chemical compounds that have recently been detected in freshwater fish in the United States (Lombardo, 1979). These classes cover a wide potential range of individual compounds and there are likely to be industrial chemicals produced within these general classes which have yet to be recognized as environmental contaminants. Rapid progress in the detection of low-level residues of chemicals in food is possible in the near future and this could enable environmentally persistent compounds to be detected well before they are likely to be of significance to public health.

TABLE II
Some Classes of Chemical Compounds Detected in Freshwater Fish in the U.S.

Aromatic amines
Brominated aromatics
Chlorinated aliphatics
Chlorinated benzenes
Chlorinated benzo trifluorides
Chlorinated nonaromatic cyclics
Chlorinated toluenes
Triaryl phosphates

Inorganic Contaminants

A brief glance at the periodic table of the elements gives an idea of the range of potential inorganic contaminants of food. Many of the elements (for example copper, zinc, and calcium) are known to be essential for life in plants and in animals, including man. Yet it is possible for food to become contaminated by an essential element, for example, abuse of a copper-containing fungicide might possibly cause unacceptably high levels of copper in food. With essential elements, there is always difficulty in quantifying the bodily requirement in order to maintain the health of man and to distinguish how much in excess of this requirement is undesirable. This chapter is not the appropriate forum to consider each and every inorganic contaminant in detail. The discussion therefore centers on the three contaminants, mercury, cadmium, and lead. These contaminants are not essential elements and are of current concern because their levels in food have increased in most industrialized countries as a result of industrial usage. Other inorganic contaminants are dealt with in less detail at the end of this section.

Mercury

The major source of mercury in the environment is the natural degassing of the earth's crust and this amounts to an annual release of 25–125 kt (World Health Organization, 1976). The annual world production of mercury from mining is about 10 kt, and about half of this is eventually lost to the sea, soil, and atmosphere (World Health Organization, 1972). The major uses of mercury are in the chloralkalai industry, in electrical equipment, in paints, measurement and control equipment, and in agriculture. The danger of industrial losses of mercury to the environment is that they are often localized. Because of this, the natural levels of mercury in the environment, including food, are often increased by the action of man.

By far the most important source of mercury in the human diet is fish. Fish concentrate mercury present in water (lakes, rivers, and seas). In normal circumstances this presence is of natural (geological) origin. Elevated levels of mercury in fish tissue can be detected when industrial activities result in increased levels of mercury in sediments or water (Ministry of Agriculture, Fisheries and Food, 1971). Mercury in fish is mainly in the form of methyl mercury (Haxton *et al.*, 1979), which is absorbed by man far more readily than inorganic mercury (World Health Organization, 1976). Methyl mercury is produced by the methylation of inorganic mercury compounds by methanogenic bacteria present in aquatic sediments. There have been many studies on the relationship between mercury in fish and the levels of mercury in human blood; an excellent summary of many of these studies has recently been published (Department of Primary Industry, 1980). For the sake of simplicity most investigators have assumed that blood-mercury levels are linearly related to mercury intake, however, this may not be so. The results of the many studies are often strikingly different. For example, if the relationship between blood-mercury levels (y,μg/liter) to mercury intake (x,μg/day) is represented by the equation $y = ax + b$, then the values of a vary from 0.04 (Haxton *et al.*, 1979) to 1.0 (Kershaw *et al.*, 1980). The significance of these very different results is best illustrated by way of an example. Suppose that each day a person consumes 100 g of fish containing 1 mg/kg of methyl mercury (values as high as this are often encountered in predatory fish, such as pike, trout, and shark taken from unpolluted waters). If a value for a of 0.04 is taken, then the corresponding blood mercury level is about 4–5 μg/liter; on the other hand if a value for a of 1.0 is taken, the corresponding value of blood mercury is 100 μg/liter. Although a blood mercury level of 4–5 μg/liter may not give cause for concern, a blood mercury level of 100 μg/liter is uncomfortably close to the level of 200 μg/liter where adverse health effects have been noted (World Health Organization, 1972). It is clear that resolution of the relationship between intake of methyl mercury and the level of mercury in blood remains an important area of research. Some communities consume far more fish than the "average man"; health authorities have to decide how much fish people in these communities may consume without damaging their health. For example, recommendations on fish consumption have been made to Canadian Indians, who relied upon fish as a major source of protein (Methyl Mercury Study Group, 1980); the recommendations included advice to limit the consumption of predatory fish such as pike. Shellfish also contain elevated levels of mercury; however, shellfish are unlikely to be eaten in sufficient quantities to make a substantial contribution to the body burden of mercury derived from food.

It is rare for humans to be exposed to organo-mercury compounds other than from fish, although people have consumed grain treated with organomercury fungicidal compounds. On occasion, animals may inadvertently be given mercury contaminated feed and humans consuming products from the animals may receive an elevated intake of mercury. For example, eggs from chickens given grain contaminated with phenyl mercury and methyl mercury contained high levels of these compounds (Englender *et al.*, 1980); people consuming 4–8 of these eggs each day had elevated blood mercury levels. Poisoning of man by inorganic mercury salts has been known for very many years. For example, workers in the felt hat industry were sometimes intoxicated by mercuric nitrate used in the treatment of the felt. After affected workers developed "hatters' shakes," these unfortunate people were immortalized by the Mad Hatter in *Alice in Wonderland.* Neurological disease due to exposure to methyl mercury has only been recognized in the past 40 years. As mentioned previously, methyl mercury is produced in nature by the methylation of inorganic mercury compounds by methanogenic bacteria present in aquatic sediments. Methyl mercury is readily absorbed by man and other animals; this is not so for inorganic mercury compounds. There are many symptoms of methyl mercury poisoning, the most common of which are the following:

1. Sensory disturbance (including paraesthesia).
2. Constriction of field of vision.
3. Impairment of hearing.
4. Impairment of speech and coordination.

Clear evidence of poisoning by methyl mercury has appeared all too often and with tragic consequences. In Japan, fish contaminated with methyl mercury from industrial sources was consumed by fishermen and their families; this resulted in deaths and permanent neurological damage (Tsubaki, 1974). A more common cause of methyl mercury poisoning is the consumption of grain treated with methyl mercury antifungal compounds (Bakir *et al.*, 1973). Grain treated with methyl mercury compounds is intended only for seeding, but people, particularly in areas of subsistence agriculture, sometimes eat such grain either because they do not know it has been treated or because they have fallen upon hard times.

Cadmium

Unlike mercury and lead, cadmium has not been used by man for many years, indeed the large scale industrial use of cadmium has taken place only in the last 50 years (Department of the Environment, 1980). Low levels of cadmium occur naturally in the environment (Kushizaki, 1977), where it is usually found in association with zinc. Since cadmium is ubiquitous, it is unlikely that any naturally occurring material will be entirely free from cadmium. Cadmium is used in electroplating, pigments, batteries, stabilizers for plastics and in alloys, especially hardened copper, brazing alloys and bearing alloys. Cadmium is transferred to the environment from mines, smelters (especially lead, copper and zinc smelters), the electroplating industry, power plants burning fossil fuels, and the ferrous metal industry. Emissions of cadmium to the air will eventually settle on the ground where it may be transferred to food and hence, to man.

Cadmium is absorbed from the soil by roots of plants and translocated from the roots to other parts of plants. In general, the levels of cadmium in cereals, fruits, and vegetables grown in uncontaminated soil are less than 0.2 mg/kg (Ministry of Agriculture, Fisheries and Food, 1973), although quick growing leafy vegetables such as spinach may take up more cadmium than other vegetables. The amounts of cadmium absorbed by plants from soils are influenced by many factors, including the soil characteristics. For example, increasing the pH of a soil tends to reduce cadmium uptake (Lucas *et al.*, 1978); there is also good evidence to show that cadmium uptake tends to be less in calcareous soils than in acid soils (Mahler *et al.*, 1978). It is evident that for a given type of soil under constant growing conditions, increasing the soil cadmium level will increase the level of cadmium in many of the plants grown on it. Cadmium levels in plants may be increased by the application of sewage sludge to agricultural land (Wood *et al.*, 1978) and there is some evidence to show (Kjellström *et al.*, 1975) that even the use of phosphatic fertilizers may increase the levels of cadmium in cereals. Clearly the extensive disposal of sewage sludge contaminated with cadmium on land used for producing food can lead to exposure to increased levels of cadmium. For this and other reasons the effects of disposal of sewage sludge to land are currently the subject of extensive research programs in many countries.

Animals used for human food always contain some cadmium. Although the levels in muscle tissue are generally less than 0.05 mg/kg (Ministry of Agriculture, Fisheries and Food, 1973) these levels can be elevated. Cadmium is present in animal tissue for the same reasons that it is present in man; the food consumed by man and animals inevitably contains some cadmium. As mentioned previously, cadmium concentrates in the kidneys and, to a lesser extent, the liver; this is true for both man (Brune *et al.*, 1980) and animals (Heffron *et al.*, 1980). In general, the higher the intake of cadmium by animals the higher will be the level of cadmium in all their tissue (Sharma and Street, 1980). There is clear evidence (Heffron *et al.*, 1980) to show that the tissue of sheep fed corn grown on land amended with cadmium-contaminated sewage sludge contained elevated levels of cadmium. The indications are that milk from cows receiving high intakes of cadmium does not contain elevated cadmium levels (Sharma *et al.*, 1979). There is no room for complacency about cadmium in the environment, and various Governments are introducing measures to ensure that cadmium contamination from various sources, such as sewage sludge, are controlled (Environmental Protection Agency, 1979; Department of the Environment, 1977).

In general, fish contain low levels of cadmium, however, this is not so for shellfish (Ministry of Agriculture, Fisheries and Food, 1973). Cadmium levels greater than 1 mg/kg are not uncommon in shellfish and much higher levels can be found in the "brown meat" from crabs and lobsters. These high cadmium levels are probably natural, although pollution of water by industrial activity has been shown (Wiedow *et al.*, 1982) to increase cadmium levels in crabs. Before an assessment of risk can be made it is necessary to know the intakes of cadmium by people who regularly consume brown crab meat; this information is very difficult to obtain.

Subacute intakes of lead and mercury from food produce adverse effects in a relatively short time, but this is not so for cadmium. Cadmium is a chronic toxin which accumulates in man's organs, especially the kidney, and may not manifest its invidious effects for many years; kidney damage caused by cadmium is irreversible. It has been recommended (World Health Organization, 1972) that weekly intakes of cadmium should not exceed 400–500 μg/individual (equivalent to about 1 μg/kg body weight/day). This recommendation was, in part, based on evidence from a study of people living in Japan who were affected by Itai-Itai disease, which is the only instance where there is clear evidence of harm to man from the ingestion of cadmium present in food. The people suffering from Itai-Itai disease were generally multiparous women above 40 years of age. The symptoms of the disease are osteomalacia and kidney dysfunction with proteinuria (proteinuria is commonly found in workers who have been industrially exposed to cadmium). It is estimated (World Health Organization, 1972) that the critical cadmium concentration in the renal cortex leading to proteinuria is about 200 μg/g wet weight.

There is a need to put a perspective on the influence of contamination of the environment with cadmium. It is right for action to be taken to control the dietary intake of cadmium by man, but it should always be remembered that in addition to food, smoking cigarettes is a major source of cadmium in man. There is evidence (Ellis *et al.*, 1979) to show that the total body burden of cadmium and the level of cadmium in the liver and kidneys of smokers is significantly higher than for nonsmokers.

Lead

There can be little doubt that lead has been the subject of more scientific reports, long standing concern and strong emotions, than any other environmental contaminant. There is ample evidence to show that all food contains some lead (Ministry of Agriculture, Fisheries and Food, 1972, 1975), although the question of whether lead in food is partially derived from natural sources or almost solely derived from pollution caused by man is the subject of recent debate (Settle, 1980).

Lead is released to the environment during the production of lead—that is, during mining, smelting and refining—and during the use of lead by man, for example in antiknock additives in gasoline, in batteries, in leaded paints, and in pesticides containing lead derivatives such as lead arsenate. Lead is also emitted during the smelting of other metals in which it is a contaminant such as copper, iron, and tin, as well as from power stations burning fossil fuel. Lead has been used for domestic water pipes and dissolves into the water, especially where the water is soft and acidic. In addition to these sources of lead, lead from solder used in the fabrication of tinplate cans is a major source of lead in canned food. Lead from gasoline and lead from solder are probably the two sources of lead contamination of most general concern in respect to food.

Average levels of lead in fresh fruit and vegetables and cereals rarely exceed about 0.2 mg/kg, although dried foods, such as herbs, and concentrated foods, such as tomato puree, contain higher levels of lead (Ministry of Agriculture, Fisheries and Food, 1975). Although the levels of lead in the muscle and fat of animals average less than 0.2 mg/kg, organs from such animals, especially the liver, contain considerably higher levels of lead.

The side seams of three-piece tinplate cans are generally soldered with an alloy containing 98% Pb and 2% Sn. During the soldering operation in can manufacture, solder penetrates the side seam to the inside of the can and fine solder particles are deposited on the internal surface of the can. The lead from the soldering operation dissolves into food contained in the cans and in consequence canned food generally contains higher levels of lead than does the equivalent fresh food. There is no doubt that the higher lead levels in canned food are associated with the use of lead-rich solders (Mitchell and Aldos, 1974). The development of new technologies which have resulted in the introduction of the two-piece food can, already used for carbonated beverages, and the welded can will eventually eliminate the contamination of canned food with lead, but it will be some years before these new technologies are in widespread use.

The concentration of lead in air varies enormously from area to area; areas remote from heavy traffic and industry may have air-lead levels of the order of 14 ng/kg (about 0.01 $\mu g/m^3$), whereas more populated areas will have air-lead levels of about 200 ng/kg (Cawse, 1977). These values are relatively small compared with air-lead concentrations near busy roads or smelters, where air-lead concentrations of 1000–2000 ng/kg often occur (Department of Health and Social Security, 1980). Airborne lead is deposited onto soil and onto growing crops and fodder, and at least some of this lead must be ingested by man. The rate at which lead is deposited from the air obviously depends upon the location under consideration. In remote uninhabited areas, deposition rates are generally less than 0.1 $\mu g/m^2$ per day (National Academy of Sciences, 1980), but in rural and industrial areas deposition rates of 70 to more than 700 $\mu g/m^2$ per day have been recorded (Chamberlain *et al.*, 1977; Bryce-Smith and Stephens, 1980). Recent work in Denmark (Tjell *et al.*, 1979) has shown that most of the lead in grass has originated from the atmosphere. The area-to-mass ratio of grass is very high, which means that grass is very susceptible to contamination by aerial deposition, and the Danish work confirms this observation. Animals who eat grass will ingest the lead deposited on the grass, so that at least some of the lead in the tissues of grazing animals has originated from the air. Evidence in support of this conclusion has been produced by studies on sheep grazing near busy roads (Ward *et al.*, 1978; Ward and Brooks, 1979). An equally important source of lead ingested by animals is the soil that they eat during grazing, and some of the lead in the soil will have originated from man-made emissions.

The contribution made by lead from air to lead in crops intended for human consumption is by no means easy to assess. The edible parts of some vegetables (such as the inner leaves of cabbages and cereals such as corn kernels) are naturally protected from contamination by lead deposition (Page *et al.*, 1971). There is no doubt that soft fruit, such as blackberries, grown near busy roads contains elevated lead levels and that less than half of this lead is removed by washing (Farmer, 1979). Lead from air undoubtedly makes a contribution to lead in food, but it is difficult to assess the magnitude of this contribution at the present time.

Vegetables cooked in water containing lead pick up lead from the water. The amount of lead transferred from the water to the food is proportional to the concentration of lead in the water (Smart *et al.,* 1981). This could be a major source of lead in food for those people having lead pipes and living in an area with plumbo-solvent water. This is, however, most unlikely to be a general problem. Lead has no known physiological function (Department of Health and Social Security, 1980), and if ingested or inhaled in sufficient quantities, can cause death or permanent harm to man. Lead has been recognized as a toxin since Roman times and in England in the early eighteenth century was recognized as the cause of "Devonshire colic," namely lead poisoning resulting from the consumption of cider produced using equipment containing lead. Lead interferes with the normal heme synthesis in man; in particular, exposure to relatively low levels of lead causes a decrease in the activity of δ-aminolevulinic acid dehydrase in red cells. It has been recommended (World Health Organization, 1972) that weekly intakes of lead should not exceed 3 mg/individual: *this recommendation does not apply to infants and children.* There is some evidence that exposure of children to relatively low levels of lead can result in an impairment of their mental development (Needleman *et al.,* 1979; Hrdina and Winneke, 1978).

Miscellaneous

It is well known that arsenic is a toxin. There is also some evidence (Nielson *et al.,* 1977) to show that arsenic may be an essential element; the same is undoubtedly true of selenium. In these instances it is a matter of judgment for the appropriate health experts to decide how much of an essential element is needed to maintain health and how much is likely to have adverse effects. The subject of fluoride in food and its addition to water is a classic example of considerable disagreement existing among the experts about the level intake of an element which has both beneficial and detrimental effects on the health of man. There is clear evidence (Singer and Ophang, 1979) to show that infants living in areas supplied with fluoridated water have elevated intakes of fluoride, but the implications of this for the general health of the infants are far from clear. It is certain, however, that their teeth benefit from the known anticaries properties of fluoride. The way in which excessive fluoride intake affects man and animals is well documented (Hodge and Smith, 1977; Suttie, 1977); however, the more subtle effects of chronic low-level intake of fluoride are far from simple to unravel. There is evidence (Marrier, 1977) that people with inadequate intakes of calcium or vitamin C or both are likely to be more at risk from low-intake, long-term ingestion of fluoride. Fluoride is present in all food (Kumpulainen and Koivistoinen, 1977), but in the U.K., at least, it has been shown (Jones *et al.,* 1971) that tea infusions are a major source of dietary fluoride. Emissions of fluoride from industry increase the levels of fluoride in vegetables and other crops, but the impact that this has on dietary intake of fluoride is difficult to assess. There is a need for more research in this area.

As mentioned earlier, there are a multitude of potential and known inorganic contaminants of food, but frequently the limitations of epidemiology, the lack of toxicological data, and the lack of information about normal levels of occurrence in food make it difficult to assess the likely impact of these contaminants on man. In some instances the effects are all too clear; for example, people have died after consuming apricots containing cyanide produced by the action of enzymes on cyanogenetic glycosides in the stone of the fruit. In other instances, food has been contaminated by a rare metal about which little is known, other than the fact that it is very toxic to man and animals. In these instances authorities responsible for public health must act in the interest of the consumer and prevent the consumption of the contaminated food. For example, food contaminated by thallium emissions from a cement factory was considered unfit to eat and was destroyed (Tooze, 1980). Inorganic contaminants, the biological effects of which are little known, are dispersed into the environment in ever increasing quantities in today's world. The use of sewage sludge on agricultural land has beneficial effects on crop growth, but it can also introduce contaminants into food grown on dressed land (Sterrit and Lester, 1980). These contaminants will not disappear once they are in the soil, but will remain for years. The application of nitrate fertilizers has beneficial effects on crop growth, but increases the level of nitrate in food and water. The effect of dietary nitrate and nitrite on man is little understood, but any increase in the dietary intake of nitrate might lead to a greater *in vivo* formation of nitrosamines after its reduction to nitrite by gastrointestinal flora. Nitrosamines are potent carcinogens and are known to be formed in the human stomach by the reaction of nitrite with secondary amines present in food.

CONCLUSIONS

Environmental contaminants of food have, at elevated levels, been the cause of human illness. The effects of this contamination have been identified by means of clinical observation and the use of epidemiological methods. The possibility that low levels of these contaminants may also be responsible for some long-term chronic effects in man cannot be discounted convincingly using existing methods, although it would appear unlikely. Nevertheless, in today's rapidly developing world, a continual surveillance of the effects of new chemicals on the environment and on the food chains favored by man and wildlife provides not only ammunition for risk-assessment calculations, but also some assurance that the pattern of human exposure to such contaminants does not change significantly during the lifetime of the human genotype.

In part, the problems of food contamination which have occurred have been the result of the lack of systematic control over the manufacture, use, and disposal of industrial chemicals. Human illness associated with such industrially derived chemicals has been the impetus for the introduction of registration schemes for new and existing chemicals in a number of countries. Such schemes seek to acquire knowledge and to forecast any possible effects, both short-term and long-term, before manufacture, distribution, and utilization occurs on any wide scale. It is to be hoped that the application of such schemes may be effective in minimizing food contamination problems in the future. Nevertheless, no control scheme will be fully effective unless it operates in conjunction with some form of continuous surveillance or food monitoring program, because no registration scheme can avoid the occurrence of accidents.

REFERENCES

Allen, J. R., and Barscotti, D. A., 1976, The effects of transplacental and mammary movement of PCB's on infant rhesus monkeys, *Toxicology* **6:**331.

Anonymous, 1976, FDA backs New York move to ban Hudson river fishing, *Food Chemical News* **17(50):**30.

Bakir, F., Damiuji, S., Amin-Zaki, L., Mrutadha, M., Khalidi, A., Al-Rawi, N., Tikriti, S., Dhakir, H., Clarkson, T., Smith, J., and Doherty, R., 1973, Methylmercury poisoning in Iraq: An inter-university report, *Science* **181:**220

Bekesi, J. G., Holland, J. F., Anderson, H. A., Fischbein, A. S., Rom, W., Wolff, M. S., and Selikoff, I. J., 1978, Lymphocyte function of Michigan dairy farmers exposed to polybrominated biphenyls, *Science* **199:**1207.

Brune, D., Nordberg, G., and Wester, P. O., 1980, Distribution of 23 elements in the kidney, liver and lungs of workers from a smeltery and refinery in North Sweden exposed to a number of elements and of a control group, *Sci. Total Environ.* **16(1):**13.

Bryce-Smith, D., and Stephens, R., 1980, *Lead or Health,* The Conservation Society Pollution Working Party, London.

Cam, C., 1960, Une nouvelle dermatose epidemique des enfants, *Ann. Dermatol. Syphiligr.* **87:**393.

Cam, C. and Nigogosyan, G., 1963, Acquired porphyria cutanea tarda due to hexachlorobenzene, *J. Am. Med. Assoc.* **183:**88.

Cawse, P. A., 1977, A survey of atmospheric trace elements in the UK: Results for 1976, Atomic Energy Research Establishment, Harwell, Report No. 8869, HMSO, London.

Chamberlain, A. C., Heard, M. J., Little, P. and Wiffen, R. D., 1977, The dispersion of lead from motor exhausts, Environmental and Medical Sciences Division, Atomic Energy Research Establishment, Harwell, Report No. 9198, HMSO, London.

Courtney, K. D., 1979, Hexachlorobenzene (HCB): A review, *Environ. Res.,* **20:**225.

Dejonckheere, W., Steurbaut, W., and Kips, R. H., 1975, Residues of Quintozene, hexachlorobenzene and pentachloroanaline in soil and lettuce, *Bull. Environ. Contam. Toxicol.* **13:**720.

De Mateeis, F., Prior, B. E. and Rimington, C., 1961, Nervous and biochemical disturbances following hexachlorobenzene intoxication, *Nature* **191:**363.

Department of the Environment, 1977, Report of the working party on the disposal of sewage sludge to land, *Standing Technical Committee Report No. 5,* HMSO, London.

Department of the Environment, 1980, Cadmium in the environment and its significance to man, *Pollution Paper No. 17,* Central Directorate on Environmental Pollution, HMSO, London.

Department of Health and Social Security, 1980, Lead and Health, Report of a DHSS working party on lead in the environment, HMSO, London.

Department of Primary Industry, 1980, Report on mercury in fish and fish products, Working Group on Mercury in Fish, Australian Government Publishing Service, Canberra.

Ellis, K. J., Vartsky, D., Zanzi, I., Cohn, S. H., and Yasumura, S., 1979, Cadmium: *in vivo* measurement in smokers and nonsmokers, *Science* **205:**323.

Englender, S. J., Greenwood, M. R., Atwood, R. G., Clarkson, T. W., and Smith, J. C., 1980, Organic mercury exposure from fungicide-contaminated eggs, *Arch. Environ. Health* **35:**224.

Environmental Protection Agency, 1979, Criteria for classification of solid waste disposal facilities and practices, *Fed. Regist.* **44(179):**53438.

Farmer, J. G., 1979, Lead in wild blackberries from suburban roadsides, *J. Sci. Food Agric.* **30:**816.

Food Safety Council, 1978, Proposed System for food safety assessment, *Food Cosmet. Toxicol.* **16,** suppl. 2.

Gatti, G. L., 1975, Report of a group of experts on pesticide residues in human fat and human milk, in: *European Colloquium on Problems Raised by the Contamination of Man and His Environment by Persistent Pesticides and Organohalogenated Compounds* (EUR 5196), Commission of the European Communities, Luxembourg.

Grant, D. L., Phillips, W. E. J., and Hatina, G. V., 1977, Effect of hexachlorobenzene on reproduction in the rat, *Arch. Environ. Contam. Toxicol.* **5:**207.

Haxton, J., Lindsay, D. G., Hislop, J., Salmon, L., Dixon, E. J., Evans, W. H., Reid, J. R., Hewitt, C. J., and Jeffries, D. S., 1979, Duplicate diet study on fishing communities in the United Kingdom: Mercury exposure in a critical group, *Environ. Res.* **10:**351.

Heffron, C. L., Reid, J. T., Elfving, D. C., Stoewsand, G. S., Haschek, W. M., Telford, J. N., Furr, A. K., Parkinson, T. F., Bache, C. A., Gutenmann, W. H., Wszolek, P. C., and Lisk, D. J., 1980, Cadmium and zinc in growing sheep fed silage corn grown on municipal sludge amended soil, *J. Agric. Food Chem.* **28:**58.

Hodge, H. C., and Smith, F. A., 1977, Occupational fluoride exposure, *J. Occup. Med.* **19(1):**11.

Hrdina, K., and Winneke, G., 1978, Paper delivered at the Working Conference of the German Association of Hygiene and Microbiology, Oct. 2–3, Mainz, West Germany.

International Association for Research on Cancer, 1973, *Certain Polycyclic Aromatic Hydrocarbons and Heterocyclic Compounds,* Vol. 3, IARC, Lyon, France.

Jelineck, C. F. and Corneliussen, P. E., 1975, *National Conference on Polychlorinated Biphenyls,* Chicago, 1975, U.S. Environmental Protection Agency Office of Toxic Substances, Washington, D.C..

Jones, C. M., Harries, J. M., and Martin, A. E., 1971, Fluoride in leafy vegetables, *J. Sci. Food Agric.* **22:**602.

Karppanen, E. and Kolho, L., 1973, The concentration of PCB in human blood and adipose tissue in three different research groups, in: *PCB Conference II, National Swedish Environment Protection Board Publication 4E,* pp. 124–127.

Kershaw, T. G., Dhahir, P. H., and Clarkson, T. W., 1980, The relationship between blood levels and dose of methyl mercury in man, *Arch. Environ. Health* **35(1):**28.

Khera, K. S., 1974, Teratogenicity and dominant lethal studies on hexachlorobenzene in rats, *Food Cosmet. Toxicol.,* **12:**471.

Kimbrough, R. D., 1974, The toxicity of polychlorinated polycyclic compounds and related chemicals, *CRC Crit. Rev. Toxicol.* **2:**442.

Kjellström, T., Lind, B., Linnman, L., and Elinder, C. G., 1975, Variation of cadmium concentration in Swedish wheat and barley, *Arch. Environ. Health* **30:**321.

Kumpulainen, J., and Koivistoinen, P., 1977, Fluorine in foods, *Residue Rev.* **68:**37.

Kuratsune, M., and Masuda, Y., 1972, PCB pollution, *Kosei no Shihyo,* **19:**11.

Kushizaki, M., 1977, Studies on soil pollution by cadmium, a heavy metal, *Jpn. Agric. Res. Quart.* **11(2):**89.

Laboratory of the Government Chemist, 1981, Report of the Government Chemist, 1977, HMSO, London.

Lillis, R., Anderson, H. A., Valcinkas, J. A., Freedman, S., and Selikoff, I. J., 1978, Comparison of findings among residents of Michigan dairy farms and consumers of produce purchased from these farms, *Environ. Health Perspec.* **23:**105.

Lombardo, P., 1979, FDA's chemical contaminants program: The search for the unrecognized pollutant, *Ann. N. Y. Acad. Sci.,* **320:**673.

Lucas, J. B., Pahren, H. R., Ryan, J. A., and Dotson, G. K., 1978, The impact of metals present in municipal sludges upon the human food chain—a risk assessment, U.S. Environmental Protection Agency, Cincinnati, Ohio.

Luquet, F. M., Goursaud, J., and Casalis, J., 1975, Pollution of mothers' milk with residues of organochlorine pesticides in France, *Lait* **55:**207.

McNulty, W. P., 1975, Primate study, National Conference on Polychlorinated Biphenyls, Chicago, 1975, U.S. Environmental Protection Agency, Office of Toxic Substances, Washington, D.C.

McNulty, W. P., Becker, G. M., and Cory, H. T., 1980, Chronic toxicity of 3, 4, 3′, 4′, and 2, 5, 2′, 5′-tetrachlorobiphenyl in *Rhesus macaques, Toxicol. Appl. Pharmacol.* **56:**182.

Mahler, R. J., Bingham, F. T., and Page, A. L., 1978, Cadmium-enriched sewage sludge application to acid and calcareous soils: Effect on yield and cadmium uptake by lettuce and chard, *J. Environ. Qual.* **7(2):**274.

Manske, D. D., and Johnson, R. D., 1977, Pesticide and other chemical residues in total diet samples, *Pestic. Monit. J.* **10:**134.

Marier, J. R., 1977, Some current aspects of environmental fluoride, *Sci. Total Environ.* **8:**253.

Marr, J. W., 1971, Individual dietary surveys: Purpose and methods, *World Rev. Nutr. Diet.* **13:**105.

Masuda, Y., 1980, Polychlorinated dibenzofurans and related compounds in patients with "Yusho," Workshop on *Impact of Chlorinated Dioxins and Related Compounds on the Environment,* Istituto di Superiore Sanita, Rome, Oct. 1980.

Methyl Mercury Study Group, 1980, McGill Methyl Mercury Study, Feb. 26, McGill University, Montreal, Canada.

Ministry of Agriculture, Fisheries and Food, 1971, Survey of mercury in food, Working Party on the Monitoring of Foodstuffs for Mercury and Other Heavy Metals, First Report, HMSO, London.

Ministry of Agriculture, Fisheries and Food, 1972, Survey of lead in food, Working Party on the Monitoring of Foodstuffs for Heavy Metals, Second Report, HMSO, London.

Ministry of Agriculture, Fisheries and Food, 1973, Survey of cadmium in food, Working Party on the Monitoring of Foodstuffs for Heavy Metals, Fourth Report, HMSO, London.

Ministry of Agriculture, Fisheries and Food, 1975, Survey of lead in food, first supplementary report, Working Party on the Monitoring of Foodstuffs for Heavy Metals, Fifth Report, HMSO, London.

Mitchell, O. G., and Aldos, K. W., 1974, Lead content of foodstuffs, *Environ. Health Perspect.* **19:**59.

Nagayama, J., Kuratsune, M., and Masuda, Y., 1976, Determination of chlorinated dibenzofurans in Kanechlors and "Yusho Oil," *Bull. Environ. Contam. Toxicol.* **15:**9.

National Academy of Sciences, 1980, *Lead in the Human Environment,* NAS, Washington, D.C.

Needleman, H. L., Gunnoe, C., Leviton, A., Reed, R., Peresie, H., Maher, C., and Barrett, P., 1979, Psychological performance of children with elevated lead levels, *N. Engl. J. Med.* **300:**689.

Nielson, S. H., Myron, D. R., and Uthus, E. D., 1977, Proceedings of the 3rd International Symposium on *Trace Elements Metabolism in Man and Animals* (M. Kirchgessner, ed.), p. 244, Tierernahrungsforschung, Weihenstephan, West Germany.

Ockner, R. K., and Schmid, R., 1961, Acquired porphyria in man and rat due to hexachlorobenzene intoxication, *Nature* **189:**499.

Page, A. L., Ganje, T. J., and Jashi, M. S., 1971, Lead quantities in plants, soil, and air near major highways in Southern California, *Hilgardia* **41:**1.

Pallotti G., Bencivenga B. and Simonetti T., 1979, Total mercury levels in whole blood, hair and fingernails from a population group from Rome and its surroundings, *Sci. Total Environ.* **11:**69.

Polishuk, Z. W., Ron, M., Wassermann, M., Cucos, S., Wasserman, D., and Lemesch, C., 1977, Pesticides in people. Organochlorine compounds in human blood plasma and milk, *Pestic. Monit. J.* **10:**121.

Portmann, J. E., 1979, Chemical monitoring of residue levels in fish and shellfish landed in England and Wales during 1970–73, *Aquatic Environment Monitoring, Report No. 1,* Ministry of Agriculture, Fisheries, and Food Directorate Fish Res. Lowestoft.

Roach, J. A. G., and Pomerantz, I. H., 1974, The finding of chlorinated dibenzofurans in a Japanese polychlorinated biphenyl sample, *Bull. Environ. Contam. Toxicol.* **12:**338.

Sartor, R. and Rondia, D., 1980, Blood lead levels and and age: A study in two male urban populations not occupationally exposed, *Arch. Environ. Health* **35(2):**110.

Settle, D. M., and Patt, C. C., 1980, Lead in albacore: Guide to lead pollution in Americans, *Science* **207:**1167.

Sharma, R. P. and Street, J. C., 1980, Public health aspects of toxic heavy metals in animal feeds, *J. Am. Vet. Med. Assoc.* **177(2):**149.

Sharma, R. P., Street, J. C., and Verma, M. P., 1979, Cadmium uptake from feed and its distribution to food products of livestock, *Environ. Health Perspect.* **28:**59.

Singer, L., and Ophang, R., 1979, Total fluoride intake of infants, *Pediatrics* **63(3):**460.

Smart, G. A., Warrington, M., and Evans, W. H., 1981, The contribution of lead in water to dietary lead, *J. Sci. Food Agric.* **32:**129.

Sterrit, R. M., and Lester, J. N., 1980, The value of sewage sludge to agriculture and effects of the agricultural use of sludges contaminated with toxic elements: A review, *Sci. Total Environ.* **16:**55.

Suttie, J. W., 1977, Effects of fluoride on livestock, *J. Occup. Med.* **19(1):**40.

Tjell, J. C., Hovmand, M. F., and Mosback, H., 1979, Atmospheric lead pollution of grass grown in a background area in Denmark, *Nature,* **280:**425.

Tooze, S., 1980, A thallium affair, *Ecologist* **10:**163.

Tsubaki, T., 1974, Clinical aspects of Minimata disease, in: *Minimata Disease* (T. Tsubaki and K. Irukayama, eds.), Kodansha, Tokyo.

UNSCEAR, 1977, Sources and effects of ionising radiations, United Nations Report to the General Assembly (E77.1X.1), p. 583.

Urabe, H., 1974, Foreward, *Fukuoka Acta Med.* **65:**1.

Venitt, S., 1980, Bacterial mutation as an indicator of carcinogenicity, *Br. Med. Bull.* **36:**57.

Villeneuve, D. C., Van Logten, M. J., Den Tonkelaar, E. M., Greve, P. A., Vox, J. G., Specijers, G. J. A., and Van Esch, G. J., 1977, Effect of food deprivation on low level hexachlorobenzene exposure in rats, *Sci. Total Environ.* **3:**179.

Ward, N. I., and Brooks, R. R., 1979, Lead levels in wool as an indication of lead in blood of sheep exposed to automotive emissions, *Bull. Environ. Contam. Toxicol.* **21:**403.

Ward, N. I., Brooks, R. R., and Roberts, E., 1978, Blood lead levels in sheep exposed to automotive emissions, *Bull. Environ. Contam. Toxicol.* **20:**44.

Wiedow, M. A., O'Connor, J. M., Hazen, R., Sloan, R., and Korcher, R., 1982,Cadmium concentrations in tissues of Hudson river blue crabs (*Calinectes sapides*), *Bull. Environ. Contam. Toxicol.* (in press).

Wogan, G. N., Paglialunga, S., and Newberne, P. M., 1974, Carcinogenic effects of low dietary levels of aflatoxin B_1 in rats, *Food Cosmet. Toxicol.* **12:**681.

Wood, L. B., King, R. P. and Norris, P. E., 1978, Some investigations into sludge amended soils and associated crops and the implications for trade effluent control, *EEC Conference on the Utilisation of Sewage Sludge on Land,* Oxford, England.

World Health Organization, 1972, Evaluation of mercury, lead, cadmium and the food additives amaranth, diethylpyrocarbonate and octyl gallate, *WHO Food Additives Series No. 4,* World Health Organization, Geneva.

World Health Organization, 1976, *Environmental Health Criteria. 1. Mercury,* World Health Organization, Geneva.

World Health Organization, 1977, *Environmental Health Criteria. 3. Lead,* World Health Organization, Geneva.

Yagamuchi, A., Yoshimura, T., and Kuratsune, M., 1971, A survey on pregnant women having consumed rice oil contaminated with chlorobiphenyls and their babies, *Fukuoka Acta Med.* **62:**112.

Yoshimura, T., 1971, Epidemiological analysis of "Yusho" patients with special reference to sex, age, clinical grades and oil consumption, *Fukuoka Acta Med.* **62:**109.

Zitko, V., 1971, Polychlorinated biphenyls and organochlorine pesticides in some freshwater and marine fishes, *Bull. Environ. Contam. Toxicol.* **6:**464.

10

AGRICULTURAL CHEMICALS

Mohammad G. Mustafa

INTRODUCTION

Intentional food additives are natural or synthetic substances added to the original foods or food products for specific purposes. Although those substances are generally considered safe, questions nonetheless arise as to their toxicities or hazards, and those will be dealt with in another chapter of this book. Of particular concern in this chapter are the nonintentional chemical additives or contaminants of foods. Based on the mode of occurrence of these chemicals in foods they may be broadly categorized as agricultural chemicals. They are nonintentional additives because they occur rather inconspicuously or insidiously, and are not meant for human consumption. For example, pesticides are applied for eliminating or limiting the growth of selected species of plants or animals, but most of them are not very selective; they find access to foods and water and harm other nontarget species, including humans. Exposure to these chemical substances may occur directly, such as through drinking water and eating processed foods, or indirectly as a result of their bioaccumulation in the food chain.

In the U.S. and other developed countries the application of a wide variety of pesticides is routine practice. Considering the economic benefits, the developing countries have begun to follow the practice, and are likely to face some typical environmental problems. Although a great deal of economic and public health benefits are derived from the use of pesticides, their widespread occurrence as environmental contaminants raises the question whether their continued, unrestricted use is indeed worth the risks of injury to humans. For example, DDT was introduced in 1942, and it became virtually a "cure-all" for vector-borne diseases in the fields of public health and crop damage. Its application spread worldwide for decades without regard to potential environmental effects and injuries to human health.

Mohammad G. Mustafa • Division of Environmental and Nutritional Sciences, School of Public Health, University of California, Los Angeles, California 90024.

TABLE I
Some Important Pesticide Chemicals and their Biological Effects[a]

Type of use and chemical category	Common name	LD_{50}, mg/kg (oral)[b]	Biological effects or mode of action
Insecticides			
Organophosphorus compounds	TEPP	1.1	Organophosphorus compounds are neurotoxic; inhibit cholinesterase in nerve tissue; do not accumulate in tissues. Muscular dysfunction and respiratory failure occur in acute poisoning.
	Mevinophos	6.1	
	Disulfoton	6.8	
	Azinphosmethyl	13	
	Parathion	13	
	Methylparathion	13	
	Chlorfenvinphos	15	
	Dichlorvos (DDVP)	80	
	Dimethoate	215	
	Trichlorfon	630	
	Chlorothion	880	
	Malathion	1375	
	Ronnel	1250	
	Abate	8000	
Carbamate compounds	Aldicarb (Temik)	0.8	Carbamate compounds are neurotoxic; inhibit cholinesterase in nerve tissue; do not accumulate in tissues.
	Baygon (Propoxur)	83	
	Zectran	37	
Organochlorine compounds	DDT	113	Being lipid soluble, organochlorines accumulate in tissue lipid; they are bioconcentrated in food chain pyramids. They are neuroactive compounds and inhibit electrolyte transport. They adversely affect reproductive systems in birds.
	DDE	217	
	DDA	—	
	Methoxychlor	5000	
	Aldrin	7000	
	Dieldrin	39	
	Endrin	18	
	Heptachlor	100	
	Chlordane	335	
	Lindane	88	

Chlorophenoxy compounds	2,4-D	100–1000	Phenoxy compounds produce systemic toxicity.
	2,4,5-T	100–1000 (several species)	They are neuropoisons; cause muscular dysfunction. They are teratogenic, but this effect may be due to contaminating dioxin compounds.
Dinitrophenols	DNOC	30	They are systemic poisons.
Bipyridyl compounds	Paraquat	30–50	They are systemic poisons; cause pulmonary edema and congestion.
	Diquat	30–50	
Carbamate compounds	Propham	5000	These compounds are allergenic.
	Barbane	600	
Urea derivatives	Monuron	3000	
	Diuron	3000	
Triazines	Simazine	100–250 (cattle and sheep)	Triazines are carcinogenic.
	Amitrole	—	
Amides	Propanil	—	They are neurotoxic.
Fungicides			
	Captan	480	They are teratogenic.
	Folpet	—	
Phenols	Pentachlorophenol (PCP)	30–100 (several species)	They are systemic poisons.
Rodenticides			
Coumarin derivative	Warfarin	200	It is an anticoagulant.
Glycoside	Red squill	—	It causes cardiac failure.
Fluoroacetates	Compound 1080	—	They inhibit the citric acid cycle.
	Compound 1081	—	
Thiourea	ANTU	3–10 (several species)	It causes massive pulmonary edema.

[a] Adapted from Murphy, S.D., 1975.
[b] LD_{50} applies to rat studies, except where noted otherwise.

CATEGORIES OF AGRICULTURAL CHEMICALS

Agricultural chemicals include herbicides, insecticides, fungicides, nematocides, rodenticides, growth regulators, fruiting agents, and defoliants grouped as pesticides (Table I); toxic metals and metallic compounds from pesticides; mineral nutrients, salts and radionuclides from fertilizers; and chemical wastes from polyethylene tarpaulin and petroleum mulches used in agriculture. Others that can be included are antibiotics and other drugs used for prevention and control of disease in cattle, sheep, and poultry; growth promoting substances in animal feeds; chemicals from packaging materials; and toxic substances from external sources that reach agricultural land, such as by-products of fossil fuel combustion and radioactive fallout.

Of these agricultural pollutants, the opportunities for exposure to pesticides are of major concern because of their multiple and extensive usage. Various types of pesticides are used in the control of insects, rodents, and other pests that are involved in either crop damage or carrying vector-borne diseases such as malaria, yellow fever, and typhus. Although by far the largest amounts of pesticides are used in agriculture, the amounts used in industries and homes are also considerable. In addition to agricultural applications, other forms of operations such as those of individual homeowners, gardeners, and fruit growers may contribute significantly to contamination of food.

PESTICIDE CHEMICALS IN FOODS

Introduction of synthetic, organic pesticides has led to control of agricultural, household, and industrial pests, but it has also brought about serious contamination of the environment. The estimated worldwide usage of pesticides during the mid-1970s exceeded 4 billion lb, and approximately half of this amount was used in agricultural crop production. The increasing dependence of agriculture on chemicals, however, has led to a buildup of chemical residues in the environment.

In the early days of insect control with pesticides, persistence was considered an important quality since the residual activity of the applied chemical would control insects for extended periods of time. These persistent pesticides, i.e., those that are slowly degraded or altered to nontoxic forms, accumulate in soils after repeated applications. They may then be transported to contaminate surface-water and groundwater, and thus find access to various food chains; or, they may remain in the soil to be taken up by plant roots, and subsequently translocated to edible portions of the crops, fruits or vegetables (Table II). Some of the classical studies have been carried out with chlorinated hydrocarbon (organochlorine), such as DDT (Table III). In orchards, for example, soil samples have shown considerable amounts of DDT residue. More seriously, samples of fruits (apples and peaches) and vegetables have been shown to contain residues of DDT, aldrin, and dieldrin (Sheets, 1967).

TABLE II

Chemical Residues Found in 360 Total-Diet Food Composites in the U.S.[a]

Metal or pesticide	Composites with residue	Composites with traces of residue[b]	Range[c] (ppm or mg/kg)
Zinc	360	0	0.1–35.5
Cadmium	211	0	0.01–0.31
Lead	180	0	0.02–1.30
Selenium	97	34	0.05–0.40
Mercury	34	17	0.01–0.04
PCB	14	13	0.050
Dieldrin	93	17	0.0006–0.0330
DDE	81	20	0.0006–0.0380
BHC	76	12	0.0004–0.0070
DDT	54	15	0.002–0.020
Malathion	53	6	0.003–0.115
Lindane	52	18	0.0003–0.0120
Diazinon	50	20	0.0007–0.0270
Heptachlor epoxide	46	19	0.0005–0.0040
TDE	3	23	0.001–0.005
Arsenic	18	2	0.03–0.60
Endosulfan	17	10	0.003–0.012
HCB	17	8	0.0003–0.0070
Parathion	17	10	0.003–0.022
CIPC	12	0	0.005–0.467
PCA	10	1	0.004–0.050
PCP	10	0	0.010–0.033
Carbaryl	8	4	0.05–0.50
TCNB	8	2	0.001–0.284
Methoxychlor	7	2	0.004–0.009
Methyl parathion	7	6	0.008
PCNB	7	4	0.002–0.005
Ethion	6	3	0.003–0.012
Orthophenylphenol	5	2	0.05–0.20
Leptophos	5	1	0.013–0.090
Perthane	4	0	0.030–2.28
Botran	3	0	0.006–0.067
Toxaphene	3	2	0.163
DCPA (dacthal)	2	0	0.003–0.013
Dicofol (kelthane)	2	1	0.010
Aldrin	1	0	0.001
Captan	1	0	0.178
Chlordane	1	1	[d]
Heptachlor	1	0	0.004
Phosalone	1	0	0.171
Ronnel	1	0	0.001
Nitrofen	1	0	0.039

[a] Composited from samples from 30 different markets in 30 different cities, Aug. 1973–July 1974 (Manske and Johnson, 1977).
[b] Detected, but concentration too low to be quantified by the analytical method used.
[c] At and above limit of quantitation.
[d] Concentration too low to be quantified.

TABLE III
Occurrence of DDT and Related Chemical Residues in the Environment, 1960–1966[a]

Material	Residue (ppm or mg/kg)	Remarks
Apple	0.20	Typical, ready-to-eat
Beef steak	0.10	Typical, ready-to-eat
Milk (cow)	0.02	Typical, ready-to-eat
Milk (human)	0.37	San Francisco
Alfalfa hay	0.10	—
Wheat-grass	211	Salmon Nat. For., Idaho
Pheasant (fat)	8	Richvale, Calif.
San Joachin River	0.000066	Vernalis, Calif.
Rainwater	0.00015	Ripley, Ohio
Ice	0.0003	Mt. Olympia, Wash.
Abalone	0.0087	Ensenada, Mexico
English sole	0.57	San Francisco Bay
Adelie penguin (fat)	0.024	Antarctica
Pear-orchard soil	38	Medford, Ore. top 6 in.
Bald eagle (breast muscle)	5	Canada (average)
Human (fat)	6.7	U.S. (average, 1963)
Human (fat)	27.2	India (average, 1964)
Human (fat)	1131	U.S. (occupational)
Human (fat)	3.0	Eskimo

[a] Adapted from American Chemical Society, 1978.

Translocation

During pesticide application humans and nontarget animal and plant organisms may be directly exposed to the chemicals. Food crops and leafy vegetables may be contaminated directly at the site of application. However, pesticides may be translocated and become a source of contamination at a site far removed from that of application. In this regard, relatively persistent chemicals are of particular concern. A number of pesticides comprised of chlorinated hydrocarbons or heavy metals are notoriously persistent. For example, DDT is only slowly metabolized by biological systems, and some of its metabolites remain toxic and resistant to further degradation. It is taken up and stored in the lipids of various organisms. Thus, bioconcentration occurs through the food chain, with increasing concentration at the top of the pyramid.

There are other means of pesticide translocation. They may be airborne as dusts or vapors, or they may accumulate as residues in soil from repeated applications.

Toxicity

Although pesticides are designed for their toxicity to selected species of animals or plants, they are not without hazards to humans. Many of these chemicals are tested for toxicological hazards, but many others are merely screened for safety prior to their application. Table IV shows relative toxicities

TABLE IV
Relative Toxicities of Some Pesticides in Male Rats[a]

Pesticide	Average LD_{50}[b] (mg/kg body weight)	Range of values
Aldrin	39	34–44
Chlordane	335	299–375
DDA	740	607–903
DDE	880	727–1065
DDT	113	87–147
Dieldrin	46	41–51
Endrin	17.8	14.7–21.5
Heptachlor	100	74–135
Toxaphene	90	67–122
Parathion	13	10–17
Dichlorvos (DDVP)	1375	1206–1568
Carbaryl (Sevin)	850	733–986

[a] Adapted from Gaines, 1960.
[b] Acute oral dose.

TABLE V
Incidents of Poisoning by Pesticide Chemicals[a]

Kind of accident	Pesticide involved	Material contaminated	Number affected	Number died	Location
Spillage during transport or storage	Endrin	Flour	159	0	Wales
	Endrin	Flour	691	24	Qatar
	Endrin	Flour	183	2	S. Arabia
	Dieldrin	Food	20	0	Shipboard
	Diazinon	Doughnut mix	20	0	U.S.
	Parathion	Wheat	360	102	India
	Parathion	Barley	38	9	Malaya
	Parathion	Flour	200	8	Egypt
	Parathion	Flour	600	88	Colombia
	Parathion	Sugar	300	17	Mexico
	Parathion	Sheets	3	0	Canada
	Mevinphos	Plants	6	0	U.S.
Eating formulation	Hexachlorobenzene	Seed grain	3000	3–11%	Turkey
	Organic mercury	Seed grain	34	4	West Pakistan
	Organic mercury	Seed grain	321	35	Iraq
	Organic mercury	Seed grain	45	20	Guatemala
	Warfarin	Bait	14	2	Korea
Improper application	Toxaphene	Collards and chard	7	0	U.S.
	Nicotine	Mustard	11	0	U.S.
	Parathion	Used as treatment for body lice	17	15	Iran
	Pentachlorophenol	Nursery linens	20	2	U.S.

[a] Adapted from the U.S. Department of Health, Education, and Welfare, 1969.

of some pesticides. Toxicological tests are generally carried out in experimental animals using biochemical, physiological, or pathological parameters that serve as models for humans. Despite an awareness of toxicity and hazard, fatal and nonfatal injuries to humans from pesticides are not uncommon. These apparently result worldwide not only from occupational exposure (by industrial workers, formulators, applicators, and nonapplicator agricultural workers) but also from misuse or mishandling of pesticides, including contamination of foods. The U.S. Department of Health, Education, and Welfare (1969) has reported some of the data on fatalities from pesticides (Table V).

Aside from the fatal and nonfatal injuries, chronic and relatively subtle effects of pesticide intoxication by man have become of great concern. Two lines of evidence particularly add to this concern: accumulation of pesticide residues in the environment including soil, water, and plant and animal organisms; and potential neurological, teratogenic and carcinogenic effects of some pesticides. Table I lists some of the common pesticides and summarizes their biological effects.

Monitoring

Chronic exposure of humans to pesticide residues is of concern. The overuse of chlorinated hydrocarbon pesticides (such as DDT) has resulted in detectable amounts of residues in virtually every living being on earth. Because of slow metabolism detectable amounts of pesticide residues entering human tissues are likely to persist for decades. Tables VI and VII show pesticide residues in human fat worldwide and in the U.S.

Regulatory measures controlling residues of pesticide chemicals in foods have been adopted in the U.S. and other developed countries, and since the 1960s the measures have become rigorous. Two international groups, the Food and Agricultural Organization (FAO) and the World Health Organization (WHO), have developed guidelines on pesticide residues for the developing nations. Such measures help to prevent the consumption of pesticide

TABLE VI

Concentration of Organochlorine Pesticide Residues in Human Body Fat[a]

		Residue concentration (ppm or mg/kg)			
Country	Year	DDT + DDE	BHC isomers	Dieldrin	Heptachlor epoxide
England	1965–1967	3.0	0.3	0.21	0.04
Denmark	1965	3.3	—	0.20	—
Canada	1966	4.3	0.1	0.22	0.14
U.S.	1968	7.7	—	0.11	—
France	1961	5.2	1.2	—	—
Italy	1966	15.4	0.1	0.68	0.23
India	1964	26.0	1.4	0.04	—

[a] Adapted from Hayes, 1975.

TABLE VII
Concentrations of Organochlorine Pesticide Residues in Human Body Fat in the U.S.[a]

Pesticide	Residue concentration (ppm or mg/kg)				
	1970	1971	1972	1973	1974
Total DDT equivalent	11.65	11.55	9.91	8.91	7.83
BHC (beta-isomer)	6.60	0.48	0.40	0.37	0.32
Dieldrin	0.27	0.29	0.24	0.24	0.20
Heptachlor epoxide	—	—	0.15	0.15	0.15
Sample size	1412	1612	1916	1092	898

[a] Adapted from U.S. Environmental Protection Agency, 1975.

residues in foods that exceed the assigned safety levels (Table VIII). Although regulatory protection is offered to people in the U.S. and other developed nations, much of the world population may not receive organized governmental supervision in this regard. Pesticide use is significantly increasing in the developing countries, but the extent of hazard from consumption of pesticide residues is not known because of the lack of sensitive monitoring.

In the U.S., the Environmental Protection Agency (EPA) in collaboration with other federal agencies establishes residue tolerance in foods for various pesticides. Based on scientific evidence, a safe residue level is established in foods, considering consumption over a lifetime (Table VIII). Most residue tolerances are expected to have adequate safety margins over the no-effect levels observed in test animals. In addition, the Food and Drug Administration (FDA) has the authority to monitor food shipments in interstate commerce, and most states have pesticide residue laws.

The pesticide-residue monitoring activities in the U.S. include sampling in nonhuman populations (such as fish, birds, and certain mammals), water,

TABLE VIII
Total Dietary Intake of Some Pesticides in the U.S.[a]

Pesticide	WHO–FAO Acceptable daily intake[b] (ppb or μg/kg)	U.S. daily intake (ppb or μg/kg)
Aldrin–dieldrin	0.1	0.08
Carbaryl	10[c]	0.5
DDT–DDD–DDE	5	0.7
Lindane	10[c]	0.05
Bromide	1000	300
Malathion	20	0.1
Parathion	5	0.01

[a] Adapted from Duggan and Corneliussen, 1972.
[b] Acceptable Daily Intake (ADI) is "the daily intake which, during an entire lifetime, appears to be without appreciable risk on the basis of all the known facts at the time." World Health Organization, 1968.
[c] Adapted from World Health Organization, 1974.

air, foods, and humans. In human studies, the sampling involves foods, fat portion of foods, and blood serum. Although pesticide residues have always been detected in U.S. foods, the levels appear to be within the limits established by FAO and WHO (Table VIII).

ANIMAL FEED RESIDUES IN FOODS

Much of the meat, milk, and eggs produced in the U.S. come from animals that have received drugs or chemicals either as injections or as additives in their feed. Of the various chemicals, hormone supplements, antibiotics and related drugs are the most extensively used in this regard.

Two major types of growth promoters are used, hormones and arsenicals. For the past few decades, synthetic estrogen diethystilbestrol (DES) had been used as an additive in cattle feed to promote rapid weight gain. This has become a controversial issue because DES has a serious carcinogenic and teratogenic potential in both humans and animals. The use of this hormone is now restricted to certain animal species for a specific period in their development, so that the hormone residues disappear from the tissues prior to marketing the animals or their products. Currently, the use of other hormones as feed supplements is also under scrutiny.

During the last few decades, the poultry and swine raised in the U.S. have been fed various organic arsenic compounds. Although the 2 ppm tolerance limit has been exceeded in poultry and swine meat rather frequently, no health problems have been reported at that level of contamination.

Antibiotics, particularly penicillins and tetracyclines, have been used in animal feeds as a preventive measure against disease. Whereas such a practice brings tremendous economic benefits, many human health problems arise due to ingestion of the drug or antibiotic residues. Individuals become sensitized by repeated exposures to these residues, which is an immunological problem that may lead to susceptibility (e.g., an allergic reaction) to the drugs. Another problem is the development of antibiotic- or drug-resistant strains of flora in the system, which renders treatment of individuals with similar antibiotics or drugs ineffective. Thus, the practice has become a controversial issue, but remains unresolved.

Other supplements of animal feed include antimicrobial and antiparasitic drugs. The levels of these drug residues occurring in human foods, and their long-term effects on human health, have yet to be fully assessed.

FERTILIZER CHEMICALS IN FOODS

Enormous quantities of mineral fertilizers are used annually in the U.S. and in the world at large. Without the use of some kind of fertilizers it would be impossible to produce enough crops to keep up with the food requirements of the population in both developed and developing nations. It is projected that by the year 2000 the worldwide usage of fertilizers for crop production will be nearly triple the levels of the early 1970s. However, the land area under cultivation is expected to increase only slightly, suggesting a massive application of fertilizers per hectare. The increased fertilizer usage will intensify certain environmental problems, including eutrophication and nitrate contamination of drinking water supplies (National Academy of Sciences, 1977).

Three critical nutrients—phosphorus, potassium, and nitrogen—constitute the bulk of chemical fertilizers. Of these, nitrate and phosphate are major contributors of water pollution. More than 70% of the fixed nitrogen entering the surface water is from nonpoint agricultural sources. A nitrate level of 10 mg/liter of water may be hazardous to human health. The population at risk comprises infants who consume synthetic milk formula with nitrate-contaminated water and suffer from variable degrees of methemoglobinemia. In the U.S., nitrate concentrations above 10 mg/liter in water are rare, but with the increased use of fertilizers both surface waters and groundwater are likely to accumulate toxic levels of nitrate (Barney, 1980). The risks may become even greater in third world countries where the need for increased crop production is crucial, and the use of chemical fertilizers will be virtually unrestricted.

Other means by which fertilizers contaminate foods include the uptake of high levels of nitrate by corn and increased uptake of radioactive potassium in crops and vegetables. There is a concern in Latin America that excessive intake of nitrate may occur through the simultaneous consumption of corn (the major food crop known to contain high levels of nitrate) and nitrate-rich water. Likewise, an increased application of phosphate and potassium fertilizers, which raise the concentration of radioactive elements in soil, may result in an increased radioactivity in crops, vegetables, and water supplies (further discussed below).

In addition to chemical fertilizers, the waste from farm animals is an important source of nitrates in water. In developing countries animal waste is extensively used as a fertilizer. Animal feed lots and large-scale use of manure as fertilizer pollute water supplies. If the trend continues, the potential for contamination of both surface waters and groundwater with nitrate will increase. High levels of nitrate–nitrogen (1100–2700 kg/hectare) have been found in soil within 100 m of old feed lots, and groundwater in the same area had as much as 70 mg/liter (Sheets, 1980).

RADIONUCLIDES IN FOODS

Radionuclides may pose a hazard to human health through their introduction into foods. Some radionuclides occur naturally in the environment. However, manmade additions during the last few decades have significantly increased their levels in the environment.

All soils contain alpha-, beta-, and gamma-radiation emitters. The relative biological effectiveness of these types of radiation is summarized in Table IX. Because some of the isotopes emitting radiation occur naturally in soil, they have not been seriously regarded as contaminants. The manmade fission products, however, have generated much greater concerns. Radionuclides from nuclear weapon tests are generally airborne, and are ultimately deposited on plants and soil surfaces. The plants can retain the radioactive substances deposited on them, or take them up from the soil via the root system.

Other sources of radioactive substances are nuclear power plants and waste disposal sites, and various commercial and noncommercial users of radioactive materials. The total global amounts of nuclear wastes generated by reactor operations have increased steadily since the early 1960s. Nuclear waste products are toxic and have long-lived radioactivity. Such sources may contaminate the soil as well as both groundwater and surface waters.

Crop plants and vegetation normally contain measurable amounts of radionuclides, but the amounts vary depending upon the types of crops, vegetation, and soils (Hansen *et al.,* 1960). The commonly occurring radionuclides are shown in Table X. The nuclear fission process produces a variety of radionuclides, none of which are innocuous. Some radionuclides have such short half-lives that the radioactivities disappear instantaneously. However, others pose a special concern because of their long half-lives, and their biochemistry and metabolism in living organisms. Radionuclides that are ingested with food and drink or are inhaled become the "internal emitters." Their toxicity depends upon their energy of radiation, and their physical and biological half-lives. The latter can be defined as the retention time for half of the original radioactive isotopes introduced to the body prior to their excretion in urine or feces. In this regard, the high energy alpha-emitters are of concern because of their ability to cause tissue damage. The major manmade radionuclides derived from soil that occur in human food supplies are the two

TABLE IX
Types of Radiation and Their Relative Biological Effectiveness

Radiation type	Penetration range into biological tissue	Relative biological effectiveness	Remarks
Alpha	0.005 cm	10–20	Corpuscular (size of a helium atom without the electrons); does not travel far but produces an intense ionization within a short path; particularly harmful if ingested.
Beta	3 cm	1	Corpuscular (a high speed electron); intermediate in penetration.
Gamma	20 cm (approx.)	1	Noncorpuscular; highly penetrating.

TABLE X
A List of Radionuclides Commonly Found in the Environment

Radionuclides (with radiation type and half-life)	Major site of biological effects	Remarks
Hydrogen-3 (beta, 12.3 years)	Whole body	Produced naturally, and also by fission and fusion reactions; present in effluents of nuclear reactors and detonated nuclear weapons. In the environment it remains combined with oxygen of water by isotope exchange reaction with hydrogen; easily enters the plants and food chain. Its biological half-life is 12 days.
Carbon-14 (beta, 5770 years)	Whole body	Occurs both naturally, and as a result of nuclear weapon tests. It exchanges with a carbon atom in organic compounds.
Potassium-40 (beta, 1.3 billion years)	Whole body muscle	Naturally occurs in soil to the extent of 20,000 millicurie per square mile to a depth of 1 foot (Alexander et al., 1960). All potassium-containing fertilizers add the radioisotope to soil. Forages (alfalfa, clover, etc.) take up large amounts of potassium including the radioisotope, and become a source for grazing animals.
Rubidium-87 (beta, 47 billion years)	Whole body	Naturally occurs but is much less abundant than potassium-40.
Strontium-90 (beta, 28.8 years)	Bone, teeth	Nuclear fission product, abundant in the environment; occurrence in soil 150 millicurie per square mile. It is in radioactive equilibrium with its ytrium-90 daughter, which emits high-energy beta particles. It can replace calcium; it is readily taken up by plants; it enters the food chain: grass-cow-milk-man. Children are particularly at risk because their developing bone will readily concentrate strontium-90 from affected milk. Its biological half-life is 50 years.
Ruthenium-106 (beta, 1 year)	Bone, liver	Produced in nuclear reactors and denotated nuclear weapons. It occurs in soil in trace quantities. It emits energetic beta particles and its daughter, rhodium-106, also emits beta particles.
Antimony-125 (beta, 2 years)	Bone, liver	Produced in nuclear reactors and detonated nuclear weapons. It occurs in soil in trace quantities.
Iodine 129 (beta, 17.2 million years)	Thyroid	Both isotopes are selectively concentrated in thyroid. Iodine-131 is of less concern as a soil contaminant because of its short half-life.
Iodine-131 (beta, gamma, 8 days)	Thyroid	

(continued)

TABLE X *(continued)*
A List of Radionuclides Commonly Found in the Environment

Radionuclides (with radiation type and half-life)	Major site of biological effects	Remarks
Cesium-137 (beta, gamma, 30.2 years)	Whole body	Nuclear fission product, most abundant in the environment; occurrence in soil 240 millicurie per square mile. It decays to barium-137, which emits gamma radiation. Being an alkali metal, it can replace sodium or potassium in the biological system. Its biological half-life is 50 days.
Lanthanum-139 (beta, unknown)	Whole body	These four alkaline earth elements occur in trace amounts, totalling 15 millicurie per square mile.
Samarium-146 (alpha, 50 million years)	Whole body	
Lutetium-176 (beta, 21 billion years)		
Rhenium-187 (beta, 70 billion years)	Whole body	
Cerium-144 (beta, 285 days)	Bone, liver	Produced in nuclear reactors and detonated nuclear weapons. It occurs in soil in trace quantities. It emits energetic beta particles and its daughter, praseodymium-144, also emits energetic beta particles.
Promethium-147 (beta, 2.5 years)	Bone, liver	Produced in nuclear reactors and detonated nuclear weapons. It occurs in soil in trace quantities.
Radium-226 (alpha, gamma, 1622 years)	Bone, liver	It is a product of uranium-238 decay chain; occurs naturally as an alkaline earth element. Because of similarity with calcium it is taken up in the bone. It constantly emits energetic alpha particles and its daughter radon-222, also emits energetic alpha particles. It is a classical cancer causing radioisotope.
Thorium-232 (alpha, 13.9 billion years)	Bone, liver	It is a product of uranium-238 decay chain; significantly contributes to soil radioactivity.
Uranium-238 (alpha, 4.51 billion years)	Bone, liver	Occurs naturally, and also as an effluent of nuclear reactors; significantly contributes to soil radioactivity.
Plutonium-239 (alpha, 24,360 years)	Bone, liver, lung	A transuranic element; fissionable; occurs naturally and also from nuclear weapon tests. Its deposition on U.S. soil surface averages 2 millicurie per square kilometer. It is rapidly fixed in soil. It enters human body by ingestion and inhalation. Transfer of plutonium from

TABLE X *(continued)*
A List of Radionuclides Commonly Found in the Environment

Radionuclides (with radiation type and half-life)	Major site of biological effects	Remarks
		ingested food to body is considered low because of slow absorption by the alimentary tracts. Bioaccumulation is limited as it moves up the food chain in both terrestrial and aquatic systems. Intake through food in the U.S. is estimated to be 2.5 picocurie per year during the mid-1960s, but is lesser now because of decrease in fallout (American Chemical Society, 1978). Nonetheless, biologically plutonium is most hazardous; its biological half-life is 200 years.

fission products strontium-90 and cesium-137, but others may occur depending upon the fallout conditions and half-lives of radioisotopes (Alexander *et al.*, 1960; American Chemical Society, 1978).

After nuclear weapon tests by the U.S., U.K., and U.S.S.R. during the late 1950s, the radioactivity in the atmosphere and rainwater increased significantly (Thatcher *et al.*, 1965). Alfalfa samples collected in several states of the U.S. during this period showed an average beta radioactivity of 10 nCi/kg dry weight (U.S. Food and Drug Administration, 1960).

The atmospheric nuclear fission products, upon deposition on plants and soil surfaces, may become redistributed by rainfall and runoff. Rainfall bringing down atmospheric radioactivity can contaminate crops. A study of Menzel *et al.* (1963) has shown that the strontium-90 concentrations in sweet corn, snap beans, potatoes, and cabbage grown in Florida was particularly high during the season of low, but frequent, rainfall. Abundant rainfall, on the other hand, will remove radioactivity from plants and soil surfaces.

In addition to plants and food crops, there is radionuclide contamination of animal species that provide food. Worldwide animal products constitute a major part of human food supply. Cattle, which supply most of the animal products in the form of meat and milk, are subject to radioactive contamination. Sources are grazing lands with natural radioactive elements, such as potassium-40, effluents of fossil fuel combustion, and particularly the fallout from nuclear weapon tests. The grazing animals usually ingest the radionuclides along with forage. Some radionuclides are consumed with soil, such as under conditions of overgrazing. Ingested radionuclides are translocated or deposited in animal tissues, particularly potassium-40 in muscle and strontium-90 in bone.

MISCELLANEOUS CHEMICAL RESIDUES IN FOODS

A number of chemicals may reach foods or food products from miscellaneous sources, originating in packaging and storage. With increased urbanization and growth of food industries, the packaging of foodstuffs has become an essential means by which transportation and storage processes are facilitated before reaching the consumers. There is often a slow release of chemicals from the packaging materials which are ingested along with foodstuffs. A prolonged storage increases the contact time for interaction between packaging materials and food constituents.

Plastic containers and wrapping papers made of polyvinyl chloride (PVC) may slowly release vinyl chloride into wines, beverages, and various food materials; in addition, PVC pipes may contaminate the drinking water. The polymer PVC itself is considered to be nontoxic. However, the food materials in PVC containers may be contaminated by the residual monomer. It has been estimated that the daily uptake of vinyl chloride from food wrappings may vary from 0.1 to 1 μg/person (Centro Europeen des Federations de l'Industrie Chimique, 1976). Phthalic acid esters (PAEs), which are extensively used as plasticizers in food wrapping films, have been detected in blood stored in plastic bags (Jaeger and Rubin, 1970). Metal containers used for canning foods release lead (from solder flux) into the food materials.

PREVENTIVE MEASURES AGAINST CHEMICAL CONTAMINATIONS IN FOODS

The dilemma posed by agricultural chemicals is obvious. Enormous economic benefits are brought about by the application of chemicals; superior agricultural productivity provides for the hungry people of the world. Yet there remains the problem of chemical residues in foods—residues that are bioconcentrated through the food chain and ultimately reach humans. The long-term consequences of continued exposure to chemical residues are a subject of conjecture at the present.

The dilemma cannot be easily resolved. Agricultural productivity must continue, but an increased reliance on agricultural chemicals should be reassessed for ultimate consequences, and a great deal of caution must be exercised to prevent undue contamination of foods by chemical residues or their accumulation in food chains. Control measures might ultimately lead to reduced consumption of potentially harmful agricultural chemicals.

For public safety consideration, foods, crops, edible plants, and animal feeds must be monitored systematically for residues. This should be done by food industries and governmental agencies. In the U.S. and other developed nations specific regulations exist for each major chemical used in agriculture, e.g., registration of pesticides. The developing nations must adopt similar measures if they intend to rely on chemicals to accelerate their agricultural productivity.

Regulation and restricted use of toxic, persistent chemicals would be particularly important. Switching to chemicals that are biodegraded in soil or catabolized in plants would be an attractive alternative. The timing of application of chemicals should be such that the residues disappear prior to harvesting the crops or processing and marketing the food products.

Although it is difficult to assess the real human hazards of agricultural chemicals from animal toxicology studies, the relative toxicities of chemicals must be determined before decisions are made about their application in crop or food production. Reported episodes of illness or fatalities caused by various agricultural chemicals must be a lesson to safeguard against future tragedies. There should be a tolerance limit set for each chemical applied in agriculture. Long-term effects of chemical residues may not be obvious, but established tolerance limits should not be exceeded in foods and food products in order to prevent long-term exposures with potential overt effects. The use of agricultural chemicals will probably continue to proliferate in order to guarantee adequate food supplies; however, research and education emphasizing human safety in these chemicals must receive a high priority. It should also be borne in mind that a continued or excessive application of chemicals (fertilizers and pesticides) may eventually harm the land and cause a decline in crop yield.

In addition to the regulatory methods, there are technical means which may be utilized to reduce the occurrence of chemical residues in foods. These include improved management of pesticide use, optimal rather than a maximal use of a chemical, modification of application methods (e.g., aerial vs. soil application), and modification of chemical formulation to encourage the development of efficient, but nonpersistent pesticides. Efficient management of pesticide use remains to be a key factor in both receiving the potential economic benefits and providing the necessary protection to humans and the environment at large.

REFERENCES

Alexander, L. T., Hardy, E. P., Jr., and Hollister, H. L., 1960, Radioisotopes in soils; particularly with reference to strontium-90, in: *Radioisotopes in the Biosphere* (R. S. Caldecott and L. A. Snyder, eds.), pp. 3–22, University of Minnesota Center for Continuation Study, Minneapolis.

American Chemical Society, 1978, Radiation in the environment, in: *Cleaning Our Environment: A Chemical Perspective,* A report by the Committee on Environmental Improvement, pp. 378–452, Washington, D.C.

American Chemical Society, 1978, Pesticides in the environment, in: *Cleaning Our Environment: A Chemical Perspective,* A report by the Committee on Environmental Improvement, pp. 322–377, Washington, D.C.

Barney, G. O., Study Director, 1980, *The Global 2000 Report to the President of the U.S., Vol. I: The Summary Report,* pp. 94–120, Pergamon Press, New York.

Centro Europeen des Federations de l'Industrie Chimique (CEFIC), 1976, *Vinyl Chloride Toxicity and the Use of PVC for Packaging Foodstuffs,* CEFIC Committee for the Toxicity of Vinyl Chloride, Brussels.

Duggan, R. E., and Corneliussen, P. E., 1972, Dietary intake of pesticide chemicals in the United States, June 1968–April 1970, *Pestic. Monit. J.* **5:**331–341.

Gaines, T., 1960, The acute toxicity of pesticides to rats, *Toxicol. Appl. Pharmacol.* **2:**88–89.

Hansen, R. O., Vidal, R. D., and Stout, P. R., 1960, Radioisotopes in soils: Physical–chemical composition, in: *Radioisotopes in the Biosphere* (R. S. Caldecott and L. A. Snyder, eds., pp. 23–36, University of Minnesota Center for Continuation Study, Minneapolis.

Hayes, W. J., Jr., 1975, *Toxicology of Pesticides,* Williams & Williams Co., Baltimore, Maryland.

Jaeger, R. J., and Rubin, R. J., 1970, Plasticizers from plastic devices: Extraction, metabolism, and accumulation by biological systems, *Science* **170:**460–462.

Manske, D. D., and Johnson, R. D., 1977, Residues in food and feed, *Pestic. Monit. J.* **10:**134–148.

Menzel, R. G., Roberts, H., Jr., Stewart, E. H., and MacKenzie, A. J., 1963, Strontium-90 accumulation on plant foliage during rainfall, *Science* **142:**576–577.

Murphy, S. D., 1975, Pesticides, in: *Toxicology: The Basic Science of Poison* (L. J. Casarett and J. Doull, eds.), pp. 408–453 (and references therein), Macmillan Publishing Co., New York.

National Academy of Sciences, 1977, *Drinking Water and Health,* pp. 411–425, NAS Washington, D.C.

Sheets, T. J., 1967, The extent and seriousness of pesticide buildup in soils, in: *Agriculture and the Quality of Our Environment* (N. C. Brady, ed.), AAAS Publication 85, pp. 311–330, American Association for the Advancement of Science, Washington, D.C.

Sheets, T. J., 1980, Agricultural pollutants, in: *Introduction to Environmental Toxicology* (F. E. Guthrie and J. J. Perry, eds.), pp. 24–33, Elsevier, New York.

Thatcher, L. L., Payne, B. R., and Cameron, J. F., 1965, Trends in the global distribution of tritium since 1961, in: *Radioactive Fallout from Nuclear Weapons Tests* (A. W. Klement, ed.), pp. 646–674, U.S. Atomic Energy Commission.

U.S. Department of Health, Education and Welfare, 1969, *Report of the Secretary's Commission on Pesticides and Their Relationship to Environmental Health,* Government Printing Office, Washington, D.C.

U.S. Environmental Protection Agency, 1975, *DDT: A Review of Scientific and Economic Aspects of the Decision to Ban its Use as a Pesticide,* prepared for: Committee on Appropriations, U.S. House of Representatives, Washington, D.C.

U.S. Food and Drug Administration, 1960, Strontium-90 analyses of human and animal food collected in 1958 and 1959, in: *Radiation Health Data* **1:**36–40. 1960.

World Health Organization, 1968, *WHO Technical Report Series 391,* p. 22, WHO, Geneva.

World Health Organization, 1974, *WHO Technical Report Series 545,* pp. 28–33, WHO, Geneva.

11

THE FIREMASTER INCIDENT

Charles L. Senn

NEW CHEMICALS AND THEIR POTENTIAL HEALTH EFFECTS

Medical and environmental specialists have become increasingly concerned with the rapid development and subsequent marketing of newly produced chemicals. The U.S. Surgeon General's report *Healthy People* (1979) says, "more than 4 million chemical compounds are now recognized; more than 60,000 are commercially produced; about 1,000 are introduced each year."

Unfortunately, many of those new chemicals have been produced, sold, and used before their health effects have been adequately evaluated. Often their potential to produce serious adverse health effects are not recognized until persons who have been exposed for long periods have developed symptoms that are then traced to exposure to a specific chemical. That was the case with vinyl chlorines, which affected many workers in the plastic industries, and with polychlorinated biphenyls, which are used as a coolant in transformers and for other purposes. The subject of this chapter is another member of the biphenyl family, polybrominated biphenyls (PBBs), which were accidentally mixed into animal and poultry feeds.

BACKGROUND OF "FIREMASTER INCIDENT"

The "incident" took place in Michigan, a state with a population of about 9.15 million. The state has a large number of large and small manufacturing and chemical plants as well as many moderate sized farms for raising dairy cattle, poultry and eggs, and other farm animals and crops.

The first warnings of the problem came in early 1973 when dairy farmers began noting serious health abnormalities among their cows. The early symptoms were described in a statement made on April 30, 1976, to a subcommittee of the Committee on Agriculture of the U.S. Congress, by Sam D. Fine,

Charles L. Senn • Senn Environmental Consulting Associates, Inc., Los Angeles, California 90032.

Associate Commissioner of the Food and Drug Administration of what was then the U.S. Department of Health, Education and Welfare. Mr. Fine said,

> The tragic story begins in May 1973, when Michigan dairy farmers began experiencing problems with their dairy herds. Throughout the summer of 1973, and extending into 1974, Michigan dairy farmers found their cattle refusing to eat manufactured feed; milk production decreased; cows lost weight, developed abnormal hoof growth and lameness, lost hair, and in general, were sick. Both cattle and swine aborted, and farmers reported that heifers were not breeding. A high neonatal death rate was observed in cattle. Some 100 cattle sent to slaughter during 1974 were found to have enlarged livers.

The *Michigan State Journal,* a newspaper of the state capital, Lansing, on November 28, 1976, described the incident and stressed public concerns for effects on the health of humans who consumed meat, milk, and eggs from animals and poultry fed on a large quantity of food to which PBB was inadvertently added. The headline was "Michigan, A PBB Guinea Pig" and the story began, "One of the world's great medical mysteries continues to unfold in Michigan, with no one in the science community sure it will ever be resolved."

OFFICIAL REPORT ON INCIDENT BY U.S. FOOD AND DRUG ADMINISTRATION

Associate Commissioner Fine of the U.S. Food and Drug Administration (FDA) told a Congressional subcommittee, "At the outset it should be recognized that this incident represents the most severe agricultural contamination problem experienced to date in the United States." His written statement continued,

> The acronym 'PBBs' stands for a class of industrial chemicals called polybrominated biphenyls. PBBs were manufactured by the Michigan Chemical Company at St. Louis, Michigan, and sold under the trade name *Firemaster* (as a fire retardant). Michigan Chemical manufactures another chemical product at the same site—magnesium oxide, an animal feed supplement—sold under the trade name *Nutrimaster.* In 1971 and 1972, the company produced several experimental batches of Firemaster which resembled *Nutrimaster* in physical appearance; both products were packaged in similar 50-lb brown paper bags. The trade name *Firemaster* and *Nutrimaster* were stenciled on the bags of PBBs and magnesium oxide, respectively.
>
> . . . Subsequent FDA investigations revealed that in May of 1973 an unknown quantity of the experimental *Firemaster* (PBB) was inadvertently substituted for *Nutrimaster* (magnesium oxide). Michigan Chemical Company shipped *Firemaster,* which was then used by several feed mills operated by the Farm Bureau Services in the manufacture of animal feed and as an ingredient in animal protein supplements in lieu of the desired magnesium oxide. These feed products containing *Firemaster* were shipped to other mills and farms for use in preparing finished feeds. The PBB has now been traced to over 1200 cattle, poultry, swine and sheep farms throughout the lower peninsula of Michigan. . . .

> Analysis disclosed the presence of PBBs in feed in amounts up to 7000 ppm, levels as high as 275 ppm in the fat portion of milk, and over 4600 ppm in fat from cattle. PBBs were also detected in the fat of poultry, swine, and sheep that had consumed the contaminated feed, as well as in eggs. State officials began quarantining affected herds and flocks in May of 1974.

Commissioner Fine reported that the incident resulted in

> the quarantine of 538 Michigan farms; it eventually resulted in the destruction of over 23,000 cattle, 5000 swine and sheep, 1.5 million chickens, 2600 lb of butter, 34,000 lb of dry milk products, 1500 cases of evaporated milk, 18,000 lb of cheese, about 5 million eggs, and 865 tons of feed. By the end of 1975, lawsuits for damages suffered totaled millions of dollars.
>
> . . . In May of 1974, our laboratory analytical capability to detect and confirm PBBs in these types of products translated to practical enforcement limits for PBB residues at 1.0 ppm on a fat basis in meat, milk, and dairy products, 0.3 ppm in animal feeds, and 0.1 ppm in eggs. These limits, which were based solely on analytical methodology available at that time, were also followed by the Michigan Department of Agriculture and the U.S. Department of Agriculture.
>
> In the summer and fall of 1974, we began to hear reports, mainly through the media, of illnesses being suffered by farm families who had consumed PBB-contaminated food products on their farms. FDA investigators visited 70 of the highly contaminated farms during this period. No unusual human health problems were reported by the persons interviewed or were noted in the medical records obtained on persons who sought medical treatment. In fact, about 60% of the families reported no health problems of any type. The Michigan Department of Public Health, which had been studying possible acute effects of PBB ingestion by Michigan residents during this same period, was also unable to document any occurrences of human disease attributable to the PBB incident. . . .

Later in his statement, Commissioner Fine said,

> PBBs are extremely stable and persistent chemicals, and because waste products from livestock exposed during the original incident would also contain PBBs, the feeding areas and the environment of these farms also became contaminated and probably will remain so for some time in the future. Thus, livestock introduced on these farms could be exposed to this indirect environmental source of PBBs and food derived from these animals could contain low levels of PBBs.

Authorities on carcinogenicity note "that fire retardants similar to PBBs (containing chlorine atoms instead of bromine atoms) are capable of causing enormous injury to the human nervous system." They noted that

> a few years ago, when the wrong spigot was turned at a Japanese factory, polychlorinated biphenyls (PCBs) contaminated a supply of cooking oil and caused irreversible brain damage in many consumers who used the oil. More recently, animal tests have shown PCB's to be carcinogens.

Authorities note, however, that the possible carcinogenic effects of the incident on humans

> will not be known for a decade or more because of the long latency period typical of chemical carcinogens. And what harm, if any, may be done by the presence of either PBBs or PCBs in human breast milk is another imponderable.

MICHIGAN DEPARTMENT OF PUBLIC HEALTH'S STATUS REPORT—PBB AND HUMAN HEALTH (1977)

This report lists a number of toxic, or potentially toxic, industrial chemicals which have entered the environment during the past 30 or 40 years, including PCBs. It notes that consumption of PBB-contaminated foods in Michigan has gradually declined since 1974, so exposure to foods highly contaminated with PBB has passed. However, "almost all residents of the Lower Peninsula probably have had some exposure to PBB through food. . . . "

PRELIMINARY ASSESSMENT OF EFFECTS ON HUMANS

A short-term study by the Department involved 165 farm residents whose animals and poultry were not fed PBB-containing feed (exposed group) and 133 persons living on farms where PBB-contaminated feed was given to animals (control group). The study included a history of illness of all persons, and physical examinations and selective laboratory work on as many as possible. PBB blood levels were obtained from 110 of 165 exposed persons and from 104 of 133 control persons. Levels were significantly lower in control persons. (In a study completed in October 1976, 96% of all nursing mothers in the Lower Peninsula had at least a trace of PBB in their milk. The highest value was 1.22 ppm with half of the women having 0.68 ppm or lower.) The Department "was unable to detect a significant difference in symptoms between those with greater and those with lesser exposures. The conclusion was that no acute disease due to the toxic effects of PBB could be demonstrated by the study . . ."

SPECIAL HEALTH EFFECT STUDY

The *Status Report* told of a study funded by the National Cancer Institute through the

> Federal Center for Disease Control for the Michigan Department of Public Health to evaluate the possibility of long-term effects of PBB on human health that might be occurring at the time or might emerge over the next ten or fifteen years. The U.S. Food and Drug Administration has also contributed for special studies. The study originally called for enrolling the following two groups of people:
>
> 1. 2000 members of farm families who lived on quarantined farms;
> 2. 2000 people who received and consumed products directly from quarantined farms.
>
> Presently the Department is formulating plans to follow 2000 farm people with no known exposure to PBB in a similar manner.
>
> The Department also established the following programs:

1. Providing PBB analytic services to physicians of people in Michigan exposed to PBB or suspected of such exposure;
2. Conducting a detailed evaluation of such persons;
3. Disseminating to the public and to physicians in Michigan any information that is gathered concerning the nature, therapy, or prevention of any possible PBB effects.

After an extensive sampling and analysis of mothers' milk that showed wide occurrence of the PBB chemical, a group of experts was called together to evaluate the question of whether mothers should be advised to discontinue breast feeding. Their report, in summary, said that in weighing the benefits of breast feeding against the anticipated hazards, it would not be good policy to advise against breast feeding, even by mothers who lived on farms where animals, milk, and eggs had been previously heavily contaminated.

FUTURE CONSIDERATIONS

The Michigan State Department of Public Health's *Status Report* said,

> Plans for the coming year include sending a one-page questionnaire to participants in the long-term study. The questionnaire will include a series of questions asking about the general state of health and place of residence of the participants. Subjects not responding to the questionnaire will be actively traced through local inquiries and phone calls to relatives.
>
> In the coming year, field workers will visit a number of participants in their homes and will administer a questionnaire asking detailed health questions comparable to those asked in the initial enrollment questionnaire with emphasis on information about significant illnesses. The field workers will draw blood from some of these people.
>
> The Health Department will continue to monitor health problems which participants report. Periodic evaluation of medical data collected on questionnaires and from hospital/physician contacts will be conducted in close cooperation with the Center for Disease Control and physicians so that appropriate medical inquiries can be pursued without delay.
>
> At this time, the Health Department has not found any answers regarding PBB and human health. The department will continue to do research and in the coming years will follow closely those people who were most highly exposed to PBB and who are enrolled in the long-term study.

SOURCES OF INFORMATION ON "FIREMASTER INCIDENT"

The information utilized in this chapter was from experiences of the author, acting as news editor at a Milwaukee Conference where the topic was discussed; and from discussions with and a letter from Mr. John Ruskin, Environmental Health Director of the County, which includes Michigan's capital, Lansing; and from the following references.

The following are all unpublished reports, except as noted:

Citrin, T., Dec. 1979, *Report on Environmental Health Focal Point,* Health Code Implementation Project, Lansing, Michigan.

Fine, S. D., *Statement of Poly-Brominated-Biphenyls (PBB) to Subcommittee on Conservation and Credit, Committee on Agriculture,* U.S. House of Representatives, April 30, 1976.

McCracken, W. E., June 3, 1980, *Overview of Act 64, The Rule Development Process and the Proposed Rules Presentation at the Hazardous Wastes Management Planning Committee Meeting,* State Department of Natural Resources, Lansing, Michigan.

Michigan Department of Public Health, 1980, *Of Chemicals and Health,* Lansing, Michigan (printed publication).

Michigan Department of Public Health, July, 1977, *PBB and Human Health—Status Report,* Lansing, Michigan.

Michigan State Department of Public Health (MSDH), Jan. 22, 1979, *Chemicals and Health Center,* MSDH, Bureau of Environmental and Occupational Health, Lansing, Michigan.

Moskal, J., Michigan: A PBB Guinea Pig, *The State Journal,* Nov. 28, 1976, Lansing, Michigan.

Randal, J., Can PBB Cause Cancer? Gannett News Service in *The State Journal,* Nov. 28, 1976, Lansing, Michigan.

U.S. Department of Health, Education and Welfare (Public Health Service), *Healthy People,* The Surgeon General's Report on Health Promotion and Disease Prevention in 1979, DHEW (PHS) Pub. No. 79-55071 (U.S. Superintendent of Documents, Washington, D.C.).

12

MINAMATA DISEASE
Organic Mercury Poisoning Caused by Ingestion of Contaminated Fish

Masazumi Harada

INTRODUCTION

Minamata disease is unique in its mechanism of outbreak. Organic mercury discharged with the wastewater from an industrial plant polluted the environment, which in turn contaminated fish and shellfish. People who ingested these marine products then became poisoned. Such widespread methyl mercury (MeHg) poisoning was the first occurrence in the history of mankind. The incident was instructive in the following sense. The seas, rivers, and atmosphere had long been regarded as places where people could throw away and dump wastes without a second thought. However, it was demonstrated that a certain substance in the environment would accumulate through the food chain and finally cause people to suffer from its toxicity. Not only persons who ingested the contaminated food were affected; embryos in the uterus suffered even more severe damages. It also became apparent that once the environment was polluted, restoring it to an acceptable state was almost impossible.

It was in May, 1956, that Minamata disease was first "discovered" in Japan (McAlpine *et al.,* 1958). At that time, its cause was held to be unknown, and it was called a "strange disease." Only three years later, in October, 1959, was the cause verified by a Kumamoto University team (Study Group of Minamata Disease, 1968). Later in 1965, the second Minamata disease outbreak occurred in Niigata prefecture, Japan (Tsubaki, 1968). Minamata disease is an incurable, manmade illness. It is organic mercury poisoning, but there is no guarantee that a similar incident may not be induced by other chemical substances.

Masazumi Harada • Department of Neuropsychiatry, Institute of Constitutional Medicine, Kumamoto University, Kumamoto City, Kuhonji 4-chome, Japan.

BACKGROUND OF THE MINAMATA DISEASE OUTBREAK

Minamata disease (abbreviated as MD hereinafter) was first observed in the area along the Sea of Shiranui, a quiet inland sea surrounded by Uto Peninsula in the north, the islands of Nagashima in the south, and Amakusa in the west (Fig. 1). A great variety of fish live abundantly in this sea. Prior to the outbreak of MD a fishermen's union in Minamata city alone annually caught *ca.* 500 tons of such species as scabbard fish, yellowtail, sea bream, gray mullet and sardine (Harada, 1972a). In addition, crabs, octopus, cuttlefish, sea cucumbers, oysters and other shellfish were caught in abundance. The main source of protein for people living in this area was the marine life, because fish and shellfish were abundant. The farm products were limited—sweet potatoes, some vegetables, and mandarin oranges—because there was little arable land behind the steep coast line. Hence, marine products constituted a major proportion of the people's diet. Japanese people, in general, are known to eat relatively large amounts of fish; the average intake of fish by Japanese is said to be 86.3 g/day. However, a survey showed that the fishermen in this area ate even more fish. That is, the husband ingested on the average 316–459 g/day, with 622 g/day as the maximum, while even the wife ate an average of 69–251 g/day. In addition, the survey revealed that fishermen's families ingested 2–4 times more fish than Japanese households in general, and nonfishermen's families in the fishing villages ate 1.5 times more than households in general (Futatsuka *et al.*, 1974). At the time the Minamata disease broke out, the population was around 200,000 along the coast of the Sea of Shiranui. Fishermen, and those who lived in the fishing villages but were not engaged in fishing, were estimated to be about half the population or 100,000 persons (Fig. 1).

Minamata is one of the towns facing the Sea of Shiranui, and Chisso Company built a plant there in 1902. The plant used to produce lime nitrate and ammonium sulfate fertilizers, but in 1932 Chisso began production of acetaldehyde using mercury as a catalyst. The initial annual output of 210 tons increased rapidly after World War II to reach 2300 tons in 1946, and 45,200 tons in 1960, the peak year (Fig. 2). Meanwhile, mercury was discharged into the sea. The exact amount is unknown, but it is estimated to have been about 600 tons (Harada, 1978b). The plant employed about 5000 workers during its peak production period—in fact, it was the biggest chemical plant in this area. The effluent from the plant was discharged and accumulated at the innermost part of the Minamata Bay, a two-branched inlet surrounded by Koiji Islet, Myojin Point, and Cape Modo (Fig. 3). In 1959, when MeHg was first suspected as the cause of the "strange" disease, an environmental survey of mercury pollution in the Bay was conducted. Chemical analyses showed that an extraordinary high level of mercury contamination existed in the Minamata Bay. In the sludge near the drainage channel, the level reached 2010 ppm (total mercury, wet weight) and declined gradually in proportion to the distance from the channel (Fig. 3) (Kitamura *et al.*, 1960b; Study Group of Minamata Disease, 1968).

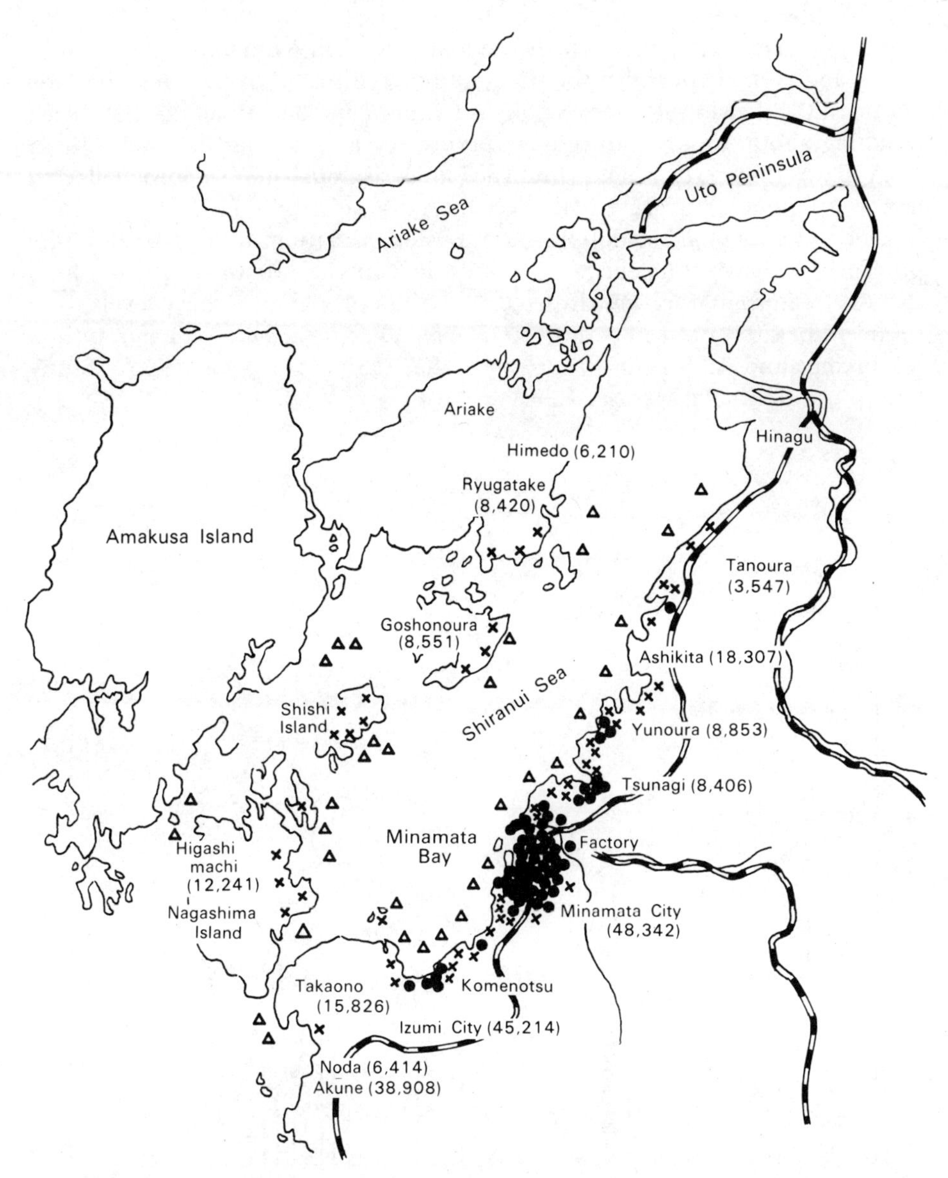

FIGURE 1. Map of the Shiranui sea. (●) Acute and subacute patients designated through 1962 (121 cases); (x) where Minamata-diseased cats have been discovered; (△) where fish were found floating; numbers in parentheses are population of 1960. From Harada (1975).

Fish lived most abundantly at this point. It was not until about 1950 that abnormalities began to appear within the bay. Catches of fish declined, shellfish died, and poisoned fish floated to the surface. In 1953 water birds, unable to fly, fell to the sea, and by 1955 poisoned fish were seen surfacing all over the Sea of Shiranui. In fishing villages, cats began to go mad and die inexplicably. In February, 1957, cats were brought from Kumamoto city to Minamata and were kept there for observation. Within 32 to 65 days, they too became ill. Likewise, 40 very small fish caught in the Minamata Bay, each weighing about 10 g, produced symptoms in 51 days when fed three times a day to cats (Sera *et al.*, 1957). The 1957 analysis of fish and shellfish collected in the Minamata Bay revealed high mercury contents; 11.4–39.0 ppm in Hormomya mutabilis, 5.61 ppm in an oyster, 35.7 ppm in a crab, and 14.9 ppm in a scioena schlegelii. Near the estuary of the Minamata River where the wastewater channel was diverted in the fall of 1958, mercury levels were found to be 20.0 ppm in a short neck clam *(Venus Japonica),* 24.1 ppm in a sea bream, and 10.6 ppm in a gray mullet (Kitamura *et al.,* 1960b; Study Group of Minamata Disease, 1968).

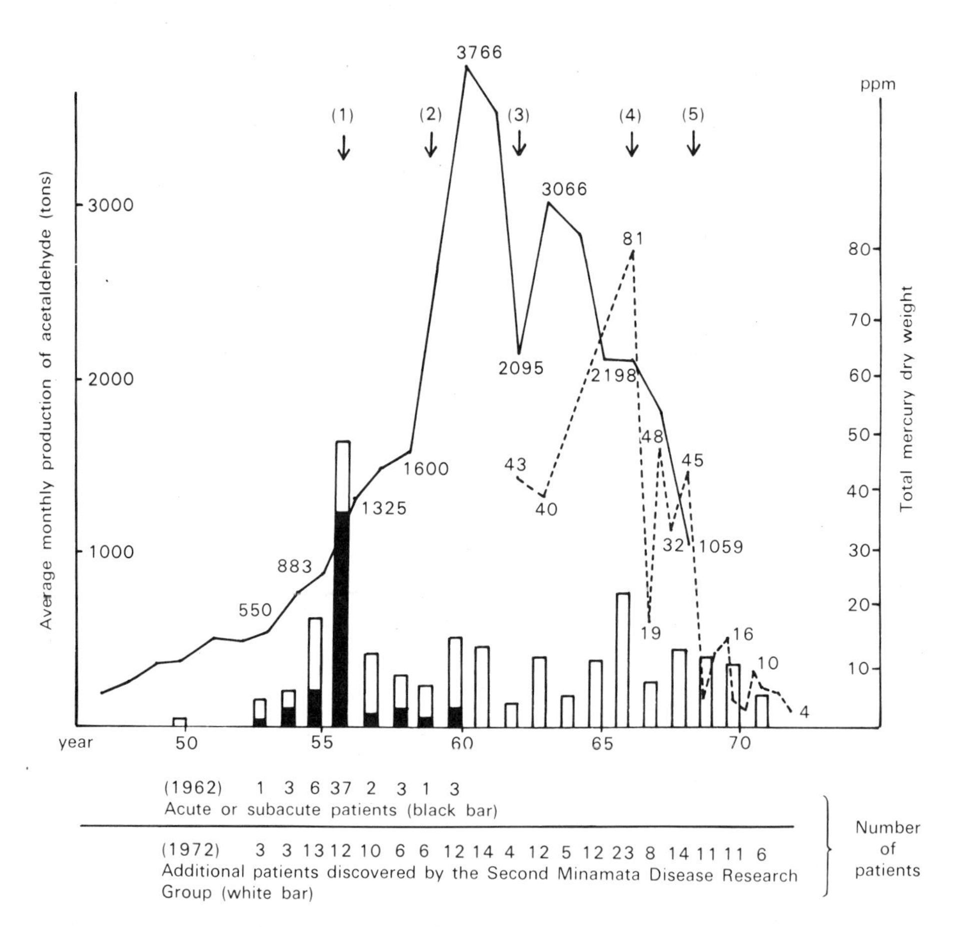

FIGURE 2. Comparative chart of acetaldehyde production rate (—), onset of Minamata disease, and mercury content in the shellfish of Minamata bay (----). Patients are limited to the Tsukinoura, Detsuki, and Yudo areas. Shellfish are *Venus japonica* of Koiji Island (Fujiki, 1972). (1) Official discovery of patients; (2) cause proved to be methyl mercury; (3) a year-long labor dispute; (4) circulatory wastewater system installed; (5) stoppage of production. From Harada (1975).

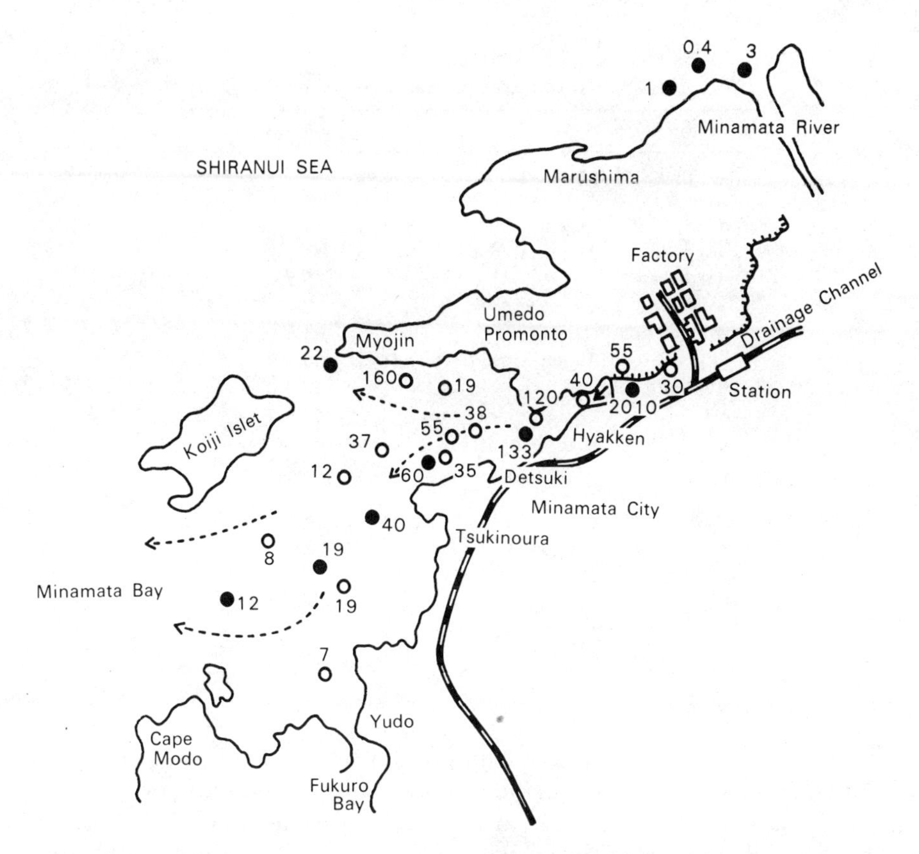

FIGURE 3. Minamata bay and mercury content of sludge at the bottom of the bay. (●) Hg = ppm wet weight (Kitamura, 1959); (○)Hg = ppm wet weight (Irukayama, 1963). From Study Group of Minamata Disease (1968).

ACUTE OR SUBACUTE MINAMATA DISEASE

Minamata disease uncovered in groups in May 1965 was considered to be acute or subacute poisoning when classified by type of poisoning. Of 34 cases found at this time, 16 died within 3 months after the onset of clinical manifestations. In four cases, death came within 6 months, and one patient died within a year (Study Group of Minamata Disease, 1968). Thus, the course of disease was short and patients' conditions deteriorated rapidly. Symptoms began with numbness of limbs, loss of strength in the limbs, and poor coordination. That is, patients dropped things held in their hands and stumbled on a flat ground. Then their speech became unclear and excessive salivation was observed. Patients were unsteady on their feet as if they were drunk, and had difficulty in writing or taking meals. In the meantime, hearing deteriorated, and tinnitus, headaches, pains in the limbs, and forgetfulness were observed. Some patients were unable to stand up and in serious cases there

TABLE I
Clinical Symptoms of Acute and Subacute Minamata Disease[a]

Symptoms	% of total examined
Disturbance of sensation	
superficial	100
deep	100
Constriction of the visual field	100
Dysarthria	88.2
Ataxia	
adiadochokinesis	93.5
finger–finger, finger–nose tests	80.6
Romberg's sign	42.9
ataxic gait	82.4
Impairment of hearing	85.3
Tremor	75.8
Tendon reflex	
exaggerated	38.2
weak	8.8
Pathological reflexes	11.8
Salivation	23.5
Mental disturbances	70.6

[a] Thirty-four adult cases of acute and subacute MD (Tokuomi, 1960).

were general convulsions, deformation of limbs, and contraction of muscles, followed by death. The combination of the clinical symptoms and signs were most characteristic, particularly in that many patients manifested identical symptoms, i.e., sensory disturbances, 100%; constriction of visual field, 100%; ataxia, 80.6–93.5%; dysarthria, 88.2%; auditory disturbances, 85.3%; tremor, 75.8%; and mental disturbances, 70.5% (Table I) (Tokuomi, 1960; Tokuomi et al, 1961; Study Group of Minamata Disease, 1968).

It became apparent from the autopsy that pathological changes were also quite unique. Lesions in the neurons of cerebral cortex were extensive. In the visual center (calarine region), motor and sensory centers (precentral and postcentral cortex), and auditory centers lesions were severe and selective. Morphological damage of cerebellar cortex was also observed. There was degeneration of granular cells, but Purkinye cells were relatively intact, indicating granular-cell type cerebellar atropy. In the peripheral nerves, destruction and demyelination in sensory nerve fibers were most conspicuous, and selective destruction of the posterior root and column also were observed (Takeuchi *et al.,* 1959; Study Group of Minamata Disease, 1968; Takeuchi, 1970).

The mercury levels detected in the patients' organs were also quite high. In 1961, 22.0–70.5 ppm of mercury was detected in the liver of MD patients, 21.2–140.0 ppm in the kidney, and 2.6–24.8 ppm in the brain. The highest value of 705 ppm was obtained from the hair of patients, and a maximal level (191 ppm) of mercury was detected in an inhabitant of the same area who did not show any symptoms (Kitamura *et al.*, 1960b; Study Group of Minamata Disease, 1968).

CONGENITAL (FETAL) MINAMATA DISEASE

In 1962 it was formally established that many cases of MD found in this area resulted from poisoning of the embryo through the placenta (Kitamura *et al.*, 1959, 1960b). The fetus was poisoned by MeHg when their mothers ingested contaminated fish (Harada, 1964; Takeuchi *et al.*, 1964; Matsumoto *et al.*, 1965).

Symptoms of patients were serious, and extensive lesions of the brain were observed. Mental retardation, primitive reflex, ataxia, growth disturbance, chorea, athetosis, dysarthria, and hypersalivation were recognized at high rates (Table II). In addition, paroxysmal symptoms, deformities of the limbs, strabismus, and pyramidal symptoms were commonly observed. Presently, 40 people born between 1955 and 1962 are officially recognized as victims of congenital MD (Fig. 4) (Harada, 1964, 1977).

Naturally, outbreaks of congenital MD had both geographic and chronological correspondence with adult cases of acute, acquired MD (Harada, 1964). In the areas where patients were most numerous, the incidence was 6.4%. The rate is considered abnormally high because the incidence of cerebral paralysis in Japan generally is 0.2–0.3%. The figure, however, includes only the serious cases and not the mild cases. It is assumed that MeHg affects embryos in varying degrees. In mild cases, motor dysfunction is light and the main symptom is retarded intelligence. In the areas where the disease occurred most frequently, the incidence of mental retardation (other than those

TABLE II
Frequency of Symptoms in Congenital Minamata Disease[a]

Symptom	1962		1971		1974	
	No. cases	%	No. cases	%	No. cases[b]	%
Mental retardation	17	100	25	100	37	100
Dysarthria	17	100	24	96	34	92
Extrapyramidal hyperkinesia	16	95	23	92	34	92
Cerebellar symptom	17	100	19	76	29	78
Deformities of limbs	17	100	21	84	23	62
Primitive reflexes	17	100	18	72	25	67
Strabismus	13	77	18	72	25	67
Hypersalivation	17	100	18	72	27	72
Paroxysmal symptom	14	82	9	36	13	35
Pathological reflex	12	75	12	48	17	45
Inhibited bodily growth	17	100	17	68	22	59

[a] Adapted from Harada (1976).
[b] Based on a total of 40 cases of congenital MD of which 3 died.

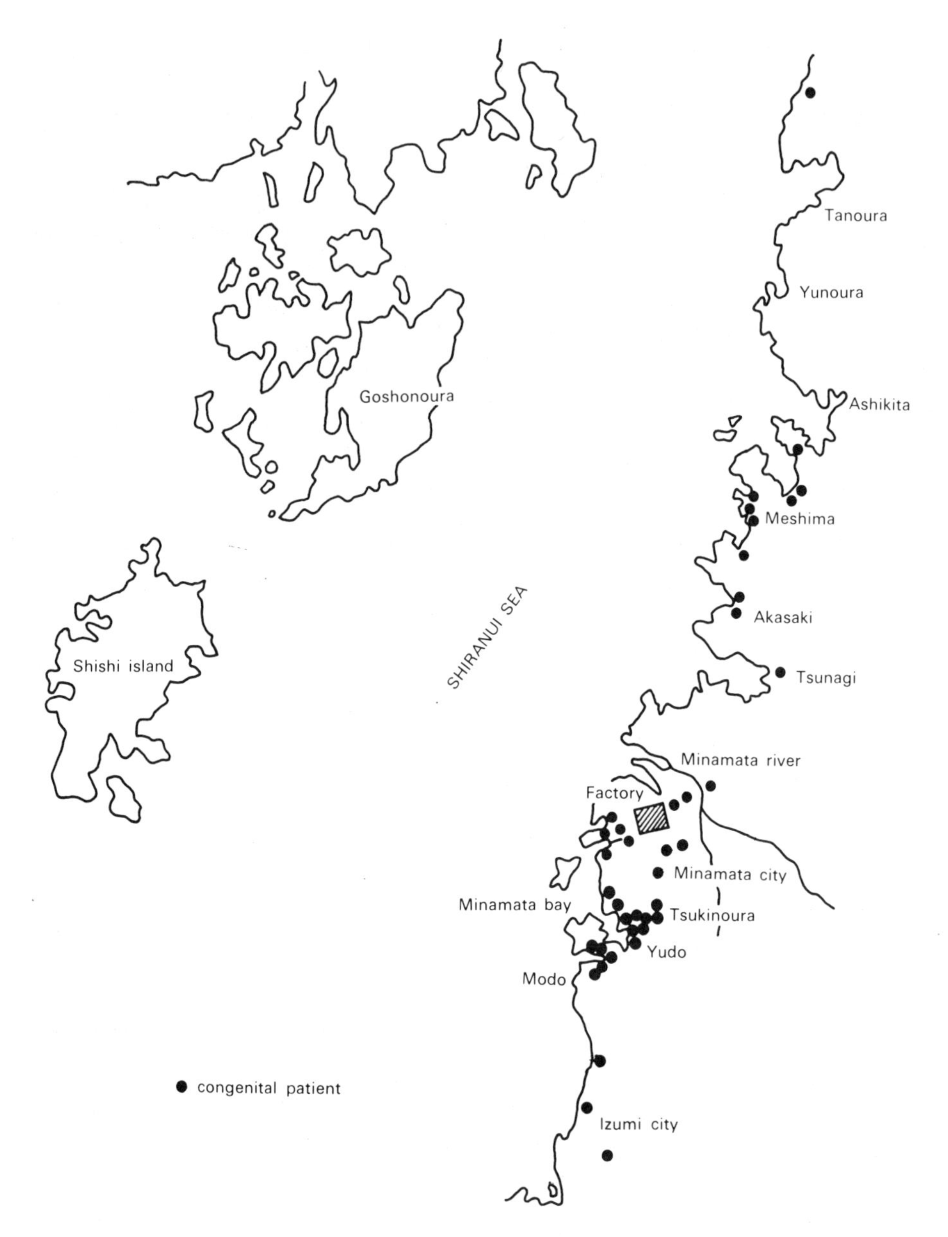

FIGURE 4. New distribution of congenital Minamata disease patients.

officially recognized as congenital MD patients) was 29.1% (Harada, 1964). Among school children in this area, 18% show impaired intelligence and 21% have sensory disturbances (Harada, 1976). Such incidence is clearly higher than the control area.

Symptoms in mothers of these patients are relatively mild compared to those of the children. However, some of the signs and symptoms of MeHg poisoning are seen in such mothers: sensory impairment in all; mild ataxia and hearing loss, 76%; and concentric constriction of vision, 48% (Table III) (Harada, 1964, 1976). Formerly, it was thought that the placenta had a protective role against MeHg. However, the accumulation of MeHg in the embryo is more marked than in the mother. Even if it does not cause serious damage to a mother, it may greatly affect the embryo. Judging from these facts, one can see that the contamination of food supply poses a grave problem for the embryo. It has also been established that MeHg affects the infant brain via mothers' milk (Deshimaru, 1969). Animal experiments have demonstrated that MeHg causes congenital deformities (Fuyuta *et al.*, 1978), although this has not been verified in studies on human beings. This is an important subject for further investigations.

According to an old custom in Japan, umbilical cords of each newborn infant are preserved by the parents. Noting this fact, the author analyzed the concentration of MeHg in the umbilical cords (Nishigaki *et al.*,1965). By means

TABLE III

Neurological Symptoms Observed Among Family Members of Acute and Subacute Minamata Disease Patients[a]

	No. of cases (% of total observed)	
Symptoms	Family members	Mothers
Sensory disturbance	115 (79)	21 (100)
extremities	109 (75)	21 (100)
perioral	44 (30)	4 (19)
Incoordination	94 (65)	16 (76)
ataxic gait	47 (32)	10 (48)
finger–nose test	69 (48)	11 (52)
adiadochokinesia	90 (62)	16 (76)
Romberg's sign	20 (14)	6 (29)
Dysarthria	65 (45)	9 (43)
Constriction of the visual field	53 (37)	10 (48)
Auditory disturbance	80 (55)	16 (76)
Tremor	51 (35)	6 (29)
Chorea and athetosis	2 (1)	0
Muscular weakness	37 (26)	7 (33)
Muscular rigidity or spasticity	43 (30)	7 (33)
Hyperreflexia	42 (30)	7 (33)
Pathologic reflexes	11 (8)	2 (10)
Muscular atrophy	16 (11)	0
Hypersalivation	8 (6)	0
Vegetative symptoms	43 (30)	12 (57)
Deformities of the extremities	27 (19)	1 (5)
Pain	52 (36)	8 (38)
Intelligence disturbance	67 (46)	14 (67)
Total examined	145 (100)	21 (100)

[a] Adapted from Harada (1972).

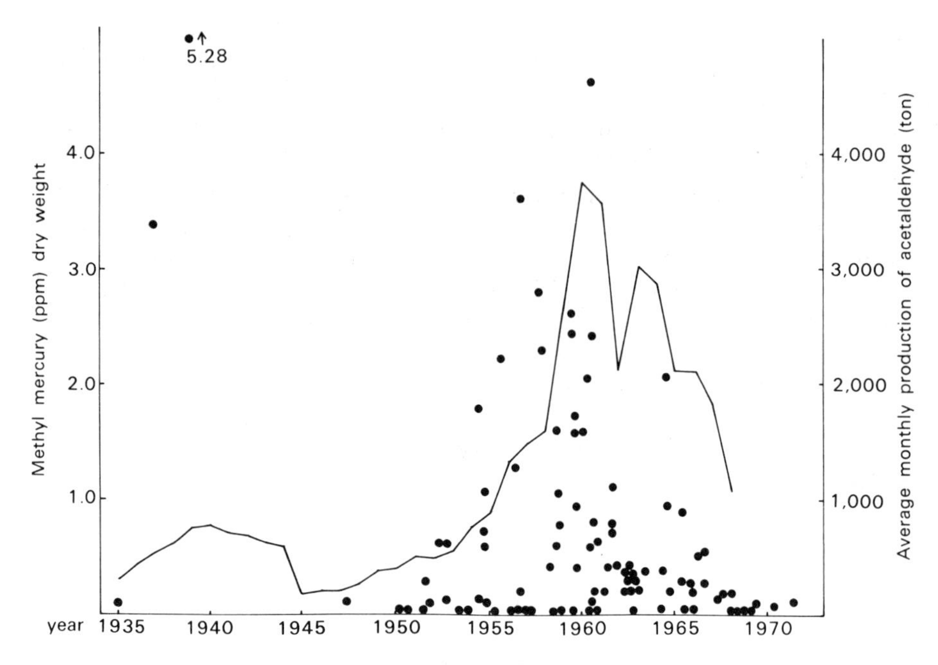

FIGURE 5. Comparative chart of acetaldehyde production rate and methyl mercury in umbilical cords. (—) Average monthly production of acetaldehyde (ton); (●) month of childbirth and methyl mercury content (ppm) dry weight. From Inshigaki and Harada (1975).

of this study, it was possible to learn the degree of MeHg contamination in patients prior to the discovery of MD. The result of the study indicated that MeHg content in the umbilical cords coincided with acetaldehyde production at Chisso Company and the degree of environmental contamination, as well as the amount of fish ingested by mothers (Fig. 5) (Harada, 1976). It is quite probable that similar phenomena would be observed in connection with other chemical substances such as PCB and BHC. In Japan, congenital PCB poisoning cases are recognized examples (Funatsu and Yamashita, 1972; Harada, 1976).

CHRONIC MINAMATA DISEASE

For 10 years the author and his associates have conducted the follow-up studies of MD for acute and typical cases. In one of the studies we found that certain symptoms were alleviated while others worsened (Harada, 1972b). There were also some cases where symptoms became atypical. Since the patients were selected according to very strict criteria in the beginning, mild and atypical cases as well as chronic or gradually deteriorating patients had been excluded from the diagnosis of MD. In order to depict the actual conditions, surveys on mothers of congenital MD patients and on family members of acute and typical cases were conducted. They are considered to be the most suitable subjects since they satisfy the identical epidemiological conditions as known MD patients. Among the subjects of the investigation, we observed a high incidence of neurological symptoms similar to those observed in MD (Tablc III) (Harada, 1972b, 1975).

In contrast to the patients with acute MD in the early stage of the outbreak, those whose symptoms deteriorated slowly (over a 10-year period) tended to show atypical signs. The author applied the term *chronic Minamata disease* to such patients. Where there was pollution of the environment (as in the case of Minamata) it was suspected that all of the inhabitants in the polluted area were contaminated. Starting in 1972, we attempted an extensive health examination of the inhabitants of a heavily polluted area (Minamata) and those of a slightly polluted area (Goshonoura). Among the inhabitants of both areas we found a number of people with such symptoms as sensory disturbances, ataxia, and constriction of visual field. In the most polluted area, one-third of the inhabitants (who had not been diagnosed as suffering from MD) were found to exhibit various neurological signs which are seen in patients with MD (Table IV) (Harada, 1975, 1978a). In addition, 50% of the inhabitants

TABLE IV

Frequency of Symptoms in Inhabitants of Minamata, Goshonoura, and Ariake Area

	No. of cases (% of total observed)					
Symptoms	Minamata (Tsukinoura, Yudo, Modo)		Goshonoura		Ariake (control)	
Sensory disturbance	260	(28.0)	132	(7.6)	94	(10.3)
Incoordination	228	(24.7)	193	(11.2)	122	(13.4)
ataxic gait	84	(9.0)	50	(2.9)	20	(2.2)
adiadochokinesia	171	(18.4)	101	(5.8)	50	(5.5)
finger–nose test	106	(11.4)	28	(1.6)	11	(1.2)
Dysarthria	114	(12.2)	63	(3.6)	18	(1.9)
Auditory disturbance	272	(29.2)	156	(9.0)	135	(14.9)
Constriction of the visual field	127	(13.7)	47	(2.7)	9	(0.9)
Tremor	94	(10.1)	87	(5.0)	27	(2.9)
Pathologic reflexes	56	(6.0)	34	(1.9)	21	(2.3)
Hemiplegia	17	(1.8)	9	(0.5)	10	(1.1)
Pain (limbs)	128	(13.7)	92	(5.3)	74	(8.1)
Epileptic seizure	26	(2.8)	19	(1.1)	8	(0.8)
Muscular atrophy	55	(5.9)	13	(0.7)	7	(0.8)
Parkinsonismus	23	(2.5)	13	(0.7)	7	(0.8)
Deformities of limbs	81	(8.7)	104	(6.0)	73	(7.8)
Intelligence disturbance	211	(22.7)	178	(10.3)	98	(10.8)
Hypertension	218	(23.4)	237	(27.5)	180	(19.1)
Total	928	(100.0)	1723	(100.0)	904	(100.07)

[a] Minamata percentages exclude the 3.9% of patients already verified in the Tsukinoura, Yudo, and Modo areas of Minamata.

[b] Second Minamata Disease Research Group were examined in 1973 (Harada, 1978b).

of the heavily polluted area complained of various symptoms seen in MD such as weakness, numbness, pains of the limbs, leg cramps, unsteady gaits, clumsiness in movement of fingers, dizziness, forgetfulness, and tinnitus. Compared to control subjects in nonpolluted area, there was a much higher incidence of these neurological signs/symptoms that lead us to believe that MeHg was a causative factor. MeHg poses a health hazard of varying severity and there are various types of MD patients besides those acute typical cases. Some patients have been confirmed to gradually manifest clinical symptoms long after the mercuric contamination had been stopped. The delayed effects of residual MeHg remaining in the body are significant clinical problems. Based on the results of such clinical studies, the governmental authorities have expanded the criteria for MD to salvage "less typical cases." The number of officially recognized MD patients today totals 1700, and about 300 of them are dead.

SAFETY OF MERCURY

According to calculations based on data from clinical cases in Minamata, Niigata, and Iraq, when an adult accumulates 100 mg of MeHg in his body, he starts manifesting signs of MeHg poisoning (Bakir *et al.*, 1973; Rustam and Hamdi, 1974). If the accumulated amount exceeds 150 mg/51 kg, the manifestations of serious typical MD begin to appear; and with 25 mg/51 kg, early symptoms of MD, i.e., sensory disturbances, appear (Bakir *et al.*, 1973; Kitamura, 1974). A value of more than 50 ppm of mercury detected in the hair, or more than 200 ppb in the blood, is regarded as a critical point in a provisional clinical criterion beyond which MeHg poisoning may manifest as clinical symptoms (Friberg, 1971; Kitamura, 1974).

A joint committee established by the Food and Drug Administration (FDA) and the World Health Organization (WHO) has put the provisional tolerable weekly intake of MeHg at 0.2 mg based on the above-mentioned clinical studies. For a person of 50 kg body weight, the tolerance level would be 0.17 mg; thus the provisional tolerable daily intake would be 0.025 mg (0.5 μg/kg). This value is defined as the *no-effect level.* Based on these results, the regulation values for mercury contained in fish was set at 1.0 ppm in Sweden, 0.5 ppm in Canada and the U.S., and 0.4 ppm in Japan. In Sweden, it is recommended that the intake of fish containing 1.0 ppm mercury be restricted to 200–300 g a week (Friberg, 1971). In reaching decisions on such values of the safety level, the following problems were encountered. Calculations were based on the typical and acute cases, and the biological half-life of MeHg was set at 70 days (Åberg, 1969; Bakir *et al.*, 1973; Kitamura, 1974). Therefore, in the event of heavy pollution over a relatively short period, the safety levels may be appropriate for acute cases. But where the contamination continued over a long period of time involving the chronic type, congenital or suckling cases, or the cases accompanied by complications, the standard safety level is not necessarily applicable. In contrast to an assumption that there is a threshold value for the onset of the disease, some advocate a *no-threshold* theory:

> There may be a certain threshold value for already recognized signs and symptoms to appear, but there is practically no threshold for other subtle and often unrecognized toxic effects of MeHg to affect an individual in ways which escape currently employed methods of evaluation (Löfroth, 1969).

According to the latter judgment, even a small amount of contamination may induce a disease such as malignant neoplasma if exposure lasts for a long period of time. It does not set the accumulation of a certain value as an absolute prerequisite of the disease and in this point the effect of MeHg is quite similar to that of radiation. This may be very useful for clarifying the mechanism by which chronic MD occurs, even though it may not have been proven sufficiently.

As explained earlier, the major symptoms of MD are observed in the central nervous system. However, many inhabitants in the polluted areas developed complications of the liver, kidney, and heart as well as hypertension. It is not clear whether such ailments are merely complications or deuteropathy caused by MeHg. In the future, the effect of MeHg on organs other than the nervous system, particularly in cases of a long-term contamination, should be studied. Recently a few reports have appeared on the systemic disease caused by MeHg (Shiraki *et al.,* 1971; Iesato et al, 1977; Harada, 1978a).

REFERENCES

Åberg, B., Ekman, L., Falk, R., Greitz, U., Persson, G., and Snihs, J. O., 1969, Metabolism of methyl mercury (^{203}Hg) compounds in man, *Arch. Environ. Health* **19:**478.

Bakir, F., Damluji, S. F., Amin-Zaki, L., Murtadha, M., Khalidi, A., Al-Rawi, N. Y., Tiriti, S., Dhakir, H. I., Clarkson, T. W., Smith, J. C., and Doherty, R. A., 1973, Methyl mercury poisoning in Iraq, *Science* **181:**230.

Deshimaru, M., 1969, Electron microscopic studies on experimental organic mercury poisoning in nursling rat brain (Jpn. ed.), *Psychiatr. Neurol. Jap.* **71:**506.

Friberg, L. (ed.), 1971, *Methyl mercury in fish, Nord. Hyg. Tidskr. Suppl. 4,* Stockholm, Sweden.

Fujiki, M., 1972, The transitional condition of Minamata Bay and the neighbouring sea polluted by factory wastewater containing mercury, Reprinted at the 6th International Water Pollution Research, June 18–23.

Funatsu, I., and Yamashita, F., 1972, Polychlorbiphenyls (PCB) induced fetopathy, I. Clinical observation, *Kurume Med. J.* **19:**43.

Futatsuka, S., Ueda, A., Teruya, H., Ueda, T., Nomura, S., and Hirano, T., 1974, Eating habits and labor of fishermen's families in Shiranui area, *Jpn. J. Pub. Health* **21:**825.

Fuyuta, M., Fujimoto, T., and Hirata, S., 1978, Embryotoxic effects of methylmercuric chloride administered to mice and rats during organogenesis, *Teratology* **18:**353.

Harada, M., 1964, Neuropsychiatric disturbances due to organic mercury poisoning during the prenatal period (Jpn. ed.), *Psychiat. Neurol. Jpn.* **66:**426.

Harada, M., 1972a, Minamata Byo (Jpn. ed.), Iwanami-Shoten, Tokyo.

Harada, M., 1972b, Clinical studies on prolonged Minamata disease (Jpn. ed.), *Psychiat. Neurol. Jpn.* **74:**668.

Harada, M., 1975, *Minamata Disease,* words and photographs by E. Smith and A. M. Smith, p. 51, Holt, Reinhardt and Winston, New York.

Harada, M., 1976, Intrauterine poisoning: Clinical and epidemiological studies and significance of the problem, *Bull. Inst. Constit. Med. Kumamoto Univ. Suppl.* **25:**1.

Harada, M., 1977, Congenital Minamata disease. Intrauterine methyl mercury poisoning, *Teratology* **18:**285.

Harada, M., 1978a, Methyl mercury poisoning due to environmental contamination "Minamata disease," in: *Toxicity of Heavy Metals in the Environment,* Part 1 (F. W. Oehme, ed.), p. 261, Marcel Dekker, New York and Basel.

Harada, M., 1978b, Minamata disease as a social and medical problem, *Jpn. Q.* **25:**20.

Iesato, K., Wakashi, M., Wakashi, Y., and Togo, S., 1977, Renal tubular dysfunction in Minamata disease, *Ann. Intern. Med.* **86:**731.

Kitamura, S., Hirano, Y., Noguchi, Y., Kojima, T., Kakita, T., and Kuwaki, H., 1959, The epidemiological survey on Minamata disease (No. 2) (Jpn. ed.), *J. Kumamoto Med. Soc.* **33(Suppl. 3):**569.

Kitamura, S., Kakita, T., Kojoo, J., and Kojima, T., 1960a, A supplement to the results of the epidemiological survey on Minamata disease (No. 3) (Jpn. ed.), *J. Kumamoto Med. Soc.* **34(Suppl. 3):**476.

Kitamura, S., Ueda, K., *et al.,* 1960b, Chemical examination of Minamata disease (No. 5) (Jpn. ed.), *J. Kumamoto Med. Soc.* **34(Suppl. 3):**593.

Kitamura, S., 1974, Accumulation of mercury in human body, (Jpn. ed.), *Adv. Neurol. Sci.* **18:**825.

Löfroth, G., 1969, Toxicity of mercury compounds in the environment, *Kagaku (Science)* **39:**592. (Jpn. ed., translated by J. Ui)

Matsumoto, H., Koya, G., and Takeuchi, T., 1965, Fetal Minamata disease. A neuropathological study of two cases of intrauterine intoxication by methyl mercury compound, *J. Neuropathol. Exp. Neurol.* **24:**563.

McAlpine, D., and Araki, S., 1958, Minamata disease. An unusual neurological disorder caused by contaminated fish, *Lancet* **20(Sept.):**629.

Nishigaki, S. M., and Harada, M., 1975, Methyl mercury and selenium in umbilical cords of inhabitants of the Minamata area, *Nature* **258:**324.

Rustam, H., and Hamdi, T., 1974, Methyl mercury poisoning in Iraq. A neurological study, *Brain* **97:**499.Study Group of Minamata Disease, 1968, *Minamata Disease,* Kumamoto University, Kumamoto City, Japan.

Sera, K., Sato, A., *et al.,* 1957, An experimental study on unknown disease of Minamata area (Jpn. ed.), *J. Kumamoto Med. Soc.* **31(Suppl. 2):**307.

Shiraki, H. and Takeuchi, T., 1971, Minamata disease, in: *Pathology of the Nervous System (II)* (J. Minckler, ed.) McGraw-Hill, Inc., New York.

Study Group of Minamata Disease, 1968, *Minamata Disease,* Kumamoto University, Kumamoto City, Japan.

Tokuomi, H., 1960,Clinical observation and pathogenesis of Minamata disease (Jpn. ed.) *Psychiat. et Neurol. Jap.* **62:**1816.

Tsubaki, T., 1968, Organic mercury poisoning in Agano area, (Jpn. ed.) *Clin. Neurol.* **8:**511.

13

MERCURY IN FOOD

Laman Amin-Zaki

INTRODUCTION

All foodstuffs contain small amounts of mercury, the level of which depends on the environment and varies from one area to another (Swedish Expert Group, 1971). But food may contain mercury in levels very much higher than the basal levels through contamination. Mercury-containing fungicides are used to treat seeds that can erroneously be used for food, or can be responsible for accumulation of mercury in celestial food chains through seed-eating birds and animals. Even more important is the methylation of inorganic mercury in the sediment in bodies of water, and its accumulation by ascending the food chain until it enters the fish eaten by man. Fish consumption is the main source of human exposure to methyl mercury (MeHg) (World Health Organization, 1976), and is the reason why mercury as an environmental contaminant has become a major concern of modern public health.

Metallic mercury and its organic and inorganic compounds are all toxic to the human being, though there are substantial differences between and within these groups (Swedish Expert Group, 1971). Poisoning with mercury compounds used to be considered mainly as an occupational hazard or a sporadic phenomenon in children until the Minamata disaster. In the 1950s, around Minamata Bay in Japan, fishermen and their families were stricken with a mysterious epidemic of neurological disease which caused structural damage to the brain (Katsuna, 1968). A number of fetal afflictions were reported (Harada, 1968). Fish, sea birds, and household cats showed signs of a similar disease. In the 1960s Minamata disease was discovered to be due to the consumption, by local residents and animals, of fish and shellfish from the bay contaminated by MeHg (Tsubaki *et al.*, 1967; Katsuna, 1968). The source of mercury was traced to the effluent from a local factory that used mercuric chloride as a catalyst to manufacture vinyl chloride, and routed its

Laman Amin-Zaki • Ministry of Health, Central Hospital, P.O. Box 233, Abu Dhabi, United Arab Emirates.

TABLE I
Some Outbreaks of Alkylmercury Poisoning

Geographical location	Year	Vehicle of exposure	No. of cases	Reference
Minamata (Japan)	1953–1960	Fish	134	Katsuna, 1968
Niigata (Japan)	1965	Fish	520	Tsubaki *et al.*, 1976
Pakistan	1960	Seed grain	100	Haq, 1963
Guatemala	1963–1965	Seed grain	45	Ordonez *et al.*, 1966
U.S.S.R.	1970	Sunflower seed	70	Shostow and Tsegeiowa, 1970
Iraq	1971–1972	Seed grain	6530	Bakir *et al.*, 1973

discharge into the bay. The attention of scientific circles was alerted to this important hazard. Yet careless discharge of mercury into the environment was allowed to continue until relatively recently. A second, larger outbreak occurred at Niigata in Japan (Tsubaki and Irukayama, 1977). Many other outbreaks have occurred in different parts of the world from similar or different sources (Table I). Sweden recognized the danger earlier than most industrial countries and took measures to minimize contamination. In 1967 the Swedish Medical Board banned the sale of fish from about 40 lakes and rivers because of the high content of MeHg (Goldwater, 1971). In the 1970s the full significance of the problem was recognized and actions were taken to control the discharge of mercury-containing wastes into lakes and streams.

SOURCES OF MERCURY IN THE ENVIRONMENT

Natural Sources

Like other elements, mercury is a natural component circulating between rocks,soil, water, atmosphere, and biosphere, including the tissues of plants and animals (Swedish Expert Group, 1971). The earth's crust contains approximately 50 ng/g, while soil, especially organic soil, can contain much higher levels (Löfroth, 1972). Volatilization and natural weathering processes are responsible for the major significant transfer of mercury between atmosphere and ground. Mercury is also released into water by natural erosion of soil. Local variations in mercury levels may occur due to uneven geochemical distribution (Löfroth, 1972).

Agricultural Sources

Mercury compounds are used for seed dressing to protect grain from fungi. At first inorganic mercury compounds were used, but since 1940 alkylmercury compounds were introduced and used on a large scale (Swedish Expert Group, 1971). The dressings were supposed to leave only small residues in the crop when the pesticide had been used according to good agricultural practice (Löfroth, 1972). But the use of these dressings has been responsible for accumulation of mercury in celestial food chains through seed-eating birds and animals, raising the mercury levels in their predators (WHO, 1976). Human poisoning, due to ingestion of treated seeds or products of animals fed on treated seed, has occurred on a limited scale in highly developed countries like the U.S. (Pierce *et al.,* 1972), and disastrously in some developing countries as Iraq (Damluji and Al-Tikriti, 1972; Bakir *et al.,* 1973). In Sweden the use of alkylmercury seed dressings has been prohibited since 1966. As a result, food levels of mercury fell by a factor of three (World Health Organization, 1976); for example, the mercury level in hen eggs fell from 29 mg/kg prior to the ban to 9 mg/kg following the ban. In Alberta the hunting of pheasants and partridges was prohibited in 1969 due to these birds being contaminated with mercury residues, and in 1973 the registration of mercurial fungicides for treatment of cereal grains was discontinued in Canada (Wheatley, 1979).

Industrial Sources

Mercury and its compounds are commonly used in industry. Mercury-containing effluent from industries such as the pulp and paper industry and the chloralkaline industry has been a major contributor to the mercury content of extensive water areas. These industries, using the mercury-cell process, discharge inorganic mercury in liquid effluent, in air emissions, and in solid waste. It used to be thought that inorganic mercury compounds disposed in the water would remain relatively harmless lying on the sediment beds. Jensen and Jernelov (1969) have demonstrated that inorganic mercury is methylated by the action of microorganisms that are richly present in river and lake sediments. The MeHg thus produced is then incorporated into marine organisms and undergoes bioaccumulation up the food chain to reach a significant level, almost entirely in the MeHg form (Swedish Expert Group, 1971), in the predatory fish and sea mammals often used for human consumption. This is believed to be a key step in the transport of mercury to man (Fig. 1; WHO, 1976). This finding has alerted the interest of toxicologists in industrial countries who started to examine fish for their mercury content from different water bodies. In Sweden, it was found that pike caught from noncontaminated waters had a level that ranged from 0.5–0.2 mg/kg, while in many contaminated waters the level reached 1.0 mg/kg, and in ten bodies of waters levels of over 5 mg/kg were reported (Swedish Expert Group, 1971). In the fish of Minamata Bay, where the first Japanese epidemic of MeHg poisoning occurred, the level of MeHg was 10–30 ppm/g wet weight (Irukayama, 1977). In some industrial parts of Canada levels up to 24 μg/g were found. Health authorities in the U.S. have defined the permissible limit for the commercial

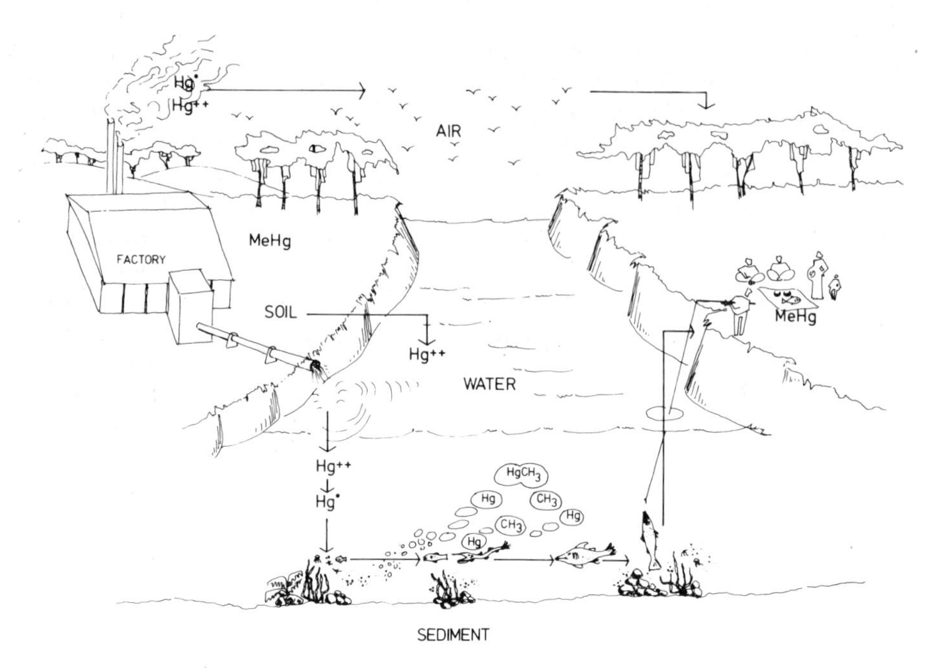

FIGURE 1. Environmental cycling of mercury.

sale of edible fish at 1.0 ppm MeHg, and in Canada the limit is 0.5 ppm (Wheatley, 1979).

Other industrial activities that account for the release of mercury to the environment include burning of fossil fuel, production of cement and phosphate, and smelting of metals (World Health Organization, 1976). Another role that industry plays in raising the level of mercury in the environment is through increasing acidity in a water body, which enhances mercury uptake by fish. In addition, the production of air-borne products by distant industries results in "acid rain," which may contribute to the rise of mercury levels in fish in water systems remote from sources of contamination (Wheatley, 1979).

METABOLISM OF METHYL MERCURY

Mercury compounds can be absorbed by inhalation, through the skin, and by ingestion. In man, the degree of absorption in food depends upon the type of mercury compound. Of ingested inorganic mercury, only 15% is absorbed, compared with an absorption of 95% of MeHg from the intestinal tract. The absorption is rapid and peak levels are reached within a few hours (WHO, 1976). In the blood the greater proportion of MeHg is carried in the red cells, the cell to plasma ratio being 10 (World Health Organization, 1976). It is then distributed almost uniformly through the body, crossing the blood–brain barrier and placenta rapidly. There is evidence to indicate that later there will be a further distribution of MeHg to the brain (Swedish Expert Group, 1971), the blood–brain concentration ratio estimated to be 1 : 6.

Transplacental transfer of MeHg takes place readily. The infant's blood levels taken at birth and up to 4 months after birth are substantially higher than corresponding maternal levels. Figure 2 shows the mercury concentrations in an infant–mother pair observed during the Iraq epidemic of 1971–1972.

MeHg is secreted in breast milk, the average level being 5% of the mean maternal blood concentration with individual variations (Fig. 3). Continued ingestion of milk containing MeHg will tend to retard the rate of decline of body mercury in infants prenatally exposed to MeHg (Amin-Zaki *et al.*, 1974a). Suckling infants may accumulate high blood levels of MeHg when their mothers are heavily exposed. Infants exposed only through mother's milk during the Iraq outbreak of 1971–1972 have accumulated MeHg levels in excess of 1000 ppb (Amin-Zaki *et al.*, 1974b).

About 90% of elimination of MeHg takes place through the biliary system, the gastrointestinal tract and the feces, and only a small part is excreted via

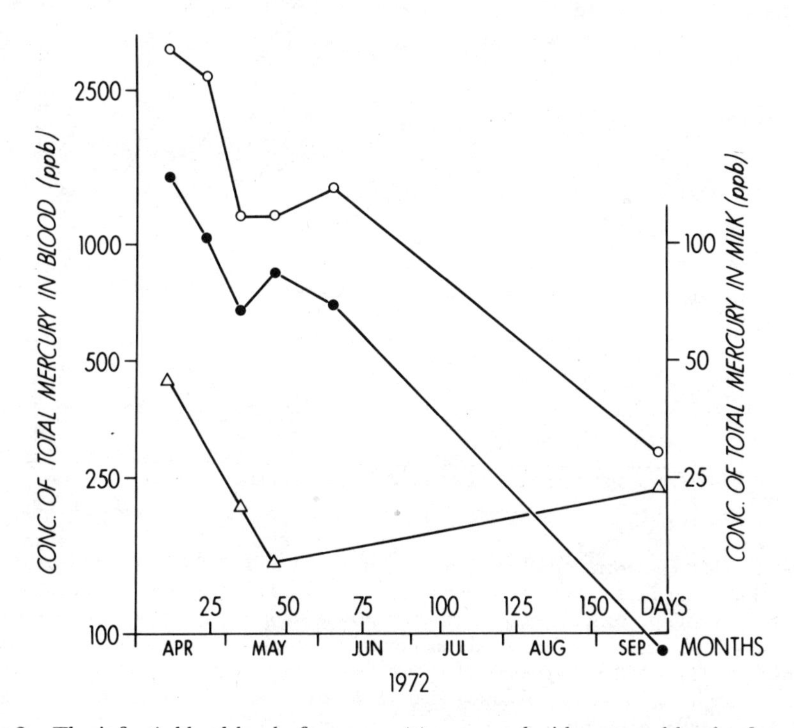

FIGURE 2. The infant's blood level of mercury (○) compared with maternal levels of mercury (●) and milk (△). The infant was born February 11, 1972. Adapted from Amin-Zaki *et al.* (1974a).

the urine (World Health Organization, 1976). Elimination via the hair may also contribute to total excretion (Swedish Expert Group, 1971). The concentration of mercury in the hair and blood of people exposed to MeHg declines in a closely parallel fashion indicating a constant relationship (Fig. 3). Thus, from logitudinal analysis of hair as depicted in Fig. 3, the blood level may be recapitulated. The average ratio of total mercury in hair to total mercury in blood concentration is 250 (World Health Organization, 1976). In man, the biological half-time of MeHg, which means the time taken for the body burden to fall by one-half, is approximately 70 days.

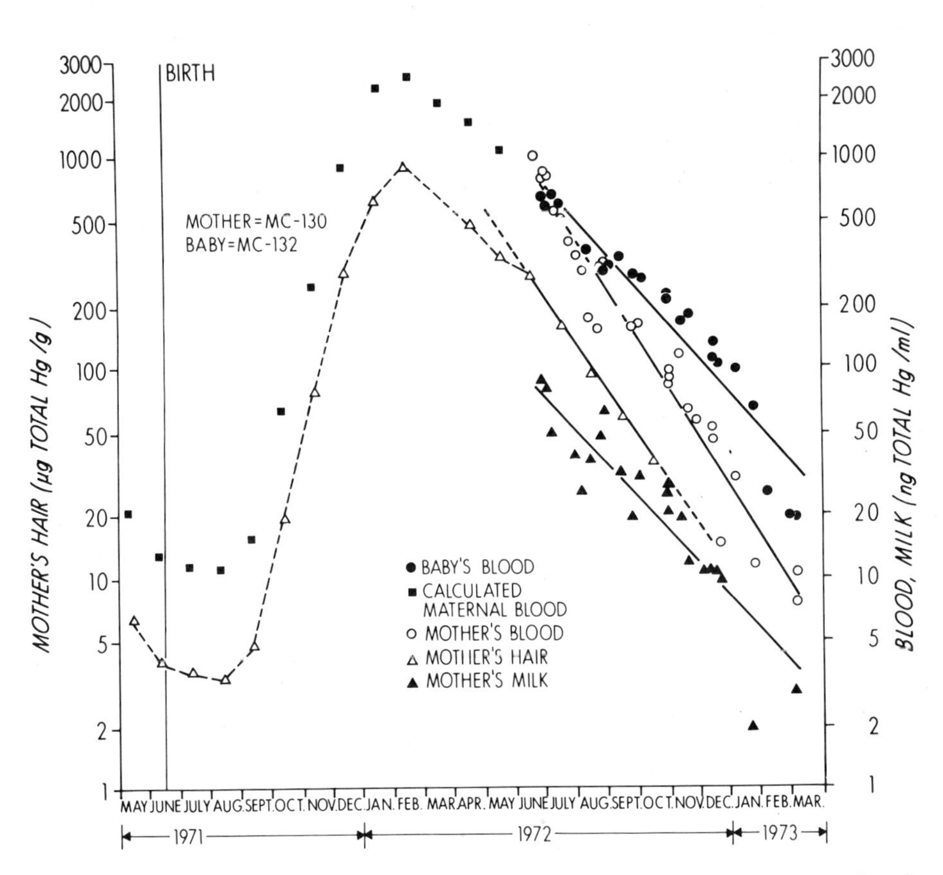

FIGURE 3. The concentration of total mercury in breast milk in relation to concentrations in maternal blood and head hair and in the infant's blood. The hair samples were analyzed in cm sections as measured from the scalp. Each cm represents approximately one month's growth of hair. Thus, the mercury concentration was plotted according to the month when the segment was formed. The calculated maternal blood values (solid squares) were obtained from the corresponding hair concentrations assuming a hair to blood ratio of total mercury of 250 : 1. Adapted from Amin-Zaki *et al.* (1976).

EFFECTS OF METHYL MERCURY ON MAN—SOME EPIDEMIOLOGICAL ASPECTS

A World Health Organization (WHO) expert group (WHO, 1976) reviewed available data on exposed populations and concluded that signs and symptoms of MeHg poisoning will appear in the most sensitive individual adults at blood concentrations of 200–500 ppb. The same group identified three major patterns of population exposure based on the level of mercury in the diet and the duration of exposure. The first category is defined by high exposure for a relatively brief period, for example, a population in Iraq having a daily intake of mercury reaching over 200 μg/kg for a period of 1–2 months (Bakir *et al.,* 1973). The second category is defined by exposure to lower daily doses of about 30 μg/kg for several months or years as in a population in Japan (Kitamura, 1971). The third category consists of fish-eating populations having a daily intake of mercury usually below 5 μg/kg with exposure lasting for a life time, for example, populations eating freshwater fish in Sweden (Birke *et al.,* 1972), Peru (Turner *et al.,* 1974), and Samoa (Marsh *et al.,* 1974). In Canada, a fourth category of seasonal exposure was described where a population had an intermittent but repetitive exposure to a wide range of daily intakes (Wheatley, 1979). Symptoms prevalent in the four categories of exposure were mainly those of affliction of the central nervous system that are not specific to poisoning with MeHg.

The damage caused by MeHg could be either reversible or irreversible (WHO, 1976). This is dependent on the dose and duration of exposure. At higher doses and longer periods of exposure, the damage can surpass the stage of reversibility (WHO, 1976). During the disastrous outbreak of MeHg poisoning in Iraq in the winter of 1971–1972 Damluji *et al.* (1972) noticed that, unlike the popular belief at the time, many of the clinical manifestations in the patients were gradually improving and milder cases were losing the disease completely.

The pathological picture is that of diffuse involvement of the central nervous system. The picture first described by Hunter and Russell (1954) was that of a bilateral cerebral cortical atrophy affecting particularly the visual cortex and the cerebellar cortex.

CLINICAL MANIFESTATIONS

Older Children and Adults

The main symptoms and signs of MeHg poisoning observed during the Iraqi outbreak were: sensory disturbances, ataxia and dysarthria, constriction of visual fields, and varying degress of motor weakness. In spite of individual variations, the severity of the poisoning is related to the body concentration of mercury. Clinical symptoms and signs seem to be associated with a mercury concentration in the brain of 5–10 mg/kg (Löfroth, 1972).

The onset is insidious following a latent period of several weeks. During the Iraqi outbreak of 1971–1972 the severity of each patient's clinical manifestation was classified into mild, moderate, severe, and very severe as adopted from Damluji (1976). Mild cases were those in whom the findings were mainly subjective and included such symptoms as malaise, headache, paresthesia, motor weakness, blurring of vision, and gastrointestinal disturbances. Moderate cases had mild ataxia and dysarthria, paresis, tremor, and some visual and auditory impairment. Severe cases had stupor, gross ataxia and dysarthria, spastic paralysis, severe and bilateral visual and auditory impairment, and mental impairment. Very severe cases were physically and mentally incapacitated. They had a combination of the following: blindness, deafness, loss of speech, decerebrate posture, severe spastic paralysis, and coma sometimes leading to death. After several weeks of the initial illness, the consumption of MeHg having stopped, the patient's condition began to improve. The improvement is inversely related to the severity of signs and symptoms (Amin-Zaki *et al.*, 1978). Many patients could resume their work or studies but they were usually left with some evidence of permanent affliction of the central

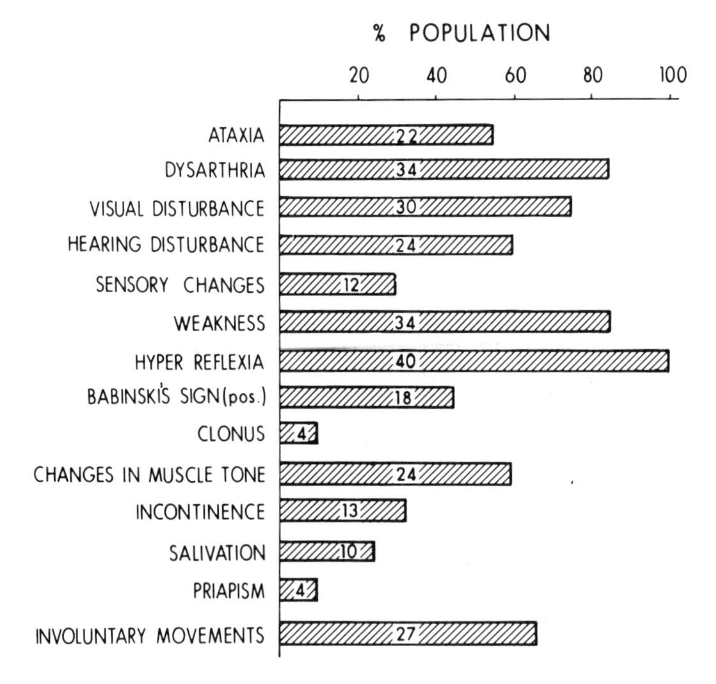

FIGURE 4. Prevalence of neurological manifestations on admission to hospital in 40 children aged 2–16 years who experienced high exposure to methyl mercury during the Iraqi epidemic of 1971/72. Adapted from Amin-Zaki *et al.* (1978).

nervous system, the least being brisk deep tendon reflexes and a bilaterally positive Babinski sign. Figure 4 gives the clinical manifestations in 40 children aged 2–16 years at the outset of their illness.

Prenatal Exposure and Exposure of the Sucklings

The Japanese and Iraqi experiences have demonstrated the extreme susceptibility of the human fetal brain to the neurotoxic effects of MeHg. MeHg concentration in cord blood is about 20% higher than maternal blood, probably because fetal hemoglobin binds more readily with MeHg than adult hemoglobin (MacGregor and Clarkson, 1974). The fetal brain contains higher concentrations of mercury than the maternal brain and displays a different pattern of distribution. Clinically, Amin-Zaki *et al.* (1974a) and Marsh *et al.* (1980) have demonstrated that high levels of MeHg in the mother can be the cause of cerebral palsy in the offspring. Marsh *et al.* (1979) have proved that Iraqi children with exposure to levels of maternal hair as low as 67.6 μg/g (approximately 200 ppb blood level) showed significant mental and motor abnormalities, while their mothers had mild transient symptoms. This indicated that the developing nervous system is specially sensitive to the effects of MeHg.

MeHg is secreted in mother's milk and infants on breast feeding will have raised blood levels throughout the period of suckling. In a group of 30 infants thus exposed in Iraq, follow-up examinations for several years have demonstrated a delay in achieving the developmental milestones, especially language development, with or without neurological deficit (Amin-Zaki *et al.*, 1980). Some experiments have indicated the potential for genetic effect of MeHg in man (Skerfving *et al.*, 1970). If proved, this and the known fetal sensitivity could be a source of great concern in communities who have an endemic elevation of body burdens of MeHg levels.

REFERENCES

Amin-Zaki, L., Elhassani, S. B., Majeed, M. A., Clarkson, T. W., Doherty, R., and Greenwood, M. R., 1974a, Intrauterine methyl mercury poisoning in Iraq, *Pediatrics* **45:**582.

Amin-Zaki, L., Majeed, M. A., Clarkson, T. W., Elhassani, S. B., Doherty, R., and Greenwood, M. R., 1974b, Studies of infants postnatally exposed to methyl mercury, *J. Pediatr.* **58(1):**81.

Amin-Zaki, L., Elhassani, S. B., Majeed, M. A., Clarkson, T. W., Doherty, R. A., Greenwood, M. R., and Giovanoli-Jakubczak, T., 1976, Perinatal methyl mercury poisoning in Iraq, *Am. J. Dis. Child.* **130:**1070–1076.

Amin-Zaki, L., Majeed, M. A., Clarkson, T. W., and Greenwood, M. R., 1978, Methyl mercury poisoning in Iraq children: Clinical observations over two years, *Br. Med. J.* **1:**613.

Amin-Zaki, L., Elahssani, S., Majeed, M. A., Clarkson, T. W., Doherty, R. A., and Greenwood, M. R., 1980, Methyl mercury poisoning in mothers and their suckling infants, in: *Mechanisms of Toxicity and Hazard Evaluation* (B. Holmstedt, R. Lauwerys, M. Mercier, and M. Roberfroid, eds.), pp. 75–79, Elsevier/North Holland, Amsterdam.

Bakir, F., Al-Damluji, S. F., Amin-Zaki, L., Murtadha, M., Khalidi, A., Al-Rawi, N., Tikriti, S., Dhahir, H. I., Clarkson, T. W., Smith, J. C., and Doherty, R. A., 1973, Methyl mercury poisoning in Iraq, *Science* **181:**230.

Birke, G., Hagman, D., Johnels, A. G., Plantin, L. O., Sjostrand, B., Skerfving, S., Westermark, T., and Osterdahl, B., 1972, Studies on humans exposed to methyl mercury through fish consumption, *Arch. Environ. Health* **25:**77.

Damluji-al, S. F., 1976, Intoxication due to alkylmercury-treated seed—1971–72 outbreak in Iraq: Clinical aspects, *Bull. WHO Suppl.* **53:**65.

Danluji-al, S. F., and Al-Tikriti, S., 1972, Mercury poisoning from wheat, *Br. Med. J.* **2:**804.

Damluji-al, S. F., Amin-Zaki, L., and Elhassani, S. B., 1972, Mercury in the environment, *Br. Med. J.* **4:**489.

Goldwater, L. J., 1971, Mercury in the environment, *Sci. Am.* **224:**15.

Haq, I., 1963, Agrosan poisoning in man, *Br. Med. J.* **1:**1579.

Harada, Y., 1968, Clinical investigations on Minamata disease. Congenital (or fetal) Minamata disease, in: *Minamata Disease, Study Group of Minamata Disease* (M. Katsuna, ed.), pp. 93–117, Kumamoto University, Japan.

Hunter, D., and Russell, D. S., 1954, Focal cerebral and cerebellar atrophy in a human subject due to organic mercury compounds, *J. Neurol. Neurosurg. Psychiatry* **17:**235.

Irukayama, K., 1977, Case history of Minamata, in: *Minamata Disease* (T. Tsubaki and K. Irukayama, eds.), pp. 2–56, Elsevier Scientific Pub., New York.

Jensen, S., and Jernelov, A., 1969, Biologic methylation of mercury in aquatic organisms, *Nature* **223:**753.

Katsuna, M. (ed.), 1968, *Minamata Disease Study, Group of Minamata Disease,* Kumamoto University, Japan.

Kitamura, S., 1971, as cited by Tsubaki, T., Shirakawa, Hirota, K., Kondo, K., Sato, T., and Kanabayashi, in: *Minamata Disease* (T. Tsubaki and K. Irukayama, eds.), pp. 83–84, Elsevier Scientific Pub., New York.

Löfroth, G., 1972, The mercury problem: A review at midway, in: *Trace Substances in Environmental Health-VI* (D. D. Hemphill, ed.) pp. 66–69, University of Missouri, Columbia, Missouri.

MacGregor, J., and Clarson, T. W., 1974, Distribution, tissue binding and toxicity of mercurials, in: *Protein–Metal Interactions* (M. Friedman, ed.), pp. 463–502, Plenum Press, New York.

Marsh, D. O., Myers, G. J., Clarkson, T. W., Amin-Zaki, L., Tikriti, S., Majeed, M. A., and Dabbagh, A. R., Dose–response relationship for human fetal exposure to methyl mercury, paper presented at the International Congress of Neurotoxicology September 27–30, 1979, Varese, Italy.

Marsh, D. O., Turner, M. D., Smith, J. C., Choi, J. W., and Clarkson, T. W., 1974, Methyl mercury in human populations eating large quantities of marine fish. II. American Samoa, in: *Proceedings of the First International Congress on Mercury, Barcelona, Vol II,* pp. 235–239, Fabrica National de Moneda y Timbre, Madrid, Spain.

Marsh, D. O., Myers, G. J., Clarkson, T. W., Amin-Zaki, L., Tikriti, S., Majeed, M. A., and Dabbagh, A. R., 1980, Fetal methyl mercury poisoning: Clinical and toxicological data on 29 cases, *Ann. Neurol.* **7:**348.

Ordonez, J. V., Cavillo, J. A., Miranda, M., and Gale, J. L., 1966, Epidemiological study of a disease in the Guatemalan Highlands believed to be Encephalitis, *Bol. Ofic. Sanit. Panam.* **60(6):**510.

Pierce, P. E., Thompson, J. F., Lifofsky, W. H., Nickey, L. M., Barthel, W. F., and Hinman, A. R., 1972, Alkylmercury poisoning in humans: Report of an outbreak, *J. Am. Med. Assoc.* **220:**1439.

Shostow, W. J., and Tsegaiowa, S. I., 1970, Acute Granosan intoxication, *Kazan. Med. Zh.* **2:**78 (translated by San Zare Assoc., Philadelphia).

Skerfving, S., Hansson, A., and Lindstem, J., 1970, Chromosome breakage in human subjects exposed to methyl mercury through fish consumption, *Arch. Environ. Health***21:**133.

Swedish Expert Group, 1971, Methyl mercury in fish. A toxicologic evaluation of risks, *Nord. Hyg. Tidskr. Suppl.* **4** (Stockholm).

Tsubaki, T., and Irukayama, K. (eds.), 1977, *Minamata Disease: Methyl mercury Poisoning in Minamata and Niigata, Japan,* Elsevier Scientific Pub., New York.

Tsubaki, T., Shirakawa, K., Hirota, K., Kondo, K., Kambayashi, K., Kirota, K., Yamada, K., and Murone, I., 1967, Outbreak of intoxication by organic mercury compounds in Niigata Prefecture: An epidemiological and clinical study, *Jpn. J. Med.* **6**:32.

Turner, M., Marsh, D. O., Rubio, C. E., Chiriboga, C. C., Chiriboga, J., Smith, J. C., and Clarkson, T. W., 1974, Methyl mercury in a population eating large quantities of marine fish. I. North Peru in: *Proceedings of the First International Congress on Mercury, Barcelona, Vol. II,* pp. 229–234, Fabrica National de Moneda y Timbre, Madrid, Spain.

Wheatley, B., (1979), *Methyl mercury in Canada: Exposure of Indians and Inuit Residents to Methyl mercury in the Canadian Environment,* Ministry of National Health and Welfare, Medical Services Branch, Ottawa, Canada.

World Health Organization, 1976, *Environmental Health Criteria. I. Mercury,* WHO, Geneva.

III

DIETARY ADDITIVES

14

COMMERCIAL ADDITIVES

Johanna Dwyer

INTRODUCTION

Scope

In the U.S. today some 2300 different substances are classified as direct additives that are deliberately added to foods as ingredients to achieve particular objectives. It is the need for and use of these substances that are the subjects of this chapter.

The process of translating assessments of risks posed by food additives into decisions about what society should do about food safety involves difficult choices and dilemmas that lack clear-cut solutions.

One difficulty in making decisions about additives is that it is not possible to choose between risk and no risk. Rather, society must decide how much risk will be allowed, and for what purpose. At one end of the spectrum is the goal of achieving well-nigh absolute safety. At the other end is the goal of determining acceptable levels of risk and then deciding usage after calculating the trade-offs between risks and benefits, with ancillary consideration of controls and costs. Social attitudes toward irreversible toxic effects and risks such as cancer, mutagenesis, and teratogenesis are probably quite different than those toward certain other less debilitating or long-lasting toxic effects. Attitudes are also likely to vary when sensitive and dependent population sectors are considered, such as children, the aged, and the ill.

Second, it has become increasingly apparent that we lack good information on present consumption levels for these substances and the means for predicting what future consumption of new additives is likely to be. Thus estimates of hazard are imprecise.

A third difficulty is for society to strike an appropriate regulatory balance

Johanna Dwyer • Frances Stern Nutrition Center, New England Medical Center Hospital, and Departments of Medicine and Community Health, Tufts Medical School, Boston, Massachusetts 02111.

between the issues involved with freedom of choice and freedom from risk. Some express concern that government food safety regulations protect our health but limit our freedom. Others believe that present regulations are appropriate in that the freedom of the manufacturer to act is appropriately balanced by the freedom of consumers to avoid risks to health.

This chapter will review these major problems and proposals for resolving the issues they involve for commercial additives.

Reasons for Interest

First, while testing of additives for safety has long been required, the concepts and methods used for risk assessment have changed dramatically since 1958, when the last major piece of legislation was enacted.

Second, exposures to these substances are rising. Commercial additives are relative newcomers to the field of food environment control and preservation techniques. The volume of their use has risen dramatically in the past few decades and it is likely to continue in the future as the trend toward purchase of ready-prepared or convenience foods continues to expand (Miller, 1977). Whereas the introduction of new additives took from decades to centuries in the past, today exposure to new additives may become widespread in only a matter of years. Thus society has had less time for experience alone to serve as the basis for judging hazard, and recourse to other techniques, including careful testing, for estimating it.

Third, methods for assessing some types of toxic potency and for estimating exposure to potentially toxic substances in foods are poorly developed, although the major items that are required for assessing hazard to human beings involve potency and exposure. An acceptable manner of combining these factors into an estimate of hazard is also lacking (Farber and Friess, 1980).

Fourth, direct additives play an important and growing role in our food system. Some of the commercial additives are necessary aids to processing if our highly urbanized population is to be fed efficiently, since they help to minimize loss between harvest or slaughter and consumption, or deterioration in quality and cost; in addition, they maximize nutritional value. Other additives have less crucial functions. In view of the importance as well as the ubiquity of various types of additives, it is important to examine the risks as well as the benefits that result from their use.

Fifth, because direct additives are deliberately put into foods during processing or formulation, they are somewhat more amenable to control than are unintentional additives, environmental contaminants, naturally occurring toxicants and microbiological hazards, which may also be hazardous. There is little doubt that some of the 7700 other substances present in foods as a result of indirect or unintentional addition and contamination pose risks to health that must be controlled. Indeed, a recent list of six broad areas of hazard and their relative priority from the standpoint of the Food and Drug Administration (FDA) named food-borne infection, malnutrition, environmental contaminants, naturally occurring toxicants, and pesticide residues as more pressing concerns than commercial food additives (Schmidt, 1975). Even advocates of stricter regulation of direct additives contend that their ill effects on health are small in comparison to alcohol, smoking, and major food ingredients and natural constituents (Jacobson, 1980); however, these latter substances are more difficult to regulate, and thus the former receive more attention, both in the law and by regulatory agencies.

Sixth, commercial food additives are of special interest because of the way they are regulated. They are covered by the Delaney Amendment to the Food, Drug, and Cosmetics Act, which states that any substance causing cancer in test animals must not be added to food in any amounts. The demonstration that they are carcinogenic in any dose in test animals is sufficient to remove them from the food supply or to prevent approval for use unless Congress takes special action. The burden for proof of safety of a new additive is on the manufacturer, with the government being responsible for evaluation of the evidence submitted. In the past few decades, advances in toxicological research and analytical chemistry have created vexing new problems in the risk assessment of additives, since some additives previously thought to be safe, such as saccharin, have been found to be carcinogenic, and residues of some other substances such as estradiol have also been found to be so. Some substances, such as nitrites and estradiol, are also produced endogenously in the human body, and it is difficult to assess the hazards associated with the small increases in usual background burden of these chemicals. Other potentially carcinogenic substances have been identified that result not from the additive itself but from its breakdown, reactions with other naturally occurring constituents, or from interactions that occur later in processing, such as during cooking. The latter include compounds found after frying, broiling, or other high temperature processing of food products. It is possible that the toxicity of some additives may be lessened by processing. Data are sparse on all of these effects, although direct additives continue to receive scrutiny in these respects.

A seventh reason for paying attention to commercial additives is that they have been publicized in the news in the past few years. Public reaction is often out of proportion with the hazards, but the pressures generated on regulatory agencies and Congress are not lessened as a result. The methods by which additives are tested for toxicity are not familiar to most consumers, and cause much confusion in their minds. The validity of the methods used in animal-model testing to calculate the effects to be expected in humans from low dose–long term exposure as compared with effects observed in high dosages in animals, and the basis for translating animal test results to estimates of safety for humans are particularly confusing to the public. Confusion is heightened by new findings of possible hazard involving popular substances such as cyclamate, saccharin, and caffeine, which are already widely used. Some consumers take these findings to indicate that a large share of some unexplainable chronic degenerative diseases such as cancer is due to the additives and that stricter regulation is necessary. Others have reacted to the same information by becoming increasingly skeptical about the whole rationale behind food-safety testing and regulations. For example, a recent cartoon in a newspaper showed an African explorer discussing food habits with a cannibal. The cannibal says, "We stopped eating Americans. Laboratory tests proved they caused cancer in crocodiles." Perhaps the prominence and publicity received by both new additives and some that were previously generally recognized as safe (GRAS) when they were subjected to more careful review and were found wanting in the 1970s is partially responsible for this confusion. As a result of the attendant publicity, the additives have sometimes served as convenient lightning rods for consumer discontents about other aspects of the food supply involving quality and safety.

The result of public confusion is that the relative hazards perceived from various foods and food ingredients often are greatly different from the ranking of problems provided by knowledgeable expert observers. This may have serious implications for public policy, since societal pressures for spending public funds may overly emphasize some topics for research or regulation and ignore others, with a distorted end result in terms of useful knowledge for improving health.

Eighth, food safety regulation is currently receiving a great deal of attention because of shortcomings and inconsistencies in the present regulatory system. The same risks (e.g., threats to life and health) occurring in different classifications of foods are treated differently; naturally occurring risks and those created by the addition of a substance are treated inconsistently, and the basis for regulatory decisions is often unclear. In order to understand this problem, it is necessary to review briefly the various regulatory categories that exist for different kinds of additives under present laws.

Let us consider the legal distinctions between different categories of additives. First are those that were considered to be generally recognized as safe (GRAS) or for which prior sanction or approval had already been given under the 1958 Food Additives Amendment.

When the GRAS list was first established, it included many additives that had been a part of the food supply for many years and that were viewed by an expert panel that reviewed evidence on their safety to offer little concern. Thousands of food ingredients, including additives such as sucrose, corn syrup, salt, and dextrose are included in this category. However, until the late 1960s, cyclamates and saccharin were also considered GRAS. After new studies put their safety in question and they were removed from the GRAS list, the FDA began to update its reviews of other substances. For only a small number of these substances were tests available that were comparable to those required for the approval of new food additives. Of the 351 substances reviewed, 71% were found to be without hazard if used at current levels or levels that might reasonably be expected in the future, and another 15% were found to be without hazard if use were limited to levels of addition that were then current.

More action was required for the remaining 14% of the substances reviewed. For 6% it was suggested that specific studies be conducted promptly because of uncertainties in the data, even though they were found to be without hazard when used at current levels. Four percent had inadequate data and more data was called for. Four percent were found to exhibit adverse effects when used in food at current levels, and it was urged that safer usage conditions be established (Code of Federal Regulations, 1977).

Other food additives, such as antioxidants, stabilizers, and similar materials, are covered under prior sanctions or approvals. Some chemicals used in paperboard products or food packages that may become unintentional additives by migration into the food are also included under this clause.

Additives that are not listed as GRAS or covered by prior sanctions or approvals require long and complex scientific testing and review before they are permitted in foods. Once they have been found to be safe, a decision may be made to permit only restricted use or limited marketing. Usually use is permitted only at a level of a hundredth of the highest dose which produces no toxic effects in animals. Thus, this group of substances is the most closely controlled of all permitted additives in the food supply; indirect additives and environmental contaminants are less rigorously restricted.

In addition to the above procedures, commercial additives are automatically declared to be unsafe under the Delaney Clause of the Food Additive Amendment if they are found to induce cancer in test animals. This causes them to be banned completely from the food supply. But such complete prohibitions do not exist for naturally occurring substances, including toxicants or contaminants in the food supply, even those causing cancer in animals, if they are considered to be unavoidable (Office of Technology Assessment, 1979). Tolerances are set at levels considered to be necessary for protection of the public health. For example, current tolerances set by the FDA for aflatoxin in peanuts and corn are at levels that, extrapolating from animal experiments, suggest present risks of 66 lifetime cancers per 100,000 persons.

It would be more logical to treat all ingredients consistently, regardless of their source, by using uniform criteria for evaluating risk, while recognizing the practical limitations that may sometimes make it impossible to regulate all substances equally. The goal of such efforts should not be to raise the level of all potentially hazardous substances up to the highest level now permitted, but rather to concentrate on evaluating and lessening the risks from the most hazardous substances, regardless of where they occur. The history of food safety shows that while remarkable progress has been made, the risks resulting from the use of some commercial additives as well as from naturally occurring toxicants and environmental contaminants were not always apparent after the kinds of testing that were then required. Thus, new methods for assessing risk are needed. Also, more attention needs to be paid to finding incentives for discovering more satisfactory alternatives for substances that are found to be hazardous.

Need for Direct Additives

If there were no need for commercial additives it would be unnecessary to investigate their risks, since they could be dispensed with immediately. Traditional American public policy as reflected in legislation and regulations partly reflects the utilitarian notion that additives are needed if they serve a useful purpose and are safe. But moralistic elements are also evident, in that it charges government with the duty of protecting health through the evaluation of the safety, not only of additives, but of the entire food supply.

Societal consensus continues to be that commercial additives do *not* serve useful purposes when they are used to disguise faulty processing and handling, to reduce nutritive value, or to deceive the consumer. However, the consensus on an appropriate definition of what is a useful purpose for an additive is presently in flux. Broad definitions of usefulness are no longer being accepted without question. Direct additives are developed and used because they are believed by the manufacturer to be potentially useful and that they will provide benefits of one sort or another that will ultimately result in greater consumer acceptance. From most food technologists' standpoints, useful purposes of food additives include improving or maintaining a food's nutritional quality; enhancing the stability or quality of a food so as to reduce food waste and cost, and increase food availability; providing a technological aid in processing and preparation; and making food more attractive to enhance consumer acceptability in a nondeceptive manner (Walker, 1976; Darby, 1979). However, other groups hold more limited views as to what constitutes the need for an additive. A cosmetic rationale (such as coloring) is frequently attacked as being insufficient justification for use, and indeed it is not considered sufficient grounds for the use of certain azo dye-type food colors in the Scandinavian countries today (Berglund, 1978). Some consumer advocates object to economic justifications for use of additives that fail to show that the advantages of lower food production costs will result in savings that will be passed on from the manufacturer to the consumer. Other groups regard food as something that should be preserved in its inviolate natural state to the greatest extent possible. Those who are the most vocal in expressing this attitude are adherents of "health," "natural," and "unprocessed" food diets, but many other American consumers as those in some other countries (such as France) appear to hold more moderate views in the same direction.

Differences in perceptions of whether a useful purpose is served by direct additives are especially pronounced for the three-fifths of the additives that are currently allowed, and that function primarily to render food more attractive to the consumer by virtue of their cosmetic properties; these are the so-called cosmetic additives. They include coloring, flavoring, emulsifying, stabilizing, and thickening agents. The other two-fifths of the commercial additives that serve other purposes in processing are less subject to criticism on these grounds.

In addition to dissension about the appropriate uses to which commercial additives are put, is the issue of the need to add to the large number of additives that already exist for certain functional purposes, especially purposes that are primarily cosmetic. Proponents of a large number of additives argue that should risks from the use of a specific substance later be demonstrated, individual exposure prior to the discovery and disruption to the food supply would be minimized if many alternatives for a given function were available. Opponents argue that the more additives there are the harder it is to monitor their use and the greater are the chances of harmful interactions between additives occurring.

RISK ASSESSMENT

The first task in the development of commercial additives is not one of determining risk but of proving the additive's efficacy for its intended purpose. The level at which it must appear as a food ingredient to do its job must also be ascertained. These tests are followed by a determination of the levels of consumption of the additive in the normal diet, based on estimates of the effectiveness and of the additive necessary to produce the technical effects and the average amount of the food or foods containing the additive that the average person would be likely to consume over the course of a year. Predicted consumption levels can only be regarded as rough estimates because knowledge of food consumption must be based on retrospective information, food practices are constantly changing, and permitted uses for additives may change over time. For example, in the case of saccharin, estimates of consumption in the 1950s were very much less than actual consumption in the late 1970s, owing to changes in food practices (especially the growing popularity of artificially sweetened, low-calorie drinks, and some other foods) and to revisions in the permitted uses of the additive. The amount of additive consumed per unit of body weight per day (mg/kg per day) is then calculated from estimates of consumption, and appropriate animal studies based on this dose are then designed to assess risk.

Two important questions must be dealt with in assessing risk. First, what types of risk are considered to be important enough to warrant investigation? Second, are methods available for assessing them? Both of these questions have been the subject of a great deal of discussion about food-additive testing in recent years.

Risks Requiring Assessment

Consumer concerns about toxicity and those of toxicologists are not always totally in synchrony, nor are the means yet available for evaluating all potential toxic effects, although their scope is growing.

At the turn of the century toxicologists were concerned mainly with preventing acute lethal effects stemming from the use of inappropriate additives. Methods for assessing these effects were developed, and a remarkable degree of safety was achieved in the food supply in these respects. Today, concern is focused on more insidious or chronic effects, such as carcinogenesis, mutagenesis, and teratogenesis. There is societal agreement that more careful consideration of these risks is warranted, not only concerning food components but also other possibly toxic substances in the environment, and scientists are beginning to develop consensus on appropriate methods for evaluating risks in these areas. However, less catastrophic and life-threatening effects are also receiving increased public attention. Consumers are growing increasingly concerned with eliminating other possible adverse effects upon physiological function and behavior, such as intolerances, adverse reactions, allergies, and effects on immune response. The psychopharmacology of additives is also receiving more attention. It is this latter group of concerns that poses the most difficulty for the toxicologist, because traditional methods for evaluating toxic effects of this sort are not yet highly developed (Miller, 1980).

Means of Assessing Risk Directly in Human Beings

Just as widespread societal consensus is often lacking concerning the need for various commercial additives, neither is there complete unanimity on the scope of the toxic effects that should be assessed and the means for assessing them. So too is there a lack of agreement concerning the best way to extrapolate findings from other species to man.

Appropriate toxicological tests provide information on what kind of injury or adverse effects are produced, how much of the test substance is required to produce the effects, and how these effects come about. The methods that are currently available for assessing potential toxic effects include studies on individuals or groups of humans, and tests involving animals or lower organisms the results of which are then extrapolated to humans. Ideally, most of us would like to have foolproof methods for assessing all aspects of toxicity that could be applied before, rather than after, substances were permitted in the food supply. Practically, the testing of substances for all potentially adverse consequences in species other than humans is impossible. Among the barriers to doing this are that methods do not exist for evaluating certain toxic effects, different effects may be seen in humans than in other species, and some substances already permitted in the food supply have not been completely tested. Methods that are currently proposed or are in use are described below. While each of the methods for determining risk have their limitations, taken together they do permit objective statements to be made about a given additive.

Epidemiological Studies

Descriptive epidemiological studies of the different circumstances under which people fall ill or die are especially helpful in providing estimates of the extent to which diseases are associated with environmental factors.

Epidemiological studies have a number of shortcomings that will be discussed shortly, but in at least one recent instance they were useful in identifying a serious toxicity stemming from the use of an intentional food additive that otherwise might not have come to light. During 1964–1967 a syndrome involving fulminating heart failure with pericardial effusion and elevated hemoglobin levels was identified among heavy beer drinkers in parts of the midwestern U.S., Quebec, and Belgium (Alexander, 1972). The extremely elevated hemoglobin levels and the association with heavy beer drinking pointed to the possibility that cobalt, used in some beers to stabilize its foaming properties in the glass, might be involved. Subsequent studies showed that indeed, cases of the syndrome had begun to appear about 6 months after breweries in the areas reporting cases had begun adding cobalt chloride to the beer, and that when the breweries stopped adding the substances to beer the epidemic stopped. However, mortality ranged from 18–47% and some after-effects were noted among survivors who had suffered from cobalt-beer poisoning. Further investigation suggested that chronic alcohol abuse, malnutrition, especially of thiamin, and preexisting cardiac damage from these or other causes also contributed to the syndrome. The doses of cobalt causing cardiomyopathy in beer drinkers were smaller than those used in medical therapy of pernicious anemia, and only after the clinical suspicion had arisen that malnutrition, lack of thiamin, and alcohol potentiated the cardiotoxic effects of cobalt were animal experiments undertaken that confirmed these observations (Berglund, 1978).

More recently, several retrospective studies of individuals with cancer of the urinary bladder have used epidemiological methods to examine the association between use of artificial sweeteners and artificially sweetened food and beverages and cancer risk (Hoover and Strasser, 1980; Wynder and Stellman, 1980; Morrison and Buring, 1980). One study also included an analysis of possible dose-related effects and also potentiation effects resulting from heavy cigarette smoking (Hoover and Strasser, 1980). None of the studies reported a clear positive association between use of these products and risk of bladder cancer, but in the latter study some possible leads between extent of use of artificial sweeteners and smoking appeared that require follow-up.

While epidemiological studies are helpful for some purposes, there are several difficulties that limit their use in assessing risks of additives before rather than after they are introduced into the food supply.

First, they can only identify substances that are presently leading to illness; they are unable to predict whether or not a new substance will have ill effects. Since chronic degenerative diseases such as some of the cancers have long latency periods (e.g., some 20 years from exposure to appearance of the cancer), with new substances it takes many years before epidemiological studies can indicate hazards. Second, epidemiological studies are best when the exposed population is clearly defined, the disease under consideration is relatively rare, and the disease is well reported. In the case of food additives, information on levels of consumption among specific groups is virtually nonexistent, and it is difficult to single out the group in which exposures were highest. The possibility also exists that the diet as a whole, or some combination of additives, rather than individual food constituents such as a particular additive, may be the problem. Statistics on the incidence of cancer and many other specific diseases are often poor, limiting epidemiological studies further.

Third, epidemiological studies are insensitive to small differences such as those likely to be observed from weak carcinogens. For example, the order of magnitude of effects associated with saccharin exposure is thought to be an increased prevalence of bladder cancer of about 2–4%. Epidemiological studies may not be sensitive enough to distinguish between differences in bladder cancer that are this small since they approach the level that could be attributed to usual variation and reporting error.

A fourth limitation of epidemiological studies is their inability to provide physically or experimentally verifiable evidence of causation. Frequently, only correlations or associations are available, and even in prospective or retrospective studies potentially confounding factors are usually present that complicate their interpretation.

Clinical Studies

Studies on human beings of the metabolism and pharmacology of food components can also provide useful information, but pose more thorny ethical questions than descriptive epidemiological studies (Goldberg, 1975; National Academy of Sciences, 1965, 1967, 1975a,b). Obviously, such studies cannot be conducted with new substances until dose levels eliciting no chronic effects have been determined in animals. Human testing at this time may be helpful in guiding selection for even more rigorous chronic testing in experimental animals. The procedures that are currently being suggested are similar to the preliminary studies used for evaluation of new drugs in humans (Food Safety Council, 1980). Cautious and limited administration of the compound at levels based on estimates of likely human dosage is performed under carefully controlled conditions using two volunteers and a single dose of the substance, in order to determine the compound's time course and major pathways of metabolism. Results from these tests help to ascertain the extent to which human metabolism and pharmacokinetics are similar to that of test animals. Assuming they are similar, further studies on humans may be carried out after additional animal studies. A slightly larger group of volunteers may be employed, doses administered for 5 days with the last dose employing radioactively labeled material, and doses may be raised to a small multiple of those used initially to obtain an idea of dose-response effects.

Obviously such studies can be carried out only after, and not before, more thorough testing has been done in animals or other species. Therefore, in spite of their shortcomings, animal tests are crucial in food safety testing.

Extrapolating the Results of Animal Toxicity Studies to Humans

Several issues are currently being debated with respect to the most appropriate toxicity tests to predict hazard in man and the best ways to extrapolate from these other species to humans. While the underlying premises are generally regarded as being valid, the extent of our certainty in making these estimates is disputed chiefly on these grounds. Therefore, these issues deserve a more careful examination.

Premises

Acute and chronic toxicity studies on laboratory animals are the most frequently used means for estimating the risks posed by an additive. The fundamental premise is that substances that are toxic in experimental animals will also pose risks to human beings.

An important second premise is that when suitable adjustments are made, it is possible to obtain fairly reliable estimates from animal tests of how much of the substances is likely to cause illness in human beings.

Pitfalls in Selecting Appropriate Animal Toxicity Tests

Toxicity tests used with animals may give different answers when used with humans for a number of reasons (Doull, 1978). First, differences in the kinetics of absorption, distribution, biotransformation, or excretion of the substance between the test species and man may exist. Second, the receptor or target organ may differ between the test species and humans. Third, some of the toxic effects of concern in man such as headache, depression, and anxiety cannot be evaluated in experimental animals. Fourth, some toxic effects that occur at a very low incidence in animals may be missed with usual toxicity studies of 25–50 animals per group even if several species are used. Fifth, animal tests usually use relatively high doses of the test substance for a shorter time than the lower doses consumed over a longer time that is more characteristic of human consumption. Most of these difficulties can be minimized by careful planning, execution, and evaluation of animal studies under the direction of expert toxicologists, but cannot be totally eliminated.

Mathematical Models

In order to derive information from acute and chronic animal tests that will be useful in estimating the effects of chronic low-dose exposure in human beings, many assumptions must be made. Mathematical models based on these assumptions are employed to perform these extrapolations. The choice of models is somewhat arbitrary and has no firm biological basis. Two major differences in existing models are the scaling factors they employ and how estimates of chronic effects at low doses are handled.

Scaling Factors. In order to compensate for differences between experimental animals and human beings in size and dietary intakes, some sort of scaling factor must be used. Three common factors are intakes stated in terms mg/kg, intakes proportional to relative body surface area in cm^2, and relative lifetime daily intakes in mg/kg body weight. The scaling factor chosen will produce very different estimates of the number of new cases of diseases which can be expected.

Low Doses. The mathematical models in use today all postulate that a proportional relationship between dose and response exists, but different assumptions are made about the magnitude of response at low doses. The question of how a model behaves at low doses is most crucial when one is considering effects in which chronic rather than acute exposures may be important. Such effects include mutagenicity, teratogenicity, and carcinogenicity. Current law requires that if a commercial (direct) additive is carcinogenic in test animals it cannot be permitted in the food supply. Statistically this is considered to be the case if the substance is estimated to exceed a lifetime risk of cancer in the animal which is greater than 10^{-6}, using the Mantel–Bryan procedure for determining a virtually safe dose (Mantel and Bryan, 1961; Federal Register, 1977).

Threshold and Single-Hit Theories of Action. The difficulty in determining the effects of low-level exposures in animal experiments is further complicated by the existence of two separate theories about how toxic substances are thought to act at the cellular level to induce carcinogenic and possibly mutagenic or teratogenic changes. The *threshold* theory holds that below a certain dose metabolic processes are sufficient to protect the body from any major harm, and that below this threshold dose there is little or no harm. The *single-hit* theory assumes that a direct relationship exists between dose and response at low as well as high levels, so that some risk or hazard, albeit very small, is always present. When one is considering cancers, some of which are believed to be caused by damage to the genetic material, even a "single hit" is thought sufficient by some scientists to cause cancer, while other investigators believe that the threshold concept is more likely to be appropriate. Carcinogens clearly exhibit a dose–response effect, just as other toxic substances do. The fundamental issue is whether there is a threshold or no-effect dose level in the dose–response curve for any or all carcinogens. Putting aside the question of whether there is a true threshold for carcinogens, some toxicologists argue that there is a "practical" or "effective" threshold, below which the body's detoxification and repair mechanisms are sufficient to make risk very small indeed (Doull, 1978). There is little doubt that a wide variety of detoxification mechanisms exist for many potentially adverse effects, and these may be operative for cancer risk as well. However, there is no surety that even if they do exist they are completely effective, or that individual differences in susceptibility do not occur. Some animals, and presumably human beings, fall ill with cancer even at very low doses of a single chemical or with small insults from many different chemicals. Also, the dose sufficient to cause one form of cancer may be insufficient to cause another. Therefore, great care in designing and conducting experiments is essential even though fundamental knowledge is not presently far enough advanced to give a definitive answer on which of these two theories is correct.

The practical significance of the single hit vs. the threshold theory controversy has to do with not only science but also politics. If the threshold notion is accepted, then regulatory policy might permit cutoff points for chemicals, while the single hit theory carried to its ultimate conclusion might lead to a regulatory policy that no amount of a proven carcinogen is acceptable or that if it were necessary to use it, efforts should be taken to find substitutes and lessen exposures to the lowest levels possible.

Extrapolation Procedures

The objective of extrapolation is to estimate either what potential risk to humans exists for certain exposure levels or to estimate a "virtually safe dose" corresponding to some small risk that is judged to be acceptable—such as 1 in 100 million as suggested by Mantel and Bryan (1961) or 1 in 1 million as proposed by the FDA (1977) (Roberts, 1980).

There are two quite different procedures that can be used to extrapolate results of animal toxicity studies to human beings: the traditional *safety-factor* approach, and the newer *risk-estimate* approach. Both procedures rely upon dose–response tests in experimental animals.

Safety-Factor Approach. This extrapolation procedure is currently used for many food components that have survived carcinogenicity testing. When the safety factor approach is used, the level of the substance that is judged to give "no effect" (i.e., no-toxic-effect level, at which treated animals are not significantly different from controls) is divided by a "safety factor" (usually 100) to obtain a predicted safe-dose level for human beings. The factor of 100 allows for an order of magnitude for greater sensitivity of man in comparison to the test animal and for another order of magnitude for variability among individual sensitivities. The use of a safety factor carries with it the implicit assumption that there is a threshold for the response at or below the no-observed-effect dose and above a hundredth of that dose. A safety factor of 100 is supported by some theoretical and experimental evidence, but the chief justification for its use rests on the fact that it has been used for many years to establish tolerances for drugs and other environmental chemicals. It has been found to be particularly useful for evaluating effects other than carcinogenicity. Once the prediction of a safe dose level for human beings is established, additional safety-testing calculations can be made on the basis of food consumption data and the likely uses of the additive to develop an acceptable daily intake for the substance.

The major difficulty with the "safety-factor" approach is that in fact there is no reason to assume that a no-effect level always actually exists; it is a function in part of the sample size employed in the experiment, the adequacy of experimental techniques, and the slope of the dose–response curve, and simply shows the level at which no statistical difference can be found. Moreover, the customary figure of 100 may be inappropriate and in any event is not always used. For example, when human toxicity data are already available, as they are for some of the fat-soluble vitamins, a factor of 10 is sometimes used, and a factor of 1000 has occasionally been used when the evaluator was especially concerned or uncertain. Finally, a threshold below which no association exists between dose and response is assumed when the safety-factor approach is used. This may not in fact be the case for all types of toxicity.

Risk-Estimate Approach. When the risk-estimate approach is utilized, it is assumed that some risk, no matter how small, exists with all levels of exposure to the substance. A small "acceptable risk" is agreed upon, such as 1 in 10^{-6} in the animal model, the dose–response information obtained from animal tests using high doses extrapolated downward to provide estimates of the risk at low dose levels. This procedure is necessary since the costs, number of animals, and time that would be required to obtain low-dose estimates by direct experimentation would be prohibitive. Various mathematical models incorporating different assumptions about the associations between dose and response at low levels of exposure are employed. The results of the extrapolation are then examined and the dose with associated risks below the acceptable level of risk is taken as the virtually safe dose (VSD). Toxicologists currently favor this approach because it appears to mirror more closely actuality and is more versatile than the safety-factor approach. It can be used for substances that are thought to have a threshold and those that do not, and can be employed without making additional assumptions about whether a single-hit or a threshold theory of toxicity with respect to carcinogenicity applies. However, consensus is lacking on which model best fits the biology of toxicity, especially when carcinogenic effects are being considered.

The mathematical methods most frequently suggested for low-dose risk assessments today are the Mantel–Bryan procedure (which is based on a probit model and leads to an S-shaped dose-response curve symmetric about the 50% response point), the one-hit model (which assumes the dose–response curve is linear in the low-dose range), the Armitage–Doll multistage model (which is also linear at low doses), the Weibull model (another generalization of the one-hit model), and the Gamma, K, or multihit model (which resembles the logit at low doses and the probit model at high doses). Recently, four of these models were compared using the same data on the toxicity of 14 substances in various species. Table I presents these findings (Food Safety Council, 1979). In general, the one-hit linear model gives the lowest virtually safe dose, the Armitage–Doll model the next lowest, the Weibull the second highest, and the K or Gamma multihit model the highest of all. The estimates of VSDs vary as much as a million times for some substances, depending upon the model used, and differences on the order of a hundred thousand times

TABLE I

Experimental Results for 14 Substances Using 4 Different Mathematical Models or Low-Dose Effects[a]

Substance	Species	Type of response	Dose unit	Estimated virtual safe dose at risk level 10^{-6}			
				One-hit	Armitage-Doll	Weibull	Multi-hit
NTA	Rat	Kidney tumor	% in diet	.00002	.000057	.52	.80
Aflatoxin B_1	Rat	Liver tumor	ppb	.000034	.00079	.04	.28
Ethylenethiourea	Rat	Fetal anomalies	mg/kg	.000045	.35	.59	2.3
2,3,7,8-tetrachlorodibenzo-*p*-dioxin	Rat	Fetal intestinal anomaly	mg/kg	.0000052	.0016	.0017	.0038
Dimethylnitrosamine	Rat	Liver tumor	ppm	.000032	.019	.019	.077
Vinyl chloride	Rat	Liver angiosarcoma	ppm	.02	.02	.0000000021	.00000000039
Hexachlorobenzene	Rat	Fetal anomaly in 14th rib	mg/kg	.00021	.00022	.00026	.00026
Botulinum toxin type A	Mouse	Death due to botulism	ng	.000000084	.0042	.0043	.013
Bischloromethyl ether	Rat	Respiratory tumor	No. of 6-hr exposures by inhalation of 100 ppb	.00016	.0004	.031	.037
Sodium saccharin	Rat	Bladder tumors	% in diet	.000043	.33	.53	1.0
Ethylenethiourea	Rat	Thyroid carcinoma	ppm	.00055	4.5	6.0	33.5
Dieldrin	Mouse	Liver tumor	ppm	.0000057	.000022	.0012	.0067
DDT	Mouse	Liver hepatoma	ppm	.00028	.00064	.017	.049
Rapeseed oil (span)	Rat	Cardiac lesion	% in diet	.000037	.000057	.0011	.0038

[a] Adapted from Food Safety Council, 1979.

are common between them. They also differ in their estimates of hazard from that predicted by the safety-factor approach. Therefore, the model chosen is a critical assumption. When the Mantel–Bryan or one-hit model is used for evaluating toxic effects other than carcinogenicity, the VSDs resulting are very much smaller than those resulting from the use of safety factors. If the Mantel-Bryan model is used for evaluating carcinogens, the VSDs are closest to the zero tolerances now imposed by the Delaney Clause.

It must be recognized that human risk estimates cannot be stated explicitly or with validity using any model, since the fundamental biological mechanisms that are operating are not apparent for all of the conditions of interest. Since extrapolation must be made across various species to humans, conservatism is justifiable.

The observed data can usually be fitted well using any of the models, and no obvious statistical basis for response behavior in the low-dose range is apparent, so that there is no firm scientific basis for choosing among alternative extrapolation techniques (Cornfield, 1977).

A recent report of the Interagency Liaison Group of the federal government (1979) concludes that there is still no way to predict an acceptable threshold below which human exposure to a carcinogen has no effect. After reviewing the Mantel–Bryan, one-hit, linear extrapolation, multistage, and multihit models it concluded that whenever quantitative risk analysis was performed, linear extrapolation (a conservative procedure) should always be included among the methods used. The single exception mentioned was that when there was reason to believe that the observed response did not fall in the convex portion of the dose–response curve, a one-hit model would be suitable. The group also emphasized that individual variability makes it difficult to have confidence that an observed no-effect level of exposure in animals could be used to estimate the total human population at risk precisely.

Another technique that has been suggested for extrapolating test data on carcinogens from animals to humans takes advantage of the fact that in experimental animals, dose is related to the median length of time it takes for the cancer to occur—that is, the "time-to-response" or latency period. Models for treating data on time-to-response have also been proposed, but because data on latency periods are not ordinarily available the results obtained using the various models with different substances cannot yet be compared. It is possible that dose is related to the latency period, but it is also possible that when doses are changed, cancer appears at the same time but in differing numbers of people. More study of this question is needed, since the issue has great practical importance.

Validity. The ideal method for extrapolating from animals to man has not yet been identified, but some data are available that enable comparisons between species to test the suitability of the models.

One recent study of environmental carcinogens concluded that dose–response relationships shown in laboratory animals are approximately similar and more sensitive than estimates of effects in human beings (National Academy of Sciences, 1975b). For the six environmental carcinogens studied by this expert group, three (benzidine, chlornophazine, and cigarette smoke) showed close associations between predictions based on animal testing and human epidemiological evidence. For the other three (aflatoxin, diethylstilbestrol, and vinyl chloride), estimates from animals were 10–100 times higher than human cancer rates. However, data on total lifetime exposures in humans was not available for either diethylstilbestrol or vinyl chloride. Also, only rare forms of cancer were surveyed and these compounds may contribute to the incidence of more common cancers in man as well. These data do suggest that, at least for these chemicals, the experimental animals appeared to be more sensitive than man. However, other scientists argue that under some circumstances experimental animals may be less, rather than more, sensitive to the effects of toxic substances than are human beings. Such effects might be due to their size (small animals often metabolizing chemicals more rapidly than large animals), age (shorter lifetimes in laboratory animals), and lack of synergism with other chemicals commonly present in the human environment but not present under highly controlled experimental conditions in the laboratory. Many carcinogens are present in actual human environments. It is argued that even the addition of a weakly active substance under such circumstances may overwhelm detoxification mechanisms in real life. Moreover, not all consumers who eat additives are normal and healthy, nor can they all be expected to consume reasonable amounts of the foods which serve as vehicles for the substances, and under such circumstances the model may be invalidated. Thus, the use of conservative models and continuing efforts to further assess their validity appears to be the best course.

Other Procedures Useful for Toxicity Screening Purposes

Even short-term tests of toxicity in animals are expensive and time consuming. Simple screening tests would eliminate poor candidates for more costly animal studies and might also provide additional information on the mutagenic and carcinogenic potential of chemicals being considered. The most rapid way of testing chemicals for their possibly hazardous properties is to observe their ability to induce mutations in bacteria. Work on the use of tissue culture techniques for screening purposes is also progressing rapidly. Some of these assays can now provide preliminary evidence on mutagenicity, others on carcinogenicity, and a third group of tests addresses both objectives. The predictive capabilities of existing tests are now being established, but data are not yet available that would permit a standardized battery of tests. Moreover, their ability to predict toxicity in animals and in humans has not yet been assessed. Therefore, their utility is limited chiefly to screening at present, and they are not considered substitutes for animal and human studies.

GAPS IN RISK ASSESSMENT TODAY

Risk Assessment for Single Constituents in Foods

The field of risk assessment is developing rapidly, but societal awareness of problems has outstripped progress. Among the most obvious limitations are the long time (at least 3 years and more usually 6) to reach conclusions when chronic toxicity studies are called for, the lack of general agreement among scientists on the most appropriate ways to estimate human hazards from exposure to low doses of potentially toxic substances on the basis of toxicity studies performed on other species with high doses, and the most satisfactory ways to translate findings on test animals to human beings in quantitative terms. A number of toxic manifestations of potential significance to humans such as behavioral effects, altered immunocompetence, intolerances, and allergies are not usually included in present procedures. Much also remains to be done to strengthen both our fundamental knowledge and procedures for risk assessments involving carcinogenicity, mutagenicity, and teratogenicity. Protection cannot be assumed when analyses are not conducted or toxicity is unknown (Farber and Friess, 1980).

Until greater agreement is reached among scientists upon how best to assess risk and to translate these findings to humans, there is little likelihood that consensus in the broader society can be reached on what level of risks will be found to be acceptable. We presently lack the means to do this as well.

Risks Arising from Interactions between Food Constituents

Generally accepted and econimical methods for determining the interactions of foods or food ingredients consumed together and how they may affect risk are also lacking. The effects of processing, storage, and home preparation techniques upon the toxicity of additives are also largely unexplored in these respects.

Human beings eat commercial additives in a wide variety of different food products. The risks of consuming an additive when it is only one of several commercial additives or multiple potential toxic agents of other sorts (including environmental contaminants, naturally occurring toxicants, pesticide residues, and toxic substances formed during processing and cooking) may be quite different from those extrapolated from test conditions concerning only a single additive. The ability to estimate risk from a single potential toxic agent is progressing rapidly, but little has been done to generate risk estimation theory for the toxic effects of multiple chemical stresses from foods. This issue needs attention, as does the broader question of estimating risks from the total body burden of all environmentally occurring carcinogens (Farber and Friess, 1980; Mantel and Bryan, 1961).

Risks Associated with Variations in the Host

The influence of diet, age, disease state, and other factors which might influence response to substances being evaluated have been given little attention, either in animal testing or in human studies; they are only now beginning to receive the attention they deserve.

Biologically each individual is unique and "special." However, some individuals and groups have special characteristics that are particularly salient when matters of food safety are considered. Examples include pregnant women, young infants, the aged, those whose food consumption habits are markedly different from those of the population as a whole, those who suffer from particular diseases or conditions that may alter their susceptibility, and others who have life style habits that alter the risks to which they are exposed. Improvements in food safety testing can better take into account some, but not all, of these characteristics. Additional measures may be necessary for coping with other problems.

In the past decade it has become apparent that information on fetal exposures and potential hazards to neonates and infants was lacking for many of the GRAS substances. Methods for testing these as well as new additives for their effects on reproduction, such as mutagenicity and teratogenicity, were found to be in need of improvement (National Academy of Sciences, 1977; Miller, 1977; Food Safety Council, 1980). In traditional multigenerational studies, test animals are usually killed when they reach 6 months of age, so that it is not possible to observe long-term effects in succeeding generations. Moreover, attention to effects usually involves observations of decreased fertility, perinatal survival and growth. But test animal dams often cannibalize malformed or dead pups at birth, and these effects are often not assessed. Studies of placental transfer of the test substances are also infrequent. Now it is recognized that placental transfer plays a critical role in fetal development and later survival, and additional study is warranted (Joint FAO/WHO Expert Committee on Food Additives, 1972).

The most appropriate tests for teratologic and mutagenic observation are still a matter of debate, although there is general consensus that such tests are called for. The validity and resolving power of the various tests for genetic damage that are currently available vary, but methodologies are also rapidly being refined.

Because young infants sometimes differ in the handling of toxic substances owing to their physiological immaturity, their large amount of actively metabolizing tissue in comparison to their size, and their relatively simple diets consisting of a few foods, their risks may be quite different from those of adults (Goldenthal, 1971). Many substances pass readily into human milk and information on this is also needed (Catz and Giacola, 1972). In order to increase our knowledge of these effects, toxicological studies on newly born animals continuing through weaning have been urged (Joint FAO/WHO Expert Committee on Food Additives, 1972).

Very little attention has been devoted to the possibility that handling of toxic substances may also be different among the aged. Toxicologists need to investigate this possibility further since the population is longer-lived today than ever before.

Consumption of a food or groups of food containing certain additives is not uniformly distributed across the population. Some individuals eat a great deal while others consume little or none. For example, Chinese–Americans consume very much higher amounts of monosodium glutamate than do those from Western European backgrounds. Attention needs to be paid to obtaining more data on the spectrum of intake values for specific food ingredients. Surveillance of possible adverse effects of existing substances would be most fruitfully concentrated among those whose consumption levels were very high (e.g., $\geq$90%). Those sometimes include persons on special diets. Diabetics in the past few decades often consumed higher amounts of artificially sweetened foods and beverages because of the belief that by so doing they would cut down on calories or sugar. More attention must be paid to developing ways for identifying those who consume large amounts of certain additives if realistic estimates and studies of exposure are to be conducted.

A final group of consumers consists of those whose metabolism is altered by a disease, condition, or use of certain drugs. Their risks from certain food components, including commercial additives, may be altered. There are so many different diseases that it is virtually impossible to ascertain all of the possible adverse reactions that might conceivably occur among such individuals. However, possible cues may be obtained during food-safety testing that deserve to be followed up in later surveillance efforts or by educational efforts such as labeling and patient-package inserts, and the like. Many otherwise healthy people receive chronic medications such as oral contraceptive agents, mood altering drugs, and analgesics. A large proportion of the population drinks and smokes. The possibility that these substances may affect the handling of some additives needs to be considered, especially for substances that are used by very large segments of the population.

Identification and protection efforts for especially sensitive population sectors (such as the fetus, infants, children, the aged, overindulgers, and the ill) need greater attention.

Finally, since absolute safety is impossible to achieve, education programs directed to high-risk groups is needed to alert them to real risks and to alleviate the perceived, but unjustified concerns of the population.

Risk and Biological Realities

The axiom of Paracelsus, father of toxicology, that "all things are poisons, all are safe, it depends on the dose" still applies today. Just as absolute freedom of choice in eating is limited by the practicalities of living, so is freedom from all eating-associated risks limited, no matter how small those risks may be.

The major obstacle militating against absolute safety is biological reality. First, while we can ban a commercial additive if it is suspected of having a potent toxic effect, and other foodstuffs might be banned for reasons of filth, for naturally occurring toxicants, or environmental contaminants, zero levels are not so easy to achieve, and such an objective may be impossible. Second, our knowledge of the effects of interactions between substances is limited. The effects of several substances together may be greater than, equal to, or less than the toxicity of any one substance acting alone, and we lack ways of measuring this. Third, all of the potential hazards that may result from substances in food may not yet be identified. After all, only in the past few decades have methods for testing substances for their potential teratogenicity and mutagenicity been developed. As fundamental knowledge grows, additional tests may be called for as well. Even today, toxicity testing does not examine all possible adverse reactions; it concentrates on the most serious risks. Risks of a milder nature may be overlooked, especially if their incidence is low. Finally, testing and extrapolation procedures employing experimental animals, which constitute the bulk of our food safety testing systems, are not the perfect models for humans. Human beings differ in their susceptability to suffering ill effects from toxic substances even when they are in good health; and when they are in poor health, or if they have special conditions of a hereditary or acquired nature, this variation in susceptibility may be even more pronounced. Other factors, such as alcohol, drug, and tobacco use, as well as age, may also influence metabolic responses. Thus, human biological variation is such that it is impossible to be certain in advance that some subgroup will not react adversely to a substance that generally has no ill effects, be it a direct additive or some other food constituent.

While absolute safety is therefore unattainable, improvement is possible. More careful testing of new additives can give us a better idea of the hazards that are likely to be involved if a new substance is permitted in the food supply. More thorough review and testing of potentially toxic substances already permitted or appearing in the food supply can be initiated, and major hazards can be eliminated. Better monitoring and surveillance systems for tracking consumption can be instituted. Finally, special precautions can be taken to assure that consumers who are at risk for one reason or another are protected or can protect themselves from potentially harmful substances.

CONSUMPTION

Risk assessments in animals can provide us with important information about the potency of particular substances and the exposures in experimental animals that result in toxic effects. Various extrapolation procedures are available that provide different estimates of the hazards to humans that may result from consumption of these same substances at levels that are assumed to be equivalent for man. Human hazard, therefore, must be estimated from animal extrapolation on the bases of average daily consumption, duration of exposure, and a variety of other factors including individual susceptibility that are difficult to mimic in experimental animals. Human food consumption studies provide additional information that may be useful for several purposes. First, they permit us to guess the likely consumption levels of a new additive. Assuming that present food patterns will continue, it is possible to calculate likely exposure levels. Second, likely consumption levels of groups with special needs can be estimated. Third, present consumption levels of substances that are permitted in the food supply and are known to be toxic (such as cyclamates and saccharin) can be monitored. Finally, as methods become more refined, estimates of the range of total intakes of various additives by different groups may be possible. All of this information may be of public health importance. Therefore, the methods presently used to assess intakes are of importance (Irving, 1978).

Present Methods for Estimating Intakes of Direct Additives and Their Problems

In order to determine exposure to a given food component, it is necessary to know its concentration in each item in the diet, the amount of those items containing the substance that is consumed, and the frequency at which they are consumed. Unfortunately, this information is not always available.

During the review of the potential hazards of 400 food ingredients that had been defined to be GRAS and permitted to remain in the food supply after the 1958 amendments to the Food, Drug, and Cosmetic Act, it became apparent that estimates of consumption using the methods then current were very difficult to establish with certainty (Siu *et al.*, 1977). Table II illustrates the wide disparity in estimates of daily intakes of several GRAS substances by

TABLE II
Comparative Estimates By Different Methods of Daily Intakes of GRAS Substances[a]

Substance	Projected consumption by persons 2 or more years old (mg)	Quantity used in food processing (mg)	Estimates based on production and import (mg)
Ammonium carbonate	741	1	Not available
Calcium gluconate	2,665	3	Not available
Modified food starch	11,834	696	Not available
Gum arobic	2,470	108	187
Monosodium glutamate	1,106	1,435	285
Nutmeg	336	11	25
Sorbitol	30,191	79	654

[a] Siu *et al.*, 1977.

the available methods. The errors involved in these estimates are considerable and have been well reviewed (Filer, 1976; Gumner and Kirkpatrick, 1979). Only their major problems will be mentioned here.

Let us consider the projected estimates of consumption compiled directly from food consumption data. The concentration of the additive in each item in the diet is difficult to ascertain and depends upon the accuracy of information provided by manufacturers of processed foods. In a National Academy of Sciences (NAS) study, estimates could only be obtained on 60–70% of the processed foods consumed in the U.S. (National Academy of Sciences, 1972–1973). Manufacturers provided data on the usual and maximal levels at which each substance was added to 28 broad categories of usual foods (e.g., baked goods, soups, and fats and oils) as well as 13 infant food categories. However, levels in each individual food product or brand within these broad categories were not obtained, so that the amount consumed of each individual item containing a given component could not be calculated. Portion size estimates were derived from a separate set of data—the 1965 USDA food consumption survey. The frequency with which each item was consumed was based on yet another set of data—the Market Research Corporation of America's menu census of the eating habits of nearly 13,000 persons over a 2-week period. Consumption estimates were obtained by multiplying usage levels in the categories by portion size for individual items by mean frequency of eating. Estimates of total daily intakes of the substances obtained by these calculations were probably quite high because of a number of conservative assumptions. First, a given ingredient was assumed to be used in all foods within a specified category if it was reported for any food in the category, even though it may have only been used in a few foods. Second, it was assumed that all categories of food were eaten each day. Finally, losses in home processing or preparation before eating were not accounted for.

Other shortcomings of the data are also unsettling and point to the need for more precise information. It was impossible to closely approximate daily intakes for populations other than infants that might be at special risk, such as pregnant women, the aged, those with specific diseases or conditions requiring special diets, or the poor. Estimates pieced together from incomplete data sets collected in different ways at different times also leave much to be desired from the standpoint of validity and reliability, especially if usage levels are changing rapidly.

Estimates of consumption based on total poundage used in processing or in production and imports divided by the total population are limited in that they are averages. If the food were eaten by only a small proportion of the total population, averages might underestimate intakes.

More valid, reliable, and up-to-date data on average daily consumption rates and duration of intakes of the total population, by vulnerable groups and by the 1% of the population with the greatest intakes of the substance in question, are needed in order to better assess the potential for hazard. Federal officials have begun developing the means for surveying the population's food consumption habits so as to obtain more accurate and timely information on consumption, and to better link these with information from commercial sources on levels of additives actually consumed. However, they are not yet in place, and we must still rely on makeshift methods to estimate consumption today. Problems involving overestimates of consumption and determination of extremes in intakes can be alleviated by using probabilistic techniques that assess the likelihood that a particular GRAS substance will be present in an individual serving of food, but the method involves expert judgments or guesses as to whether the component appears in each specific food (National Academy of Sciences, 1977). Better information from manufacturers is needed before real progress can be made.

Changes in Intakes of Additives

Food consumption patterns are constantly changing, and as these occur intakes of additives and other food components may change as well. Intakes of commercial additives may rise sharply for a specific substance when new food preferences stimulate rapid growth in a particular product line (as occured in the 1960s when low-calorie foods and beverages containing the artificial sweeteners cyclamate and saccharin rocketed in popularity). Intakes also change when a new additive is brought onto the market, or when a specific additive is banned and another is substituted in its place (as has occured with food colors in the past decade as red #2 was banned and the safety of red #40 was questioned, stimulating many manufacturers to substitute red #3 in products containing the former). Intakes may decline when manufacturers alter product formulations (as was the case with monosodium glutamate, some modified starches, and sodium chloride in baby foods), or consumer preferences for "nonprocessed" foods rise (as witnessed by the growth of so-called natural and organic foods in the 1960s and 1970s). Because of this constant state of flux, population exposure to chemicals in food and elsewhere in the environment should be constantly monitored. Current systems are inadequate and uncoordinated. When consumption increases markedly for substances previously judged to be safe, chronic toxicity tests are not automatically required to be revised. Methods must be improved and better integrated so that consumption data on commercial additives, naturally occuring toxicants, nonintentional additives, and environmental contaminants are constantly surveyed and updated.

IMPROVING FOOD-SAFETY DECISION MAKING

Regulatory Dilemmas

Decisions to determine whether and at what levels a substance can be intentionally added to foods are made by federal regulatory agencies that administer the laws passed by Congress concerning food safety. Regulatory decisions are inherently controversial because the decisions that must be made always involve trade-offs between such factors as food availability, cost, acceptable levels of risk, and freedom of choice. Stakeholders differ in their judgments and in their views of the appropriate balance between these factors. In the past two decades difficulties in making these decisions have increased. Among the regulatory dilemmas which are increasingly being debated are considerations of "acceptable" levels of risk, cost–benefit assessment, and consumer information in food safety decision-making. The relevance of these issues is highlighted by a consideration of some of the more controversial issues in the past few years and their antecedents (Fusfield, 1979; Winspringer, 1979).

Legislation passed in 1958 prohibited the use in any amount whatsoever of intentional additives that were found to induce cancer in animals or man. Reviews of food safety data on substances already permitted in food supplies also revealed lacunae with respect to information on mutagenicity, teratogenicity, and intergenerational effects which merited attention. Both the affected industries and the regulatory agencies found that these considerations complicated their tasks. In spite of intensified efforts to improve food safety testing, the level of public concern began to rise. In the late 1960s the artificial sweetener cyclamate was banned when new tests in animals indicated that the substance was carcinogenic. Because cyclamates were widely used as artificial sweeteners in popular foods, this regulatory action caused a good deal of disruption in the food industry, but the regulation was eventually put into effect. In the next few years other permitted additives, such as some of the food colors, came under suspicion, and actions to ban their use were also taken, focusing further public attention to the issue. In the late 1970s when the FDA announced that it intended to ban the use of saccharin because new animal tests had shown that saccharin was carcinogenic, great public furor ensued, because no other noncaloric artificial sweetener was available. Congress acted to place a moratorium on the ban. Pending further evaluation of the evidence and the issues, the agency began a campaign to inform consumers on its possible hazards using warning labels and announcements. Shortly thereafter, on the basis of another new study indicating the carcinogenicity of nitrites, the Food Safety and Quality Service and the FDA jointly proposed that their use as additives would also be gradually restricted. Because nitrites are widely used in food processing to prevent the growth of *Clostridium botulinum,* which produces a deadly toxin, no readily available substitute existed with the same desirable charcteristics, alternative storage measures would be required in its absence, and costs for such a changeover would be considerable, the trade-offs involving risks and benefits to health in this case were vigorously argued. To add to the confusion, the validity of the study upon which the regulatory decision was based was subsequently questioned and shortcomings in procedures were discovered, resulting in a reversal of the original phase-out, although the matter still remains under review.

More recently, evidence on the adverse effects of alcohol and caffeine on fetal development has led the regulatory agencies to consumer information efforts directed toward women who may be at risk.

In the present food safety decision-making system, additives are categorized as either safe or unsafe on the basis of risk assessment techniques usually involving experimental animals. Those that are found to be carcinogenic in test animals are regarded unsafe and are not permitted as intentional additives at any level in human food. The levels permitted for other additives are based on estimates of the hazard they may involve, but usually do not involve outright bans. These regulatory interpretations of safety for food additives have come under increasing criticism in the past decade from private industry, scientists, and consumers alike. Industry critics and scientists complain that it is illogical to treat the same risks of carcinogenicity occurring in disparate classifications of foods and additives differently and inconsistently. Consumers fail to understand the relevance of animal tests to human hazards, especially when the substances declared to be hazardous are already in the food supply and are widely used.

Acceptable Levels of Risk

In previous sections the methods for assessing risks from direct additives were reviewed and found to provide useful but usually incomplete and imprecise information. Studies in species other than human beings are the most practical, but to date they have often been limited in the scope of the effects that they assessed, and views on what should constitute standard test batteries vary. Both the safety-factor approach and the risk-assessment approach for extrapolating the risks observed in high-dose experiments in animals to exposure levels in humans suffer from shortcomings. The estimate of a VSD corresponding to some small "acceptable" risk such as 1 in 1 million as proposed by the FDA (1977) or 1 in 100 million as originally suggested by Mantel and Bryan (1961) varies by as much as 100,000-fold depending upon the extrapolation technique employed. Little data is available on whether estimates of VSDs correspond to human morbidity and mortality estimates for diseases such as cancer, which are the very ones of greatest concern. At present they can only be considered to be conservative guesses, but direct epidemiological and clinical studies on human beings also have serious limitations. Thus, the ability of science to provide conclusive judgments on risks is limited. Issues involving what hazards are acceptable to society and the appropriate balance between hazards, costs, and choices concerning which are to be tolerated in the food supply involve not only scientific data but social, economic, legal, and philosophical considerations. While improvements in risk assessment can clarify estimates of human hazard, they are simply adjuncts to, rather than substitutes for, societal decisions on what is regarded as an acceptable hazard (Food Safety Council, 1980).

Cost–Benefit Analysis

Many criticisms of the current food safety decision-making system have to do with its failure to consider social benefits in a formal, direct, and systematic manner, the lack of procedures for considering benefits from additives that have been found to bear risk, and the inadequacy of information on the nature of the risks, benefits, and rationale for regulatory decisions that is provided to producers and consumers (Food Safety Council, 1980). Since benefits are already considered in making regulatory decisions on allowable tolerance levels for some indirect additives, natural toxicants, and environmental contaminants today (including some that are carcinogens) some observers have suggested that they should be included on a more explicit level when regulatory decisions about direct additives are made. Others have suggested that cost–benefit analyses might be employed as an alternative to the absolute prohibitions of the Delaney Clause in decision making on low-level carcinogens.

While few dispute the need for more open and explicit procedures, consensus is lacking on whether formal, analytical techniques for weighing costs and benefits (such as cost–benefit analysis) are sufficiently valid and reliable to justify their use as the major or sole means for making food-safety decisions. The difficulties that remain unresolved are several. First, no generally acceptable means for comparing risks in terms of human morbidty and mortality to benefits in terms of dollars is available. A second shortcoming is that definitions of benefits and the values to be placed on them are also ambiguous. Cost–benefit analysis, which casts all these factors into the same unit, usually money, may thus be inappropriate because the monetary value of a human life or the costs of disease are difficult to determine. Some critics believe that risks, especially if they are life threatening, should take absolute precedence over benefits. Benefits are also difficult to estimate in dollar terms. Third, both risks and benefits usually have considerable margins of error. Fourth, risks and benefits may accrue to different individuals. Finally, an important ethical question remains unresolved, regardless of the outcome of any single cost–benefit assessment. Those who hold a moralistic or intuitionist philosophy believe that the greatest good to society will come from preserving life, and define this end as the goal. Cost–benefit analysis is based on the utilitarian notion that the common good consists of the greatest benefit to the greatest number. The notion underpinning this analytic approach may thus be unacceptable, especially when the risks, costs, and benefits are unequally distributed across the population. In practice, American food safety decisions have usually involved a combination of moralistic and utilitarian approaches. Some implicit, if not explicit, reckoning of risks and benefits is done to clarify the trade-offs, but consideration is also given to ethical issues, the differences between social and private costs, special interests, and social values, norms, and opinions during the political adversary process which then reifies or modifies these proposals (Ostenso, 1979; Reed, 1979; Throdahl, 1979).

It is likely that the validity and quality of food safety decision-making can be increased by the use of consistent frameworks which structure the early part of the process and permit fuller identification and consideration of the costs, benefits, and consequences of proposed regulations. Among the useful principles of analysis which might be employed are these: definition of the problem, statement of objectives, identification of private and social costs and benefits, discounting of future costs and benefits to their present value, analysis of the importance of uncertainties with respect to the key variables upon the results of the analysis, assessment of ethical issues involved, and discussion of results in line with the above considerations (Office of Technology Assessment, 1980). In contrast, decision-making systems that attempt to remove most aspects of regulatory action from adversarial and political processes by substitution of formal cost–benefits analysis for these time honored procedures are unlikely to be acceptable.

Consumer Information

Food safety regulation involves maintaining a delicate balance between consumer protection and consumer choice. Theoretically a variety of regulatory actions could be employed which emphasized one or the other to a greater extent. On one extreme are outright bans of ingredients found to have carcinogenic, mutagenic, or teratogenic effects, along the lines of the Delaney Clause's proscriptions of additives found to be carcinogenic. On the other extreme are redefinitions of acceptable risk to permit the use of weak carcinogens or other toxicants, especially if they had other virtues. The present trend in regulatory law appears to emphasize providing consumers with information so that they can protect themselves if they wish to do so while giving them more freedom of choice than would be possible if outright bans were employed. However, a number of practical problems with this approach have already become evident that should cause us to pause before contemplating drastic overhauls and revisions of existing legislation. The first problem is that adequate information simply does not exist on many food safety questions. All risks related to direct food additives are not known, measurable, or detectable today, and the situation is even more uncertain when indirect additives and environmental contaminants are considered. The process of consumer education on some of these issues cannot begin until information is available. Classical toxicological techniques used until a few years ago provided little information on fetal exposures, intergenerational effects, or neonatal and infant risk levels, for example, and are only now being updated.

The second difficulty lies in the individual consumer's ability to understand information on hazards, which is often difficult for laymen to interpret. It is based largely on studies of risks in animals, and is highly vulnerable to public misinterpretation and skepticism. In addition, news stories on hazards often tend to be sensationalistic and uncritical rather than analytical in nature, and such information sources compete for the consumer's attention.

A third and even more serious difficulty involves problems consumers may have in understanding the warnings and acting upon them in their own best interest. Most of the information involves statistical statements of probabilities involving small and ambiguous risks. Many consumers find such information difficult to understand. Even general warnings such as those that appear on cigarette packets cannot be comprehended by consumers who are illiterate, preliterate, or who have other difficulties in reading or understanding label information. Consumers may also become confused by the conflicting messages provided by advertising of the product's benefits, and refuse to believe that a food or beverage eaten by the "beautiful people" in a television commercial could actually be risky. Also, some argue that more conscious choices will be made when consumers must take a specific step, such as adding saccharin to food rather than simply buying food that they may or may not be aware contains saccharin.

A fourth difficulty involving consumer information as an alternative regulatory measure is that in the case of certain additives (such as those causing intolerances in a small proportion of consumers) the information is irrelevant to all but these individuals and thus, the costs of providing such information are very high (Vanderveen, 1977). Moreover, in spite of the popularity of informational measures as an alternative to more restrictive legislation among many in the food industry, few companies have voluntarily labeled their food products with information on the content of additives such as tartarazine, which is suspected of causing adverse reactions in some individuals.

A fifth difficulty is that many consumers simply may be uninterested in making decisions about food safety, preferring to make their decisions on the basis of price, taste, freshness, and the like, and to assume that regulations have already taken care of the food safety question.

A sixth difficulty is that much of the information about the effects of food additives comes from food studies produced or sponsored by industry. These studies have been criticized on several grounds, the most common being the dangers of potential conflicts of interest, the lack of peer review before results are released to the public, and the fact that few of the scientists who have found adverse effects have made an effort to stimulate public action on the problem. While there is little reason to suspect that these criticisms are relevant in most cases, they do little to enhance the credibility of information efforts sponsored by either industry or government.

A final difficulty is that it is not clear that simply because current laws already permit several very dangerous substances from being consumed that consumers favor similar regulations with respect to hazardous substances in the food supply even if the risks involved are somewhat smaller. Those who smoke and drink are well aware of what they are doing and these behaviors are not essential to life. However, eating is essential, and most of us assume that the array of foods we are choosing from is *already* wholesome and safe, so that our choices at the table can be based on individual tastes and preferences rather than further complicated by health hazard appraisal estimates.

CONCLUSION

It remains to be seen whether in the next few years existing food safety legislation will be radically overhauled or reinterpreted and adjusted to deal with these problems in a more systematic and timely manner. Regardless of the outcome, the issues discussed in this chapter are likely to play an important part in whatever resolution is finally arrived at. If prophecy is permissible, we predict that more rigorous methods of risk assessment, less emphasis on zero risk, and more explicit decision-making procedures will receive emphasis. Cost–benefit analysis will founder on the shoals of uncertainty and the rocky reefs of politics.

ACKNOWLEDGEMENTS. This chapter was written with the assistance of Christine Larsen. J. Dwyer acknowledges the support of Career Development Award K 04 AM 00273-03 from the National Institutes of Health which permitted the publication of this paper. We thank Glenda Brown, Jill Williamson, and Andrea Scarpa for their assistance in the preparation of the manuscript.

REFERENCES

Alexander, C. S., 1972, Cobalt-Beer Cardiopathy: A clinical and pathologic study of twenty-eight cases. *Am. J. Med.* **53:**395–417.

Berglund, F., 1978, Food additives, in: *Toxicological Aspects of Food Safety, (Arch. Toxicol. Suppl. 1),* pp. 33–46, Springer Verlag, New York.

Catz, C. S. and Giacola, G. P., 1972, Drugs and breast milk, *Pediatr. Clin. North Am.* **19:**151–166.

Code of Federal Regulations, 1977, Title 21–121.101, Food and Drugs, Part 100–199, U.S. Government Printing Office, Washington, D.C.

Cornfield, J., 1977, Carcinogenic risk assessment, *Science* **198:**693.

Darby, W. J., 1979, Food additives, *J. Florida Med. Assoc.,* **66:**471–475.

Dill, W. R., Overview of policy issues, in: *Science and Technology Policy: Perspectives for the 1980s* (H. Fusfield, ed.) pp. 172–175, *Ann. N.Y. Acad. Sci.* **334:**172–175.

Doull, J. F., Assessment of food safety. *Fed. Proc. Fed. Am. Soc. Exp. Bio.* **37:**2594–2597, 1978.

Farber, T. M., and Friess, S., 1980, Food safety, pp. 55–68 in: *Animal Agriculture: Research to Meet Human Needs in the 21st Century,* (W. G. Pond, R. A. Merkel, L. D. McGuilliard, and V. J. Rhodes, eds.), Westview Press, Boulder, Colorado.

Federal Register, Feb. 22, 1977, **42(35):**10412.

Filer, L. J., 1976, Patterns of consumption of food additives, *Food Technol.* **30:**62.

Food and Drug Administration, Feb. 22, 1977, Criteria and procedure for evaluating assays for carcinogenic residues, *Federal Register,* **42:**16412.

Food Safety Council, 1979, *Proposed System for Food Safety Assessment,* Scientific Committee, Food Safety Council, Washington, D.C. p. 149.

Food Safety Council, 1980, *Principles and Processes for Making Food Safety Decisions,* Social and Economic Committee, Food Safety Council, Washington, D.C. (unpublished manuscript).

Fusfield, H., 1979, Science and technology policy: Perspectives for the 1980s, *Ann. N. Y. Acad. Sci.* **334:**1–279.

General Mills, Inc., 1977, *A Summary Report on U.S. Consumer Knowledge: Attitudes and Practices About Nutrition,* General Mills, Minneapolis.

Goldberg, L., 1975, Safety evaluation concepts, *J. Assoc. Off. Anal. Chem.* **58:**635.

Goldenthal, E. I., 1971, A compilation of LD50 values in newborn and adult animals, *Toxicol. Appl. Pharmacol.* **18:**185–207.

Gorman, J., 1979, *Hazards to Your Health: The Problems of Environmental Disease,* New York Academy of Sciences, N.Y., New York.

Gumner, S. W. and Kirkpatrick, D. C., 1979, Approaches for estimating human intakes of chemical substances, *Can. Inst. Food Science Technol. J.* **12**:27.

Hoover, R. N., and Strasser, P. H., 1980, Artificial sweeteners and human bladder cancer, *Lancet* **1**:837–840.

Interagency Regulatory Liaison Group, Feb. 6, 1979, *Scientific Bases for Identifying Potential Carcinogens and Estimating Their Risks,* Food and Drug Administration, Washington, D.C.

Irving, G. W., 1978, Safety Evaluation of the food ingredients called GRAS, *Nutr. Rev.* **36**:351.

Jacobson, M., 1980, Diet and cancer, *Science* **207**:258–261.

Joint FAO/WHO Expert Committee on Food Additives, 1972, Evaluation of food additives. Some enzymes, modified starches and certain other substances: Toxicological evaluations and specifications and a review of the technological efficacy of some antitoxicants, *W. H. O. Tech. Rep. Ser.* **488.**

Mantel, N., and Bryan, W. R., 1961, Safety testing of carcinogenic agents, *J. Nat. Cancer Inst.* **27**:455.

Mantel, N., Bohidar, W. R., Brown, D. C., Ciminera, J. J., and Tukey, J. W., 1971, An improved "Mantel-Bryan" procedure for safety testing of carcinogens, *Cancer Res.* **35**:759.

Miller, S. A., 1977, Additives in our food supply, *Ann. N. Y. Acad. Sci.* **300**:397.

Miller, S., 1980, The new metaphysics, *Nutr. Rev.* **38**:53–64.

Morrison, A. E., and Buring, J. E., 1980, Artificial sweeteners and cancer of the lower urinary tract, *N. Engl. J. Med.* **302**:537–541.

National Academy of Sciences, 1965, *Some Considerations in the Use of Human Subjects in Safety Evaluation of Pesticides and Food Chemicals,* Publication 1270, National Research Council, NAS, Washington, D.C.

National Academy of Sciences, 1967, *Use of Human Subjects in Safety Evaluation of Food Chemicals,* Publication 1491, National Research Council, NAS, Washington, D.C.

National Academy of Sciences, Subcommittee on Review of the GRAS List, 1972 *A Comprehensive Survey of Industry on the Use of Food Chemicals Generally Recognized as Safe (GRAS),* Phase II, National Technical Information Service Reports (PB 221–949, 1972; PB 221–925, 1973; PB 221–939, 1973), Food Protection Committee, National Research Council NAS, Washington, D.C.

National Academy of Sciences, 1975a, *Experiments and Research with Humans: Values in Conflict,* National Research Council, NAS, Washington, D.C.

National Academy of Sciences, 1975b, *Principles for Evaluating Chemicals in the Environment.* p. 126, National Research Council, NAS, Washington, D.C.

National Academy of Sciences, Subcommittee on GRAS, 1977, *Estimating Distribution of Daily Intakes of Certain GRAS Substances,* List Survey (Phase III), Food and Nutrition Board, National Research Council, NAS, Washington, D.C.

Office of Technology Assessment, U.S. Congress, 1979, *Environmental Contaminants in the Food Supply,* Government Printing Office, Washington, D.C.

Office of Technology Assessment, U.S. Congress, 1980, *The Implications of Cost-Effectiveness Analysis in Medical Technology,* Government Printing Office, Washington, D.C.

Ostenso, G. L., 1979, Overview of policy issues: Panel report, in: Science and technology policy: Perspectives for the 1980s (H. Fusfield, ed.), *Ann. N. Y. Acad. Sci.* **334**:80–84.

Reed, S. K., 1979, Food and nutrition, in: Science and technology policy: Perspectives for the 1980s, (H. Fusfield, ed.), *Ann. N. Y. Acad. Sci.* **334**:71–79.

Roberts, H. R., 1980, *Food Safety Risk Assessment,* Unpublished manuscript prepared for Peer Review Seminar of Food Safety Council, Palo Alto, California.

Schmidt, A. M., 1975, Food and drug Law: A 2000 year perspective, *Nutr. Today* **10**:29–32.

Siu, R. G. H., Borzelleca, J. F., Carr, C. J., Day, H. G., Fomon, S. J., Irving, G. W., LaDu, B. W., McCoy, J. R., Miller, S. A., Plaa, G. L., Shimkin, M. B., and Wood, J. L., 1977, Evaluation of health aspects of GRAS food ingredients: Lessons learned and questions unanswered, Fed. Proc. *Fed. Am. Soc. Exp. Biol.* **36**:2519–2562.

Throdahl, M. C., 1979, Regulatory policy, in: Science and technology policy: Perspectives for the 1980s (H. Fusfield, ed.), *Ann. N. Y. Acad. Sci.* **334:**163–171.

Vanderveen, J., 1977, Government regulatory difficulties, *Ann. N. Y. Acad. Sci.* **300:**406–410.

Walker, R., 1976, Food additives: The benefits and the risks, *J. Biosocial Sci.* **8:**211–218.

Winspringer, W. W., 1979, A labor perspective on science and technology, in: Science and technology policy: Perspectives for the 1980s (H. Fusfield, ed.), *Ann. N. Y. Acad. Sci.* **334:**264–275.

Wynder, E. L., and Stellman, S. D., 1980, Artificial sweetener use and bladder cancer: A case control study, *Science* **207:**1214–1216.

15

COMMON FOOD ADDITIVES AND SPICES IN THAILAND
Toxicological Effects

Yongyot Monsereenusorn

Of the food additives used today some are as old as recorded history; others are recent products of chemical laboratories. The Food and Agriculture Organization (FAO) originally defined an additive as a "nonnutritive substance added intentionally to food generally in small quantities to improve its appearance, flavor, texture, or storage properties." With the modernization of society, the complexity of modern life, and changes in our eating habits, the exposure to these additives is ever-increasing. Consequently there is rising awareness of the toxicological problems with food additives in almost every country. In Thailand, due to the lack of comprehensive legislation and enforcement of regulations, the abuse of food additives and even adulteration of foods by unscrupulous tradesmen are serious problems.

This chapter describes in brief some toxicological problems encountered with some selected food additives and spices in Thailand, with special emphasis on the author's own experience concerning red chili.

MONOSODIUM GLUTAMATE

Monosodium glutamate (MSG) is the sodium salt of one of the most common amino acids found in the body and in foods. It is prepared from natural sources or by chemical synthesis and is used to enhance the flavor of foods. The Food and Drug Administration (FDA) places MSG in the category termed *generally regarded as safe* (GRAS).

The assumption that MSG is an innocuous substance for human consumption has been questioned in view of its role in the Chinese Restaurant Syndrome in susceptible individuals (Schaumberg *et al.*, 1969). Three cate-

Yongyot Monsereenusorn • Department of Biology, Faculty of Science, Ramkamhaeng University, Banggapi, Bangkok 24, Thailand.

gories of symptoms can be cited by MSG: burning, facial flushes, and chest pain. Headache is also a consistent complaint in a minority of individuals. While the response is dose-related, it is difficult to define its exact mechanism. Since glutamic acid is present in large amounts in the central nervous system (CNS), it has been suggested to be a neuro-humoral transmitter. Therefore, it seems reasonable to assume a CNS mechanism for all the sensory phenomena. Reports from Olney (1969) and Olney and Sharpe (1969) of brain lesions in certain mammalian species after parenteral MSG administration suggested a multifaceted neuroendocrine disturbance. Following the administration of MSG to neonate mice (3 mg/g body weight), retinal degeneration (Lucas and Newhouse, 1957) and necrosis of the hypothalamic neurons (Olney, 1969) were observed. As adults, treated animals (5–7 mg/g body weight) showed stunted skeletal development, marked obesity, and female sterility. However, later attempts to induce lesions in primates by MSG administration have proved unsuccessful. It is probable that in primates such as man, the mean peak concentrations of glutamic acid are well below those likely to cause neurological damage. Furthermore, an increase in plasma glutamic acid concentration occurs only for a brief period (Marrs *et al.*, 1978).

It would appear that a threshold level of the neurotoxicity of MSG, when administered in diet, has yet to be established for any species, either in neonates or in mature individuals. In earlier, conventional 2-year feeding studies, diets containing up to 5% MSG for the rat or weanling mouse (Heywood *et al.*, 1977) and up to 10% for the dog (James *et al.*, 1978), administered *ad libitum* have not been associated with any clinical or histopathological evidence of CNS damage. Glutamic acid did not accumulate in liver, kidney, duodenum, or brain tissue, and to date there is indeed no conclusive evidence from any dietary study that would suggest a lack of safety of MSG as a food additive. It appears, therefore, that the transitory clinical symptoms of the Chinese Restaurant Syndrome may be the sole consequences of excessive MSG consumption.

BORAX

Borax, with a chemical name of sodium borate decahydrate or disodium tetraborate ($Na_2B_4O_7 \cdot 10\ H_2O$), and a molecular weight of 381.38, is highly soluble in water.

In Thailand, borax is frequently used as food preservative and sometimes to disguise, and therefore render marketable, food that has deteriorated. The Science Department of the Ministry of Industry initially informed the public of its harmful effects in 1955. In 1974, the Ministry of Health officially banned borax as a food additive. Consequently, survey of borax contamination in various foodstuffs in the market has shown that its occurrence and use seem to be restricted in recent years (Table I).

Borax is absorbed directly by oral route or through the skin (George,

TABLE I
Incidences of Borax Contamination as Surveyed by the Department of Health in Thailand[a]

Types of food inspected	Year inspected	Percentage of samples contaminated with borax
meat	1960	78.75
pork	1973	56.89
shrimp	1974	91.18
seafood	1975	37.89
fish	1976	7.14
dessert	1977	few

[a] Modified from Chertchewasart, 1980.

1965). The absorbed portion is excreted through urine, and in minute quantities through sweat (Valdes-Dapena and Arey, 1962). If high amounts are consumed, the substance may be found accumulated in liver and brain (Pfeiffer *et al.*, 1945).

The toxicological evaluation of borax is summarized in Table II. It can be seen that borax has a wide range of actions on various enzyme activities and physiological functions. The importance of borax in producing testicular dysfunction is suggested by the preliminary evidence (Lee *et al.*, 1978) that it can accumulate in the testes and that this accumulation is both time- and dose-dependent. Furthermore, the testicular lesion can persist long after the toxic response to borax has occurred. The relevance of this observation to humans depends on whether such chronic exposure could occur in man.

TABLE II
Summary of Various Toxicological Effects of Borax

LD_{50}:	
acute	4.5–4.98 g/kg (rat)
chronic	525 ppm for 90 days (rat)
	175 ppm for 90 days (dog)
Enzyme activities:	
urate oxidase	
alcohol dehydrogenase	
glyceraldehyde phosphate dehydrogenase	
urease, arginase, cholinesterase, pepsin, phosphodiesterase, chymotrypsin hexokinase	
Dermal effect	
Effects on growth:	
growth suppression, decreased food utilization efficiency (5250 ppm)	
Effects on reproduction:	
1170 ppm for 90 days—testicular atrophy	
5250 ppm for 30 days—degeneration of gonads	
morphologic alteration of some organs	

Although there is little data on borax levels in food in Thailand, concentrations have been reported to range from 0.007% to 1.185% (Medical Science Dept, Ministry of Health, 1977, from Chertchewasart, 1980). It is not known whether such levels in food can induce any toxicological effects in humans. On the other hand, results obtained from the studies of Weir and Fisher (1972) indicated that borax is relatively nontoxic by oral route and there is a reasonable margin of safety between the toxic dose in animals and the amount actually consumed by man.

RED PEPPER

Red pepper, of the *Capsicum* species, is a popular spice in Thailand. The active ingredient responsible for the irritating and pungent effects in the fruits is classified within a compound group called the *capsaicinoids,* which are acid amides of vanillylamine and C_9 to C_{11} branched-chain fatty acids with five analogues: capsaicin, dihydrocapsaicin, nordihydrocapsaicin, homocapsaicin, and homodihydrocapsaicin. Capsaicin is the major component, occupying 60% of the total capsaicinoid, followed by dihydrocapsaicin. The proportion of nordihydrocapsaicin is less than 1%. Only traces of homodihydrocapsaicin and homocapsaicin are detected (Iwai *et al.,* 1979; Buranawuti and Glinsukon, 1980). Most pharmacological and toxicological actions of red pepper have been attributed to capsaicin. Acute toxicity of capsaicin indicates a high susceptibility in guinea pigs, rats, and mice (Glinsukon *et al.,* 1980). It also has a wide and interesting profile of biological activity (see reviews by Virus and Gebhart, 1979; and Monsereenusorn *et al.,* 1982).

Capsaicin (or red pepper) has well-documented pharmacological effects on the cardiovascular and respiratory systems. Acute intravenous or intraventricular administration of capsaicin on anesthetized cats and dogs resulted in hypotension, bradycardia, and apnea with reflex bronchoconstriction (Russell and Lai-fook, 1979). Upon local application, it has also been shown to cause neurogenic inflammation and pain. However, chronic or pretreatment of capsaicin, either topically or parenterally, produces an anesthesia-like condition (Jancso-Gabor, 1980) and subsequent insensitivity to almost all types of chemical pain that persists for several months, while responses to light touch, and mechanical stimulation are unimpaired. This inhibition of neurogenic pain and inflammation produced by capsaicin has been related to the change in morphology and number of fibers in the dorsal root ganglion (Lawson and Nickels, 1980) and subsequent blockage of impulse generation (Porszasz and Jancso, 1959) of the peripheral nerve endings (Holzer *et al.,* 1979); it might also cause the release (Gamse *et al.,* 1979) and depletion (Yaksh *et al.,* 1979) of substance P from the spinal cord and the skin (Hayes and Tyers, 1980a,b; Gamse *et al.,* 1980). Recently, the stimulation of prostaglandin synthesis (Collier *et al.,* 1976) and its enhanced release (Juan *et al.,* 1980) have

TABLE III[a]

Effect of Capsicum annuum and Capsaicin on Rat's Intestinal Enzymes (μM Substrate Hydrolyzed/hr)

Duration of treatment (days)	Lactase			Sucrase			Maltase		
	Control	Capsicum annuum[b]	Capsaicin[c]	Control	Capsicum annuum[b]	Capsaicin[c]	Control	Capsicum annuum[b]	Capsaicin[c]
0	2.6 ± 0.3	2.4 ± 0.3	2.5 ± 0.1	22.5 ± 2.4	24.1 ± 2.6	24.1 ± 2.5	136.0 ± 14.0	140.1 ± 15.0	134.1 ± 14.1
1	2.8 ± 0.3	2.9 ± 0.3	2.9 ± 0.3	21.0 ± 2.5	22.0 ± 2.4	21.0 ± 2.8	156.7 ± 15.8	148.2 ± 16.0	149.6 ± 16.2
2	2.2 ± 0.3	2.4 ± 0.4	2.4 ± 0.3	17.5 ± 1.9	19.2 ± 3.2	15.4 ± 3.2	126.0 ± 13.2	132.1 ± 14.2	122.1 ± 13.4
3	2.0 ± 0.1	2.2 ± 0.2	2.4 ± 0.4	16.7 ± 1.2	12.5 ± 1.1[d]	13.1 ± 1.2[d]	120.0 ± 11.2	99.0 ± 8.9[d]	101.2 ± 9.0[d]
4	2.3 ± 0.2	2.7 ± 0.4	2.5 ± 0.5	17.7 ± 1.1	10.1 ± 0.9[d]	12.2 ± 2.1[d]	131.0 ± 10.4	95.4 ± 10.2[d]	105.4 ± 12.4[d]
5	2.6 ± 0.4	2.4 ± 0.5	2.9 ± 0.4	19.5 ± 1.8	10.2 ± 1.8[d]	12.4 ± 1.2[d]	162.1 ± 19.1	88.2 ± 10.2[e]	106.2 ± 12.1[e]

[a] Presented at the Eighth International Congress of Pharmacology (July 19–24, 1981), Tokyo, Japan.
[b] The whole fruit contains 257.0 ± 18.6 mg capsaicin/100 g dry wt capsicum. Dosage was 10 g/100 g body wt per day.
[c] Dosage was 25.7 mg/100 g body wt per day.
[d] $P < 0.02$
[e] $P < 0.005$

also been implicated as one of the explanations for the subsequent functional impairment of chemosensitive fibers by capsaicin pretreatment.

A single dose of capsaicin administered subcutaneously results in an immediate dose-dependent fall in body temperature of rats. When the capsaicin injections are repeated, the hypothermic effect decreases and finally vanishes. A desensitized condition arises that is characterized by the rats' inability to protect themselves from overheating in a warm environment. It has been suggested that the decreased tolerance of capsaicin-treated rats for elevated ambient temperatures might be due to an impaired peripheral heat-dissipating mechanism, such as reduction of saliva secretion, decreased grooming activity and inability to escape to a cooler environment (Obal *et al.*, 1979), or desensitization of the warmth receptors (Hori, 1980). This malfunction of the thermoregulatory mechanisms then lead the body thermostat to set at higher levels (Szekely and Szolcsanyi, 1979). The molecular basis of such desensitized hyperthermia is not clear. Stimulation of Ca^{+2}-dependent adenyl cyclase in the brain has been suggested (Horvath *et al.*, 1979).

Capsaicin has been blamed as the etiological factor of various gastrointestinal disorders. These include stimulation of gastric secretion (Limlomwongse, *et al.*, 1979) and gastrointestinal motility. According to the recent concept (Szolcsanyi and Bartho, 1978, 1979, 1980), activation of capsaicin-sensitive sensory nerve endings results in a discharge of a transmitter, which in turns excites cholinergic neurons at the myenteric plexus. The result is a local chemoreflex in which coordinated peristaltic reflexes take place. The mediator that is released from the capsaicin-sensitive nerve endings of the gut (Bartho and Szolcsanyi, 1978) has not yet been identified. It cannot be substance P, as in the case of thermoregulating and sensory pharmacology, since desensitization of substance-P receptors does not interfere with the neurogenic contraction elicited by capsaicin (Bartho and Szolcsanyi, 1978), and capsaicin treatment has no effect on intestinal substance-P concentration (Holzer *et al.*, 1980).

Excess consumption of red chili or its pungent substance capsaicin affects intestinal structure and functions. Some examples are inhibition of glucose (Monsereeunsorn and Glinsukon, 1978), fat (Buranawuti, 1980) and electrolyte (Monsereenusorn, 1980) absorption. Recently it was found that the intestinal disaccharidase activities in rat fed with capsaicin and red pepper were also reduced. Table III shows that the toxicity of capsicum extract on intestinal sucrase, maltase, and lactase activities was higher than its pungent substance, capsaicin, suggesting that other capsaicin derivatives in red pepper have a synergistic effect.

CONCLUSION

The toxicological actions of various food additives and spices described are based on animal experiments. It should be emphasized that in attempting to evaluate their risks, it is important to recognize that the oral dose is but one among several determinants. Other factors such as the maturity of enzyme systems, absorption by the gastrointestinal tract, and individual metabolic capabilities could act in concert to produce the toxicity.

ACKNOWLEDGEMENT. The author would like to express his sincere appreciation to Dr. Heinz Palla for his criticism and suggestions.

REFERENCES

Bartho, L., and Szolcsanyi, J., 1978, The site of action of capsaicin on the guinea-pig isolated ileum, *Naunyn-Schmiedeberg's Arch. Pharmacol.* **305:**75–81.

Buranawuti, T., 1980, *Effect of Capsaicin on Lipid Absorption,* presented in the 9th Physiological Society Meeting of Thailand (April, 10-11, 1980), Chulalongkorn University, Bangkok, Thailand.

Buranawuti, T., and Glinsukon, T., 1980, Determination of capsaicin in various species of capsicum fruits and its toxicity in mice, *R. Thai Army Med. J.* **33**(2)**:**85–99.

Chertchewasart, V., 1980, Borax, *Witayasart* **34**(8)**:**624–631 (in Thai).

Collier, H. O. J., McDonald-Gibson, W. J., and Saeed, S. A., 1976, Stimulation of prostaglandin biosynthesis by drugs: Effects *in vitro* of some drugs affecting gut function, *Br. J. Pharmacol.* **58:**193–199.

Gamse, R., Molnar, A., and Lembeck, F., 1979, Substance-P release from spinal cord slices by capsaicin, *Life Sci.* **25:**629–636.

Gamse, R., Holzer, P., and Lembeck, F., 1980, Decrease of substance-P in primary afferent neurons and impairment of neurogenic plasma extravasation by capsaicin, *Br. J. Pharmacol.* **68:**207–213.

George, A. J., 1965, Toxicity of boric acid through skin and mucous membrane, *Food Cosmet. Toxicol.* **3:**99–101.

Glinsukon, T., Stitmunnaithum, V., Toskulkao, C., Buranawuti, T., and Tangkrisanavinont, V., 1980, Acute toxicity of capsaicin in several animal species, *Toxicon* **18:**215–220.

Hayes, A. G., and Tyers, M. B. 1980a, Effect of capsaicin on nociceptive heat, pressure, and chemical thresholds and on substance P levels in the rat, *Brain Res.* **189:**561–564.

Hayes, A. G., and Tyers, M. B., 1980b, Capsaicin depletes substance P from dorsal horn and skin and discriminates heat from chemical and pressure nociceptive stimuli in the conscious rat, *J. Physiol.* **300:**25P.

Heywood, R., James, R. W., and Worden, A. N., 1977, The *ad-libitum* feeding of monosodium glutamate to weanling mice, *Toxicol. Lett.* **1:**151–155.

Holzer, P., Jurna, I., Gamse, R., and Lembeck, F., 1979, Nociceptive threshold after neonatal capsaicin treatment, *Eur. J. Pharmacol.* **58:**511–514.

Holzer, P., Gamse, R., and Lembeck, F., 1980, Distribution of substance-P in the rat gastrointestinal tract: Lack of effect of capsaicin pretreatment, *Eur. J. Pharmacol.* **61:**303–307.

Hori, T., 1980, The capsaicin desensitized rat: Behavioral thermoregulation and thermosensitivity of hypothalamic neurons in: *Thermoregulatory Mechanisms and Their Therapeutic Implications,* Fourth International Symposium on the Pharmacology of Thermoregulation (Oxford, 1979), pp. 214–215, Karger, Basel.

Horvath, K., Jancso, G., and Wollemann, M., 1979, The effect of calcium on the capsaicin activation of adenylate cyclase in rat brain, *Brain Res.* **179:**401–403.

Iwai, K., Suzuki, T., and Fujiwake, H., 1979, Formation and accumulation of pungent principle of hot pepper fruits, capsaicin and its analogues, in: *Capsicum annuum* var. *annuum* cv. Karayatsubusa at different growth stage after flowering, *Agric. Biol. Chem.* **43(12):**2493–2498.

James, R. W., Heywood, R., and Salmona, M., 1978, Uptake of glutamate in beagle dogs after oral gavage with MSG, *Toxicol. Lett.* **2:**305–311.

Jancso-Gabor, A., 1980, Anaesthesia-like condition and/or potentiation of hexobarbital sleep produced by pungent agents in normal and capsaicin-desensitized rats, *Acta Physiol. Acad. Sci. Hung. Tomus* **55(1):**57–62.

Juan, H., Lembeck, F., Seewann, S., and Hack, U., 1980, Nociceptor stimulation and PGE release by capsaicin, *Naunyn-Schmiedeberg's Arch. Pharmacol.* **312:**139–143.

Lawson, S. N., and Nickels, S. M., 1980, The use of morphometric techniques to analyse the effect of neonatal capsaicin treatment on rat dorsal root ganglia and dorsal roots, *J. Physiol.* **303:**12P.

Lee, I. P., Sherins, R. J., and Dixon, R. L., 1978, Evidence for induction of germinal aplasia in male rats by environmental exposure to boron, *Toxicol. Appl. Pharmacol.* **45:**577–590.

Limlomwongse, L., Chaitauchawong, C., and Tongyai, S., 1979, Effect of capsaicin on gastric secretion and mucosal blood flow in the rat, *J. Nutr.* **109(5):**773–777.

Lucas, D. R., and Newhouse, J. P., 1957, The toxic effects of sodium-L-glutamate on the inner layers of the retina, *Am. Med. Assoc. Arch. Opthalmol.* **58:**193–201.

Marrs, T. C., Salmona, M., Garattini, S., Burston, D., and Matthews, D. M., 1978, The absorption by human volunteers of glutamic acid from monosodium glutamate and from a partial enzyme hydrolysate of casein". *Toxicology* **11:**101–107.

Monsereenusorn, Y., and Glinsukon, T., 1978, Inhibitory effect of capsaicin on intestinal glucose absorption *in vitro, Food Cosmet. Toxicol.* **16:**469–473.

Monsereenusorn, Y., 1980, Effect of capsaicin on intestinal fluid and Na^+-absorption, *J. Pharmacol. Biodyn.* **3(12):**631–635.

Monsereenusorn, Y., Kongsamut, S., and Pezalla, P., 1982, Capsaicin: Literature survey, *CRC Crit. Rev. Toxicol.* (in press).

Obal, F., Benedek, G., Jancso-Gabor, A., and Obal, F., 1979, Salivary cooling, escape reaction, and heat pain in capsaicin-desensitized rats, *Pflugers Arch.* **382:**249–254.

Olney, J. W., 1969, Brain lesions, obesity, and other disturbances in mice treated with monosodium glutamate, *Science* **164:**719–721.

Olney, J. W., and Sharpe, L. G., 1969, Brain lesions in an infant rhesus monkey treated with monosodium glutamate, *Science* **166:**386–388.

Pfeiffer, C. C., Hallmann, L. F., and Gersh, I., 1945, Boric acid ointment: A study of possible intoxication in the treatment of burns, *J. Am. Med. Assoc.* **128:***266–273.*

Porszasz, J., and Jansco, N., 1959, Studies on the action potentials of sensory neurons in animals desensitized with capsaicin, *Acta Physiol. Acad. Sci. Hung.* **16:**299–306.

Russell, J. A., and Lai-fook, S. J., 1979, Reflex bronchoconstriction induced by capsaicin in the dog, *J. Appl. Physiol.* **47(5):**961–967.

Schaumburg, H. H., Byck, R., Gerstl, R., and Mashman, J. H., 1969, Monosodium-L-glutamate: Its pharmacology and role in the Chinese Restaurant Syndrome, *Science* **163:**826–828.

Szekely, M., and Szolcsanyi, J., 1979, Endotoxin fever in capsaicin treated rats, *Acta Physiol. Acad. Sci. Hung. Tomus* **53(4):**469–477.

Szolcsanyi, J., and Bartho, L., 1978, New type of nerve-mediated cholinergic contractions of the guinea-pig small intestine and its selective blockage by capsaicin. *Naunyn-Schmiedeberg's Arch. Pharmacol.* **305:**83–90.

Szolcsanyi, J., and Bartho, L., 1979, Capsaicin-sensitive innervation of the guinea pig taenia caeci, *Naunyn-Schmiedeberg's Arch. Pharmacol.* **309:**77–82.

Szolcsanyi, J., and Bartho, L., 1980, The capsaicin-sensitive non-parasympathetic excitatory innervation of the small intestine, in: *Symposium on Modulation of Neurochemical Transmission* (J. Knoll and E. S. Vizi, ed.), Akademiai Kiado, Budapest, pp. 311–320.

Valdes-Dapena, M. A., and Arey, J. B., 1962, Boric acid poisoning, *J. Pediatr.* **61:**531–546.

Virus, R. M., and Gebhart, G. F., 1979, Pharmacologic actions of capsaicin: Apparent involvement of substance P and serotonin, *Life Sci.* **25:**1273–1284.

Weir, R. J., and Fisher, R. S., 1972, Toxicologic studies on borax and boric acid, *Toxicol. Appl. Pharmacol.* **23:**351–364.

Yaksh, T. L., Farb, D. H., Leeman, S. E., and Jessell, T. M., 1979, Intrathecal capsaicin depletes substance-P in the rat spinal cord and produces prolonged thermal analogesia, *Science* **206:**481–483.

16

THE HYPERKINESIS CONTROVERSY

Morris A. Lipton and Jeanine C. Wheless

Current evidence suggests that the syndrome of hyperkinesis occurs in a heterogeneous group of children and results from multiple etiologies. Hyperkinesis can occur in children with normal intelligence who have emotional disorders such as anxiety, depression, phobias, psychoses, and personality disorders. It is frequently found in children with minimal brain dysfunction (MBD) where it may be associated with "soft neurological signs" such as impairment of laterality and spatial orientation, and evidence of motor incoordination (Wender, 1971). Hyperactivity in children has been related to organic brain syndromes, sensory disorders like blindness or deafness, endocrine disorders like hyperthyroidism and possibly hypoglycemia, and toxicity from environmental pollutants like lead. The fact that the syndrome is from four to nine times more prevalent in males than females suggests that there may be a genetic factor as well. In addition to those factors that have been unequivocally demonstrated to contribute to etiology, claims have been made that the hyperkinetic syndrome may be associated with food allergy, vitamin deficiency, stress from fluorescent lighting and television cathode tubes, and the ingestion of food additives.

Accepted treatments of hyperactivity are both pharmacological and psychological. It should be emphasized that, while both drugs and psychological treatments diminish symptoms and improve the child's behavior and learning ability, there is no "cure" for hyperactivity. Consequently, it is not surprising that some parents, dissatisfied with conventional treatments, seek and accept new treatments that claim to be more effective, even though these methods have not been scientifically confirmed. Treatments based on the elimination of fluorescent lighting, avoidance of possible food allergens, the ingestion of

Morris A. Lipton • Department of Psychiatry and Biological Sciences Research Center, University of North Carolina School of Medicine, Chapel Hill, North Carolina 27514. *Jeanine C. Wheless* • Biological Sciences Research Center, University of North Carolina School of Medicine, Chapel Hill, North Carolina 27514.

mega-quantities of the water-soluble vitamins, and the elimination of synthetic food additives from the diet are among those therapies that have been advocated by different groups of quasi-scientific investigators and accepted by different segments of the lay public.

The most popular of the novel treatments in the U.S. is based upon a hypothesis developed by Dr. Ben Feingold, a pediatrician and allergist from San Francisco. Dr. Feingold has proposed that some children have a central nervous system "variation" that predisposes them to sensitivity to synthetic food additives, particularly to food colors and antioxidants [butylated hydroxyanisole (BHA) and butylated hydroxytoluene (BHT)]. In such children, ingestion of these additives results in hyperkinetic behavior; elimination of these items from the diet leads to dramatic improvement and even cure. Feingold (1975) claims that 50–70% of hyperactive children placed on a diet devoid of additives will have a complete remission and that 75% of children who have been treated with stimulants can discontinue medication. The diet, he says, becomes effective in several days to several weeks, and the younger the child, the quicker and more complete the response. One hundred percent adherence to the diet is mandatory, Feingold insists, because any infringement—for example, one bite of a food or sip of a beverage containing additives—causes a return of symptoms within about 3 hr, and persisting as long as 72 hr. Successful treatment usually requires the entire family to be on the diet, and an individual sensitive to artificial food colors, flavors, and antioxidants must, according to Feingold (1975), avoid them for life. Although Feingold claims that dramatic successes have occurred, neither he nor his advocates have conducted controlled, double-blind trials to test his hypothesis, and his claims are based entirely on anecdotal reports and findings from open clinical trials. Nevertheless, the Feingold diet has been widely publicized and has gained considerable popularity among the lay public. A recent letter in *Science* (Morrison, 1978) reports that more than 20,000 families of hyperkinetic children adhere to the diet and advocate its use.

Based upon his hypothesis, Feingold has argued that food additives should be eliminated from federally funded school lunch programs, that the quantity and nature of food additives in all commercial foods should be specifically designated, and that a "logo" should be placed on foods that contain no additives. Since color additives appear in such diverse substances as toothpaste, perfume, cough drops, and vitamin preparations, as well as in most processed foods, it is no simple task to prepare an additive-free diet. Home preparation of all foods is generally required, which precludes eating school lunches, at restaurants, or with friends. Furthermore, since total compliance is mandatory, Feingold recommends that the entire family adhere to the dietary program in order to ensure compliance by the hyperactive child. The dramatic nature of the therapeutic claims, the attendant publicity, and the apparent popular acceptance of this dietary treatment have made the question of the validity of the Feingold claims both a scientific and a political issue. If, indeed, the brains and minds of many children are adversely affected by common food additives, governmental regulation may be necessary. On the other hand, when there is a high level of expectation, when the whole family unifies and sacrifices to follow this therapeutic diet, when the child's role in the family becomes dominant insofar as he is the one who determines what the remainder of the family can eat, there is ample opportunity for placebo effect.

Many open studies, in which parents or physicians place children on the additive-free diet and then anecdotally report their results, have been conducted, and these generally support Feingold's claims. Cook and Woodhill (1976) reported that parent and teacher ratings indicated that 13 of 15 children had a reduction in hyperactive behavior while on the elimination diet and a deterioration of behavior when infractions of the diet occurred. In another study (Salzman, 1976), 14 of 15 children were reported to have statistically significant improvement in overactivity, excitability, distractibility, and impulsiveness while on the diet. Of 32 families who participated in a study conducted by Brenner (1977), 34% reported an excellent response to the diet, 25% reported an equivocal response, and no change occurred in 40%. Parents observed detrimental effects on behavior when "challenge" foods containing additives were ingested, and deterioration of behavior when the diet was terminated. Harper and co-workers (1978) stated that 50% of 54 children demonstrated significant improvement in behavior while on the Feingold diet.

Rigorous, double-blind, controlled clinical trials have also been conducted and these studies have generally yielded negative results. Based on these findings, it appears that the beneficial results of the diet reported in open studies are largely due to placebo effects.

Two research strategies have been employed in the double-blind experiments. Because of the difficulty and expense of conducting controlled, blind trials, only one study has employed the strategy of comparing the additive-free diet with an ordinary diet containing additives (Harley *et al.*, 1978a). All of the food for the families participating in the study was provided for the duration of the trial and neither the investigative team nor the family knew which diet was being consumed at any particular time; diet phases were switched so all families ingested both diets. In 36 school-age boys, no significant differences in behavior occurred under the two dietary conditions, according to teacher and objective ratings. With these school-age children, some parents reported statistically significant improvement on the additive-free diet, but this occurred only when the additive-free diet followed the control diet; parents were unable to detect differences in behavior when the Feingold diet preceded the additive-containing diet. The failure to find deterioration when the additive-containing diet was introduced after a week or more on the additive-free diet differs from Feingold's assertion that even a minor infraction will cause deterioration in behavior that can last for several days. Furthermore, the *diet × order effect* weakens the evidence of the parental ratings since the results should have been independent of the order in which the two diets were given. Ten preschool boys were studied under the same dietary conditions with the same battery of tests and evaluations, with the exclusion of teacher ratings. Parents of these young children were able to detect differences with the additive-free diet, and this was true whether the Feingold diet preceded or followed the additive-containing diet. The authors (Harley *et al.*, 1978a, p. 827) conclude: "While we feel confident that the cause–effect relationship asserted by Feingold is seriously overstated with respect to school-age children, we are not in a position to refute his claims regarding the possible causative effect played by artificial food colors on preschool children."

Because the design is simpler, most of the double-blind studies have used a "challenge" strategy. In this type of trial, children who are reported by their parents to respond to the additive-free diet are blindly "challenged" by the addition of food color additives to the additive-free diet. Weiss *et al.* (1980) conducted a study of 22 preschool children reported to improve on the Feingold diet. Only one child, a 3-year-old girl, had a consistent and significant adverse response to challenge with 36 mg/day of food color. Of 9 putative diet responders studied by Harley *et al.* (1978b), only one had an adverse reaction to ingestion of 26 mg/day of colors. In a study of 22 children by Levy *et al.* (1978), no positive response to the additive-free diet nor deterioration in behavior following active challenge with 5 mg/day of tartrazine was detected by teachers, clinicians, or objective ratings. Although mothers reported significant improvement with the diet, they were unable to detect detrimental effects of active challenge. Thirteen children from this group were tested further and significant challenge effects were noted only when the ratings were done a few hours after administration of the challenge, a finding quite different from Feingold's claim that behavioral effects can persist for several days. Mattes and Gittelman-Klein (1978) conducted a 10-week, double-blind challenge trial of a single child whose parents had volunteered him because they were certain his responses would unequivocally confirm the efficacy of the Feingold diet. In this child, neither maternal nor teacher ratings revealed a significant difference between active (39 mg/day food color) and placebo challenges, and both the teacher and the child guessed that only placebo had been administered throughout the trial. However, the mother correctly guessed challenge phases 8 out of 10 times, despite the fact that her objective ratings did not differentiate the dietary periods.

In the course of investigating the effects of food additives on the clinical symptoms of hyperkinesis, several investigators have encountered an interesting behavioral phenomenon whose relevance to hyperkinesis is uncertain but that is nonetheless worth noting. The first observation was made by Conners, Goyette, and coworkers (Conners *et al.*, 1976; Goyette *et al.*, 1978), who found that hyperkinetic children administered active challenge who showed no differences in their overall behavior in school or over a full day, nonetheless showed some decrements in performance in a laboratory task requiring attention and concentration. The effect was noted shortly after ingestion of the additive and persisted only about an hour. Because of this laboratory finding, the investigators then asked parents to focus on the 3-hour period immediately following ingestion of the challenge substances. In preliminary experiments, parents reported a change in behavior during this brief period. However, the investigators were unable to repeat this finding on a larger sample of children.

Swanson and Kinsbourne (1979) replicated the Conners *et al.*, (1976) experiment with several modifications. The most important change was that in addition to using 26 mg of food colors, 100 mg of color administered as a bolus in a chocolate-covered capsule, was also tested. Second, in order to ensure compliance, the hyperkinetic children were admitted as inpatients and placed on an additive-free diet for 4–6 days. Finally, the investigators employed a learning task in a laboratory setting, measuring the errors before and for several hours after ingestion of the challenge. They had previously found that this learning task could measure the favorable effects of stimulant medication (Swanson *et al.*, 1978). In this study, Swanson and Kinsbourne were unable to detect an effect of 26 mg of additive, but reported a decrement in performance after ingestion of 100 mg. The effect, though not large, seemed to be statistically significant and occurred 1.5 and 3.5 hr after administration of the active challenge. They attribute their positive findings of the adverse effects of food colors to the use of 100 mg of these agents, and feel that there is a dose–response curve for the food colors just as there is for any toxic substance.

The relationship of the findings of Conners *et al.* (1976; Goyette *et al.*, 1978) and of Swanson and Kinsbourne (1979) to the Feingold hypothesis is by no means clear. On the one hand, these data suggest that the mixture of food color additives is not biologically inert. On the other hand, these data do not support Feingold's contention that a minor infraction of the diet produces a dramatic and prolonged behavioral change. Instead, the most positive data suggest that a major infraction produces a minor, transient effect. Swanson and Kinsbourne (1979) submit that the relationship between the effects of the long-term exposure to low levels of the toxic substance and the short-term exposure to high levels of the same substance is complex. Thus, they consider it possible that even though a large, single dose of color may not produce an acute change in behavior, the chronic daily exposure to a small amount may have adverse long-term effects on behavior. This possibility must be considered, but there is currently no evidence to support it.

In all of the behavioral "challenge" studies, the eight most common food colors have been lumped together and administered in a mixture calculated to resemble the probable percentage intake of each dye (see Table I). There is, of course, the distinct possibility that only one of the dyes might be toxic while the others are biologically inert. This possibility has been tested. Logan and Swanson (1979) used an homogenate of whole rat brain to investigate the effect of food dyes on the uptake of neurotransmitters or neurotransmitter precursors. Of the 7 food dyes tested, 5 had been contained in the mixture used in challenge experiments with hyperkinetic children. The investigators

TABLE I
Artificial Food Colors in the Nutrition Foundation "Challenge" Cookies[a]

Color	% of total
FD & C Blue #1	3.12
FD & C Blue #2	1.70
FD & C Green #3	0.13
FD & C Red #3	6.08
FD & C Red #40	38.78
FD & C Yellow #5	26.91
FD & C Yellow #6	22.74
Certified Orange B	0.54

[a] Challenge cookies were prepared by the Nutrition Foundation to contain a mixture of the eight most common food dyes in the proportions estimated to be ingested on ordinary diets. Thirteen mg of this artificial color mixture was contained in each active challenge dessert. (Dr. William Darby, President, The Nutrition Foundation, personal communication).

found that a mixture containing 1 μg/ml of each of 7 food dyes inhibited uptake of 9 neurotransmitters or neurotransmitter precursors to an average of about 50%. When the dyes were tested individually, it was found that only one dye, Red #3, also called Erythrosin B, erythrosine, or tetraiodofluorescein, inhibited the uptake of all of the neurotransmitters by about 50%. The other 6 dyes were inert. The lack of specificity for any single neurotransmitter lead Logan and Swanson (1979) to suggest that Red #3 has a generalized membrane effect. Another laboratory (Lafferman and Silbergeld, 1979) confirmed the *in vitro* findings of Swanson and Logan, although the degree of uptake inhibition differed in the two studies. However, later biochemical studies (Mailman *et al.*, 1980) demonstrated that the ability of Red #3 to inhibit dopamine uptake into brain synaptosomes was dependent upon the concentration of tissue present in the assay mixture. Thus, the finding that Red #3 inhibits dopamine uptake may actually be an artifact that results from nonspecific interactions with brain membranes. This finding suggested that because the body would be a large reservoir of tissue (a "buffer") for nonspecific interactions, the color would be relatively inert biologically. Preliminary studies using rat tests of two behavioral symptoms similar to those of MBD children ("hyperactivity" and "impulsivity") showed that intraperitoneal doses of 50 mg/kg or larger were needed to cause behavioral effects. The issue of the *in vivo* neurotoxicity of Red #3 is not yet settled.

This then is the state of the art with respect to the behavioral toxicity of food colors. Several conclusions can be drawn. First of all, it seems clear enough that the Feingold claims and those from other open trials have been grossly exaggerated and that, at best, only a tiny fraction of hyperactive children may be "sensitive" to the common food additives that have been studied.

Although the results from double-blind studies have been largely negative, a number of trends are weakly detectable. Younger children may be more sensitive to the detrimental effects of additives, and parents (but not objective ratings) have frequently been able to detect improvement in behavior in preschool hyperkinetic children who are placed on the additive-free diet (Harley *et al.*, 1978a). Transient decrements in attention and concentration have also been reported following challenge with 26–100 mg of color. Thus, hyperactive children who showed no adverse reaction to active challenge in their overall behavior in school or over a day, nonetheless showed some decrements in performance in specific laboratory tasks requiring concentration and attention when rated for brief periods immediately following ingestion of the challenge (Conners *et al.*, 1976; Goyette *et al.*, 1978; Swanson and Kinsbourne, 1979). There may possibly be a dose–response curve for the food colors just as there is for any toxic substance, but this has not yet been demonstrated. Finally, the controversy as to whether or not Red #3 is the specific culprit remains to be resolved. These findings raise some doubts that the food colors are completely innocuous for all children.

It is our opinion, however, that the existing data are such to preclude any major legislative or administrative action to remove food additives or to limit their use severely. Further intensive studies of those very few children who seem to respond adversely to the dyes seem warranted, though they may be difficult to perform. Finally, additional *in vivo* animal research and *in vitro* biochemical research, dealing with the questions of whether these food additives have any biological activity in the central nervous system, and what the mechanism of this activity might be, are needed.

REFERENCES

Brenner, A., 1977, A study of the efficacy of the Feingold diet on hyperkinetic children, *Clin. Pediatr.* **16:**652–656.

Conners, C. K., Goyette, C., Southwick, D. A., Lees, J. M., and Andrulonis, P. A., 1976, Food additives and hyperkinesis: A controlled double-blind experiment, *Pediatrics* **58:**154–166.

Cook, W. S., and Woodhill, J. M., 1976, The Feingold dietary treatment of the hyperkinetic syndrome, *Med. J. Aust.* **2:**85–90.

Feingold, B. F., 1975, *Why Your Child Is Hyperactive,* Random House, New York.

Goyette, C. H., Conners, C. K., Petti, T. A., and Curtis, L. E., 1978, Effects of artificial colors on hyperkinetic children: A double-blind challenge study, *Psychopharmacol. Bull.* **14:**39–40.

Harley, J. P., Ray, R. S., Tomasi, L., Eichman, P. L., Matthews, C. G., Chun, R., Cleeland, C. S., and Traisman, E., 1978a, Hyperkinesis and food additives: Testing the Feingold hypothesis, *Pediatrics* **61:**818–828.

Harley, J. P., Matthews, C. G., and Eichman, P., 1978b, Synthetic food colors and hyperactivity in children: A double-blind challenge experiment, *Pediatrics* **62:**975–983.

Harper, P. H., Goyette, C. H., and Conners, C. K., 1978, Nutrient intakes of children on the hyperkinesis diet, *J. Am. Diet., Assoc.* **73:**515–520.

Lafferman, J. A., and Silbergeld, E. K., 1979, Erythrosin B inhibits dopamine transport in rat caudate synaptosomes, *Science* **205:**410–412.

Levy, F., Dumbrell, S., Hobbes, G., Ryan, M., Wilton, N., and Woodhill, J. M., 1978, A double-blind crossover trial with a tartrazine challenge, *Med. J. Aust.* **1**:61–64.

Logan, W. J., and Swanson, J. M. L., 1979, Erythrosin B inhibition of neurotransmitter accumulation by rat brain homogenate, *Science* **206**:363–364.

Mailman, R. B., Ferris, R. N., Tang, F. L. M., Vogel, R. A., Kilts, C. D., Lipton, M. A., Smith, D. A., Mueller, R. A., and Breese, G. R., 1980, Erythrosine (Red no. 3) and its nonspecific biochemical actions: What relation to behavioral changes? *Science* **207**:535–537.

Mattes, J., and Gittelman-Klein, R., 1978, A crossover study of artifical food colorings in a hyperkinetic child, *Am. J. Psychiatry* **135**:987–988.

Morrison, M., 1978, The Feingold diet (letter to the editor), *Science* **199**:840.

Salzman, L. K., 1976, Allergy testing, psychological assessment and dietary treatment of the hyperactive child syndrome, *Med. J. Aust.* **2**:248–251.

Swanson, J. M., and Kinsbourne, M., 1979, Artificial color and hyperactive behavior, in: *Treatment of Hyperactive and Learning Disordered Children: Current Research,* (R. M. Knights and D. Bakker, eds.), pp. 131–149, University Park Press, Baltimore.

Swanson, J. M., Kinsbourne, M., Roberts, W., and Zucher, K., 1978, A time–response analysis of the effect of stimulant medication on the learning ability of children referred for hyperactivity, *Pediatrics* **61**:21–29.

Weiss, B., Williams, J. H., Margen, S., Abrams, B., Caan, B., Citron, L. J., Cox, C., McKibben, J., Ogar, D., and Schultz, F., 1980, Behavioral responses to artificial food colors, *Science* **207**:1487–1489.

Wender, P., 1971, *Minimal Brain Dysfunction in Children,* Wiley-Interscience, New York.

17

THE CHINESE RESTAURANT SYNDROME

Martin Gore

INTRODUCTION

In 1968 a letter from a Chinese physician named Kwok was published in the New England Journal of Medicine, describing certain symptoms he had experienced after eating Chinese food (Kwok, 1968). Since then there has been much research attempting to place what was a random observation onto a more scientific base. Although 13 years have passed since that first report, this attempt has failed; indeed, if anything the situation has become more confused. There is argument over the very symptoms that make up the syndrome, let alone an understanding of their cause and time course.

In this chapter I will review the different studies that have been carried out in the effort to evaluate the Chinese Restaurant Syndrome with, particular reference to symptomatology, cause, and possible pathophysiology. Before I do this however I think it is appropriate to look at the historical background that surrounds the syndrome.

HISTORY

The symptoms that Kwok first described in 1968 were numbness at the back of the neck radiating down both arms and back, palpitations, and general weakness. These occurred 15–20 min after eating in Chinese restaurants, especially when northern Chinese food was served. Kwok had never experienced such symptoms before he went to live in the U.S. and then only when he ate out in restaurants. He suffered similar but milder symptoms as part of a hypersensitivity reaction to acetylsalicylic acid. He is Chinese, and he found that Chinese friends of his also complained of similar reactions. Al-

Martin Gore • University College Hospital, London, United Kingdom.

though he had done no research concerning this phenomenon he suggested four possible causes: soy sauce, which he discounted as he used to eat this at home without adverse effect; the wine in which the food was cooked; the high sodium content of the food, with consequent hypernatremia and intracellular hypokalemia; and monosodium glutamate, which is used extensively in the preparation of food in Chinese restaurants. This first communication by Kwok was quickly followed up by further correspondence and editorial comment in the New England Journal of Medicine. Two correspondents reported work on the symptomatology and cause of what Kwok had described, suggesting that his symptoms were caused by the oral ingestion of monosodium glutamate, which can be present in large quantities in the food served in Chinese restaurants, (Schaumburg and Byck, 1968; Ambos *et al.,* 1968). Indeed, Schaumburg and Byck said that the reaction had been known for many years and was referred to by the term *The Chinese Restaurant Syndrome.*

Since all along it has been suggested that monosodium glutamate was responsible for the syndrome, it is interesting to trace the history of this nonessential amino acid that is found so widely in mammalian tissues. The ancient Japanese used seaweed (*dashikombu* or *Laminara japonicum*) in their cooking to give food "umami" or tastiness, but it was not until 1908 that Ikeda isolated glutamate from this seaweed and recognized that it was the cause of the flavor enhancing properties of the seaweed. Glutamate itself had been first isolated many years before in Europe in 1866 by Ritthausen, although it was not realized then that it possessed any properties that might be of use in food preparation. However, during the nineteenth century work was in progress to determine the cause of the taste left by the products of certain chemical reactions, and through the work of Justus von Liebig and Julius Maggi the meaty taste resulting from the hydrolyzation of vegetable protein was described. At first Maggi thought that he had actually reproduced meat extract but soon found that the quality of his products depended very much on the choice of protein used, and he developed a very successful industrial process for the hydrolyzation of protein. The hydrolysate that he was using is now known to contain glutamate, but it was not until the work of Ikeda, Liebig, Maggi, and others was brought together that monosodium glutamate became an additive in the food industry. This first occurred in 1910, and since then increasing amounts of monosodium glutamate have been used by industry.

The effects of monosodium glutamate and glutamic acid on human subjects was studied long before Kwok's letter in 1968. Albert *et al.* in 1946 reported the effect of glutamic acid administration on mentally retarded subjects. Oral administration caused fullness of the stomach, vomiting, and diarrhea, while intravenous injection caused flushing, vomiting, dizziness, and general malaise. They also suggested that its administration might have a beneficial effect on mentally retarded subjects and in the same year Goodman *et al.* suggested the use of glutamate in the treatment of epilepsy. In 1954 Himwich *et al.* studied the effects of high-dose oral administration of monosodium glutamate on human males and found no adverse effects. The following year, Himwich and his colleagues (1955) suggested the use of monosodium glutamate in the treatment of various psychiatric disorders, and in the same paper reported incidences of hyperglycemia and hypotension following its administration. In the same year Alexander *et al.* (1955) proposed a possible use for glutamic acid in hepatic coma. Asten and Ross (1960), while studying the effects of glutamic acid, administered large amounts of both monosodium glutamate and glutamic acid to their subjects for a 2–6 month period without any record of adverse reactions. In addition, Gasster (1961) suggested that glutamate could be given to elderly patients for behavioral and memory problems.

In 1957 came the first report of toxicity of glutamate in laboratory animals (Lucas and Newhouse, 1957). Researchers showed that glutamate caused retinal damage in mice after subcutaneous injections, and this was followed by a report of damage to the arcuate nucleus of the hypothalamus after monosodium glutamate administration (Olney, 1969). During the 1960s and 1970s, a large amount of research had been conducted on the toxicity of glutamate with some conflicting results. At the same time, its central metabolic role was firmly established. More recently, work has centered around the role of glutamate in the nervous system as a possible central neurotransmitter.

The relationship between the symptoms of the Chinese Restaurant Syndrome and monosodium glutamate was not suspected until 1968 with Kwok's letter. It is the study of this relationship during the 1970s that forms the basis of this chapter.

THE SYNDROME

The Chinese Restaurant Syndrome consists of a triad of symptoms occurring about 15 min after the ingestion of a Chinese meal and lasting approximately 1 hr. The symptoms are facial pressure, a burning sensation over the upper trunk, and chest pain. As has already been suggested, there is argument over whether these are the only symptoms and their precise time course.

The symptoms that have been ascribed to the Chinese Restaurant Syndrome can be divided into four groups. First, there are different perceptions of the basic triad and although this means that different words are used, it is accepted that a variety of different descriptions represent the same phenomenon. Second, there are general symptoms of a central nature such as headache and light-headedness that have been widely reported in association with the syndrome and need separate discussion. Third, there are general gastrointestinal symptoms that, although claimed by some to be part of the syndrome, occur in many different situations and have been shown not to be specifically associated with the Chinese Restaurant Syndrome. Finally, there are many anecdotal reports of more unusual symptoms usually occurring in doctors or their immediate family. I will deal with each of these groups separately.

The Classic Triad

The facial pressure and chest pain may be felt just as a tightness or tingling. The chest pain may be very severe, crushing, and central, and indeed may simulate that of a myocardial infarction. The burning sensation has also been described as tingling but may be felt as a warmth, numbness, tightness, or stiffness. The area affected by this sensation is the upper trunk both anteriorly and posteriorly, the back of the neck, and the shoulders. Some sufferers complain of a weakness in their arms and this is presumably due, at least in part, to the sensation in the shoulders.

General Central Symptoms

Headache has been part of many descriptions of the syndrome and is often described as migraine-like. It has been shown (Schaumburg *et al.*, 1969) that subjects who experience headache after the ingestion of monosodium glutamate have a history of either migraine or vascular tension headaches. Support for the view that headache is not part of the syndrome comes from placebo-controlled trials, where headache was reported after the ingestion of both placebo and monosodium glutamate (Rosenblum, 1971; Tidball, 1972; Kenney, 1979; Gore and Salmon, 1980). Although these trials show that headache is reported more frequently after monosodium glutamate ingestion than placebo, this difference never reaches statistical significance.

Feelings of light-headedness or dizziness have been reported in most trials, but only in one (Rosenblum *et al.*, 1971) has its frequency reached statistical significance when compared with the placebo. Also, in this particular trial, the only other symptom that occurred more frequently after monosodium glutamate ingestion when compared with placebo was tightness of the face, but there were no other features of the Chinese Restaurant Syndrome found in any of the subjects tested.

Similarly nobody has been able to show that there is an increased incidence of thirst after the ingestion of glutamate. However, when glutamate is taken in high concentrations it may have a local effect in the mouth causing a sensation of dryness, but this is a very subjective sensation and there is no statistical evidence that glutamate ingestion is responsible for subsequent thirst.

Gastrointestinal Symptoms

Some case reports list nausea, vomiting, and abdominal discomfort as part of the syndrome. Once again, when placebo-controlled trials are analyzed there is no statistical difference between the frequency of these symptoms and the ingestion of monosodium glutamate or placebo (Kenney, 1979). Even the trials that strongly suggest that monosodium glutamate causes the Chinese Restaurant Syndrome do not record these general gastrointestinal symptoms as part of the syndrome (Schaumburg and Byck, 1968; Schaumburg *et al.*, 1969; Kenney and Tidball, 1972; Ambos *et al.*, 1968).

Although monosodium glutamate may enhance the flavor of some foods, when taken in large quantities it has an unpleasant taste and thereby may cause nonspecific gastrointestinal symptoms. Furthermore, virtually all foodstuffs can cause such symptoms in individuals under differing circumstances and if taken in large enough quantities.

Case Reports

There have been several reports in which the Chinese Restaurant Syndrome has taken on some rather bizarre and alarming forms. All these reports are anecdotal accounts and therefore lack the authority of placebo-controlled trials, and they also very often deal with doctors and their families. For both these reasons such reports are sometimes ridiculed or worse still, completely ignored. Although the interpretation of such case histories must be guarded, they can be very useful in pointing the way for future research. Indeed, it should be remembered that it was just such a communication that rekindled interest in the clinical effects of monosodium glutamate. There is also a further point to consider: if a person suffers from a mysterious set of symptoms and these symptoms can be alleviated or abolished by a change in diet without detriment to the nutritional status of the patient, then such a change must be worthwhile, even if the scientific foundation for such treatment turns out to be bogus. The following cases are not in any way representative but have merely been chosen as a random cross section.

The first case concerns a physician's wife and son who both suffered the classic symptoms of the Chinese Restaurant Syndrome (Colman, 1978). The boy also developed bowel and bladder incontinence as well as hyperactivity, from which he had previously suffered. His mother developed psychological symptoms some 48 hr after her acute reaction to glutamate, when she displayed signs of a depressive illness associated with outbursts of rage and paranoia. Following this she was challenged with monosodium glutamate on two occasions, first with wanton soup and then with "Accent" and on each occasion her symptoms returned. Both mother and son were symptom free after being put on a glutamate-free diet. A similar case was reported in the Swedish literature (Friberg, 1978). In this case, a quarter of an hour after eating soup in a Chinese restaurant a mother and eight-year-old son were suddenly taken ill. Both mother and son suffered identical symptoms of intense flushing in the face, pain in the temples, vertigo, nausea, stiff neck, and tachycardia. The other diners were unaffected and the son was symptom free within 3 or 4 days. His mother, however, suffered loss of sensation in her cheeks and tongue and did not become symptom free for 4 weeks. There is no report that either the mother or the son in this case suffered any psychiatric disturbance. The suggestion that there might be some such disturbance following ingestion of monosodium glutamate is not altogether surprising in the light of glutamate's possible role as a central nervous system neurotransmitter. It was suggested that there might be a genetic basis to monosodium glutamate reactions, but this has been studied in three families and shown not to be so, although the number of families studied was obviously very small (Schaumburg *et al.,* 1969).

In another report (Reif-Lehrer, 1976) a child suffered "shudder" attacks that were not typical of the minor seizures sometimes seen in childhood, and did not respond to diphenylhydantoin therapy. These attacks were studied in some detail with the result that the child was put on a monosodium-glutamate-free diet and his attacks ceased. He was later challenged with monosodium glutamate on more than one occasion and his attacks returned. Among a group of volunteers in whom the effects of monosodium glutamate was studied (Ghadimi *et al.,* 1971; Ghadimi and Kumar, 1972) one volunteer was observed to undergo a mood change after the ingestion of monosodium glutamate; from being lively and talkative the subject became rather subdued, quiet and withdrawn.

It has been noted that some people only develop symptoms in certain Chinese restaurants or on certain occasions, a fact that Kwok (1968) noted in his own case. I would like to relate here a similar though slightly different experience of the Chinese Restaurant Syndrome related to me by one sufferer. This particular person gets classic symptoms but not until several Chinese meals have been eaten. The interesting point is that the meals do not have to be taken on the same day but are often several days, a week, or even 10 days apart. After about the third meal the symptoms begin to appear and become more severe with each successive meal. After a break of several weeks in which no Chinese meals are eaten she remains symptom free, and may then eat a further Chinese meal with no effect, but the symptoms can be induced again with repeated ingestion of Chinese food as has just been described. This particular sufferer is not only a distinguished research scientist and therefore accustomed to making careful observations, but she also noted all these facts before she even knew about the existence of the Chinese Restaurant Syndrome or monosodium glutamate reactions. I have not published this case and only record it here as an example of one of the many different stories that research workers in this field are told by a variety of individuals in all walks of life.

THE EVIDENCE FOR AND AGAINST THE EXISTENCE OF THE CHINESE RESTAURANT SYNDROME

Several reports of placebo-controlled trials indicate that monosodium glutamate ingestion can cause classic symptoms of the Chinese Restaurant

Syndrome (Schaumburg and Byck, 1968; Schaumburg *et al.*, 1969; Ambos *et al.*, 1968; Kenney and Tidball, 1972; Reif-Lehrer, 1976 1977; Kenney, 1979). The incidence of reaction to monosodium glutamate differs from trial to trial, but is generally accepted that the reaction rate is about 30% of subjects tested. Schaumburg *et al.* (1969), however, showed that virtually all their subjects would experience classic symptoms of the Chinese Restaurant Syndrome if large enough oral doses of monosodium glutamate were given. They found in one trial that 55 out of 56 subjects complained of symptoms, and in another that 35 out of 36 subjects could likewise be made to complain of symptoms. They used oral doses of up to 12 g of monosodium glutamate. They also found that it was the monosodium L-glutamate that caused the symptoms and not monosodium D-glutamate, monosodium L-aspartate, sodium chloride or glycine. Ghadimi *et al.* (1971), and Ghadimi and Kumar (1972) likewise showed that histidine does not cause the symptoms.

It is also considered that women are more commonly affected than men, approximately in the proportion of three to five (Kenney, 1979).

There does, however, seem to be a threshold for the amount of glutamate ingested below which even the most vigorous reactor will show no symptoms. This level is considered to be about 3–4 g monosodium glutamate (Schaumburg et al., 1969; Kenney, 1979). There is approximately 3 g monosodium glutamate in a 200 ml portion of wanton soup (Schaumburg *et al.*, 1969).

In contrast to the above studies there are some placebo-controlled trials that suggest monosodium glutamate is not the cause of the Chinese Restaurant Syndrome (Morselli and Garattini, 1970; Bazzano *et al.*, 1970; Himwich *et al.*, 1955; Gore and Salmon, 1980). Morselli's study can be excluded from the argument, as only 3 g monosodium glutamate were used and this obviously does not reach the recognized threshold. The study of Bazzano *et al.* was only performed on male subjects as was the study by Himwich *et al.*, and they have thus both been critized for this reason. However it must be realized that in both these studies, subjects were given high doses of monosodium glutamate for a prolonged period of time with no apparent adverse effect. In fact in Bazzano's study eleven subjects were given between 25 and 147 g of monosodium glutamate daily for 14–42 days with no apparent effect. Gore and Salmon's work, however, takes into account both the threshold level and the suggested female preponderance, in that 25 out of 55 subjects were women. All subjects took 6 g of monosodium glutamate double blind and placebo controlled and there was not one classic reaction.

There has been argument over the vehicle in which monosodium glutamate is given, whether the subject has been fasting or not, and whether a high carbohydrate or protein load is given at the same time as the monosodium glutamate. Rosenblum *et al.* (1971) tested monosodium glutamate with chicken stock and sodium chloride, giving the monosodium glutamate in high dosage up to 12 g, and found no reaction at all. Kenney and Tidbull (1972) did not find any difference in the number of reactors to monosodium glutamate after fasting or after different types of breakfasts. They did, however, suggest the

30% incidence of reaction to monosodium glutamate. It has been suggested that when monosodium glutamate is taken with a meal the plasma level of glutamate is lower than when the monosodium glutamate is taken dissolved in water (Stegink *et al.,* 1979), but Kenney and Tidball (1972) found no difference in the plasma levels of glutamate between reactors and nonreactors. This would suggest that, even if the absorption of glutamate is affected by its vehicle or circumstance ingestion, as there is no difference in the plasma glutamate levels of reactors and nonreactors, this difference is probably unimportant.

Gore and Salmon (1980) tested 55 subjects with 6 g monsodium glutamate double blind and placebo controlled and did not find one classical reactor. The difference between this trial and those that strongly suggest that monosodium glutamate causes the Chinese Restaurant Syndrome (Schaumburg *et al.,* 1969; Kenney and Tidball, 1972; Kenney, 1979; Reif-Lehrer, 1977) is that the subjects not only were unaware that they were receiving monosodium glutamate, but also had never heard of the Chinese Restaurant Syndrome or glutamate reactors, and indeed were led to believe that the trial concerned the taste of the test samples rather than any symptoms they might experience. The bias in the other trials may be considerable because of the awareness of the subjects of the existence of the syndrome. This has been shown to be so by Kerr *et al.* (1979) when they studied the symptoms caused by various foods in over 3000 patients. They showed that not only was there an extremely low incidence of classic Chinese Restaurant Syndrome symptoms (1–2%) but also that the symptoms were associated with only Chinese food in 0.19% of cases. However when the results were analyzed by whether or not the subjects had heard of the Chinese Restaurant Syndrome, then 12% of those familiar with the syndrome complained of characteristic symptoms. Kerr *et al.* (1979) also point out that in a previous survey 90% of the "health conscious" population associated unpleasant symptoms with food as compared with 43% of the general public. In addition, 31% of the subjects aware of the Chinese Restaurant Syndrome believed that they had experienced the syndrome, compared to the 2.3% of the general adult population, most of whom had reported nonspecific symptoms (Kerr *et al.,* 1979). Kenney (1980) has recently agreed that bias will be brought into any trials if the subjects used are aware of the existence of the syndrome or monosodium glutamate reactors.

Finally, a recent report (Kenney, 1980) has strengthened the belief of Gore and Salmon (1980) that monosodium glutamate is not responsible for the syndrome or is only responsible for the syndrome in certain circumstances. In this double-blind, placebo-controlled trial Kenney found that typical symptoms of the Chinese Restaurant Syndrome occurred with other foodstuffs. He showed that out of 60 subjects, 14 developed classic symptoms of burning, tightness, or pain in the chest, neck, face, and arms. Six reported these symptoms after drinking coffee, six after spiced tomato juice, and only two after a 2% solution of monosodium glutamate. This latest piece of evidence seems to suggest very strongly that monosodium glutamate is not solely responsible for symptoms of the Chinese Restaurant Syndrome.

THE PATHOPHYSIOLOGY OF THE SYNDROME

There have been many suggestions as to the pathophysiology of the Chinese Restaurant Syndrome and the first of these has concerned the role of monosodium glutamate as a central nervous system neurotransmitter. There is little work directly aimed at this explanation and at present it is only a hypothesis.

Reif-Lehrer (1976) suggested that if monosodium glutamate is a neurotransmitter or is in some way involved in neurotransmission then it might "excite" nerve endings. This is based on the knowledge that monosodium glutamate has a distinctive taste that is separate from the basic tastes of saltiness, bitterness, sourness, and sweetness (Yamaguchi and Akimitsu, 1979), and that because taste involves the chemical excitation of nerve endings monosodium glutamate must have neuro-excitatory properties. The Chinese Restaurant Syndrome would therefore be induced by such a neurological mechanism.

Ghadimi and his co-workers (1971) suggested that the symptoms of the syndrome are very like those experienced when acetylcholine is given. They therefore proposed that the syndrome might be caused by transient acetylcholinolysis and showed that reactors to monosodium glutamate, when given anticholinergic drugs such as atropine, experienced less severe symptoms and those symptoms increased when cholinergic drugs such as prostigmine were given. Reif-Lehrer (1976) suggested that there might be a "benign" inborn error of metabolism that is present in monosodium glutamate reactors. She supported this hypothesis by citing other benign errors such as those responsible for the "asparagus effect," whereby thio-compounds appear in the urine after eating asparagus (Alison and McWhirter, 1956), the appearance of red urine after eating leeks (Alison and McWhirter, 1956) and the headache that some people get after eating cheeses or other foods containing tyromine (Blackwell and Mabbitt, 1965; Hannington, 1967).

Schaumburg *et al.* (1969) suggested that there may be arterial receptors sensitive to monosodium glutamate and the evidence for this was that two of his subjects experienced a burning sensation traveling down their abdomens and into their legs after the intravenous injection of monosodium glutamate. He suggested that there were arterial receptors in the walls of the aorta causing sensation to be felt over this particular distribution. He also demonstrated that when monosodium glutamate was injected intravenously into an arm that had a pressure cuff on it and hence no venous return, a burning sensation was felt only in the arm. He also said that the monosodium glutamate reaction was not an allergic response since he found that the administration of antihistamine and aspirin did not alter the reaction. As has already been mentioned, he went on to demonstrate that the reaction was specific to monosodium L-glutamate since monosodium D-glutamate, monosodium L-aspartate, and glycine did not provoke the typical reaction. Levey *et al.* in 1949 had shown that out of several amino acids administered parenterally only glutamate and aspartate caused any ill effects, nausea and vomiting. Ghadimi *et al.* (1971) showed that there was also no reaction to histidine administration.

Several trials have shown no objective changes in pulse or blood pressure after the administration of monosodium glutamate (Rosenblum *et al.*, 1971; Morselli and Garattini, 1970) and also no such objective changes of pulse, blood pressure, or respiration rate during actual glutamate reactions (Kenney, 1979; Kenney and Tidball, 1972). Kenney also showed in these trials that there was no electrocardiograph change and no change of skin temperature over the areas where flushing was felt. He was also unable to demonstrate any change in the levels of serum transaminases or routine blood chemistry except for glucose and LDH (lactate dehydrogenase), both of which differed in reactor and nonreactor subjects, but in neither case did this difference reach statistical significance.

Kenney (1979) has suggested that these symptoms of the Chinese Restaurant Syndrome arise from the esophagus. He showed this by making known reactors first gargle with monosodium glutamate, and then take monosodium glutamate in a gelatine capsule thus avoiding contact between monosodium glutamate at the nerve endings of the oro-pharanx and esophagus. In both these experiments fewer people reacted and he therefore concluded that the nerve endings in the esophagus were being stimulated by the monosodium glutamate either after oral ingestion or intravenous injections. He also felt that people who suffer from gastroesophageal reflux are more likely to be monosodium glutamate reactors than those who do not. This hypothesis is strengthened by his latest report (Kenney, 1980), which has already been discussed and shows that more people reacted with classical Chinese Restaurant Syndrome symptoms to coffee and spiced tomato juice than after monosodium glutamate ingestion.

THE TOXICITY OF GLUTAMATE

An evaluation of whether or not monosodium glutamate is toxic in man, based on the research discussed, is well beyond the scope of this chapter. However, it is important to point out that although there are many reports in the literature suggesting histological damage, there are also many reports that suggest it is very hard to demonstrate any toxicity. A general view of this argument is that although toxic effects have been demonstrated, most of the experiments involve massive doses of monosodium glutamate usually administered in very unphysiological ways. Nobody has ever shown that there are any histological or biochemical toxic sequelae in man. There has been much argument as to whether it is possible to damage the fetal brain following the maternal administration of glutamate, but experiments with the rhesus placenta, which is very similar to the human placenta in many ways, has shown that glutamate does not transfer across it (Pitkin *et al.*, 1979).

There has been some worry about the presence of monosodium glutamate in baby foods, but when one considers that glutamate is present in human breast milk this worry seems to be unreasonable. Indeed a breast-fed infant takes in more glutamate/kg body weight per day during infancy than at any other time in its life (Baker *et al.*, 1979). In addition ingestion of monosodium glutamate failed to increase the levels of glutamate in the milk of lactating women, even though their plasma levels could be raised.

It has also been demonstrated that infants fed oral or parenteral feeding formulas have normal glutamate levels and show no ill effects (Filer *et al.*, 1979).

There has recently been a review of some of the work on animals that were given monosodium glutamate orally, and none of the papers cited demonstrated any serious clinical behavioral, biochemical, hematological, or histopathological changes (Cooper, 1979).

CONCLUSION

Although early work on the toxicology of glutamate suggested that it might be a dangerous substance, this has been seriously called into question and now there is a good deal of evidence to show that it is in fact quite safe. Indeed, it appears on the FDA listing of GRAS substances. It has not yet been demonstrated that monosodium glutamate, when taken orally by humans, results in any biochemical or histological change.

Although the initial evidence of both Schaumburg and Kenney for the classic syndromes of facial pressure, burning sensation over the trunk, and chest pain or tightness was extremely strong and its causation fairly certain—that is, the ingestion of monosodium glutamate—recent work has thrown considerable doubt over these observations. Rosenblum *et al.* (1971) were unable to show that such a syndrome existed after the ingestion of monosodium glutamate, and their work is now strongly supported by that of Gore and Salmon (1980). Kenney's latest report (1980) is of major importance for two reasons. First, he demonstrates convincingly that monosodium glutamate is not the only cause of the syndrome and second, this view represents a slight but subtle change in one who has probably done more work than anyone else on the evaluation and significance of the Chinese Restaurant Syndrome.

This new suggestion that there is a ubiquity of causative agents was in fact first noted by the editor of the New England Journal of Medicine in 1968 when he reported the correspondence he received following the first report of the syndrome by Kwok (1968) a few weeks earlier. Incriminating foods included matzo and split pea soup in kosher delicatessens, mustard, tea, poisonous fish (*Ciguatera*), and many others. A letter following this editorial comment was published suggesting that Dr. Kwok's original letter was a hoax and congratulating the editor on his sense of humor (Porter, 1968). This author must have been somewhat taken aback with the subsequent furor over the Chinese Restaurant Syndrome and possibly even slightly embarassed by his suggestion. However it now appears that there may be some truth in his initial skepticism.

REFERENCES

Albert, K., Roch, P., and Waelsch, N., 1946, Preliminary report on the effect of glutamic acid administration in mentally retarded subjects, *J. Nerv. Ment. Dis.* **104:**263–274.

Alexander, R. S. W., Berman, E., and Balfour, D. C., Jr., 1955, Relationship of glutamic acid and blood ammonia to hepatic coma, *Gastrointerology* **19:**711–718.

Alison, A. C., and McWhirter, K. G., 1956, Two unifactorial characters for which man is polymorphic, *Nature (London)* **178:**748–749.

Ambos, M., Leavitt, N. R., Marmorek, L., and Wolschina, S. B., 1968, (Letter), *N. Engl. J. Med.* **279:**105.

Asten, A. W., and Ross, S., 1960, Glutamic acid and human intelligence, *Psychol. Bull.* **57:**429–434.

Baker, G. L., Filer, L. J., and Stegink, L. D., 1979, Factors influencing dicarboxylic amino acid content of human milk, in: *Glutamic Acid: Advances in Biochemistry and Physiology* (L.J.Filer, *et al.,* ed.)pp. 111–125, Raven Press, New York.

Bazzano, G., D'Elia, J. A., and Olson, R. E., 1970, Monosodium glutamate: Feeding of large amounts to man and gerbils, *Science* **169:**1208–1209.

Blackwell, B., and Mabbitt, L. A., 1965, Tyramine in cheese related to hypertensive crises after monoamine-oxidase distribution, *Lancet* **1:**938–940.

Colman, A. D., 1978, Possible psychiatric reactions to monosodium glutamate, *N. Engl. J. Med.* **299:**902.

Cooper, P., 1979, MSG—Food for thought, *Food Cosmet. Toxicol.* **17:**541–542.

Editor, 1968, *N. Engl. J. Med.* **279:**106.

Filer, L. J., Baker, G. L., and Stegink, L. D., 1979, Metabolism of free glutamate in clinical products fed infants, in: *Glutamic Acid: Advances in Biochemistry and Physiology* (L. J. Filer, ed.) pp. 353–363, Raven Press, New York.

Friberg, S., 1978, The Chinese Restaurant Syndrome, *Lakartidningen* **75:**3564.

Gasster, M., 1961, Clinical experience with L-glutamate in aged patients with behavior problems and memory defects, *J. Am. Geriatr. Soc.* **9:**370–375.

Ghadimi, H., and Kumar, S., 1972, Current status of monosodium glutamate, *Am. J. Clin. Nutr.* **25:**643–646.

Ghadimi, H., Kumar, S., and Abaci, F., 1971, Studies on monosodium glutamate ingestion, Part I, Biochemical explanation of the Chinese Restaurant Syndrome, *Biochem. Med.* **5:**447.

Goodman, L. S., Swinyard, E. A., and Toman, J. E. P., 1946, Effects of L-glutamic acid and other agents on experimental seizures, *Arch. Neurol. Psychiatry* **56:**20–29.

Gore, M. E., and Salmon, P. R., 1980, Chinese Restaurant Syndrome: Fact or fiction? *Lancet* **1:**251–252.

Hannington, E., 1967, Preliminary report on tyramine headache, *Br. Med. J.* **2:**550–551.

Himwich, W. A., and Petersen, I. M., 1954, Ingested sodium glutamate and plasma levels of glutamic acid, *J. Appl. Physiol.* **7:**196–199.

Himwich, E. H., Wolff, K., Hunsicker, A. L., and Himwich, W. A., 1955, Some behavioral effects associated with feeding sodium glutamate to patients with psychiatric disorders, *J. Nerv. Ment. Dis.* **121:**40–49.

Kenney, R. A., 1979, Placebo controlled studies of human reaction to oral monosodium L-glutamate, in: *Glutamic Acid: Advances in Biochemistry and Physiology*,(L. J. Filer, *et al.*,)pp. 363–373, Raven Press, New York.

Kenney, R. A., 1980, Chinese Restaurant Syndrome, *Lancet* **1:**311–312.

Kenney, R. A., and Tidball, C. S., 1972, Human susceptibility to oral monsodium L-glutamate, *Am. J. Clin. Nutr.* **25:**140–146.

Kerr, G. R., Wu-Lee, M., El-Lozy, M., McGandy, R., and Stare, F. J., 1979, Prevalence of the Chinese Restaurant Syndrome, *J. Am. Diet, Assoc.* **75:**29–33.

Kwok, R. H. M., 1968, Chinese Restaurant Syndrome, *N. Engl. J. Med.* **278:**796.

Levey, S., Harroun, J. E., and Smyth, C. J., 1949, Serum glutamic acid levels and occurrence of nausea and vomiting after intravenous administration of amino acid mixtures, *J. Lab. Clin. Med.* **34:**1238–1248.

Lucas, D. R., and Newhouse, J. P., 1957, The toxic effects of sodium L-glutamate on the inner layers of the retina, *Arch. Ophthalmol.* **58:**193–201.

Morselli, P. C., and Garattini, S., 1970, Monosodium glutamate and the Chinese Restaurant Syndrome, *Nature* **227:**611–612.

Olney, J. W., 1969, Brain lesions, obesity and other disturbances in mice treated with monosodium glutamate, *Science* **164:**719–721.

Pitkin, R. M., Reynolds, W. A., Stegink, L. D., and Filer, L. J., 1979, Glutamate metabolism and placental transfer in pregnancy, in: *Glutamic Acid: Advances in Biochemistry and Physiolog* (L. J. Filer, *et al.*, ed.) pp. 103–111, Raven Press, New York.

Porter, W. C., 1968 (Letter), *N. Engl. J. Med.* **279:**106.

Reif-Lehrer, L., 1976, Possible significance of adverse reactions to glutamate in humans. *Fed. Proc. Fed. Am. Soc. Exp. Biol.* **35:**2205–2212.

Reif-Lehrer, L., 1977, A questionnaire study of the prevalence of Chinese Restaurant Syndrome, *Fed. Proc. Fed. Am. Soc. Exp. Biol.* **36:**1617.

Rosenblum, I., Bradley, J. D., and Coulston, F., 1971, Single and double blind studies with oral monosodium L-glutamate in man, *Toxicol. Appl. Pharmacol.* **18:**367–373.

Schaumburg, H. H., and Byck, R., 1968, Sin-Gib-Syn: Accent on glutamate, *N. Engl. J. Med.* **279:**105.

Schaumburg, H. H., Byck, R., Gerstyl, R., and Mashman, J. H., 1969, Monosodium L-glutamate: Its pharmacology and role in the Chinese Restaurant Syndrome, *Science* **163:**826–828.

Stegink, L. D., Reynolds, W. A., Filer, L. J., Baker, G. L., Daabes, T. T., and Pitkin, R. M., 1979, Comparative metabolism of glutamate in the mouse, monkey, and man, in: *Glutamic Acid: Advances in Biochemistry and Physiology*, (L. J. Filer, pp. 85–103, Raven Press, New York.

Yamaguchi, S., and Akimitsu, K., 1979, Psychometric studies on the taste of monosodium glutamate, in: *Glutamic Acid: Advances in Biochemistry and Physiology*, (L. J. Filer, pp. 35–55, Raven Press, New York.

IV

SOCIAL TOXICANTS

18

SOCIAL TOXICANTS AND NUTRITIONAL STATUS

Mark L. Wahlqvist

If a man will be sensible and, one fine morning, while lying in bed, count on his fingertips how many things in life truly give him enjoyment, invariably he will find food to be the first.
Chinese Proverb

Throughout history man has been prepared to take some risk with the things he ingests since they have been to him more than nutrition: pleasure, communication, curiosity, and creativity. The functions of food and beverages are several (Table I) (Mead, 1955; Lowenberg *et al.*, 1979; Schack, 1978: Turner 1980). It is the ratio of risk to total benefit that must be borne in mind by nutritionists.

To the nutritionist a social toxicant may represent either a social factor that can adversely affect nutritional status, or an item ingested primarily for social rather than nutritional reasons and that may lead to harm.

NUTRITIONALLY ADVERSE SOCIAL FACTORS

Ethnic Background

The way in which cultural diversity is associated with variation in nutritional status is particularly evident in North America and Australasia where immigrant European, Asian, and African as well as indigenous cultures are found together. The hunter–gatherer life-style of Australian Aborigines (Hetzel and Frith, 1978: Wahlqvist, 1980, 1981) and the semiagriculturalist practice of the American Indians (Lowenberg *et al.*, 1979) were interfered with and, for urban fringe dwellers of these ethnic groups, nutritional prob-

Mark L. Wahlqvist • Department of Human Nutrition, Deakin University, Geelong, Victoria, Australia 3217.

TABLE I
Social Functions of Food and Beverages

Mood alteration
Providing security
Symbolic
Establishing roles and status
Socializing
Celebration
Religious or philosophical expression
Art form (food preparation, viniculture)
Economic

lems are frequently seen. These problems reflect not simply a change in the food system, but extensive societal change for the ethnic group concerned.

In general, minority food cultures tend to move toward the majority food culture. This is already evident in those from southern Europe who migrated to Australia after the Second World War (Hopkins *et al.*, 1980; Kosmidis *et al.*, 1980), and appears to be reflected in an increase in the diseases associated with affluent societies such as colorectal cancer (McMichael *et al.*, 1979).

Traditional food culture is also lost as urbanization occurs, and with the loss appear health problems more characteristic of industrialized society. Such changes are under way in many underdeveloped countries. In the Pacific Island of Nauru, prosperous through phosphate exports, abundant imported foodstuffs are overeaten and obesity and maturity-onset diabetes have emerged (Ringrose and Zimmet, 1979). For Orang Asli or indigenous people of peninsular Malaysia, the more contact they have with the general population, the greater their coronary disease risk by way of serum lipids and blood pressure (Chong, 1976; Chong and Pang, 1978). Urban Ethiopian men also show higher blood lipids when their diets become more westernized (Ostwald and Gebre-Medhin, 1978).

Socioeconomic Status

The marked differences in prevalence of protein energy malnutrition (PEM) between developed and underdeveloped countries clearly demonstrate the importance of socioeconomic factors in achieving adequate nutritional status. The further pressures within an industrialized society of economic influences on food choice have been demonstrated by McKenzie in the U.K. (1980). In Australia, there is evidence that children from lower socioeconomic groups are more likely to have inadequate or no breakfasts (Storey, 1979), with implications for physiologic and scholastic performance (Tuttle *et al.*, 1954). The way in which smoking affects body weight also appears to depend on social class (Ashwell *et al.*, 1978).

Family Relationships

Food purchase, preparation and consumption are most often family matters so that problems within the unit can adversely affect nutritional status. In Western society, an increasing number of family units have only one parent or two working parents, so that ready-to-eat food is more in demand (Krupinski and Stoller, 1978). In this situation, the family has less control over and probably less knowledge of the components of the food it eats. The food outlet has, through its contribution, decreased the working knowledge of food that the family used to have. The education system will presumably be required to help with this change in consumer knowledge, so that the potentially adverse effects of a rapidly changing food system can be minimized.

Not only does the child pattern food behavior after the parent, but the child with information obtained from school, the media, or peers can in turn influence the parent. Close parent–child nutrient-intake interrelationships have been observed insofar as factors affecting serum lipids are concerned (Laskarzewski *et al.*, 1980).

Accommodation

Food storage and preparation facilities can affect nutritional status and these vary greatly according to the kind of accommodation an individual has.

Whether a household has a designated place for eating or encourages eating in a number of venues can also be important as far as the restraint of food consumption is concerned.

Occupation

The workplace has become for some, through related health services, a place for identification of nutritionally related problems (Kornitzer *et al.*, 1980; Rose *et al.*, 1980; Stewart *et al.*, 1980). When provision of food by the employer becomes a benefit of employment, the relative contribution of the canteen to daily needs tends to increase. There is, therefore, a growing need for nutritionists to be involved in assessment of food supply at the place of employment.

Time Management

Food consumption patterns appear to be influenced by both under- and overutilization of time. Both the bored and those with little time to spare may overeat, the latter because they eat too fast or without planning. The extent to which leisure time is used for physical activity and the extent to which work tasks expend physical energy will also affect energy balance.

Physical Activity

Appetite control is optimal when regular exercise is undertaken (Åstrand and Rodahl, 1970). With a greater energy requirement, a wider choice of food items is possible inasmuch as more of the energy-dense items can be incorporated. The basic nutritional principle that the wider the variety of foods the more likely is nutrient intake to be adequate and the less likely any toxic effect, is also better served with a greater energy requirement.

Education

To some extent it is difficult to separate education from socioeconomic factors affecting nutritional status. The down-turn in mortality from ischemic heart disease in North America may have occurred because of reduced exposure to risk factors (Stallones, 1980). Whether the down-turn actually represents better access to information about risk factors, more ability to respond to recommendations, more resources to accomplish change in risk, greater access to medical care, or a combination of these is difficult to assess.

However, knowledge of nutritional needs or problems may not be enough to effect behavioral change. Motivation, techniques for change, and problem-solving skills are also required (Caliendo, 1979).

The print and electronic media can be used to affect food habits in underdeveloped (Wood-Bradley *et al.,* 1980) as well as developed countries (Senate Standing Committee on Education and the Arts, 1978). Food advertisers, if not nutritionists, are convinced about the media possibilities. In the Australian summer of 1978–1979, owing to a concerted advertising campaign, flavored sweetened milk displaced carbonated sweetened beverages as the principal drink purchased by Melbourne children. In the Australian capital, Canberra, a survey showed that the percentage of advertisement time devoted to fast-foods, (i.e., biscuits, sweets, and drinks) was 44% during 4:00–6:00 PM, compared with only 6% during 6:00–7:00 PM (Senate Standing Committee on Education and the Arts, 1978).

Transportation

Although road, rail, water, and air transport contribute to the movement of food from grower to processor to user, the user himself may not always be well-served by modern transport systems.

In many ways, the automobile serves as a model for an examination of some of the contemporary factors adversely affecting nutrition in the West. In the first place, it reduces the level of physical activity. Next, it allows an increase in the range of places in which it is possible to eat and thereby some of the constraints on overconsumption are reduced. It allows the purchase of food in larger quantities because transport of the food is easier. It also allows access to ready-to-eat food outlets which are beyond ordinary walking distance from home or employment and often not related to a public transport system. The car itself provides a relatively private place for food consumption and driving may encourage chewing to stay awake, to avoid tension, or to counter boredom. The car adds to the risks of alcohol abuse. The automobile illustrates how an apparently unrelated technological change can affect the food system and nutritional status.

The Food System

The mode of food production, processing, marketing, and pricing can all affect nutritional status (Abelson, 1975; Caliendo, 1979).

Personal Habits

The adverse effects of cigarette smoking on vitamin C status are now recognized (Pelletier, 1975). The magnitude of effect of cigarette smoking on vitamin C is related to the number of cigarettes smoked. For those who persist with cigarette smoking, it may be necessary to recommend a higher vitamin C intake than for nonsmokers. Smoking may have a more general effect on nutritional status by impairment of taste and smell.

There is also evidence that those who ingest more than 80 g ethanol daily have low plasma and platelet vitamin C concentrations (Strauss *et al.*, 1981). The effect may in part be due to reduced vitamin C absorption (Fazio *et al.*, 1981). Often, alcohol abusers are also cigarette smokers and the effect of both on vitamin C status may be additive.

The wide-ranging nutritional consequences of alcohol abuse are reviewed elsewhere (Thomson, 1978; Hurt *et al.*, 1979). As far as cardiovascular disease is concerned, it would appear that alcohol induces competing risk phenomena (Kozararevic, *et al.*, 1980). Less than one standard drink a day is associated with more ischemic heart disease, perhaps because high-density lipoprotein cholesterol is lower. For all alcoholic beverages except beer their use is associated with higher blood pressure and therefore probably more death from stroke.

Health Care System

The inadequate training of health-care professionals in nutrition is one of the ways in which the health-care system itself can adversely affect nutritional status (Wahlqvist, 1981). The health-care system requires facilities for prevention, identification, and management of nutritional problems. Even in the practice of clinical nutrition and dietetics, the consideration of nutritional problems in isolation from other health problems and of a particular nutrient rather than intact food or food patterns can present problems (Wahlqvist *et al.*, 1981). Where management of medical problems is drug oriented, drug–nutrient interactions may be seen (Mueller, 1980).

RISK OF SYMBOLIC FOODS

In food cultures, a particular item may symbolize food as a whole. In the Greek meal it is bread (Mead, 1955); for most Chinese it is rice; for Papua New Guineans it is the sweet potato. The potential problem with the dominance of one food item is that it diminishes the likelihood of variety in the diet and makes the community vulnerable should the symbolic food be unavailable.

At times, the desire for a culturally characteristic food or beverage may override a knowledge of possible death. This happens, for example, with pufferfish—potentially contaminated with tetrodotoxin—yet eaten by Chinese and Japanese (Committee on Food Protection, 1973), and with the fermented Javanese food bongkrek, which may be contaminated with *Pseudomonas cocovenenans* (Winarno, 1979).

BEVERAGES

Like food, beverages are used as a means of communication, entertaining others, and for mood alteration. Where a water supply is safe microbiologically and toxicologically, water itself would generally be the nutritionally preferred drink to quench thirst. Yet it occurs less and less to children and adults in affluent societies to drink water, sweetened carbonated beverages being preferred by children and caffeinated or alcoholic beverages by adults.

The potential hazard or hazardous component in commonly used beverages as well as the possible nutritional benefits are shown in Table II (Zarembski and Hodgkinson, 1962; Graham, 1978; Siegel, 1979). Caffeine occurs in tea, coffee, guarana, maté, kola nuts, and cocoa beans. Guarana yields beans for roasting, and maté provides tealike leaves used to make stimulating beverages in South America. The kola nut is used in the manufacture of the cola group of soft drinks. The average caffeine contents of common beverages are (Burg, 1975; Bunker and McWilliams, 1979);

1. Brewed Coffee, 85 mg/150 ml cup
2. Instant Coffee, 60 mg/150 ml cup
3. Decaffeinated Coffee, 3 mg/150 ml cup
4. Black Tea, 50 mg/150 ml cup
5. Green Tea, 30 mg/150 ml cup
6. Cocoa Drinks, 6–42 mg/150 ml cup
7. Cola Drinks, 22–65 mg/360 ml container

The effects of caffeine need to be considered on a dose per kg basis which means that effects in children will be relatively greater for a given ingested amount. The lethal dose for adult men is probably in excess of 10 g (Timson, 1978; Graham, 1978). The various potential adverse effects of caffeine are

TABLE II
Orally Ingested Items That Usually Serve Little Nutritional Purpose

Item	Potential hazard or hazardous component	Possible nutritional benefit
Beverages		
Tea	Caffeine, oxalate, tannin	Fluoride source; vehicle for water ingestion; added milk and sugar as protein and energy sources
Coffee	Caffeine, tannin	Potassium; nicotinic acid; vehicle for water; added milk and sugar
Cocoa	Theobromine, caffeine, tannin	—
Alcohol	Ethanol, amines, aldehydes, ethers, fusel oils	Local noncommercial beverages may contribute vitamins and minerals; energy source; accompany and encourage use of food
Carbonated, sweetened	Dental caries	Energy source; fluid replacement
Ginseng	Saponin glycosides	—
Confectionery	Dental caries, coloring agents	Energy source
Chewables		
Gums	Dental caries if sweetened	Dental and oral hygiene
Betel nut	Psychotropic factor	Stimulates salivary flow; medicinal
Tobacco	Oral cancer, abnormal fetal development	—

summarized in Table III (Anonymous, 1979, 1980; Cohen, 1980; Greden *et al.*, 1980). Of most concern has been its possible mutagenicity and teratogenicity (Mau and Netter, 1974; Weathersbee *et al.*, 1977; Borlee *et al.*, 1978).

Animal studies indicate the threshold of caffeine for no fetal effect is about 50 mg/kg as a single dose. This would be about 25–40 cups for man. At this stage it is probable prudent for pregnant women to reduce caffeinated beverage intake (Goyan, 1980).

A recent case-control study by MacMahon *et al.* (1981) in Boston has shown an association between drinking coffee and cancer of the pancreas, such that the relative risk associated with up to 2 cups per day was 1.8 after adjustment for cigarette smoking. Whether or not the relationship is causal will depend on further investigation.

TABLE III
Putative Adverse Effects of Caffeine

Symptoms	Grades of evidence[a]
Cardiovascular	
ischemic heart disease	3
arrhythmias	2
lipoprotein	1
Central Nervous System	
adenosine antagonist	2
behavioral changes	1
withdrawal headache	2
Mutagenic	3
Teratogenic	3
Breast	
fibrocystic breast disease	3
Gastrointestinal	
heartburn	3

[a] Evidence for effects has been graded on a scale of 1 to 3. Assignment of 1 indicates good, 2 fair, and 3 weak evidence that effects might be seen in man.

Any possible effect of coffee on ischemic heart disease might be through its interaction with smoking on lipoprotein cholesterol (Heyden *et al.*, 1979; Anonymous, 1981).

It must be remembered that not all effects of coffee may be mediated by caffeine and this appears to be the case as far as heartburn is concerned (Cohen, 1980).

Teas made from a variety of plant sources are currently popular in the west amongst those who are interested in 'health foods'. These are not all as safe as might be thought (Editorial, 1979). For example, teas made from juniper berries can cause gastrointestinal irritation; those from shave grass or horsetail plants, adverse effects may result from the nicotine and the thiaminase that they contain; those from buckthorn bark, senna leaves, dockroots and aloe leaves are cathartics; teas from catnip, juniper, hydrangea, lobelia, jimson weed and wormwood are anticholinergic, can produce euphoria and hallucinations; nutmeg is hallucinogenic and hepatotoxic; chamomile can cause sensitivity reactions; licorice root can cause sodium and water retention; pennyroyal oil and devil's claw foot contain oxytocics; sassafras root bark is hepatotoxic and carcinogenic; Indian tobacco contains lobelline which can cause paralysis and hypothermia; mistletoe contains phytotoxins with muscle depolarizing properties; apricot kernels contain the cyanogenetic glycoside amygdalin.

Ginseng, used for a long time in oriental medicine, is now used in a variety of forms including teas in the West. The active constituents, saponin glycosides, have now been reported to cause the ginseng abuse syndrome (GAS) (Siegel, 1979). GAS includes mood alteration, anorexia, hypertension and hypotension, edema, amenorrhea, diarrhea, and skin eruptions.

CONFECTIONERY

The adverse effects of confectionery relate chiefly to the increased likelihood of dental caries (Newbrun, 1979; Lee, 1981). There may also be a problem where energy needs have been met, but all nutrient needs have not. Nevertheless, the role of confectionery use in childhood and adult life requires better description. An Oxford study indicates that confectionery plays an important part in creating a children's world separate from that of the adult (James, 1979).

CHEWABLES

Man has long chewed for nonnutritional reasons (Table II). Oral satisfaction is considered a basic psychological need. For oral hygiene, chewing sticks were used by the Greeks and Romans; and the counterpart to the modern toothbrush is thought to have been devised by the Chinese in the fifteenth century (Lewis and Elvin-Lewis, 1976).

Although the use of chewing gum (chicle latex from *Achras zapota*) and bubble gum (latex from *Mimusops dalata*) are regarded as North American practices, gums, latex, and resins have long been used by indigenous peoples for chewing (Lewis and Elvin-Lewis, 1976). With added sugar, gums may promote dental caries.

In Arabia, the leaves and young shoots of the shrub *Catha edulis,* called kat or khat, are chewed for their stimulant properties.

Betelnut *(Areca catechu)* and lime is used widely as a chewable throughout Asia and Papua New Guinea (Wood-Bradley *et al.,* 1980). It appears to have mood-modifying properties. Additionally, where food supply is short, it may be used to stave off hunger and, indeed, when used by children who could otherwise be nibbling food, may contribute to malnutrition.

Tobacco does not have to be smoked to cause harm (Darby, 1979). Not only smoking (Simpson, 1957) but also the chewing of tobacco by pregnant women has been shown to lead to low birth weight (Krishna, 1978). Krishna also observed in Maharashtra that tobacco chewing led to an increased stillbirth rate and a low male : female infant sex ratio.

LIFESTYLE

From the Edinburgh–Stockholm comparison of risk factors for ischemic heart disease, it can be seen that several factors were more prevalent in Edinburgh, which has the higher mortality rates (Logan *et al.,* 1978). Edinburgh men were shorter and fatter, had higher blood pressures, smoked more cigarettes, drank more alcohol, had more electrocardiograph abnormalities, lower exercise tolerance, higher serum insulins, higher serum triglyceride concentrations, lower concentrations of serum high-density lipoprotein cholesterol, and lower contents of serum triglyceride and cholesterol ester linoleic acid. It is likely that at least some of these risk factors operate independently of each other. Yet they have occurred together more frequently in one place than another. This clustering of risks suggests that there may be a basic lifestyle problem.

Lifestyles with different components were examined for relationships to nutrient intake and laboratory indices of nutritional status by Baird and Schutz (1980). The study indicates that as the general quality of life improves, so the nutritional status of the population could be expected to improve.

The recognition of a particular lifestyle, which may alter several nutritional and other health-related variables, may be of considerable importance in preventive medicine (Turner, 1980). A checklist of the many components of lifestyle is available (Stanley *et al.*, 1978). Although pathogenetic mechanisms are important to understand inasmuch as they may allow the chain of events leading from cause to effect to be broken, it may sometimes be more effective overall to alter a lifestyle.

REFERENCES

Abelson, P. H., 1975, *Food: Politics, Economics, Nutrition and Research.* American Association for the Advancement of Science, Washington, D.C.

Anonymous, 1979, Workshop on caffeine, *Nutr. Rev.* **37:**124.

Anonymous, 1980, Second international caffeine workshop, *Nutr. Rev.* **38:**196.

Anonymous, 1981, Coffee: Should we stop drinking it? *Lancet* **1:**256.

Ashwell, M., North, W. R. S., and Meade, T. W., 1978, Social class, smoking and obesity, *Lancet* **2:**1466.

Åstrand, P.-O., and Rodahl, K., 1970, *Textbook of Work Physiology,* McGraw-Hill Book Co., New York.

Borlee, I., Lechat, M. F., Bouchaert A., and Mission, C., 1978, Le cafe, facteur de risque pendant la grossesse? *Louvain Med.* **97:**297–284.

Baird, P. C., and Schutz, H. G., 1980, Life style correlates of dietary and biochemical measures of nutrition, *J. Am. Diet. Assoc.* **6:**228.

Burg, A. M., 1975, The effects of caffeine on the human system, *Tea Coffee Trade J.* **147:**40, 88.

Bunker, M. L., and McWilliams, M., 1979, Caffeine content of common beverages, *J. Am. Diet Assoc.* **74:**28.

Caliendo, M. A., 1979, *Nutrition and the World Food Crisis,* Macmillan Publishing Co., New York.

Chong, Y. H., 1976, Aspects of ecology of food and nutrition in peninsular Malaysia, *Environ. Child Health* **October:** 239.

Chong, Y. H. and Pang, C. W., 1978, Serum lipids, blood pressure and body mass index of Oran Asli—possible effects of evolving dietary socioeconomic changes, *Proceedings of the International Symposium on Alterations of the Diet on Health,* U.S.–Japan Panel on Malnutrition, Osaka.

Cohen, S., 1980, Pathogenesis of coffee-induced gastro-intestinal symptoms, *N. Engl. J. Med.* **303:**122.

Committee on Food Protection, 1973, *Toxicants Occurring Naturally in Foods,* 2nd ed., Food and Nutrition Board, National Research Council, National Academy of Sciences, Washington, D.C.

Darby, W. J., 1979, Some observations concerning nutrition and dental health, Marabou Symposium, Prevention of Major Dental Disorders, *Näringsforskning 23 (Suppl.)* **17**:37.

Editorial, 1979, "High" Tea, *Med. J. Aust.* **1**:232.

Fazio, V., Flint, D. M., and Wahlqvist, M. L., 1981, Acute effects of alcohol on plasma ascorbic acid in healthy subjects, *Am. J. Clin. Nutr.* **34**:2394.

Goyan, J. E., 1980, Statement, Sept. 4, 1980, Food and Drug Administration, Washington, D.C.

Graham, D. M., 1978, Caffeine—its identity, dietary sources, intake and biological effects. *Nutr. Rev.* **36**:97.

Greden, J. F., Victor, B. S., Fontaine, P., and Lubetsky, M., 1980, Caffeine-withdrawal headache: A clinical profile. *Psychosomatics* **21**:411.

Heyden S., Heiss, G., Manegold, C., Tyroler, H. A., Hames, C. G., Bartel, A. G., and Cooper, G., 1979, The combined effect of smoking and coffee drinking on LDL and HDL cholesterol, *Circulation* **60**:22.

Hetzel, B. S., and Frith, H. J., 1978, The nutrition of aborigines in relation to the ecosystem of central Australia, Commonwealth Scientific and Industrial Research Organization, Melbourne, Australia.

Hopkins, S., Margetts, B. M., Cohen, J., and Armstrong, B. K., 1980, Dietary change among Italians and Australians in Perth, *Community Health Studies* **4**:67.

Hurt, R. D., Nelson, R. A., Dickson, E. R., Higgins, J. A., and Morse, R.M., 1979, Nutritional status of alcoholics before and after admission to an alcoholism treatment unit, in: *Fermented Food Beverages in Nutrition* (C. F. Gastineau, W. J. Darby, T. B. Turner, eds.) pp. 397–408, Academic Press, New York.

James, A., 1979, Confection, concoctions and conceptions, *J. Anthropol. Soc. Oxford* **X(2)**:83.

Kornitzer, M., DeBacker, B., Dramaix, M., and Thilly, C., 1980, The Belgian heart disease prevention project: Modification of the coronary risk profile in an industrial population, *Circulation* **61**:18.

Kosmidis, G. C., Rutishauser, I. H. E., Wahlqvist, M. L., and McMichael, A. J., 1980, Food intake patterns amongst Greek migrants in Melbourne, *Proc. Nutr. Soc. Aust.* **5**:165.

Kozararevic, D. McGee, D., Vojvodic, N., Racie, Z., Dawber, T., Gordon, T., and Zukel, W., 1980, Frequency of alcohol consumption and morbidity and mortality: The Yogoslavia cardiovascular disease study, *Lancet* **1**:613.

Krishna, K., 1978, Tobacco chewing in pregnancy, *Br. J. Obstet. Gynaecol.* **85**:726.

Krupinski, J., and Stoller, A., 1978, *The Family in Australia—Social, Demographic and Psychological Aspects,* 2nd ed., Pergamon Press, Melbourne, Australia.

Laskarzewski, P., Morrison, J. A., Khoury, P., Kelly, K., Glatfelter, L., Larsen, R. and Glueck, C. J., 1980, Parent–child nutrient intake interrelationships in school children ages 6 to 19: The Princeton school study, *Am. J. Clin. Nutr.* **33**:2350.

Lee, V. A., 1981, The nutritional significance of sucrose consumption, 1970–1980, *CRC Critical Reviews in Food Science and Nutrition* **14**:1.

Lewis, W. H., and Elvin-Lewis, M. P. F., 1976, Oral Hygiene, in: *Medical Botany: Plants affecting Man's Health,* pp. 226–270, John Wiley and Sons, New York.

Logan, R. L., Reimersma, R. A., Thomson, M., Oliver, M. F., Olsson, A. G., Walldius, G., Rössner, S., Kaijser, L., Callmer, E., Carlson, L. A., Lockerbie, L., and Lutz, W., 1978, Risk factors for ischemic heart-disease in normal men aged 40, *Lancet* **1**:949.

Lowenberg, M.E., Todhunter, E. N., Wilson, E. D. Savage, J. R., and Lubawski, J. L., 1979, *Food and People,* 3rd ed.,John Wiley and Sons, New York.

MacMahon, B., Yen, S., Trichopoulos, D., Warren, K., and Nardi, G., 1981, Coffee and cancer of the pancreas, *N. Engl. J. Med.* **304**:630–633.

Mau, G., and Netter P., 1974, Kaffee- und Alkaholkonsum-Risikofaktoren in der Schwangerschaft? *Gerburtshilfe Frauenheilkd.* **34**:1018.

MacKenzie, J., 1980, Economic influences on food choice, in: *Nutrition and Lifestyles* (M. Turner, ed.),pp. 91–103, Applied Science Publishers, London.

McMichael, A. J., McCall, M. G., Hartshorne, J. M., and Woodings, T. L., 1979, Patterns of gastrointestinal cancer in European migrants to Australia: The role of dietary change, *Int. J. Cancer* **25:**431.

Mead, M., 1955, *Cultural Patterns and Technical Change,* Mentor-Unesco, New York.

Mueller, J. F., 1980, Drug–nutrient interrelationships, in: Human Nutrition: *A Comprehensive Treatise,* Vol. 3B, *Nutrition and the Adult: Micronutrients* (R. B. Alfin-Slater and D. Kritchevsky, eds.) pp. 351–365, Plenum Press, New York.

Newbrun, E., 1979, Dietary carbohydrates: Their role in cariogenicity, *Med. Clin. North Am.* **63:**1069.

Ostwald, R., and Gebre-Medhin, M., 1978, Westernization of diet and serum lipids in Ethiopians, *Am. J. Clin. Nutr.* **31:**1028.

Pelletier, O., 1975, Vitamin C and cigarette smokers, in: Second Conference *on Vitamin C* (C. G. King and J. J. Burns, eds.), Vol. **258,**pp. 156–168, N.Y. Academy of Sciences.

Ringrose, H., and Zimmet, P., 1979, Nutrient intakes in an urbanized Micronesian population with high diabetes prevalence, *Am. J. Clin. Nutr.* **32:**1334.

Rose, G., Heller, R. F., Pedoe, H. T., and Christie, D. G., 1980, Heart disease prevention project: A radomized controlled trial in industry, *Brit. Med. J.* **1:**747.

Schack, W. A., 1978, Anthropology and the diet of man, in *Diet of Man: Needs and Wants* (J. Yudkin, ed.), pp. 261–280, Applied Science Publishers, London.

Senate Standing Committee on Education and the Arts, 1978, Inquiry into the impact of television on the development and learning behaviour of children, Australian Government Publishing Services, Canberra.

Siegel, R. K., 1979, Ginseng abuse syndrome, *J. Am. Med. Assoc.* **241:**1614.

Simpson, W. J., 1957, A preliminary report on cigarette smoking and the incidence of prematurity, *Am. J. Obstet. Gynecol.* **73:**808.

Stallones, R. A., 1980, The rise and fall of ischemic heart disease, *Sci. Am.* **243:**43.

Stanley, J. R., Lambert, R. P., Buckley, J. D., and Mackay, I. R., 1978, Social factors and clinical disease: A check list data base for correlative analysis, *Aust. N.Z. J. Med.* **8:**387.

Stewart, A., Wahlqvist, M. L., Olipant, R. C., Simons, L. A., and Jones, A. S., 1980, A study to examine the prevalence of coronary risk factors in an industrial community, *Proc. Nutr. Soc. Aust.* **5:**215.

Storey, E., 1979, Nutritional disorders of childhood, *Food Nutr. Notes Rev.* **36:**168.

Strauss, B. G., Fazio, V. A., Flint, D. M., Korman, M. G., Brodie, G. M., Hunt, P. S., and Wahlqvist, M. L., 1981, Platelet ascorbic acid concentrations in nutritionally deprived groups in the Australian community, *Recent Advances in Clinical Nutrition* **1:**68.

Thomson, A. D., 1978, Alcohol and nutrition, *Clinics Endocrinol. Metab.* **7 (2):**401–428.

Timson, J., 1978, How harmful is your daily caffeine? *New Sci.* **June 15:**736.

Turner, M., 1980, *Nutrition and Lifestyles,* Applied Science Publishers, London.

Tuttle, W. W., Daum, K., Larsen, R., Salzano, J., and Roloff, L., 1954, Effect on school boys of omitting breakfast: Physiologic responses, attitudes, and scholastic attainments, *J. Am. Diet. Assoc.* **30:**674.

Wahlqvist, M. L., 1980a, Malnutrition in Australian society *Menzies Symposium,* Vol. 1 (in press).

Wahlqvist, M.L., 1981, Nutrition and the under graduate medical curriculum: An Australian Perspective. *Recent Advances in Clinical Nutrition* 1 (in press).

Wahlqvist, M. L., Coles-Rutishauser, I., and Briggs, D. R., 1981, Food composition tables: A nutritionists view, *Food Technol. Aust.* **33:**118.

Weathersbee, P. S., Olsen, L. K., and Lodge, J. R., 1977, Caffeine and pregnancy, a retrospective survey, *Postgrad. Med.* **62:**64.

Winarno, F. G., 1979, Fermented vegetable protein and related foods of South East Asia with special reference to Indonesia, *J. Am. Oil Chem. Soc.* **56:**363.

Wood-Bradley, R., Flint, D. M., and Wahlqvist, M. L., 1980, Food and nutrition in an independent Papua New Guinea, *Search* **11(3):**73.

Zarembski, P. M., and Hodgkinson, A., 1962, The oxalic acid content of English diets, *Br. J. Nutr.* **16:**627.

19

ALCOHOL AND BODY DAMAGE

Derrick B. Jelliffe and E. F. Patrice Jelliffe

From the beginning of time man has been plagued by famine, disease, and pestilence. Thus, life has always been difficult, and any mood-altering drug found by our ancestors was absorbed into the ritual of subsequent cultures. This was particularly so with substances that could lessen stress and anxiety and induce a jovial, exaggerated "all's well with the world" feeling.

Most present-day cultures have traditional mood-changing preparations, ranging from the *kava* of Polynesia to *hashish* in some Arab countries. But undoubtedly the most widespread—and popular—of the various mood-changing drugs is ethyl alcohol (ethanol), colloquially referred to as *alcohol.*

Alcohol has been produced since time immemorial in a number of different forms based on the fermentation of sugar by yeast. Even some animals seem attracted to alcohol; wild elephants, for example, have a great liking for ripe fruits that have dropped to the ground and fermented.

For Western man, alcohol has a very strong historical and traditional background. The wine-drinking cultures of the Mediterranean depended upon fermented cereal grain drinks and alcoholic honey drinks, such as mead, from Northern Europe. Similarly, the use of alcohol has been widespread from ancient times in large areas of the Far East in the form of rice wine, and in Africa as palm wine and gruel-like beer, made from millet and other grains.

Thus, in most of the world, and certainly in Europe and North America, the simple compound of carbon, hydrogen and oxygen (C_2H_5OH), or ethyl alcohol, has always dominated the mood-changing scene. Its usage has evolved into a complex network of socially acceptable customs and practices, such as selecting "correct" drinks to serve with different foods, or for various occasions.

The deep-seated use of alcohol in Western culture is emphasized by the wealth of rituals surrounding its consumption—different glasses for different wines and beverages, the variety of complicated toasts, and the symbolic significance of red wine in the Communion. The aromas and flavor of some

Derrick B. Jelliffe and E. F. Patrice Jelliffe • Population, Family and International Health Division, School of Public Health, University of California, Los Angeles, California 90024.

alcoholic drinks, especially wines, are an important feature of gourmet cooking.

PHARMACOLOGY

Alcohol is not a stimulant, but rather is essentially a depressant of the higher brain-control centers. Thus, it permits relaxation and the temporary release from some of life's tensions. The depressing effects of alcohol on the central nervous system have been used in times past as a crude form of anesthesia—for example, during battlefield amputations in earlier centuries. During the early industrial revolution, it was commonplace for babies left at home by their working mothers to be "sedated" with brandy.

In the past, as well as now, the ill effects of alcohol have depended on the type, concentration, and amount that has been consumed. The introduction of strong distilled spirits or hard liquors has had more serious effects than those of beers or ales. This was especially noticeable in England in the early eighteenth century, when gin drinking became a major social problem, as vividly portrayed in Hogarth's savage cartoons.

Strangely, alcoholism was not regarded as a widespread, major, and increasing public health problem by the medical profession until the last 10–15 years, when increasing drunk-driving offenses (Zylam, 1964), rising deaths from cirrhosis of the liver (Anonymous 1981b,c) and other serious ill consequences have focused increasing attention on the world's largest addiction problem. Its seriousness is brought into focus even more by the rise in alcoholism among women (Saunders *et al.*, 1981) and children in all Western countries, as well as its less-recognized increase in developing countries.

SOCIALLY SANCTIONED TOXIN

Whether we care to admit it or not, alcohol is a socially acceptable toxin—hence the word *intoxicated* for someone "under the influence."

Last year in California alone, the average per capita rate of consumption of "alcoholic beverage" was reported as 26 gal, with a total consumption of 542.7 million gal. These statistics point up the necessity for understanding the effects of alcohol on the human body. Its most immediate influence is on the brain, where it acts as a depressant, rather than a stimulant, as many believe. Alcohol actually slows down the brain's control mechanisms and, depending on the dosage, can cause mild or serious mental disorganization, loss of muscular control, drowsiness, coma, or even death.

Small amounts depress the reticular-activating system of the brain, which normally alerts the cerebral cortex (the portion of the brain that integrates its activities). As a result, the functioning capacity of the cortex is decreased and activities requiring alertness and concentration are performed less efficiently. At the same time, ideas and images are stimulated (at least in the mind of the drinker) and mood changes to expanded euphoria. The effects of alcohol become noticeable when the concentration in the blood is about 5 parts of alcohol to 10,000 parts of blood. In the U.S., at 15 parts per 10,000, the drinker is legally defined as being "under the influence of alcohol" at 15 parts per 10,000. At 40 parts per 10,000, unconsciousness usually occurs, and a level of 50 parts per 10,000 can be fatal. The legally defined levels vary in different countries.

Another reaction is dilation of the capillaries, located just under the skin, causing them to carry more blood. Although the skin *feels* warmer from being flushed with blood, the body temperature actually tends to fall because this internal heat is carried to the skin's surface, where it is dissipated.

Alcohol is easily absorbed through the stomach. Most of it, however, enters the small intestine, where it is rapidly and completely absorbed. Food hinders its passage to the intestine, thus slowing down its absorption into the body. On the other hand, carbon dioxide is stated to speed up absorption of alcohol from the intestine, explaining why the bubbles in champagne make it more rapidly effective than other beverages of equal alcohol content.

When even more alcohol is imbibed, the more resistant parts of the brain become depressed. When it reaches the cerebellum, which controls muscular coordination, speech becomes confused and equilibrium is affected. The brain centers that control consciousness are the next to be involved. But the vital centers controlling the heart and respiration are only affected by extremely high doses of alcohol.

A small portion of consumed alcohol leaves the body in the urine and breath; the rest is oxidized, mainly in the liver. If alcohol is imbibed in such a quantity that the liver's ability to oxidize is exceeded, it will continue to circulate in the blood—eventually causing intoxication. When taken in combination with barbiturates and tranquilizers, alcohol has an increased toxicity and can be fatal at lower levels.

EMPTY CALORIES, OBESITY, AND MALNUTRITION

Alcohol supplies energy for the body (7.1 kcal/g) that is almost double that supplied by either protein or carbohydrate (each approximately 4 kcal/g). Because of its content of relatively "empty" calories, prolonged overconsumption of too much alcohol can result in malnutrition. Despite the energy supplied by alcohol, it contains no other required nutrients, and such deficiencies as the vitamin B complex may result.

Some persons can consume as much as 1800 kcal/day in the form of alcohol. In addition, since alcohol causes depression of the central control system, before-dinner cocktails may mean that larger amounts of food will be eaten.

There is considerable variation in the calorie content of the usual quantities of different alcoholic drinks. In an average-sized drink, 12 oz (1 glass) of beer has 150 kcal, a martini (2 oz) has 160 kcal., a "shot" of whisky (1.5 oz) has 110 kcal., and 4 oz of dry wine has 160 kcal. Since comparisons are interesting, as well as odious, it may be noted that one martini has the same calorie content as half a pint of whole milk but the latter has vastly more food value. Alcohol respects neither social level nor occupation, and extra calories can manifest themselves in the form of "beer belly" or "executive jowls."

FETAL DAMAGE

Recent work indicates that alcohol overdose is of much greater significance than previously thought. One very serious effect of alcohol consumption is the fetal alcohol syndrome (FAS). Heavy drinking during pregnancy leads to a 30–50% risk of FAS.

There are still questions about light drinkers and the susceptibility of the fetus in the early weeks of pregnancy. The present view is that appreciable risks may perhaps occur even with light drinking. Simply put, alcohol is best avoided if pregnancy is likely or certain. Recent estimates suggest that FAS is now the third largest cause of congenital mental deficiency in the U.S. and probably the leading preventable cause of mental deficiency in the world.

LIVER DAMAGE

The liver is especially susceptible to damage from alcohol. Until recently cirrhosis of the liver was thought to be the result of malnutrition, as seen in advanced alcoholics ("skid row" types), who have inadequate diets, poor absorption, and damaged nutritional body metabolisms. Now it is known that while this type of inadequate nutrition can play a role in liver damage, the main cause of cirrhosis is the direct toxic effect of alcohol on the liver. This means that cirrhosis of the liver—now the seventh cause of mortality in the U.S.—can and does occur in the well-fed executive alcoholic.

CEREBRAL DAMAGE

Another affected area of the body is the central nervous system, particularly the brain. Indeed, the main reason for using alcohol is because of its pharmacological cerebral effects. However, the recent increase in alcoholism in various parts of the world has drawn attention to the fact that alcohol can cause serious, sometimes fatal, cerebral brain damage. One of these syndromes in Wernicke's encephalopathy, which occurs in the chronic alcoholic who has had a poor diet, decreased absorption, and inadequate and deficient stores of nutrients. It is often precipitated by "binge drinking" for several weeks with little food intake, and leads to loss of consciousness, weakness of the eye muscles, and a staggering gait.

INTESTINAL DAMAGE

In adults, chronic long-term alcohol abuse can lead to body damage in many other organs and systems. The pancreas and the surface of the gastrointestinal tract can be damaged. Various other ill effects may occur, including poor absorption and diarrhea, which can lead to insufficient intake of thiamin, folic acid, and vitamin B_{12}. The bone marrow can be affected, resulting in impaired erythropoiesis. A combination of inadequate folic acid in the diet, reduced absorption, and the toxic effect on the bone marrow can cause anemia. Alcoholic cardiac myopathy is also recognized to be a not uncommon toxic effect in chronic alcoholism (St. Leger *et al.*, 1979).

CONCLUSION

Apart from its well-recognized psychosocial effects, one of the few facts supporting the value of alcohol is associated with coronary heart disease. According to one investigator, there may be a connection between the remarkable decline in coronary heart disease in the U.S. with the almost parallel spectacular increase in wine consumption.

As stressed earlier, the use of alcohol as a "social lubricant" is deeply enmeshed in the Western way of life. Screening to identify early alcoholism has been a major topic of recent medical literature, and involves the use of questionnaires and blood tests. Essentially, however, a reduction in alcoholism requires an individual awareness that dependence can develop easily and insidiously. With the multiple family, economic and world stresses, anxieties of modern life, and in some cases the frustration and boredom of routine existences and unsatisfying jobs, the temptations are obvious.

The occasional use of alcohol can do no harm, except in pregnancy. The risk is that occasional use can change to frequent and then to continuous use. Another problem is that some individuals seem more susceptible than others to alcohol and to the body-damaging effects of prolonged high alcohol intake.

All in all, the penalties of alcohol abuse and overdosage are physical and social ill effects, both acute and chronic. Alcohol is double-edged. In "moderation" it is a "social lubricant"; in excess it can be fatal—leading to cirrhosis of the liver, cerebral injury, fetal deformities, and other less-recognized types of body damage.

Basically, the ease of availability of alcohol is a critical problem. This was first shown in Canada by Seeley (1960), who demonstrated that when the price of alcohol fell by half, the death rate from cirrhosis doubled. When wine rationing was discontinued in France in 1947, deaths from cirrhosis increased no less than 5 times over the succeeding 5 years (Pequignot, 1974). A reduction in the legal age for drinking in the U.S. was followed by a rise in road accidents in which alcohol was implicated (Zylam, 1964).

A recent editorial concerning the serious recent rise in alcoholism in Britain is revealing (Anonymous, 1981a):

> The fall in relative cost of alcoholic drinks was confirmed as an important determinant, as was the increase in number of retail outlets. A major new finding was that advertising of spirits had had a highly significant effect on consumption, thus belying the advertisers' defence that their campaigns merely seek to encourage consumers to buy their particular product . . . This is further evidence that a realistic policy for preventing the continuing increase in alcohol-related problems must include some control of its availability and stricter regulations on advertising . . . the romanticised view of alcohol as portrayed in advertisements should be discouraged, and countered by effective health education programmes . . . Year by year, as politicians prevaricate, the national rise in alcohol consumption continues and the damage caused intensifies. How much longer must we wait?

REFERENCES

Anonymous, 1981a, Alcoholism: Time for action, *Br. Med. J.* **1:**1177–1178.

Anonymous, 1981b, Alcoholic liver disease: Morphological manifestations, *Lancet* **1:**707–711.

Anonymous, 1981c, Epidemiological problems with alcohol, *Lancet* **1:**762–764.

Pequignot, G., 1974, Les problemes nutritionels de la societe industrielle. *Vie Med. Can. Fr.* **3:**216–225.

Saunders, J. B., Davis, M., and Williams, R., 1981, Do women develop alcoholic liver disease more readily than men? *Br. Med. J.* **1:**1140–1143.

Seeley, J., 1960, Death by liver cirrhosis and the price of beverage alcohol, *Can. Med. Assoc. J.* **83:**1361–1366.

St. Leger, A. S., Cochrane, A. L. and Moore, F., 1979, Factors associated with cardiac mortality in developed countries with particular reference to the consumption of wine, *Lancet* **1:**1017–1020.

Zylam, R., 1964, Fatal crashes among Michigan yough following reduction of legal drinking age, *Q. J. Stud. Alcohol* **35:**283–286.

20

ALCOHOL AND THE FETUS

Melvin Lee and Joseph Leichter

The historical literature contains numerous references to the potentially adverse effects of maternal alcohol consumption during pregnancy on the pre- and postnatal growth and development of the child (Warner and Rosett, 1975). However, it was the reports of Lemoine *et al.* (1968) and Jones *et al.* (1973) that directed attention to the spectrum of abnormalities that has come to be known as the *fetal alcohol syndrome* (FAS). They and others have described a pattern of developmental abnormalities in children born to women who had consumed substantial amounts of alcohol throughout pregnancy. These include pre- and postnatal growth retardation, with more severe retardation in length than in weight at birth, microcephaly, a constellation of craniofacial anomalies (including short palpebral fissures, epicanthal folds, ptosis and strabismus, a flattened philtrum, maxillary hypoplasia, cleft palate, and micrognathia), joint anomalies, cardiac and genital anomalies, and abnormalities of the liver, kidneys, and skeleton (Jones *et al.*, 1976; Loser and Majewski, 1977, Goldstein and Arulanantham, 1978; Habbick *et al.*, 1979; Qazi *et al.*, 1979). Central nervous system abnormalities include small brain size, brain malformations, abnormal migration of developing brain cells, poor coordination, irritability, hyperactivity, and retarded postnatal mental development. IQ scores tend to be low, related to the severity of clinical signs, and do not appear to change with time (Clarren *et al.*, 1978; Jones, 1975; see Streissguth *et al.*, 1980 for a more detailed review). All of these signs are not seen in each case and, indeed, the clinical picture may vary from only low birth weight to cases exhibiting most of the signs described above. However, Jones *et al.* (1976) emphasize that, although the individual signs are not necessarily unique, the total spectrum of changes is sufficiently characteristic that FAS is a distinct clinical entity.

Fetal alcohol syndrome has been observed in children of all races studied (Streissguth, 1978). The incidence among children born to women who con-

Melvin Lee and Joseph Leichter • Division of Human Nutrition, School of Home Economics, University of British Columbia, Vancouver, British Columbia, V6T 1W5, Canada.

sume alcohol will depend on the manner in which drinking habits are classified and on the definition of FAS. However, it has been reported that the prevalence is 20–40%, increasing with greater consumption (Ouellette and Rosett, 1976; Majewski *et al.*, 1976). The spectrum of malformations is also reported to be greater among infants of heavy drinkers, although the studies of Little (1977) and others indicate that even moderate consumption of alcohol during pregnancy has adverse effects. The prevalence of FAS in the general population has been estimated to be about 1 in 750 births (Streissguth 1978), but this must vary according to the drinking habits of different population groups. Nevertheless, both Streissguth (1978) and Jones *et al.* (1976) have emphasized that this is probably one of the most common forms of mental deficiency with a known etiology.

Several studies have examined the extent to which the frequency and severity of FAS is related to the amount of alcohol consumed and the temporal pattern of consumption. Ouellette and Rosett (1976) reported that 33.3% of the babies born to heavy drinkers had congenital anomalies, compared with 13% and 7.4% of the babies born to moderate drinkers and abstainers, respectively. Furthermore, 33.3%, 10.5%, and 0% of the children born to heavy, moderate and nondrinkers, respectively, were below the 3rd percentile for birth length, and similar relationships were seen for birth weight and head circumference. Although the sample size was small, these findings are in agreement with those of other studies (Hanson *et al.*, 1978) in suggesting that the incidence of anomalies and growth retardation is greater in children born to heavy drinkers than in those born to moderate drinkers. Although the results may be confounded by contributory factors, such as maternal smoking, nutritional state, and social class, there is evidence that alcohol *per se* is responsible for many of the clinical findings.

The association of maternal alcohol consumption with retarded fetal growth, documented in both human and animal studies, has received considerable attention. Kaminski *et al.* (1976), Little (1977), and others have demonstrated that both heavy and moderate consumption of alcohol during pregnancy is associated with depressed birth weight, and Little has shown that this is independent of maternal smoking habits, a recognized birth-weight depressant. In a subsequent study Little *et al.* (1980) reported that even the offspring of women who drank prior to recognition of pregnancy, but abstained during pregnancy, exhibited a significant decrease in birth weight, intermediate between the birth weights of offspring of abstainers and those of women who continued to drink throughout pregnancy. On the other hand Rosett *et al.* (1980) reported that abstinence during the last trimester resulted in less growth retardation than was seen in the offspring of those who continued to drink during the last trimester. These studies leave open the possibility that alcohol consumption prior to pregnancy may have persistent adverse effect that are not obliterated by abstinence during pregnancy. This does not yet seem to have been examined in animal models of FAS.

There is one interesting report (Scheiner *et al.*, 1979) describing a pattern of congenital anomalies characteristic of FAS in a child born to a couple who had stopped drinking one and a half years before conception. Although one may question the reliability of the reported abstinence, and recognizing that other factors may have contributed to the anomalies seen, this case, together with the studies of Little *et al.* (1980) raise the question of whether prolonged alcohol consumption prior to pregnancy may not result in adverse metabolic alterations in the mother that contribute to the outcome of pregnancy.

Children with FAS do not appear to demonstrate postnatal catch up growth (Mulvihill and Yeager, 1976). Jones *et al.* (1973) reported on eight children with FAS, some of whom were followed for as long as 4 years. Postnatal linear and weight growth were retarded in all children, and head growth (circumference) was retarded in all but one. Three of the children had been raised in foster homes and six had been hospitalized at some point, with adequate calorie intakes during hospitalization, so the lack of postnatal catchup growth was presumably not due to unsatisfactory care during this time. Clarren and Smith (1978) have hypothesized that the persistent growth defect is due to a prenatal effect on cell proliferation, leading to diminished fetal cell number. However, there does not yet appear to be experimental evidence for this.

Several studies have been concerned with the metabolic factors responsible for FAS growth retardation. Root *et al.* (1975) failed to find significant alterations in hypothalamic–pituitary functions or in several metabolic parameters (including serum sodium, chloride, potassium, calcium, magnesium, phosphate, or bicarbonate) in four children born to a woman with a history of alcoholism, although all were retarded in physical growth and bone age, and had abnormal electroencephalograms. Tze *et al.* (1976) reported normal growth-hormone response and normal circulating somatomedin levels in five cases of FAS.

Alcohol has also been used to prevent premature labor, and Fox *et al.* (1978) have shown that acute alcohol administration near term temporarily suppresses fetal breathing movements. In studies with sheep, Mann *et al.* (1975) found that acute alcohol administration causes a significant maternal hyperlacticacidemia and hyperglycemia, presumably reflecting the increased NADH/NAD ratio resulting from alcohol oxidation. Whether either of these observations has any significance for FAS (due to chronic alcohol consumption) does not seem to have been investigated.

The FAS has been identified primarily through retrospective clinical studies. Retrospectively, it is very difficult to obtain accurate information not only about the amount of alcohol consumed during pregnancy but also with regard to other risk factors such as inadequate nutrient intakes, smoking, and use of drugs that may compound the adverse action of ethanol on the fetus. Animal experimentation has the advantage of permitting rigid and simultaneous control of diet and ethanol intake, thus reducing confounding interactions.

Some features of the FAS have been reproduced in animal models, but the results are contradictory because of the wide variety of experimental conditions employed. The quantity and route of alcohol administered differs from study to study. Some have administered alcohol prior to mating, throughout gestation, and during lactation, while others have administered alcohol only during one period or only during a portion of pregnancy. Furthermore, different species and strains of animals were used, and pair feeding was not always employed to control for nutrition and caloric intake. Nevertheless, many investigators who have employed rodents in their experiments (Chernoff, 1977; Schwetz *et al.*, 1978; Abel and Greizerstein, 1979; Abel, 1979; Abel and Dintcheff, 1978; Tze and Lee, 1975; Leichter and Lee, 1979; Henderson *et al.*, 1979; Harris and Case, 1979; Henderson and Schenker, 1977; Pilstrom and Kiessling, 1967) have reported that maternal alcohol consumption during gestation results in retarded fetal and postnatal growth. In addition to retarded postnatal growth, the offspring of animals receiving alcohol during gestation also exhibit a retardation in the rate of skeletal maturation (Leichter and Lee, 1979). Despite the fact that the mothers did not receive alcohol after delivery (Leichter and Lee, 1979) or the young were fostered at birth to control mothers (Abel and Dintcheff, 1978), catch up was not observed either in growth or in skeletal maturation during the first four weeks postnatally. The retarded growth can not be corrected postnatally by reduction of litter size (to three pups) and provision of an alcohol-free diet (unpublished data). Taken together, these observations suggest a primary effect of alcohol on the prenatal growth process.

Abel (1978, 1979) and Abel and Dintcheff (1978) administered alcohol to pregnant rats intragastrically by gavage during gestation and observed that the fetal growth retardation was dose related. Maternal doses of 1–2 g/kg body weight per day had a minimal effect, and doses of 4 and especially 6 g/kg body weight per day produced significant growth retardation. The offspring exposed to 4 g/kg body weight per day exhibited catch up growth by day 21 postnatally, while the animals exposed to 6 g/kg body weight per day exhibited growth retardation at 5 months of age.

The inclusion of pair-fed controls in animal models insures that the nutrient and calorie intake is similar in both the alcohol-treated and the control animals. However, even pair feeding does not rule out the possibility that the utilization of nutrients may be impaired by the effects of alcohol on maternal intestinal absorption and urinary excretion or on the transfer of nutrients from the maternal circulation to the fetus. Frank and Baker (1980) found decreased blood folate and thiamin in male rats when ethanol was fed chronically with nutritionally adequate diets. In man, too, folate and thiamin absorption is decreased in alcoholism (Baker *et al.*, 1975). There is also a possibility that alcohol affects the utilization of nutrients by the fetus. Rawat (1979) reported that ethanol consumption by pregnant rats resulted in a significant inhibition in the rate of [U – ^{14}C]leucine incorporation into fetal cardiac proteins. Decreased total RNA and unchanged total DNA in ethanol-exposed neonates have been reported by Henderson and Schenker (1977).

Brown *et al.* (1979) demonstrated that exposure to ethanol retards growth and differentiation in cultured rat embryos, suggesting that ethanol can exert a direct effect on fetal growth, without the confounding factors of nutrition and alterations in maternal metabolism or physiological state.

Alcohol has been shown to be teratogenic when administered to mice during the gestation period, but few obvious malformations have been reported in rats. Chernoff (1977) found that malformations of 18-day-old mouse fetuses increased with increasing maternal blood-alcohol concentrations. The most frequently observed malformations were neural and cardiac anomalies, deficient occiput ossification, and eyelid dysmorphology. Similar teratogenic effects of alcohol in mice were reported by Kronick (1979) and by Randall and Taylor (1979). Chernoff (1977) also observed that different strains of mice responded differently to the same dose of alcohol, which was related to their rate of alcohol metabolism. In a study by Schwetz *et al.* (1978) pregnant mice, rats, and rabbits were given 15% ethanol in their drinking water as the sole source of liquid. In mice the highest blood-alcohol concentration was 204 mg/100 ml, in rats 40 mg/100 ml, and in rabbits 28 mg/100 ml. The high blood-alcohol levels in mice are probably responsible for the greater teratogenic effect of alcohol observed in this species because, according to Chernoff (1977), the teratogenic potential of ethanol seems to depend upon maintenance of high maternal blood-alcohol levels. This indicates that genetic differences in ethanol metabolism affect the outcome of pregnancy.

It has been suggested that elevated maternal blood levels of the ethanol metabolite acetaldehyde may be of importance in the toxic actions of ethanol on the fetus (Veghelyi *et al.*, 1978). Such increased blood-acetaldehyde concentrations are apparently due to an inherited or acquired low maternal acetaldehyde dehydrogenase activity, which is necessary for acetaldehyde oxidation. Kasaniemi and Sippel (1975) determined the contents of ethanol and acetaldehyde in placentas and fetuses of pregnant rats given ethanol. Ethanol was present in the placentas and fetuses in concentrations similar to that in the maternal circulation. In contrast, the acetaldehyde content of the placenta was only 25% of that in the maternal aorta, and no acetaldehyde was detected in the fetus. Since there was no accumulation of acetaldehyde in the fetal tissues it is doubtful whether acetaldehyde is the causative agent in the FAS. Although many aspects of the FAS in humans have been replicated in experimental animals, the mechanism by which alcohol induces these abnormalities remains unsolved.

Animal models used to examine the impairment of cognitive function associated with the FAS have been reviewed recently by Streissguth *et al.* (1980). Offspring exposed to alcohol prenatally generally show impaired learning on some tasks that could be related to neurochemical, hormonal, and neural anomalies.

It is clear from both clinical studies and animal experimentation that the consumption of large amounts of alcohol throughout pregnancy has deleterious consequences for the physical growth and development and the mental development of the offspring. It is not yet clear whether more moderate or sporadic drinking has similar effects. In any event, alcohol consumption during pregnancy appears to represent a problem of considerable public health significance and one that deserves vigorous public health and nutrition education programs.

REFERENCES

Abel, E. L., 1978, Effects of ethanol on pregnant rats and their offspring, *Psychopharmacologia* **57**:5.

Abel, E. L., 1979, Prenatal effects of alcohol on adult learning in rats, *Pharmacol. Biochem. Behav.* **10**:239.

Abel, E. L., and Dintcheff, B. A., 1978, Effects of prenatal alcohol exposure on growth and development in rats, *J. Pharmacol. Exp. Ther.* **207**:916.

Abel, E. L. and Greizerstein, H. B., 1979, Ethanol-induced prenatal growth deficiency: Changes in fetal body composition, *J Pharmacol. Exp. Ther.* **211**:668.

Baker, H., Frank. O., Zetterman, R., Rajan, K. S., TenHove, W., and Leevy, C. M., 1975, Inability of chronic alcoholics with liver disease to use food as a source of folates, thiamine, and vitamin B6, *Am. J. Clin. Nutr.* **28**:1377.

Brown, N. A., Goulding, E. H., and Fabro, S., 1979, Ethanol embryotoxicity: Direct effects on mammalian embryos *in vitro, Science* **206**:573.

Chernoff, G. F., 1977, The fetal alcohol syndrome in mice: An animal model, *Teratology* **15**:223.

Clarren, S. K., and Smith, D. W., 1978, The fetal alcohol syndrome, *N. Engl. J. Med.* **298**:1063.

Clarren, S. K., Alvord, E. C., Sumi, M., Streissguth, A. P., and Smith, D. W., 1978, Brain malformations related to prenatal exposure to ethanol, *J. Pediatr* **92**:64.

Frank, O., and Baker, H., 1980, Vitamin profile in rats fed stock or liquid ethanolic diets, *Am. J. Clin. Nutr.* **33**:221.

Fox, H. E., Steinbrecher, M., Pessel, D., Inglis, J., Medvid, L., and Angel, E., 1978, Maternal ethanol ingestion and the occurrence of human fetal breathing movements, *Am. J. Obstet. Gynecol.* **132**:354.

Goldstein, G., and Arulanantham, K., 1978, Neural tube defect and renal anomalies in a child with fetal alcohol syndrome, *J. Pediatr.* **93**:636.

Habbick, B. F., Zalesky, W. A., Casey, R., and Murphy, F., 1979, Liver abnormalities in three patients with fetal alcohol syndrome, *Lancet* **1**:580.

Hanson, J. W., Streissguth, A. P., and Smith, D. W., 1978, The effect of moderate alcohol consumption during pregnancy on fetal growth and morphogenesis, *J. Pediatr.* **92**:457.

Harris, R. A., and Case, J., 1979, Effects of maternal consumption of ethanol barbital, or chlordiazepoxide on the behaviour of the offspring, *Behav. Neural Biol.* **26**:234.

Henderson, G. I., and Schenker, S., 1977, The effect of maternal alcohol consumption on the viability and visceral development of the newborn rat, *Res. Commun. Chem. Pathol. Pharmacol.* **16**:15.

Henderson, G. I., Hoyumpa, A. M., McClain, C., and Schenker, S., 1979, The effects of chronic and acute alcohol administration on fetal development in the rat, *Alcoholism* **3**:99.

Jones, K. L., 1975, Aberrant neuronal migration in the fetal alcohol syndrome, *Birth Defects Orig. Artic. Ser.* **11**:131.

Jones, K. L., Smith, D. W., Ulleland, C. N., and Streissguth, A. P., 1973, Pattern of malformations in offspring of chronic alcoholic mothers, *Lancet* **1**:1267.

Jones, K. L., Smith, D. W., and Hanson, J. W., 1976, The fetal alcohol syndrome: Clinical delineation, *Ann. N. Y. Acad. Sci.* **273**:130.

Kaminski, M., Rumeau-Rouquette, C., and Schwartz, D., 1976, Consommation d'alcool chez de femmes enceintes et issue de la grossesse, *Rev. Epidemiol. Med. Soc. Sante Publique* **24**:27.

Kasaniemi, Y. A., and Sippel, H. W., 1975, Placental and foetal metabolism of acetaldehyde in rat. I. Contents of ethanol and acetaldehyde in placenta and foetus of the pregnant rat during ethanol oxidation, *Acta Pharmacol. Toxicol.* **37**:43.

Kronick, J. B., 1976, Teratogenic effects of ethyl alcohol administered to pregnant mice, *Am. J. Obstet. Gynecol.* **124**:676.

Leichter, J. and Lee, M., 1979, Effects of maternal ethanol administration on physical growth of the offspring in rats, *Growth* **43**:288.

Lemoine, P., Harrousseau, H., Borteyru, J. P., and Menuet, 1968, Les enfants de parents alcooliques. Anomalies observees: A propos de 127 cas, *Ouest Med.* **25**:476.

Little, R. E., 1977, Moderate alcohol use during pregnancy and decreased infant birthweight, *Am. J. Public Health* **67**:1154.

Little, R. E., Streissguth, A. P., Barr, H. M., and Herman, C. S., 1980, Decreased birth weight in infants of alcoholic women who abstained during pregnancy, *J. Pediatr.* **96**:974.

Loser, H., and Majewski, F., 1977, Type and frequency of cardiac defects in embryo-fetal alcohol syndrome, *Br. Heart J.* **39**:1374.

Majewski, F., Bierich, J. R., Loser, H., Michaelis, R., and Lieber, B., 1976, Zur klinik und pathogenese der Alkohol-Embryopathie, Berichte uber 68 falle, *Muench. Med. Wochenschr.* **118**:1635.

Mann, L. I., Bhakthavathsalan, A., Liu, M., and Makowski, P., 1975, Placental transport of alcohol and its effect on maternal and fetal acid–base balance, *Am. J. Obstet. Gynecol.* **122**:837.

Mulvihill, J. J., and Yeager, A. M., 1976, Fetal alcohol syndrome, *Teratology* **13**:345.

Ouellette, E. M., and Rosett, H. L., 1976, A pilot prospective study of the fetal alcohol syndrome at the Boston City Hospital. Part II. The infants, *Ann. N. Y. Acad. Sci.* **273**:123.

Pilstrom, L., and Kiessling, K. H., 1967, Effect of ethanol on growth and on the liver and brain mitochondrial functions of the offspring of rats, *Acta Pharmacol. Toxicol.* **25**:225.

Qazi, Q., Masakawa, A., Milman, D., McGann, B., Chua, A., and Haller, J., 1979, Renal anomalies in fetal alcohol syndrome, *Pediatrics* **63**:886.

Randall, C. L., and Taylor, W. J., 1979, Prenatal ethanol exposure in mice: Teratogenic effects, *Teratology* **19**:305.

Rawat, A. K., 1979, Derangement in cardiac protein metabolism in fetal alcohol syndrome, *Res. Commun. Chem. Pathol. Pharmacol.* **25**:365.

Root, A. W., Reiter, E. O., Andriola, M., and Duckett, G., 1975. Hypothalamic-pituitary function in the fetal alcohol syndrome, *J. Pediatr.* **87**:585.

Rosett, H. L., Weiner, L., Zuckerman, B., McKinlay, S., and Edelin, K. C., 1980, Reduction of alcohol consumption during pregnancy with benefits to the newborn, *Alcoholism* **4**:178.

Scheiner, A. P., Donovan, C. M., and Bartoshesky, L. E., 1979, Fetal alcohol syndrome in child whose parents had stopped drinking, *Lancet* **1**:1077.

Schwetz, B. A., Smith, F. A., and Staples, R. E., 1978, Teratogenic potential of ethanol in mice, rats, and rabbits, *Teratology* **18**:385.

Streissguth, A. P., 1978, Fetal alcohol syndrome: An epidemiological perspective, *Am. J. Epidemiol.* **107**:467.

Streissguth, A. P., Landesman-Dwyer, S., Martin, J., and Smith, D. W., 1980, Teratogenic effects of alcohol in human and laboratory animals, *Science* **209**:353.

Tze, W. J., and Lee, M., 1975, Adverse effects of maternal alcohol consumption on pregnancy and foetal growth in rats, *Nature* **257**:479.

Tze, W. J., Friesen, H. G., and MacLeod, P. M., 1976, Growth hormone response in fetal alcohol syndrome, *Arch. Dis. Child.* **51**:703.

Veghelyi, P. V., Osztovics, M., Kardos, G., Leisztner, L., Szaszovszky, E., Ingali, S., and Imrei, J., 1978, The fetal alcohol syndrome: Symptoms and pathogenesis, *Acta Paediatr. Acad. Sci. Hung* **19**:171.

Warner, R. H., and Rosett, H. L., 1975, The effect of drinking on offspring. An historical survey of the American and British literature, *J. Stud. Alcohol* **36**:1395.

21

ALCOHOL AND CEREBRAL THIAMIN DEFICIENCY

A. Stewart Truswell and Frank Apeagyei

HISTORY

Last year was the 100th anniversary of the first description of Wernicke's encephalopathy. In his textbook of neurology Carl Wernicke (1881) described three cases of opthalmoplegia, disorientation, intermittent somnolence, and ataxia. All died and at postmortem examination they showed small hemorrhages symmetrically disposed in the grey matter around the third and fourth ventricles and the aqueduct. Wernicke thought the condition was an acute inflammatory disease of the oculomotor nuclei (hemorrhagic polioencephalitis). Two of his cases were alcoholic men but his first case was not. This first patient was a 20-year-old woman with sulphuric acid poisoning and persistent vomiting from pyloric stenosis. In the light of present knowledge, thiamin deficiency is the fundamental abnormality in Wernicke's encephalopathy and the associated amnesic syndrome, first described as an independent disease by Sergei Sergevich Korsakoff in Moscow in 1887 (Victor and Yakorlev, 1955).

EVIDENCE LINKING WERNICKE'S ENCEPHALOPATHY AND KORSAKOFF'S PSYCHOSIS WITH THIAMIN DEFICIENCY (TRUSWELL, 1979)

a. Wernicke's encephalopathy can develop without drinking alcohol. It has been reported with a variety of backgrounds, including pyloric stenosis, esophageal obstruction, gastric carcinoma, digitalis toxicity, malabsorption, hyperemesis gravidarum, prolonged intravenous feeding without vitamin supplements, fasting for obesity, hunger

A. Stewart Truswell and Frank Apeagyei • Human Nutrition Unit, Biochemistry Department and Commonwealth Institute of Health, Sydney University, New South Wales 2006, Australia.

strike, chronic hemodialysis, and in prisoners of war in Singapore in World War II. (De Wardener and Lennox, 1947; Cruickshank, 1950).

b. Korsakoff's psychosis is part of the same clinical complex as Wernicke's encephalopathy. Most cases of Wernicke's encephalopathy go on to Korsakoff's psychosis; most cases of Korsakoff's psychosis were preceded by Wernicke's encephalopathy (Victor *et al.,* 1971). The pathological changes in the brain are essentially the same (Gudden, 1896). Nonalcoholic cases of Wernicke's encephalopathy can go on to Korsakoff's psychosis.

c. Alcoholic patients with Wernicke–Korsakoff disease give a history of inadequate food intake for some weeks (Victor *et al.,* 1971). Alcoholism predisposes to thiamin deficiency—first, from displacement by alcohol of thiamin-containing foods, and second, because alcohol impairs the absorption of thiamin (Tomasulo *et al.,* 1968). Alcohol, like carbohydrates, requires thiamin for its metabolism. It is not the prosperous overweight social drinker, who consumes alcohol heavily yet manages to keep his job, that developes Wernicke–Korsakoff disease. This type of drinker is likely to end up with cirrhosis. Wernicke–Korsakoff disease typically occurs in the person who drinks very heavily for several weeks, so heavily that he can't work; his life is socially disorganized and he eats little or nothing. Anorexia and later vomiting are early features of thiamin deficiency, seen in experimental animals; they create a vicious cycle and can lead to diagnostic confusion. There are no stores of thiamin in the body and the diuretic action of alcohol may possibly impair renal attempts at conservation. Symptoms of thiamin deficiency occur after 3–4 weeks of very low or zero intake.

d. Patients with Wernicke–Korsakoff disease have signs of general malnutrition. Korsakoff in 1889 noted "almost always there is severe emaciation."

e. Experimental thiamin deficiency in animals resembles the pathological or clinical features of Wernicke–Korsakoff disease (Mesulam *et al.,* 1977).

f. The opthalmoplegia, disorientation, and ataxia of Wernicke's encephalopathy respond promptly to thiamin treatment. Korsakoff's psychosis takes longer.

g. Biochemical tests indicate severe thiamin deficiency in Wernicke's encephalopathy. The TPP effect (the effect from adding thiamin pyrophosphate *in vitro*) is very high in the red-cell transketolase assay (Victor *et al.,* 1977; Wood, 1979; Truswell *et al.,* 1972; Baker, 1966), provided the patient has not been given thiamin treatment before the blood is taken.

The current revision of the *International Classification of Diseases* (World Health Organization, 1975, Vol. II) indicates that Wernicke's encephalopathy should be classified under item 265.1 "other and unspecified manifestations of thiamin deficiency," while Korsakoff's psychosis is classified under two numbers, both described as mental disorders—one for alcoholic (291.1), the other for nonalcoholic (294.0).

The cardiovascular form of thiamin deficiency, beri beri, used to be a common affliction of the rice-eating people of southern and eastern Asia. The first animal model for beri beri was discovered by Eijkman in Java around 1890, and thiamin was first isolated in Java by Jansen and Donath in 1926 (Davidson *et al.*, 1979). Yet Wernicke's encephalopathy and Korsakoff's psychosis never seem to have been prominent in these sections of Asia.

While beri beri has now become uncommon, apparently, throughout Southeast Asia and the western Pacific (Davidson *et al.*, 1979; Burgess and Burgess, 1976; Donoso, 1978), there is recent concern about Wernicke's encephalopathy in the U.S. and Britain, as reflected in editorials in the New England Journal of Medicine (Weinstein, 1978), Lancet (Anonymous, 1979a) and the British Medical Journal (Anonymous, 1979b).

In Australia, it has been reported that Wernicke's encephalopathy and/or Korsakoff's psychosis are not uncommon in three of the country's seven states, Victoria (Wood, 1979; Wood and Breen, 1980) Western Australia (Harper, 1979) and Queensland (Brown, 1979; Price and Theodoros, 1979). In Western Australia, 51 cases of Wernicke's encephalopathy were diagnosed at postmortem over a period of 4 years. Only seven had been correctly diagnosed while the victims were alive (Harper, 1979). In Queensland about 10% of the chronic residents in psychiatric hospitals now have Korsakoff's psychosis (Brown, 1979), and the annual number of admissions with this diagnosis has tripled since 1968 (Price and Theodoros, 1979). The syndrome is evidently occurring in New Zealand as well, judging from four recent cases reported from the Auckland Hospital (Wallis *et al.*, 1978).

SYDNEY STUDY 1979–1980

There have been no reports from the state of New South Wales or its capital, Sydney, the largest city in Australia. The authors have made an estimate of the frequency of Wernicke–Korsakoff disease by three approaches:

1. from the discharge summaries of in-patients in the 11 major Sydney hospitals for the 15-year-period of 1964–1978,
2. from postmortem records at the Police Mortuary and the Institute of Neuropathology over the same 15-year-period, and
3. from the prevalence in mid-1979 of cases of Korsakoff's psychosis in the state mental institutions in and near Sydney.

From 1964 to 1968 there were about 35 cases per year; between 1969 and 1972 numbers rose to 82. In 1973 there were 152, and up to 1978 the numbers have stayed between 135 and 179. The great majority were alcoholic cases. The largest number, 784 of the 1370 cases over 15 years, were diagnosed as Korsakoff's psychosis, with 320 cases of Wernicke's encephalopathy and

267 of Wernicke and Korsakoff diseases combined. From inspection of the summaries the diagnosis was considered firm or probable in most of the cases.

The annual incidence of Wernicke–Korsakoff disease has thus gone up fivefold in Sydney in the past 15 years. The population of the city increased from an estimated 2.4 to 3.2 million (men, women, and children), i.e., 1.3 times, while estimated average alcohol consumption in Australia changed from 5,784 g/head per year to 7,424 g/head per year (also 1.3 times). Multiplying the two, the Sydney population times Australian alcohol consumption, less than a 2-fold increase occurred in the period.

There are two possible explanations for the discrepancy. One is a change in diagnostic accuracy or fashion. The sudden jump in cases between 1972 and 1973 suggests this and the irregularities are even greater for individual hospitals. In one hospital the annual admissions were 0, 0, 0, 0, 0, 0, 0, 0, 0, 20, 54, 24, 12, 17 and 32 cases. Wernicke's encephalopathy can be missed unless doctors are on the lookout for it.

Another explanation is a threshold effect, i.e., that medical complications of alcohol increase faster than consumption rate. Cases of alcoholic cirrhosis in Sydney were also found in our study to have gone up about 5 times, though this must be a delayed manifestation of alcoholism. Allocation of cirrhosis to alcohol or hepatitis might have been affected by a diagnostic fashion during the period.

Examination of postmortem records showed that during the 15 years, 86 cases were brought in dead to the Police Mortuary, having died at home or in the street, and were found to have the changes of Wernicke's encephalopathy at necropsy. With prior fixing of the brain and an experienced neuropathologist the changes of Wernicke's encephalopathy are characteristic. Other cases found in the postmortem study at the Institute of Neuropathology included 15 where the diagnosis had been missed and 11 where Wernicke–Korsakoff disease was really an incidental finding in a patient who died of a more serious disease such as cancer or hepatic failure.

In the seven state psychiatric institutions there was a total of 138 in-patients in mid-1980 with Korsakoff's psychosis or Wernicke–Korsakoff disease, compared with 348 diagnosed as alcoholic psychosis. The ratio between those two overlapping diagnoses in the institutions ranged from 1:5 to 1:1.

The figures obtained from all three approaches are more likely to under-estimate rather than over-estimate the frequency of Wernicke–Korsakoff disease. Records were not examined for small and private hospitals; the red-cell transketolase is not being used routinely to check the thiamin status of complex clinical pictures in alcoholics and patients with such symptoms as persistent vomiting. The prevalence study did not include patients living in the community as out-patients.

COMMENT

Diagnosed cases of Wernicke's encephalopathy and Korsakoff's psychosis have been increasing in Sydney in the past 15 years. It appears that something similar is happening in other affluent countries. Prevention will require education of the public about one more danger of alcohol and the need to eat as well as drink. In the short-term, fortification of beer and cheap wines with thiamin is being considered in Australia. A well-monitored trial of this measure in one of the states would be valuable.

Thiamin deficiency can lead to Wernicke's encephalopathy or to beri beri cardiomyopathy. The two diseases can occur together but it is surprising how often they do not. We cannot yet explain why the brain is affected in one person and the heart in another. It has been suggested that possibly the cardiomyopathy occurs in people who use their muscles for heavy work and so accumulate large amounts of pyruvate and lactate, producing intense vasodilatation in the muscles and increasing cardiac work, while encephalopathy is the first manifestion in less active people.

ACKNOWLEDGEMENTS. We are grateful to Dr. J. H. Gaha, Professor R. McLennan, Dr. R. Shureck, and Dr. J. Rankin for help and advice and to the medical superintents of 11 state general hospital and 7 psychiatric institutions in Sydney for access to medical records. This work was supported by the Sydney University Nutrition Foundation.

REFERENCES

Anonymous, 1979a, Wernicke's preventable encephalopathy (editorial), *Lancet* **1:**1122–1123.

Anonymous, 1979b, Wernicke's encephalopathy (editorial), *Br. Med. J.* **2:**291–292.

Baker, R. N., 1966, The transketolase test in neurological disease, *Bull Los Angeles Neurol. Soc.* **31:**125.

Brown, J., 1979, Alcoholic psychoses in Queenland psychiatric hospitals in: *Prevention of Alcohol-Related Brain Damage,* Report of Workshop April 5–6, p. 48, Commonwealth Department of Health, Canberra.

Burgess, H. J. L., and Burgess, A. P., 1976, Malnutrition in the Western Pacific region, WHO *Chron.* **30:**64–69.

Cruickshank, E. K., 1950, Wernicke's encephalopathy, *Q. J. Med.* **19:**327.

Davidson, S., Passmore, R., Brock, J. F., and Truswell, A. S., 1979, *Human Nutrition and Dietetics,* 7th ed., Churchill Livingstone, Edinburgh and London.

De Wardener, H. E., and Lennox, B., 1947, Cerebral beri beri (Wernicke's encephalopathy): Review of 52 cases in a Singapore prisoner-of-war hospital, *Lancet* **1:**11–17.

Donoso, G. (WHO Regional Office for S.E. Asia) (1978) personal communication.

Gudden, H., 1896, Klinische und anatomische Beitraege zur Kenntniss der multiplen Alkoholneuritis nebst Bemerkungen ueber die Regenerationsvorgaenge in peripheren Nervensystem, *Arch. f. Psych.* **28:**643–741.

Harper, C., 1979, Wernicke's encephalopathy: A more common disease than realised, *J. Neurol. Neurosurg. Psychiatry* **42:**226–231.

Korsakoff, S. S., 1889, Psychic disorder in conjunction with multiple neuritis, translated from the Russian by M. Victor and P. Yakovlev (1955), *Neurology* **5:**394.

Mesulam, M. M., van Hoesen, C. W., and Butters, N., 1977, Clinical manifestations of chronic thiamine deficiency in the rhesus monkey, *Neurology* **27:**239–245.

Price, J., and Theodoros, M. T., 1979, The prevention of Korsakoff's psychosis, *Med. J. Aust.* **1:**285.

Tomasulo, P. A., Kater, R. M. H., and Iber, F. L., 1968 Impairment of thiamine absorption in alcoholism, *Am. J. Clin. Nutr* **21:**1340–1344.

World Health Organization, 1975, *International Classification of Diseases,* vols. 1 and 2, Geneva (1975 revision).

Truswell, A. S., 1979, Thiamin and alcohol-related brain damage, in: *Prevention of Alcohol-Related Brain Damage,* Report of a Workshop, April 5–6, pp. 67–77, Commonwealth Department of Health, Canberra.

Truswell, A. S., Konno, T., and Hansen, J. D. L., 1972, Thiamine deficiency in adult hospital patients, *S. Afr. Med. J.* **46:**2079–2082.

Victor, M., Adams, R. D., and Collins, G. H., 1971, *The Wernicke–Korsakoff Syndrome,* Blackwell, Oxford.

Wallis, W. E., Willoughby, E., and Baker, P., 1978, Coma in the Wernicke–Korsakoff syndrome, *Lancet* **2:**400–401.

Weinstein, M. C., 1978, Prevention that pays for itself (editorial), *N. Engl. J. Med.* **299:**307–308.

Wernicke, C., 1881, *Lehrbuch der Gehirnkrankheiten für Aerzte und Studirende,* Vol. 2, pp. 229–242, Theodor Fischer, Kassel.

Wood, B., 1979, Paper at International Medical Advisory Conference of Brewers of Australia, Canada, UK, and USA, October 21–24, Melbourne.

Wood, B., and Breen, K. J., 1980, Clinical thiamine deficiency in Australia: The size of the problem and approaches to prevention, *Med. J. Austr.* **1:**461–464.

V

INFECTIONS

22

FOOD INFECTIONS

Shirley L. Fannin

Food infection has been the focus of much public health attention over the past century. Though many scientists had suspected food as a source of human illness, it was not until the science of bacteriology developed that those theories could be given a scientific basis.

With present-day knowledge of the bacteriology of food infections, it does not require a great deal of logic to surmise that primitive man suffered from illness and death from contaminated food. This would be especially true before the use of fire to cook food. Even the primitive mind must have noted that food kept longer after cooking, and caused less illness.

Recorded history from the Egyptians and Mesopotamians gives tantalizing glimpses of man's developing knowledge of diseases associated with environment. In ancient Egypt, at least with the pharoah and priest classes, the concept seemed well developed.

Though all ancient religions combined the concept of cleanliness into the practice of their religion, the Hebrews formalized these concepts in the books of the Torah and the laws codified in the Talmud. The concept of contagion and hygiene is discussed in great detail, and practice of these laws became a responsibility of the entire community rather than just the priests or ruling classes. The laws relating to food are especially well developed and suggest that the Hebrews recognized the dangers of some foods as sources of illness.

The Greeks developed the concept of personal hygiene to the level of religion, and contributed much to the understanding of the balance needed between man and his environment. In the third century B.C., Hippocrates of Cos, who recorded his thoughts on the nature of diseases and epidemics, included food as one of the elements that the physician must consider when looking for a cause and cure of epidemic illness.

The Romans first elevated the concept of control of disease by control of the environment to an institutional level. In Roman cities huge aqueducts

Shirley L. Fannin • Acute Communicable Disease Control, Public Health, Department of Health Services, Los Angeles, California 90012.

were built to bring clean water from the mountains and large sewage conduits were built and maintained. City officials, called *aediles,* were appointed to oversee the sewerage system, garbage disposal, and even destruction of unwholesome food (Richmond, 1935).

Following the fall of the Roman empire, during the second and third centuries A.D., many of the concepts of cleanliness and disease prevention were lost. Medieval society was marked by a return to paganism, belief in magic, and the concept of disease as a result of the wrath of God. The medieval towns were described as walled fortresses with severe overcrowding where people lived with their own filth. As the population moved from farms to crowded enclosures, the problem of providing uncontaminated food and ridding the living space of human waste was overwhelming. This situation helped create the 1000 years of darkness marked by recurrent epidemics and loss of human life in staggering numbers (Munday and Riesenberg, 1958).

In the sixteenth century, a monk called Gerolamo Fracastoro reasoned that a living contagious particle called *semenaria* was the cause of infection and could be spread by direct contact with contaminated articles, including food and water, and spread through the air. One hundred years later, Anton van Leeuwenhoek in Delft developed the first microscope and was the first man to see "animalcules of diver sorts."

In 1830, the microscope was improved both in power and precision. This advance in technology led to the discovery of the direct link between disease and microbes by Louis Pasteur. The science of bacteriology rapidly developed soon thereafter with new discoveries being reported in rapid succession. With this new knowledge came advances in methods of controlling infections.

The process of pasteurization was developed and instituted as a method of purifying milk. This led to increased shelf life of the milk and, thereby, safer milk for consumption by city dwellers.

Recognition of the link between bacterial contamination of food and diarrheal disease was a great step forward in child health. In the mid-nineteenth century, infant mortality was almost 50% of live births. Diarrheal disease was among the leading causes of this high mortality. By 1915, infant mortality had fallen to 10% of live births with the most striking decrease in deaths due to diarrheal disease and other infections. By comparison, present-day infant mortality in the U.S. is close to 1.3%.

The most common food infections are related to bacteria, viruses, and parasites. The symptoms caused by these infections are varied, usually related to a particular organism's method of attacking the gastrointestinal tract. Some bacteria damage the intestinal wall by causing death of cells in the brush border of the small intestine. The loss of this brush border interferes with normal digestive processes resulting in the loss of osmotic pressure, and accumulation of water and by-products of digestion in the intestinal lumen resulting in rapid peristalsis and evacuation. As the progress of food is speeded up through the intestinal cycle, water is not absorbed in the large intestine and diarrhea results.

Some organisms attack the large intestine. Amebiasis and shigellosis, in particular, cause characteristic damage to the wall of the large intestine. The result of this damage is to interfere with normal processes that take place in the large intestine, chiefly reabsorption of water and electrolytes. The resulting diarrhea and massive losses of both water and electrolytes lead to dehydration and metabolic imbalance.

A third result of infection of the intestine is the spread of the infecting organism to the rest of the body. This is accomplished when the organism gains access to the bloodstream through the damaged wall of the intestine. The circulatory system is a very effective mechanism for spread of the agents of food infection, resulting in infections of other organ systems remote from the initial site of invasion. Typical of this spread is liver infection, secondary to hepatitis A virus, and the infection of kidneys, bone, and the central nervous system, which follow *Salmonella typhosa* infection.

A fourth cause of pathology from food infections is secondary to toxins produced by the infecting organism. Staphylococcal food poisoning and botulism are two of the most common food infections that cause illness through toxin production. The staphylococcal toxin has a direct effect on the lining of the gastrointestinal tract while botulinal toxin has a direct effect on the central nervous system.

FOOD HANDLING

Numerous disease organisms may contaminate food under appropriate conditions. The most common organisms associated with food infections are staphylococcus, salmonella and shigella. The most common contributing factor to food contamination is poor food-handling practices. There are several steps in the process of food handling where contamination can take place. The first step is production. The source of food infections may be brought into the kitchen with the food. Green vegetables meant to be eaten raw can be contaminated in the field. Poultry or eggs often have surface contamination with salmonella organisms. *Trichinella spiralis* cysts are estimated to infest about 2% of pork sold commercially.

The second step in food processing where contamination occurs is in the storage and preparation of food. Surface contamination of food with bacteria is common. Refrigeration controls the amount of contamination by retarding the growth of bacteria that may be present. The cooking process will kill most bacteria that remain on the surface of food. Undercooking of infected meat is recognized as a major cause of trichinosis. Salmonella contamination of the cooking surfaces is a well-documented cause of subsequent salmonella food infections. Protecting food that is meant to be eaten uncooked generally depends on cleaning the surface of the food thoroughly or protecting it from contamination by packaging or minimal handling. The food preparers are possible sources of contamination. Care is taken to decrease the possibility of food-handler contamination by the practice of scrupulous handwashing and minimal touching of food before serving.

The third step in the food-handling process where contamination occurs is in the actual serving of the food. The dishes on which the food is served, as well as the person serving, are potential sources of contamination. A recent extensive outbreak of hepatitis A was associated with a doughnut shop where the infected food handler was in charge of powdering the doughnuts. Careful sanitizing of dishes and eating implements is important to prevent exposure during food serving. Training and supervision of food handlers to insure good hygienic practice is necessary in any commercial food establishment.

SALMONELLOSIS

Salmonella enteritis is an acute gastrointestinal disease frequently associated with ingestion of contaminated food. There are more than 1600 types of salmonella, many of them capable of causing enteric illness. The usual form of this illness consists of vomiting, diarrhea, and fever alone; however, spread of the disease to almost every other system by way of the bloodstream has been documented.

Salmonella is a gram-negative, motile, rod-shaped bacteria of the family Enterobacteriaceae. These organisms are biochemically distinct from genera of *Arizona* and *Citrobacter* even though they are closely related.

The salmonellae are hardy and adaptive organisms. They infect a multitude of hosts including mammals, reptiles, birds, and insects. They can remain viable in the environment even when dessicated and deprived of proper growth media. Some types are heat resistant and most are resistant to freezing. Another example of their adaptability is their genetic capability of acquiring drug resistance, which gives them an advantage in the host-parasite relationship.

Since animals are the main reservoir of salmonellae, foods of animal origin are the most frequent source of human infection. Food handlers are less likely to contaminate the food they are processing than to become infected from the food. Whereas shigellae have an infective level of 10^2 or more, salmonellae apparently must be present at levels of 10^5–10^6 before illness occurs.

Salmonella enteritis has remained the most frequent reported cause of foodborne outbreaks over more than 15 years of data collection at the Center for Disease Control (CDC). The disease is felt to be greatly underreported because of the mild self-limited nature of most illnesses. These would not come to medical attention in most instances and those that were seen might never have confirmatory stool testing. Even though large outbreaks, involving numerous guests at parties, are most often reported, the single cases or small household outbreaks are most frequent.

The salmonella organism is widespread in nature and found in virtually all species of mammals, birds, and reptiles. The species most common in human disease, however, are associated with poultry, swine, and cattle. *S. typhimurium,* the most common subtype, is frequently reported from poultry and eggs. The use of raw eggs has been discouraged because of this hazard (Ager *et al.,* 1967). *S. dublin* has been associated with disease secondary to consumption of raw milk (Werner *et al.,* 1979). *S. infantis* was recently found contaminating a special dietary supplement used for hospitalized infants and adults. The contamination evidently resulted from use of dried-egg product in its preparation.

There is an age predominance of salmonella enteritis with most cases occurring in children under age of five. This may be related to the fact that salmonella enteritis is a more severe disease in young children and thus more likely to come to medical attention and diagnosis. Males and females are equally affected.

The small intestine is the site of greatest pathology. Changes in the mucosa range from moderate inflammation and edema in mild cases to ulcerations in both large and small intestines in severe cases. Bacteremia is quite common especially during the febrile period of the illness. Hematogenous spread results in arthritis, osteomyelitis, pneumonia, meningitis, and endocarditis not infrequently seen in immuno-compromised hosts. Salmonella osteomyelitis has been seen often in children with sickle cell hemoglobinopathy.

The incubation period of this disease can vary from 6 hr to 7 days with most cases occurring at about 24 hr following exposure. The onset is rapid with fever, headache, myalgias, followed soon after by vomiting, diarrhea, and abdominal cramps. The illness, though severe, generally lasts only 2–3 days. It is not unusual for stool shedding of the bacteria to persist for 2–4 weeks. In a few cases, fecal shedding has continued for months and years. The chronic carrier state of *S. typhosa* is a well-known example of this and has been documented for the lifetime of the carrier.

Treatment is usually not necessary beyond symptomatic treatment. There is good evidence to suggest that antibiotic therapy will increase the length of fecal shedding (Dixon, 1965; Aserkoff and Bennett, 1969), which is to be discouraged, if possible. The decision to treat will depend on the toxicity of the illness, the condition of the patient, and the likelihood of suffering major complications. Patients at both extremes of age, those with immune suppression, and those with hemoglobinopathies deserve consideration for antibiotic therapy. Tetracycline, ampicillin, and trimethoprim sulfa have been used with success. The development of resistance to antibiotics has been noted, so that bacterial sensitivity patterns should be a guide to appropriate therapy.

Control of salmonella disease requires particular attention at all stages of food storage, preparation, and serving. Meat products should be thoroughly cooked with special attention to prevention of contamination of food preparation areas. The use of raw eggs should be discouraged and dried-egg products should be pasteurized. Scrupulous handwashing is necessary for all stages of food handling.

STAPHYLOCOCCUS AUREUS

Staphylococcus aureus is the second most common bacteria associated with foodborne illness. It causes almost 21% of reported episodes. This bacterium is a gram-positive, non-spore-forming coccus. It is a facultative anaerobe and grows well in a 10% salt solution that inhibits most other bacteria. The bacteria is ubiquitous in the environment. It colonizes the skin and mucous membranes and can survive for periods of time on inanimate objects, such as countertops, clothing, and other dry surfaces.

The illness related to the presence of this organism in food is due to a preformed enterotoxin produced by the bacteria. There are at least five antigenically distinct enterotoxins produced by staphylococcus—A, B, C, D, and E. All are heat resistant; boiling for 30 min will not inactivate them. A concentration of about 1 μg/100 g of food has been associated with illness.

Staphylococcal food poisoning is rarely fatal so the exact pathology caused in the intestinal tract has not been well studied. Vomiting, diarrhea, and abdominal cramping are the usual presenting symptoms. The cause of the vomiting is due to a direct toxic effect on the viscera, mediated by impluses to the vomiting center through the sympathetic nervous system and the vagus nerve. Diarrhea is attributed to a direct toxic effect on the cells lining the gut, causing a failure of water transport with resulting loss of large amounts of water into the intestinal lumen.

Phage group III staphylococci, which produces type A and type D toxins, are the most common causes of outbreaks. Prior to pasteurization, outbreaks related to mastitis in cattle and caused by phage group IV were seen more frequently. There is no seasonal incidence; and outbreaks are seen wherever food is prepared and served.

The major reservoir of staphylococcus is man. It frequently colonizes the nose, throat, and skin. As high as 40–50% of normal asymptomatic adults are found to be colonized. In a CDC survey of foodborne outbreaks, an infected lesion on the hand of a food handler associated with the outbreak was found 20% of the time (Merson, 1973).

The incubation of staphylococcal food poisoning is very short, varying between 30 min and 8 hr, with peak incidence at about 3–4 hr. The variation in the incubation period likely relates to the amount of toxin consumed. Fever is absent and the major symptoms of vomiting, diarrhea, and abdominal cramping vary from mild to severe, probably also related to the level of toxin exposure.

Diagnosis of staphylococcal food poisoning can usually be made on clinical grounds. The association with a suspect meal, the short incubation period, the absence of fever, and the relatively short course of the illness tend to differentiate it from other bacterial food poisonings. Recovery is usually complete within 24 hr and there are no sequelae.

Treatment is directed toward fluid replacement and supportive care as needed. There is no specific antitoxin available.

The only reasonable preventive measures are related to proper food handling practices, which minimize contamination with the organism initially and prevent growth of the organism in case contamination does occur. Keeping infected people out of the kitchen, proper refrigeration, and good handwashing should minimize this source of food poisoning.

SHIGELLOSIS

Shigellosis is an acute bacterial disease of the lower colon, sigmoid, and rectum. The onset is abrupt, characterized by diarrhea, abdominal cramps, tenesmus and occasionally fever. The stools may contain blood, mucus, and pus in severe cases. The incubation period of the disease is from 1 to 7 days, but is usually less than 4 days. A recent outbreak in Los Angeles, California included cases with common exposure less than 18 hr previously. The causative agent is a bacteria which is a nonmotile, gram-negative rod from the Enterobacteriaceae group. There are four groups of shigella: Group A, *S. dysenteriae;* group B, *S. flexneri;* group C, *S. boydii;* and group D, *S. sonnei.* There are more than 39 serotypes distributed between the four groups found worldwide. The most common types encountered in the U.S. are *S. sonnei, S. flexneri,* and *S. dysenteriae.* In Los Angeles, sporadic cases have been secondary to *S. sonnei,* while institution outbreaks among the developmentally disabled have more frequently been due to *S. flexneri* types 1 and 3.

The human gut is the only significant reservoir of this disease. Spread is by the fecal–oral route, and infection can occur by ingestion of as few as 200 organisms according to studies reported by Dupont *et al.* (1972).

Though the majority of cases of shigellosis are spread from person to person, food- and waterborne outbreaks have been reported. Two factors would predict these modes of transmission, one being the low infective dose and the other being the shedding of the organism in the stool of healthy-appearing convalescent cases. This shedding usually lasts for several weeks but has rarely been seen for a year or more.

The fact that food- and waterborne shigella outbreaks are not seen more frequently probably relates to the relative fragility of the organism. When shigellae are in the gastrointestinal tract, they maintain growth and replication readily; however, they are labile when removed from this environment. They are sensitive to heat and drying and are easily overgrown by competing bacteria on artificial media.

Shigellosis has a seasonal pattern with peak incidence between August and October. Susceptibility is higher in young children than adults and affects males more often than females, except in the 20–30-year-old group, where women are more often infected than men. This may relate to increased exposure of young mothers to their infected children.

Large epidemics occur in institutions, especially those housing the developmentally disabled. This is probably due to the episodes of fecal soiling and the virtual impossibility of maintaining adequate hygiene in these settings. Outbreaks have also occurred in day-care centers and nursery schools where children are not toilet trained. Because of the high secondary attack rates, a schoolwide outbreak can occur, which inevitably spreads to the families of the infected children. Secondary attack rates in families may be as high as 75%.

The pathogenesis of shigellosis is characterized by invasion of the epithelial cells of the intestinal tract. In fact, it was demonstrated by LeBrec *et al.* (1964) that without invasion, pathogenic effects did not occur. Once inside the cells, the bacteria grow and multiply, producing toxic products that diffuse into the circulation.

These toxic products may account for the convulsions and other neurologic symptoms seen with shigellosis in the absence of fever or with only low-grade fever.

Shigellosis sometimes has two phases. The first phase is characterized by abrupt onset of diarrhea, cramping, abdominal pain, dehydration, and fever. This lasts from 1 to 3 days. The second phase is characterized by persistent diarrhea, tenesmus, anorexia, and weight loss which lasts for several weeks. This prolonged illness can be cleared with antibiotic therapy.

The course of shigellosis can be shortened and fecal shedding halted with antibiotic therapy. A stool culture and sensitivities are recommended because of changing sensitivity patterns. A study in San Francisco, reported in *California Morbidity* in September, 1978, described only 78% of shigella isolates sensitive to ampicillin; 100% of these same isolates were sensitive to trimethoprim sulfa *(Bactrim, Septra).* In 1979, the Food and Drug Administration (FDA) approved trimethoprim sulfa for the treatment of shigellosis.

The prevention of shigella contamination of food is related directly to good food management. Cases and convalescent shedders should be excluded from food handling until cleared by three negative stool cultures on three successive days, starting at least 3 days after antibiotic therapy is finished.

BOTULISM

Botulism is a disease caused by the toxin produced by *Clostridium botulinum.* This disease is considered under food infections though it is a relatively rare disease at this time.

Clostridium botulinum is an anaerobic, gram-positive bacillus widely distributed in nature. It has been isolated both from the soil and from marine environments. The bacillus has a spore form that is very heat resistant, and capable of surviving boiling for many hours. Under appropriate conditions the spores germinate, producing an extremely heat-labile toxin. This toxin is one of the most potent neurotoxins known. Minute amounts of contaminated food are capable of killing adults.

There are seven different toxins designated A through G. The most common toxins associated with human illness are A, B, and E. Types C and

D have been associated with outbreaks in birds and small animals. Type F and G toxins are rare types, seldom seen in humans. The toxins are heat labile but require relatively high temperatures to inactivate them totally. Boiling readily inactivates the toxin.

Botulinal toxin is absorbed from the lymphatics of the gut wall. It acts at the myoneural junction as a potent inhibitor of acetylcholine release. The toxin evidently may be produced during bacterial multiplication in the gastrointestinal tract or in the depths of infected wounds. Postmortem findings are not classic and most often reflect complications, such as pneumonia and engorgement in the vessels of the central nervous system.

Botulism has been recognized since the late 1800s. Numerous outbreaks and single cases have been reported over the years. Type A botulism accounted for 23% of the outbreaks, type B 6%, type E 3%, and type F 0.1%. At least 68% of outbreaks were of unknown types. Almost all of the botulism outbreaks have been related to improper canning methods, both at home and commercially. The incidence of cases tends to rise during periods of increased interest in home canning.

The toxin type associated with a particular outbreak reflects the spore type found in the environment. Home-canned vegetables, fruits, and pickles have been implicated in more than 90% of cases. Meat products, jams, and jellies seldom, if ever, are implicated, probably because soil contamination is unlikely with meats and the high sugar content of jams and jellies does not provide an appropriate medium for spore germination. Type E outbreaks have been associated most frequently with fish and fish products. Commercial products, such as vichyssoise, peppers, mushrooms and smoked white fish have been the source of outbreaks. These are usually traced to a breakdown in the canning process. Identification of a contaminated commercial product initiates immediate and urgent public health action to recall the product. Because of the usually large batch and wide distribution of commercial products, the potential for widespread illness is great.

The incubation period for botulism ranges from 6 hr to 8 days, with the largest number of cases occurring between 12 and 24 hr. Early symptoms most often are gastrointestinal, with nausea, vomiting, and constipation. The symptoms that usually suggest the diagnosis however are neurologic. The onset is symmetrical weakness or paralysis of the cranial nerves progressing downward to the nerves of the limbs and trunk. Diplopia, dysarthria, dysphagia, and blurred vision are the most common symptoms. In severe cases with respiratory muscle involvement, dyspnea may be present. Symmetrical weakness of the limbs is almost always present; deep tendon reflexes are normal. There are no sensory deficits and the patient is usually mentally clear. Mucous membranes may be dry and appear inflamed, fever is absent, temperature and pulse are normal, paresthesias are usually absent and if present, suggest another diagnosis.

The diseases most commonly mistaken for botulism are Guillain-Barré syndrome, staphylococcal food poisoning, carbon monoxide poisoning, cerebrovascular accident with basilar artery involvement, myasthenia gravis, tick paralysis, atropine poisoning, trichinosis, poliomyelitis, and diphtheritic paralysis.

Diagnosis of botulism requires demonstration of the toxin in either the blood or stool of the patient or in an incriminated food. The difficulty of these diagnostic measures account for the 68% of cases being classified as unknown toxin type.

The treatment generally consists of three stages. First is the removal of unabsorbed toxin from the gastrointestinal tract by lavage or emetic if the toxin consumption has been less than 4 hr, and by use of a purgative if several more hours have elapsed. Neither of these procedures will be worthwhile if the ingestion occurred 12 hr or more before institution of therapy. The second stage is respiratory support by mechanical ventilation. Tracheotomy will likely be necessary. The third stage is neutralization of the absorbed toxin. The earlier an antitoxin is given, the better. Late administration of antitoxin will probably have little therapeutic effect, though unbound toxin has been detected as long as 30 days post ingestion. The occurrence of side effects to the antitoxin are frequent, about 20%, and can be severe. Careful evaluation of risks vs. benefits of antitoxin use should be made in all cases. Even though therapy with antibiotics and guanidine have been used, their efficacy is still unproven. Long-term, supportive care in an intensive care unit preventing pneumonia and other complications seems to offer the best chances for survival.

At present, the fatality rate is between 20 and 25%. The case fatality rate is lower with type B botulism than with types A and E. Adults die more frequently than children. Those cases with earliest onset have the worst prognosis. Severity of illness is not always related to volume of food eaten. This may be due to uneven distribution of the toxin throughout the food or to variation of susceptibility between people.

Using proper canning methods is obviously the primary method of prevention and control of botulism. Thorough heating of home-canned vegetables before tasting or consuming could decrease the risk somewhat.

INFANT BOTULISM

An interesting syndrome known as infant botulism was first described in California in 1976 (Pickett *et al.,* 1976). This is a syndrome affecting infants under the age of 6 months who present lethargy, a weak suck, a weak cry, and limpness. The illness may be progressive requiring ventilatory assistance. The affected infants are alert and have no preceding history of family illness, birth defect, or trauma.

Studies (Arnon *et al.,* 1978) have demonstrated the presence of both the organism *C. botulinum* and type specific toxin in the feces of affected infants.

Because of the young age of the cases, studies were undertaken to determine the etiology of the disease. It became clear early in the investigation that ingestion of preformed toxin had not occurred.

The most reasonable explanation for this syndrome is that *C. botulinum* spores are ingested from the environment and begin to germinate and form toxin within the lumen of the gut. The amounts of toxin are minute, probably accounting for the persistent but not rapidly progressing symptoms. The only food item that has been epidemiologically associated with the disease is honey. In several cases of type A botulism and none of the cases of type B, honey was the only food item ingested, other than milk. Laboratory studies on batches of honey demonstrated the presence of spores in a few instances.

This disease description has given rise to a great deal of speculation about its relationship to sudden infant death syndrome (SIDS). The affected age group for both disorders is the same. It is possible that a severe form of infant botulism might account for some but not all SIDS cases.

Infant botulism appears to affect boys more often than girls. It is confined to children under 6 months of age. Treatment is supportive only, sometimes requiring weeks of intensive care. Recovered cases seem to have no sequelae. Cases have occurred in both breast-fed and formula-fed infants (California State Department of Health Services, 1978).

Since 1976, cases of infant botulism have been reported from across the U.S., in England, Australia, and Czechoslovakia. Since the distribution of *C. botulinum* is worldwide, it is reasonable that this disease will also be found worldwide.

CAMPYLOBACTER FETUS (SUBSPECIES *JEJUNI, INTESTINALIS*)

Campylobacter is a recently recognized cause of human enteritis. The species *C. fetus* has long been recognized by veterinarians as a cause of infectious abortions in cattle and ewes, while other members of the genus are associated with several diseases of domestic animals including enteritis of dogs and pigs.

King (1957) was the first to report on human strains in 1957, but it was more than 10 years later, when Butzler *et al.* (1973) reported a method for successfully isolating the organism from fecal culture, that identification could become routine laboratory procedure. Skirrow (1977) modified the Butzler method and demonstrated that 7.1% of 803 isolates from patients with diarrhea contained *C. fetus,* subspecies *jejuni* or subspecies *coli.* He also demonstrated specific agglutinins in the sera of 31 of 38 patients with campylobacter enteritis; 10 had rising titers. Skirrow's findings confirmed campylobacter as a cause of enteritis and suggested that they were fairly common. Karmali and Fleming (1979) reported a method for rapid diagnosis of campylobacter using direct phase-contrast microscopy.

The syndrome related to campylobacter is marked by fever, severe abdominal pain, and profuse, watery diarrhea. Blood has been described in as high as 90% of cases in a pediatric study (Pai *et al.*, 1979). Significant fever has been found present in all patients greater than 12 weeks of age. Diarrhea persists 1–3 weeks. Four percent of untreated patients continue to shed the organism in their stool for 6 weeks even though more than half have cleared their stool by the fourth week. All treated patients became stool negative within 48 hr.

This disease apparently affects all ages. Two outbreaks have been reported associated with consumption of raw milk (Taylor *et al.*, 1979; California State Department of Health Services, 1979). Several studies failed to demonstrate mode of transmission or reservoir even though person-to-person, as well as common source, spread have been suggested by different reports. Two waterborne outbreaks have been reported from Colorado and Vermont (CDC, 1978a,b).

As more laboratories develop the special techniques necessary to isolate this organism, more will be learned about its epidemiology. Until there is a more universal diagnostic ability, the disease will probably not be included in the regulated diseases of sensitive occupations.

The drug of choice for therapy is erythromycin or gentamicin for bacteremias. Ampicillin, carbenicillin, and cephalothin-type drugs do not show good *in vitro* activity against campylobacters.

CLOSTRIDIUM PERFRINGENS

Clostridium perfringens food infection is an acute intestinal illness characterized by abdominal cramps and diarrhea. The incubation period is between 8 and 14 hr. Nausea occurs in about 50% of the cases, but vomiting is unusual. Fever and headache are not present.

Clostridium perfringens (Clostridium welchii) is a nonmotile, encapsulated, anaerobic bacillus. It is widely distributed in nature and has been isolated from soil, dust, raw meat, and the intestinal tract of animals and man. Spores may be heat labile or heat resistant. Some spores survive boiling up to 6 hr. Growth can occur at pH values from 5 to 9, and at salt concentrations higher than those used to cure meat. Studies demonstrated that at least 10^8 organisms are required to produce illness (Dische and Elek, 1957).

There are five types of *C. perfringens* designated A through E. Food poisoning is usually caused by type A organisms. More than ninety serotypes of *C. perfringens* have been identified. The same strains capable of producing alphatoxin, which is responsible for gas gangrene (myonecrosis), cause food infections.

About 7% of foodborne outbreaks reported to CDC are due to *C. perfringens*. This is thought to be a sizeable underestimation because of the technical difficulty many laboratories experience in isolating anaerobes. Many illnesses listed as unknown etiology have a clinical description compatible with *C. perfringens*.

Because *C. perfringens* is found in the intestinal tract of normal, healthy people, diagnosis of food poisoning secondary to this organism cannot be made solely on a positive stool culture from the case. It is best demonstrated by culturing the suspected food and both the ill and well persons who ate the food. Finding the same serotype in both the food and the sick person exposed to the food offers strong evidence of food-associated illness.

Since this organism is so widespread, it is almost impossible to eliminate contamination of food. Preventive measures should ensure serving meat dishes hot soon after cooking. When storing is necessary, the meat should be cooled rapidly and refrigerated until served. Large cuts of frozen meat should be thawed in the refrigerator.

BACILLUS CEREUS

Bacillus cereus is a bacterium that has been associated with food infections for at least 30 years. It is a gram-positive rod, aerobic, spore-forming, and produces at least two different toxins. The first syndrome recognized to be associated with this bacterium is characterized by diarrhea, abdominal cramps, and nausea with virtually no vomiting. This syndrome has an incubation period of 6–14 hr and lasts 10–36 hr. The symptoms and incubation period are the same as *C. perfringens,* making distinction on a clinical basis impossible. The second syndrome recognized is one characterized by nausea, vomiting, abdominal cramps, and no fever or diarrhea. This syndrome has an incubation period of 1–8 hr and a duration of 8–10 hr.

According to work reported by Terranova and Blake (1978), the two distinct syndromes are caused by two distinct toxins. The toxin associated with upper intestinal tract illness (vomiting, no diarrhea) is heat stable. The toxin associated with lower gastrointestinal illness (diarrhea, no vomiting) is heat labile. The heat-stable toxin has been associated with eating fried rice. The heat-labile toxin has been associated with consumption of various vegetables and meats (Terranova and Blake, 1978).

There have been few outbreaks reported in the U.S. compared to Europe. It has been suggested that the similarity between the upper gastrointestinal tract illness and staphylococcal food poisoning probably limits the search for a specific diagnosis. This same explanation holds for the underreporting of the illness caused by the heat-labile toxin, lower gastrointestinal tract illness, which clinically resembles *C. perfringens* food poisoning.

Since this organism is widespread in soil, it can be expected to be a contaminant of raw, dried, and processed food. A level of at least 10^5 organisms/g of food is necessary to cause illness.

Bacillus cereus spores can survive high temperatures including boiling. They multiply rapidly in food held at room temperature. The numerous episodes of poisoning related to fried rice are thought to be related to the restaurant's practice of storing cooked rice at room temperature to prevent clumping (Mortimer and McCann, 1974). To prevent the possibility of this type of food infection, there should be prompt refrigeration of food whenever serving is to be delayed.

Since this disease is self-limited and of short duration, no antibiotic therapy is indicated. Definitive diagnosis depends on demonstrating the presence of 10^5 or more of the organisms in a suspect food item. A stool culture on the case cannot be interpreted alone because *B. cereus* is sometimes found in the gut of healthy people.

VIBRIO CHOLERAE

Cholera is a self-limited, acute, dehydrating, diarrheal disease caused by an exotoxin of *V. cholerae.* Illness caused by this toxin varies from a mild diarrhea to an explosive diarrhea with extreme and rapid fluid loss leading to death in a matter of hours.

Vibrio cholerae is a gram-negative, slightly curved organism that is propelled by a single polar flagellum in a rapidly darting motion. The organism grows well from 22° to 40°C. Although it has maximal growth at body temperature, it also grows at room temperature. In the classic rice-water stool, the vibrios are present in large numbers; however, when searching for asymptomatic carriers, special laboratory media are required because of the relatively few organisms to be found.

Cholera is endemic in Asia and periodically becomes epidemic, spreading rapidly from country to country. The latest pandemic started in 1961, peaked in 1975, and has continued the downward trend in a number of countries reporting cases into 1981. This pandemic has involved almost every Asian country, the countries of the Middle East and Africa, and the European countries bordering the Mediterranean.

The U.S. reported two cases locally acquired in 1978; one was laboratory associated, while the other was traced to consumption of shellfish in Louisiana. Four cases have been reported secondary to travel; only one of these was in a U.S. citizen. The other three were found in Southeast Asian refugees during the resettlement process. There has been no secondary spread from cases in the U.S.

Epidemic *V. cholerae* are divided into three serotypes: Inaba, Ogawa, and Hikojima, classified by the somatic (O) antigen. The Inaba type has A and B somatic antigens, the Ogawa serotype has A and C antigens, while the Hikojima serotype has all three antigens—A, B, and C. The El Tor biotype has almost completely replaced the classical strain as the causative agent of pandemic cholera. This biotype was first described in 1905 and has since become the predominant epidemic strain.

Vibrio cholerae is not found in other animals besides man. The organism survives for a relatively short time, 4–7 days, when excreted into water. It does not withstand drying or an acidic environment. The El Tor strain survives for longer periods of time in both the host and the environment than the former classic strain.

Shifts of serotypes between Ogawa and Inaba have been noted in endemic areas. This leads to speculation that new epidemic strains may arise from unrecognized reservoirs in nature through a series of genetic changes under the right conditions. Supporting this speculation are the shifts *in vitro* reported by Sack and Miller (1969) and Barua (1974).

Cholera is predominantly a disease of poor people. This is most likely related to their poor hygiene, crowded living conditions, and lack of access to protected sources of water and food. Refugee camps, religious gatherings, or any large gatherings where sanitary facilities are inadequate for the number of people gathered, are prone to cholera outbreaks in endemic areas.

The disease is not easily spread from person to person. This is most likely related to the large number of organisms required to infect. Hornick, *et al.* (1971) reported that 10^6 organisms are required to infect when gastric acidity is neutralized; when no neutralization was done, at least 10^{11} organisms were required to infect.

The El Tor strain of cholera may cause a rather mild self-limited diarrheal illness or an acutely fulminating illness characterized by massive fluid loss, 10–15% dehydration, metabolic acidosis, and death within a few hours. The cause of the massive fluid loss—diarrhea—is an enterotoxin called *choleragen* produced by the *V. cholerae* either as they grow in the intestinal tract or on appropriate culture media. The enterotoxin acts directly on the lining of the gastrointestinal tract causing the mucosal cells to secrete large amounts of chloride. Secondary water and bicarbonate loss occur simultaneously.

The incubation period of cholera is 1–5 days; however, in common-source outbreaks, most cases occur within 2–3 days. The illness usually presents as an acute, profuse, watery diarrhea with no tenesmus or abdominal cramping. The diarrhea is brown in color at the outset but soon becomes gray, ricewater in appearance, and has an odor characterized as slightly fishy. Vomiting, if present, usually occurs after the diarrhea is well established. Fluid losses up to 30 liters/day have been observed, resulting in clinical signs of severe dehydration, i.e., sunken eyes, poor tissue turgor, weakness, rapid thready pulse, coldness of skin, wrinkling of the skin of the digits, and hyperpnea. There is no fever; in fact, often the temperature is subnormal. Death may occur within 4 hr without appropriate immediate therapy. Usually death during the acute illness occurs on the second or third day. Untreated cases will usually begin recovery from the diarrheal phase of the illness in 5–7 days. Use of appropriate antibiotics will shorten this phase to about 3 days.

Treatment of cholera is related to replacement of the fluid losses and correction of the electrolyte imbalances. In a conscious patient this can be assisted by giving glucose electrolyte solutions orally as well as intravenous replacement therapy. Because loss of bicarbonate is excessive, special attention to correction of the severe metabolic acidosis is a key factor in treatment.

According to a leading authority on cholera, A. H. Benenson, the complications most often seen are severe metabolic acidosis, hypokalemia, acute renal failure, pulmonary edema, hypoglycemia, convulsions or tetany, and bilateral corneal scarring in a stuperous patient (Benenson, 1970). Metabolic acidosis is directly related to loss of bicarbonate in the stool; persistent nausea and vomiting may be secondary to uncorrected metabolic acidosis. Hypokalemia may cause cardiac arrhythmia, weakness, and abdominal distention. Acute renal failure may be secondary to undertreated dehydration or prolonged hypokalemia. Pulmonary edema has been documented when fluids are rapidly replaced without correcting metabolic acidosis. The mechanism of the pulmonary edema relates to the severe peripheral vasoconstriction that causes an overload of the heart and pulmonary circulation. This can be avoided by replacing the bicarbonate ion at a faster rate. Hypoglycemia may occur when the patient has had no significant alimentation for several days. Children are especially prone to this complication. Convulsions and tetany seen in some severe cholera cases may be due either to loss of magnesium or a calcium phosphate imbalance secondary to the loss of bicarbonate ion. Corneal scarring generally occurs in patients with long periods of stupor caused by conjunctival dehydration and loss of the wink reflex.

Cholera spread can be controlled by protecting the food and water supply from fecal contamination (Fetsenfeld, 1965). This can be accomplished by boiling or chlorinating all water, pasteurizing milk, and eating only foods that are hot. Sanitary sewage disposal is critical. Where public systems are not available, emergency alternate facilities should be provided. Attention must also be given to mechanical contamination of food by flies, cockroaches, and other vermin. Fly breeding control and screening are important adjuncts to management. Food handling practices particularly in commercial food establishments must be monitored closely. This includes good handwashing practice and protection of the preparation surfaces from contamination.

Most developed countries in the world do not have endemic or epidemic cholera. When cases are imported, they seldom spread. The level of sanitation, practice of personal hygiene, and a relatively low level of poverty are factors that account for this.

TRICHINELLA SPIRALIS

Trichinosis is an acute or subacute infection caused by *T. spiralis,* a round worm. The infection develops after ingestion of improperly cooked meat containing trichina cysts. The gastric juice digests the wall of the cyst releasing the larvae. The larvae become embedded in the wall of the intestine within a week after ingestion. The females produce embryos that are deposited in the lymph spaces and mucous membrane. These embryos travel by way of the lymphatics to mesentery nodes; from there, they reach the general circulation by way of the thoracic duct. From the bloodstream, the larvae reach the striated muscle, where they encyst and remain viable for many years.

Trichinosis is prevalent wherever meat that is infested with trichina cysts is eaten under cooked. Several outbreaks of trichinosis have been reported from Eastern Europe and Russia, where homemade sausage from trichinous pork is eaten raw or partially cooked. In 1952, it was estimated (Link, 1952) that one out of every six Americans were infested with Trichina. By 1970, this prevalence had been reduced to 4.2%. Legislation requiring heat treatment of garbage fed to swine, and a greater knowledge of proper methods of cooking pork are factors most likely to have contributed to this decrease.

There is no age, sex, race, season, or climate that affects trichinosis prevalence. This disease is not transmitted from person to person.

Susceptibility is general. Neither natural or acquired immunity occurs in man. Several attacks have been documented in the same person suggesting either no acquired immunity or very weak immunity.

The incubation period of trichinosis is 2–28 days. The first group of symptoms occur during the invasive stage in the intestine. These symptoms are predominately digestive symptoms: anorexia, abdominal pain, nausea, and vomiting. During lymphatic and bloodstream dissemination, muscle pain is a dominant symptom. These pains, which may be severe, are due to penetration of the worms into the muscle fibers with attendant inflammation. Edema of the eyelids is seen quite often; sometimes the entire face is swollen. Transient rashes, either scarlatiniform or urticarial, may be present. Other symptoms include tingling, itching, or burning of the skin. Temperature may rise to 102 or 103°F. There is an elevation of the white-cell count with a significant eosinophilia. The final stage of the disease is the encystment, which occurs during convalescence. The acute illness varies in length from 10 days to 1 month.

There are many features of trichinosis that resemble allergic reactions. The skin manifestations, periorbital edema, and eosinophilia are evidence of a possible allergic phenomenon (Bohrod, 1961).

Many cases of trichinosis are mild and undiagnosed. Of those infections recognized, the mortality rate is about 5%. Absence or sudden drop of eosinophilia has been observed to be related to a poor prognosis.

Treatment is not recommended for mild disease. The drug of choice for moderate to severe disease is thiabendazole (Campbell, 1971). Corticosteroids may be used in severe illness or in those with severe allergic symptoms. Muscle pain and headache respond to aspirin.

Prevention of trichinosis is dependent on proper cooking of meat that might be infested. It has been stated that appropriate freezing of meat would kill the larvae in most meats. This may be true of the majority of Trichina-infested meat in the temperate zone; however, recent reports have suggested a cold-adapted *T. spiralis* found in Alaskan bear meat (CDC, 1979). Pasteurization of garbage fed to hogs should be continued.

HEPATITIS A (INFECTIOUS HEPATITIS)

The most common virus associated with food infection is hepatitis A. The virus has been demonstrated by electron microscopy by Feinstone *et al.*, 1973.

This virus is worldwide in distribution. The distinguishing characteristics permitting recognition of hepatitis A epidemics are cyclic occurrences, age-specific rates highest in school children, seasonal cycle, rural rates higher than urban, household transmission, immune serum globulin as an effective prophylaxis, and fatality rate is low. Humans appear to be the only important reservoir. Transmission by blood transfusion rarely occurs because the viremic stage is very short, probably less than 24 hr.

The incubation period is 3–7 weeks with the largest number of cases clustering at about 4–5 weeks after exposure. Virus is shed in the stools in great numbers in the two weeks prior to onset of jaundice. Very few viruses are present at the time of jaundice and none 3 days later.

Fecal–oral transmission is probably the only mechanism of importance. Person-to-person spread is most common; however, common source outbreaks in both food and water have been frequently reported.

Waterborne outbreaks are directly related to sewage contamination of the water. This seldom occurs with municipal water supplies in the U.S., but has been reported often with small private supplies.

Foodborne hepatitis occurs by several mechanisms. The most frequent mechanism is direct contamination of food by an infected food handler. Since the illness may be unrecognized during the 1–2 weeks before onset of jaundice, a food handler who practices poor handwashing may easily transmit via food items. A recent large outbreak in northern California occurred when two cooks at a restaurant worked during the late incubation period. One hundred cases with two deaths occurred (Schoenbaum *et al.*, 1976). A second mechanism is the use of contaminated water for washing containers for food not subsequently cooked (Murphy and Petrie, 1946). A third source is the use of sewage-contaminated water for soaking or reconstituting foods not subsequently cooked. A fourth mechanism is the growth of subsequently uncooked food in a sewage contaminated environment. Green, leafy vegetables grown where human waste is used as fertilizer is one graphic example.

Another interesting mode of transmission other than human contact is by way of raw or incompletely cooked mollusks. This has been frequently reported and usually indicates a situation where mollusks are harvested from sewage-contaminated waters. Most commercially harvested mollusks in the U.S. are protected from this contamination by strict regulation of the beds in which they are grown.

Infection with hepatitis A confers life-long immunity. The most common pathology that follows infection is liver parenchymal cell necrosis. In mild cases, this necrosis is spotty. In severe cases, necrosis becomes confluent and lobular collapse occurs. Biliary pigments are altered and biliary stasis occurs in the canaliculi of the bile collecting system. Periportal inflammation and mononuclear cell infiltration of the sinusoids also occur.

The clinical illness that occurs with hepatitis A infection is sudden. Malaise, anorexia, muscle-aching, nausea, and vomiting occur in most cases. Headache, fever, abdominal pain, or discomfort usually in the epigastrium of the right upper quadrant, occur less frequently. These symptoms precede jaundice by 3–10 days. Itching believed to be secondary to jaundice may occur both before and after jaundice appears. Hepatomegaly and hepatic tenderness are important clinical signs. Liver enzyme elevations may sometimes be extreme. The SGOT level (Serum Glutamic Oxaloacetic Transaminase) generally reaches 1000–1200 units. The fatality rate is thought to be very low; in military populations, the rate is one per 1000 of recognized cases.

There is no specific treatment for hepatitis A infection. Bed rest, though often ordered, probably does not affect the outcome of the disease. Dietary prohibitions also have no basis. Alcoholic beverage may be limited to modest amounts. Prohibiting alcohol consumption for 6–12 months has been done in the past, but has no scientific basis. Antibiotics and steroids should not be used as they do not affect the outcome of the disease.

Preventing hepatitis A infections can be accomplished by administration of pooled gamma globulin to household contacts to a case or, prior to exposure, to persons traveling or working in developing countries where sewage disposal is inadequate.

Preventing foods from becoming contaminated requires care to be taken in preparation of foods not intended to be cooked before ingestion. These include good handwashing techniques for the food handler, and proper cleaning of vegetables meant to be eaten raw.

REFERENCES

Ager, E. A., Nelson, K. E., Galton, M. M., Boring, J. R., III, 1967, Two outbreaks of egg-borne salmonellosis and implications for their prevention, *J. Am. Med. Assoc.* **199:**372.

Arnon, S. S., Midura, T. F., Damus, K., Wood, R. M., and Chin, J., 1978, Intestinal infection and toxin production by *Clostridium botulinum* as one cause of sudden infant death syndrome, *Lancet* **1:**1273–1277.

Aserkoff, B., and Bennett, J. V., 1969, Effect of antibiotic therapy in acute salmonellosis on the fecal excretion of salmonellae, *N. Engl. J. Med.* **281:**636.

Barua, D., 1974, Laboratory diagnosis of cholera, in: *Cholera* (D. Barua and W. Burrows, eds.), pp. 359–366, W. B. Saunders, Philadelphia.

Benenson, A. H., 1970, Cholera, in: *Communicable and Infectious Diseases* (F. H. Top, Sr. and P. F. Wehrle, eds.), pp. 174–182, C. V. Mosby, St. Louis.

Bohrod, M. G., 1961, Trichinosis with special reference to the allergic component, *Am. J. Gastroenterol.* **36:**67.

Butzler, J. P., Dekeyser, P., Detrain, M., and Dehaen, F., 1973, Related vibrio in stools, *J. Pediatr.* **82:**493.

California State Department of Health Services, 1978, Antibiotic susceptibility and treatment of *Shigella* infection, *California Morbidity* **38 (Sept. 29).**

California State Department of Health Services, 1979, *Campylobacter fetus* subspecies *jejuni:* Increasingly recognized public health significance, *California Morbidity* **44 (Oct. 26).**

Campbell, W. C., 1971, Anti-inflammatory and analgesic properties of thiabendazole, *J. Am. Med. Assoc.* **216:**2143.

Center for Disease Control, 1978a, Waterborne Campylobacter gastroenteritis—Vermont, *Morbidity and Mortality Weekly Report* **27(25)** (June 23), p. 207.

Center for Disease Control, 1978b, Campylobacter enteritis—Colorado, *Morbidity and Mortality Weekly Report* **27(27) (July 7):**226–231.

Center for Disease Control, 1979, Trichinosis–United States, 1978, *Morbidity and Mortality Weekly Report* **28(45)** (Nov. 16), pp. 541–543.

Dische, F. E., and Elek, S. D., 1957, Experimental food poisoning by *Clostridium welchii, Lancet* **2:**71.

Dixon, J. M. S., 1965, Effect of antibiotic treatment on duration of excretion of *Salmonella typhimurium* by children, *Br. Med. J.* **2:**1343.

Dupont, H. L., Hornick, R. B., Snyder, M. J., Libonati, J. P., Formal, S. B., and Gangarosa, E. J., 1972, Immunity shigellosis. II. Protection induced by oral live vaccine or primary infection, *J. Infect. Dis.* **125:**12.

Feinstone, S. M., Kapikian, A. Z., and Purceli, R. H., 1973, Hepatitis A: Detection by immune electron microscopy of a viruslike antigen associated with acute illness, *Science* **182:**1026.

Fetsenfeld, O., 1965, Notes on food, beverages and fomites with *Vibrio cholerae, Bull. World Health Organization* **33:**725.

Hornick, R. B., Music, S. I., Wenzel, R., Cash, R., Libonati, J. P., Snyder, M. J., and Woodward, T. E., 1971, The Broad Street Pump revisited. Response of volunteers to ingested cholera vibrios, *Bull. N. Y. Acad. Med.* **47:**1181.

Karmali, M. A., and Fleming, P. C., 1979, *Campylobacter enteritis* in children, *J. Pediatr.* **94:(4):**527.

King, E. O., 1957, Human infections with vibrio fetus and a closely related vibrio, *J. Infect. Dis.* **110:**119.

LeBrec, E. H., Schneider, H., Magnani, T. J., and Formal, S. B., 1964, Epithelial cell penetration as an essential step in the pathogenesis of bacillary dysentery, *J. Bacteriol.* **88:**1503.

Link, V. B., 1952, Trichinosis, a national problem, presented at the National Conference on Trichinosis, December 15, Chicago, Illinois.

Merson, M. H., 1973, The epidemiology of staphylococcal foodborne disease, in: *Proceedings of the Staphylococci in Foods Conference,* pp.20–37, Pennsylvania State University Park, Pennsylvania.

Mortimer, P. R., and McCann, G., 1974, Food-poisoning episodes associated with *Bacillus cereus* in fried rice, *Lancet* **1:**1043.

Munday, J. H., and Riesenberg, P., 1958, *The Medievel Town,* N. J. Van Nostrand, Princeton.

Murphy, W. J., and Petrie, L. M., 1946, Outbreak of infectious hepatitis apparently milkborne, *Am. J. Public Health* **36:**169.

Pai, C. H., Sorger, S., Lackman, L., Sinai, R.E., and Marks, M. I., 1979, Campylobacter gastroenteritis in children, *J. Pediatr.* **94(4):**589.

Pickett, J., Berg, B., Chaplin, E., and Brunstetler-Shafer, M., 1976, Syndrome of botulism in infancy: Clinical and electrophysiologic study, *N. Engl. J. Med.* **295:**770–772.

Richmond, I. A. (ed.), 1935, The aqueducts of Ancient Rome by Thomas Ashby, Clarendon Press, Oxford.

Sack, R. B., and Miller, C. E., 1969, Progressive changes in vibrio serotypes in germ-free mice infection with Vibrio cholera, *J. Bacteriol.* **99:**688.

Schoenbaum, S. C., Baker, O. J., and Jezek, Z., 1976, Common-source epidemic of hepatitis due to glazed and iced pastries, *AM. J. Epidemiol.* **104(1):**74–80.

Skirrow, M. B., 1977, Campylobacter enteritis: A "new" disease, *Br. Med. J.* **2(6078):**9–11.

Taylor, P. R., Weinstein, W. M., and Bryner, J. H., 1979, *Campylobacter fetus* infection in human subjects: Association with raw milk, *Am. J. Med.* **66:**779–783.

Terranova, W., and Blake, P. A., 1978, *Bacillus cereus* food poisoning, *N. Engl. J. Med.* **298(3):**143–144.

23

MAJOR FOOD-BORNE PARASITIC INFECTIONS

Derrick B. Jelliffe and E. F. Patrice Jelliffe

Numerous parasites can enter the body within food or as a contamination of foodstuff. The most important of these include *Ascaris lumbricoides* (roundworm), *Truchuris* (whipworm), *Taenia* sp. (tapeworms), *Echinococcus granulosus* (hydatid cyst), protozoa (*Entameba histolytica* and *Giardia lamblia*) and liver flukes (see Chapter 26).

ASCARIASIS

Roundworm infection, although cosmopolitan in distribution, is at the present time much more common and of greater importance among children in most subtropical and tropical countries, especially in those with warm, moist climates that favor the survival of ova after they have reached the ground. Infection occurs exclusively as a result of the swallowing of mature embryonated ova acquired from soil contaminated by human feces, that is, as a result of consuming dirt or raw vegetables, soiled fingers, or in water, such as is found in the shallow, muddy, polluted ponds. Fly-borne contamination of food is a further, lesser hazard.

Infection is occasionally seen even in the early months of life in areas where ascariasis is highly endemic—the baby presumably becoming infected from the mother's fingers, from contaminated feeding utensils, or from dust. However, human milk may contain antiparasitic factors, as well as protective substances against bacterial and viral infections.

Ascariasis is most common during the "preschool" age period from 1 to 5 years, when the child begins to lead a more independent life and may spend much of his time playing in the feces-contaminated compound. It is an exploratory period that is reflected by the child experimentally putting things

Derrick B. Jelliffe and E. F. Patrice Jelliffe • Population, Family and International Health Division, School of Public Health, University of California, Los Angeles California 90024.

in his mouth, so that there tends to be not only a high incidence but also heavy worm burdens.

Roundworm infection occurs most easily in warm, humid tropical regions; very hot, dry, desert climates tend to desiccate and kill ova after they have reached the ground. In general, it may be said that the incidence of ascariasis, at least among adults, is roughly inversely proportional to the number of effective sanitary latrines *actually being used* by a population (including young children), and it has even been suggested in some places to use this as a rough index of public health and sanitation. Infection rates are especially high, extending into all age groups, in regions where the custom of fertilizing the field with raw, untreated night soil is practiced widely, as in some parts of Asia.

Clinical Features

Due to Larvae

If large numbers of larvae migrate at one time, a child may show respiratory signs or an enlarged liver (Jelliffe and Jelliffe, 1978).

Due to Adult Worms

A wide variety of different clinical features may be produced by adult roundworms, depending to a great extent upon their situation, the worm burden and the host's nutrition. The clinical features include those due to (a) migration (e.g., following fever into the nose or anus), (b) intestinal obstruction (when a tangled mass of roundworms block the ileocaecal opening), (c) general, and (d) malnutrition.

General. With a small parasite load in a relatively well-nourished individual, ascariasis may be symptomless and unsuspected, unless abnormal migration occurs.

By contrast, a heavy worm burden, especially associated with poor nutrition, often appears to produce a typical clinical picture, although it is usually uncertain which features are due to the presence of roundworms and which are due to associated factors, such as an inadequate dietary intake or other intestinal parasites that are very often present. A typically affected child (as seen, for example, in the slums of Southeast Asia), is aged about 2–5 years and is listless, underweight, and somewhat anemic. The cheeks are rounded and rather pendulous, resembling the "moon-face" of *kwashiorkor.* The skin is frequently lusterless and scaly, and the hair is dry and somewhat light in color. In Indian children of this sort, Bitot's spots and follicular keratosis are frequently seen. The abdomen is distended, full, and somewhat pendulous. The outlines of the worm-filled small intestines may be visible, and occasionally cord-like ascarids can be palpated through the thin abdominal wall.

Malnutrition. A high worm burden may sometimes represent 5–10% of a stunted malnourished tropical child's body weight, so that the direct nutritional drain of the parasitic mass must often be important, in view of the child's high nutritional needs during this period of rapid growth and the often inadequate or marginal diet. It seems possible that the roundworm may affect the host's nutrition by the following means:

1. Ingestion of food from the intestinal tract directly into the parasite's alimentary canal, as can be confirmed by radiological examination of an affected child following a barium-meal, when the worm's alimentary canal can be visualized as string-like shadows. Carbohydrate and protein-splitting enzymes have been identified in the parasite's intestinal tract. It may be noted that 48% of the dried weight of the adult worm has been shown to be composed of protein, containing many of the essential amino acids; while another small, but possibly cumulatively important, protein drain is represented by the fact that a single fertile female lays about 200,000 or more ova daily.
2. Absorption of certain foods, notably sugars, can occur directly through the worm's cuticle.
3. Damage to the intestinal wall may be produced by repeated minor trauma and possibly by "toxins"; while, of greater importance in the narrow lumened intestines of small children, ascarids may interfere with food absorption as a result of direct blockage of the villous surface by the large number of parasites present.
4. The antienzyme, ascarase, which contains antiprotein splitting enzymes, may be secreted to further impair the digestion and absorption of food by the host child.

Nutritionally, heavy infections with *A. lumbricoides* have been related to stunting, protein-energy malnutrition, xerophthalmia, and kwashiorkor.

Diagnosis

Intestinal ascariasis is diagnosed by the finding of adult roundworms in the stools or vomitus, or by finding the ova in the stools. The number of worms present can, under average conditions, be roughly estimated by the number of ova present in the standard 2–3 mg smear. Under 20 eggs/smear indicates a light infection, while over 100 are present with a heavy worm burden of over 50 ascarids.

Treatment

Theoretically, ascariasis needs treatment even if the infection is so light that only one unfertilized ovum is found in the stools, as even a single ascarid is potentially capable of producing serious, or even fatal, complications, should ectopic migration occur. In addition, with modern drugs such as piperazine, the presence of pyrexial illness should not be regarded as a contraindication to therapy, but rather as an indication, in view of the tendency to migration or "balling" that seems to occur as a result of fever. Care should be taken to avoid initial treatment with such partially-ascaricidal drugs as tetrachlolethylene in children with hookworm infection who are also harboring roundworms. In fact, as a rule, in intestinal polyparasitism, ascariasis ideally should be treated first rather than risk the danger of initiating migration as a result of therapy with drugs not effective against the roundworm.

The specific therapy of ascariasis has been revolutionized in the last few years. Relatively toxic drugs, such as santonin and oil of chenopodium, are being universally superseded by less toxic remedies needing neither preparation nor purgation, notably piperazine derivatives and more expensive widespectrum anthelminthis, such as thiabendazole (Davis, 1973).

Prevention

The prevention of ascariasis will not be achieved in some tropical areas for many years and is related both to improvement of socioeconomic conditions in a region, and particularly, to the spread of general health education in a community by such means as school health services and "mothercraft" classes. Specific measures may be considered under two headings.

Adequate Disposal of Feces

The provision of cheap but effective toilets, such as bore-hole latrines, will certainly be of assistance, although the problems of enabling and persuading children to use them still remains. In Sri Lanka, "toddler-sized" squatting plates have been popularized. In addition, cultural attitudes towards stools and understandable lack of realization that intestinal parasitic infection may be acquired by the ingestion of feces-contaminated food will both present health education problems.

Removal of Worms

With the development of drugs of such low toxicity as the piperazine compounds, it should become increasingly possible to employ widespread periodical routine deworming campaigns through such organizations as Young Child Clinics ("Under Fives") and school health services.

TRICHURIASIS (TRICHOCEPHALIASIS)

Infection with the whipworm, *Trichuris trichura (Trichocephalus trichurus)* is one of the most common helminthiases of childhood in many subtropical and tropical countries and also appears to be one of the most neglected. In tropical countries whipworm infection is usually more or less ignored, with the assumption that it is harmless. That this is not always the case has been amply demonstrated.

Adult whipworm may be recognized by shape, the anterior three-fifths being much slender than their posterior portions. They most commonly live in the human caecum, but like *Entameba histolytica,* frequently also attack the appendix, rectum, and sigmoid colon. The worms are attached to the mucosa by the slender anterior portion of the body, which is threaded into the tissue.

The clinical picture of whipworm infection is closely related to the number of worms present in the bowel. With light infections, there are either no related symptoms or mild ones, of which the relationship to the infection is doubtful. The most frequently mentioned of these is right abdominal pain. That the worms are allergic is suggested by the common occurrence of peripheral eosinophilia and Charcot-Leyden crystals in the stools. Some symptoms of allergy may therefore be expected to occur occasionally even with a light infection.

The clinical picture of heavy whipworm infection, as was seen in New Orleans and in some West Indian islands, is fairly characteristic, consisting of prolonged diarrhea with blood-streaked stools, abdominal pain, tenesmus, and loss of weight, usually in children aged 1–4 years. There may be a hypochromic anemia, together with failure to gain weight, and retarded development. Chronic infection in many cases eventually results in rectal prolapse, and the worms may then be seen attached to the mucosa of the prolapsed bowel. By clinical means alone, excluding sigmoidoscopy, the dysentery of uncomplicated trichuriasis is practically undistinguishable from that of amebiasis. Children show the syndrome when the worm burden is over a fairly constant level [probably about 1000 worms or 100 eggs per 2–3 mg smear, according to the Beaver (1950) direct smear technique]. There are few satisfactory drugs. Mebendazole appears to be the drug of choice.

CESTODIASIS (TAPEWORMS)

Infections by *Taenia saginata and T. solium* result from eating undercooked beef or pork respectively, containing the cysticercal stages of the parasites. Infections often have a distinct relation to local food patterns and prejudices. In general, tapeworm infections do not usually comprise an important disease in most tropical children, and then tend to affect the older groups.

The clinical features resulting from cestodiasis are far from clear-cut or certain. In well-nourished children there are usually no symptoms until the mother notices the proglottids either in the stools, or less commonly, actually "migrating" through the anal opening.

In view of their great length and absorption surface, adult *T. solium* and *T. saginata,* especially if several are present, may sometimes cause alimentary and nutritional side effects, both in the form of increased appetite and hunger pains, and also as an additional factor in the production of malnutrition.

In tapeworm infection, ova may be present in the feces, while proglottids always are passed in the stools. Niclosamide is currently the drug of choice (Davis, 1973), while dichlorophen and oral mepacrine can be used.

Cysticercosis. Man can occasionally become infected with the larval stage of *T. solium* as a result of ingesting ova either from another infected person, or by auto-infection. This may be symptomless or with variable features depending upon where "feeding" occurs (e.g.,muscle pain, convulsion, headache, vomiting, etc., with brain involvement).

Sanitary disposal of feces will reduce the spread of ova *T. solium* from person to person, and in the case of *T. saginata* and *T. solium,* from humans to cattle and pigs respectively. Ideally, the incidence of the so-called "great" tapeworms will be minimized by adequate meat inspection at slaughterhouses. As this is not at present universally possible in most tropical regions, all meat should be very well cooked before being eaten.

ECHINOCOCCOSIS

Echinococcus granulosus is a small tapeworm that lives as an adult in the small intestines of the pig. In the intermediate host, which includes domestic ungulates such as sheep, cattle, pigs, and in some countries goats, horses, camels, and buffaloes, and occasionally man, the larval stage produces the disease called echinococcosis or hydatid disease. In some areas a wild animal reservoir cycle may be a source of infection to dogs and man.

When *E. granulosus* cysts are ingested by the dog, they develop into the adult parasite in the upper intestinal tract and eventually eggs are passed in the dog's feces. These ova, when ingested by the intermediate host, produce a slow-growing hydatid cyst. Dogs and sheep are the most common hosts of the disease. The eggs can reach the mouth of man in different ways; with water or vegetables, in dirt, in wool or fur, or by flies. Direct contact between dogs and children is considered to be of prime importance in the epidemiology of the disease. The custom, in some rural areas of South America, of allowing dogs to lick eating utensils or to clean the face and anal regions of babies when they vomit or have diarrhea may play a major role in transmission. Because of the dog's habit of cleaning his anal region by licking, eggs of this tapeworm will be found in and around the region of the mouth. And disagreeable as it may be to accept this, a person who has a hydatid cyst has ingested canine feces containing eggs of the tapeworm.

The distribution of the endemic areas coincides for the most part with the major livestock-raising zones of the world where they are many dogs and a close contact between dog and man.

Clinical Features

The pathology and symptomatology of hydatid disease are best understood if one bears in mind that it is a noninfiltrative cystic tumor, with a very slow, expansive growth process. Incidental autopsy study findings indicate that many cases of hydatid cysts occur without symptoms.

The clinical picture depends on the organs affected, the site of localization, the size of the new cyst, and the presence or absence of complicatons. The liver is the organ most frequently affected (47–76%), followed by the lungs (16–40%).

There is a relatively higher prevalence of detected hydatid cysts in the lung, brain, and orbit in children than in adults. The rigid nature of the skull and orbit explains early clinical symptoms.

Prevention

Owing to the present lack of chemotherapy and the seriousness of the surgery involved, hydatid disease has an importance out of proportion to its commoness. Prevention is therefore the only approach to control. This is primarily a problem of health education related to the cultural background and epidemiology of each locality. Factors that appear to be common among populations with high transmission rates include rural habitats where raising livestock is the primary occupation, low socioeconomic and educational levels, low sanitary standards, a relatively high ratio of dogs per inhabitant, and the very common practice of household slaughtering.

The dog-domestic animal–dog–man cycle could easily be broken if every dog owner would keep his dog under permanent control and feed it no offal. Although this approach is theoretically simple, it is difficult to carry out because it implies behavioral changes in people. More detailed consideration is given elsewhere (D'Alessandro, 1978).

AMEBIASIS

Although human infection with the amoeba (*E. histolytica*) occurs in most areas of the world, the frequency of human disease varies widely. In some parts of the tropics (e.g., Nigeria, Thailand, Mexico, parts of India, and South Africa), amebiasis is a common and serious condition. Clinically, this presents in two main ways:

1. *Diarrhea or dysentery.* Loose stools occur with or without blood. There is not a typical picture, and inspection may reveal a formed stool with just flecks of blood and mucus or something that resembles red currant jelly. Amebae are present in the stools in most cases and can be detected by microscopic examination. Special tests may be necessary, including sigmoidoscopy. Varying degrees of fever, flatulence, colic, and abdominal pain may be present. Definite diagnosis mainly rests on finding the parasite. Repeated stool examinations may be necessary. Patients with symptoms must be treated. Flagyl (metranidazole) 20–40 mg orally in three divided doses daily for 5–7 days is the first line of therapy. However, amebiasis varies considerably and an alternative treatment needs to be considered, such as a 7-day course of tetracycline or chloroquine.

2. *Liver abscesses.* Uncommonly, the liver and other organs may be invaded. Patients with a liver abscess have fever and painful liver. If suspected, further hospital investigations are needed. Initial treatment is with metronidazole. Pus may have to be aspirated with a syringe.

GIARDIASIS

While not having the severe ill effects of *E. histolytica,* infection with the *Giardia lamblia* parasite occurs worldwide, especially in less developed communities with poor hygiene. It is common in children and can be responsible for growth failure as the large number of parasites interfere with the absorption of fat, causing pale, frothy stools and flatulence. Diagnosis is suggested by the typical light-yellow, frothy "beaten egg" appearance of the stools, and is confirmed by finding cysts in the feces. Sometimes in cases of severe diarrhea in small children, active moving forms can be seen in the stools.

As with amebiasis, other symptoms are common. Metronidazole is the drug of choice in the same dosage as for amebic diarrhea. It has a cure rate of over 90%. Only if this fails should "atebrin" (mepracine, quinacrine) be given in oral dose of 5 mg/kg daily in three divided doses for 10 days. This drug with its bitter taste is less well tolerated by children.

Both amebiasis and giardiasis have long been recognized as common conditions in developing countries. However, both in the U.S. and Western Europe illness due to these parasites seems to have become more common in recent years—or possibly has become diagnosed more frequently. One contributing factor may be the large numbers of people who are symptomless cyst carriers, who have emigrated to the U.S. or Europe in recent decades from Third World countries. At the same time an increasing awareness, and the fact that they usually respond well to drug therapy, may have focused attention on these tiresome forms of parasitic "food poisoning."

Cyst carriers are usually symptomless and may inadvertently contaminate food from their own soiled fingers. Therefore, carriers are more of a problem if they are working in food preparation, and the dishes which are most likely to be affected are those which are uncooked, such as salads. Hand washing prior to food preparation is obviously of great importance, as are short finger nails.

REFERENCES

Beaver, P. C., 1950, The standardization of local smears for estimating egg production and worm burden, *J. Parasitol.* **36:**451–456.

D'Alessandro, A., 1978, Echinococcosis, in: *Diseases of Children in the Subtropics and Tropics,* Edward Arnold, London.

Davis, A., 1973, *Drug Treatment of Intestinal Helminthiases,* World Health Organization, Geneva.

Jelliffe, D. B., and Jelliffe, E. F. P., 1978, Intestinal helminths, in: *Diseases of Children in the Subtropics and Tropics,* Edward Arnold, London.

24

ADVERSE EFFECTS OF DIET IN NEW GUINEA
Kuru and Enteritis Necroticans

R. W. Hornabrook

The New Guinea Highlands provide two unique examples—Kuru and Pig-bel—of the adverse effects of dietary practice on health. In each case, disease results from a complex association of factors, cultural circumstances, food composition, and contamination, each providing essential components in pathogenesis.

One must first consider the cultural setting in which these diseases occur. Within the valleys and slopes of the Central Cordillera, between 1200 and 2400 m in altitude, dwell subsistence gardeners. The staple foodstuff is the sweet potato *Ipomoea batatas;* a sophisticated system of horticulture has been succeeded by high yields and a dense human population. The staple root crop is supplemented with various green vegetables but animal protein is a scarce commodity. Wild game, such as rodents and small marsupials, are scarce and the only domestic animal, the pig, tends to be slaughtered infrequently and as part of social ritual.

Food is often roasted on the hot cinders of an open fire, but one meal each day is usually prepared in a type of oven in which the food is, in effect, steamed. A circular pit, about 1 m in depth and of similar diameter, is excavated. The depression is then filled with red-hot stones which are manipulated until level with the surrounding soil. These are covered with a layer of green foliage. Onto this, various items of food to be cooked are piled in a mound and then the whole structure is enclosed with further green leaves and grass. A layer of mud and soil is used to roof the whole oven. Cooking is accomplished by inserting bamboos filled with water through the sides of the oven; then this water is drained so that it percolates down through the heated stones. The steam rising up through the food raises the temperature

R. W. Hornabrook • Department of Medicine, Wellington Clinical School of Medicine, Wellington Hospital, Wellington, New Zealand.

to approximately 80°C. Two and a half hours later, when the structure is dismantled, the food is ready for consumption.

Early explorers recognized the protein-deficient nature of the diet. Hipsley and Clements (1950) were the first to draw attention to the fact that there was a low intake of both energy and protein. This view was later confirmed by Oomen (1961) and Hipsley and Kirk (1965).

Venkatachalam (1962) found energy intakes within an acceptable range but significant protein deficiency. Daily intakes of less than 30 g protein for male adults were sometimes reported. The good physique of the Highlanders contrasted with their defective diet. Clinical malnutrition was rare but infant growth rates were slow and puberty delayed. Oomen (1970) even suggested that nitrogen-fixing bacteria in the intestine were responsible for correcting a negative nitrogen balance. More recently, Norgan *et al.*, (1974) have reexamined the energy and protein levels of diet, finding the deficiency less serious than earlier writers had portrayed. In relation to the causation of Kuru and Pigbel, the significant fact appears to be that protein food is scarce, and in any form is a rare indulgence for the people.

KURU

Kuru is a subacute cerebellar degeneration (Hornabrook, 1968, 1976). The disease is remorselessly progressive and a mild clumsiness advances over a period of 1 year to produce a profound disintegration of the control of muscular activity. Eventually, weakness and immobility, starvation and secondary infection result in the death of the patient within 2 years of the first symptoms. Kuru is restricted to a relatively small area of 2000 km^2 some 45 km southwest of the town of Kainantu in the Eastern Highlands of the Central Cordillera. Here dwell some 40,000 people, belonging to eight distinct ethnolinguistic groups. Shortly after this region came under government administration, the disease was discovered and its existence reported by Gajdusek and Zigas (1957). It was the most important cause of death, affecting principally the women. Among the 11,000 South Fore people, it caused an annual mortality of 32/1000 in the adult female age group but only 5/1000 among the adult men. Children constituted one-third of the cases and here male and female were equally involved (Fig. 1).

The anthropological investigation of Glasse (1967) revealed that the disease first appeared in a small hamlet towards the turn of the century. From this focus, it had spread at the rate of about one mile in five years, achieving its greatest geographical extent and maximum incidence about the time of the first European contact in the early 1950s. Striking changes in the epidemiological characteristics have occurred since, (Mathews, 1967, 1976; Hornabrook and Moir, 1970; Alpers, 1979). The geographical area has shrunk. The juvenile cases have disappeared. The overall incidence of the disease is continuing to decline. The number of female cases has diminished and the

FIGURE 1. A funeral feast after the death of a Kuru patient at Anumpa Village, New Guinea Highlands, 1964. The female relatives of the patient sit with the bundles of food which are their share of ritual mortuary payments, no men are present, and the food will be taken away to be consumed with their children.

male/female ratio approaches unity. There is every indication that Kuru will disappear from the area within the next 15–20 years.

Mathews *et al.* (1968) and Glasse and Lindenbaum (1976) drew attention to the association of Kuru with the practice of cannibalism. The people in the region had adopted the practice of cannibalism of deceased relatives. The bodies were dismembered in the gardens and prepared for cooking in the traditional earth oven. The preparation of the food was undertaken by the female relatives accompanied by their prepubescent children. The same family groups participated in the subsequent feast, the food being shared among them on a ritualistic basis. Adult men did not participate in either the butchery or the cannibalism.

Gajdusek *et al.* (1966, 1967) reported that Kuru could be transmitted to chimpanzees by the intracerebral inoculation of brain tissue obtained at autopsy from Kuru patients. A long incubation period of 2–3 years preceded the appearance of clinical disease in the ape. Hadlow (1959) had drawn attention to the similarity of the histological lesions in the brain of Kuru with those seen in the central nervous system of sheep dying of scrapie, a slow virus disease. Kuru became the first slow virus disease of the nervous system of man and at the same time, the first chronic degenerative disease of man to be related to an infectious agent (Sigurrdson, 1954). Later, Creutzfeldt–Jakob disease was also shown to be transmissible to experimental animals (Gibbs *et al.,* 1968), and became the third slow virus disease of the nervous system with clinically and histologically similar characteristics. These disorders, along with mink encephalopathy were grouped together into the spongiform encephalopathies. Creutzfeldt–Jakob disease is a rare sporadic illness appearing among all human societies, but Kuru is an epidemic disorder in a very restricted area.

Evidence that infection occurred in association with the practice of cannibalism could be adduced from the anthropological facts concerning the spread of the disease from family to family. The epidemiological features with the early involvement of the cannibalistic females and their children gave weight to the possibility, and continuing support of the theory has been gained from the observation that disappearance of the disease and changes in age and sex incidence were first apparent among villagers who had discarded the practice of cannibalism. It is possible that infected material may be eaten but also that the preparation and handling of the raw, uncooked tissue, provide abundant opportunities for the inoculation of infected material through cuts and scratches in the skin. Whatever the mode of entry, it would appear that the practice of cannibalism is a central part in the existence of Kuru.

It is possible to suggest that Kuru owes its existence in the Highlands of New Guinea to a sociocultural practice, cannibalism. This practice provided an environment in which the transmission of a slow virus infection among a restricted population could occur. It has been suggested that the origin of the Kuru epidemic was a single sporadic case of Creutzfeldt–Jakob disease, (Hornabrook and Wagner, 1975). A hypothetical case, occurring in the first hamlet affected by Kuru, and the cannibalism of the body by the family kin, provided the initial nidus from which all subsequent patients were derived.

PIGBEL

Enteritis necroticans is not unique to New Guinea. Scattered outbreaks have been reported from various parts of the world (Murrell, 1979). It was a significant problem in malnourished people in northern Europe after the Second World War. Only in New Guinea is it a common and important public health problem and there it is known in the pidgin English vernacular as "Pigbel."

In New Guinea, enteritis necroticans arises among Papua New Guineans subsisting on traditional protein-deficient diets. It does not appear in Europeans living in the same area. The disease has its main incidence in childhood but cases may be encountered in all age groups.

Symptoms usually occur within hours, although sometimes they occur up to one week, after a protein-rich meal. The symptoms are abdominal distension and pain, vomiting, and loose, usually bloody bowel motions. In severe cases, intestinal obstruction develops and is followed by collapse and often death. Many patients recover spontaneously and supportive treatment with fluid replacement is effective in others. In the most severe cases, recourse to surgical exploration with removal of gangrenous bowel may be necessary (Shepherd, 1979).

At operation, varying lengths of the small bowel are found to be dilated and alternating areas of congestion and necrosis create a mottled pattern. Fibrino purulent peritonitis causes adhesions, and yellowish spots of necrosis arise on the antimesenteric border (Cooke, 1979). Large numbers of anaerobic *Clostridium welchii* Type C organisms are found in the bowel contents. Antibodies to the β-toxin of the organism may be found in the serum of recovering patients. The β-toxin is a protein which is broken down readily by proteolytic enzymes. In the absence of intestinal proteases, it accumulates within the gut and causes necrosis of the mucous membrane, the primary lesion.

The factors leading to development of clinical disease in Pigbel have been reviewed by Lawrence (1979). He has shown that the responsible organism is common in the intestine of man and animals in the New Guinea Highlands, as well as in the soil and as a food contaminant (Fig. 2). The spores, being

FIGURE 2. Sina Sina village in Chimbu district of New Guinea Highlands. Shown is butchery of pigs; the background to "Pigbel" or enteritis necroticans. Some of the 6000 pigs killed in the course of a single morning will be ceremoniously given away to neighboring clans. The carcasses may be carried for many miles over several days before being consumed.

relatively heat stable, are unlikely to be destroyed by cooking temperatures. After heating, those still present in the food become active and a surge of growth occurs. The β-toxin liberated during this active bacterial growth causes Pigbel.

Lawrence (1979) has postulated that the protein-deficient diet of the Papua New Guineans results in diminished proteolytic enzyme secretion by the pancreas. Furthermore, frequent consumption of sweet potato, which contains trypsin inhibitors, reduces the effectiveness of the enzymes that are produced. The sporadic large meat meal provides a good medium for the growth of *C. welchii* and also substrate excess to compete with the already inadequate amounts of proteolytic enzymes. Yet other factors may be involved; many patients with this disease have heavy infestations of *Ascaris lumbricoides* and this parasite possesses antiproteolytic substances (Walker *et al.*, 1979).

In the case of Pigbel, a protein-deficient diet leads to a reduction in intestinal proteolytic enzymes. At the same time, it induces a hunger for meat and the likelihood of a massive protein intake when meat becomes available. The food is usually contaminated and *C. welchii* from this source, or from the bowel contents, proliferate. The protein toxin which is produced is not broken down normally, partly because of the enzyme deficiency and partly because of the enzyme inhibition by factors such as the effect of ascariasis and sweet potato trypsin inhibitors. The β-toxin reaches toxic levels and induces a disease state.

REFERENCES

Alpers, M. P., 1979, Epidemiology and ecology of Kuru, in: *Slow Transmissible Disease of the Nervous System, Vol. 1,* (S. B. Prusiner and W. J. Hadlow, eds.), p. 66. Academic Press, New York.

Cooke, R., 1979, The pathology of Pigbel, *Papua New Guinea Med. J.* **22:**35.

Gajdusek, D. C., and Zigas, V., 1957, Degenerative disease of the central nervous system in New Guinea: The epidemic occurrence of Kuru in the native population, *N. Engl. J. Med.* **257:**974.

Gajdusek, D. C., Gibbs, C. J., Jr., and Alpers, M. P., 1966, Experimental transmission of a Kuru-like syndrome to chimpanzees, *Nature* **209:**794.

Gajdusek, D. C., Gibbs, C. J., Jr., and Alpers, M. P., 1967, Transmission and passage of experimental "Kuru" to chimpanzees, *Science* **155:**212.

Gibbs, C. J., Jr., Gajdusek, D. C., Asher, D. M., Alpers, M. P., Beck, E., Daniel, P. M., and Mathews, W. B., 1968, Creutzfeldt–Jakob disease (spongiform encephalopathy): Transmission to the chimpanzee, *Science* **161:**388.

Glasse, R. M., 1967, Cannibalism in the Kuru region of New Guinea, *Trans. N. Y. Acad. Sci., Ser. 2* **29(6):**748.

Glasse, R. M. and Lindenbaum, S., 1976, Kuru at Wanitabe, in: *Essays on Kuru* (revised ed.), Papua New Guinea Institute Medical Research Monograph No. 3, Faringdon, Berks, E. W. Classey.

Hadlow, W. J., 1959, Scrapie and Kuru, *Lancet* **11:**289.

Hipsley, E. H., and Clements, F. W., 1950, *Reports of the New Guinea Nutrition Survey,* 1947 Expedition, Department of External Territories, Canberra, Australia.

Hipsley, E. H., and Kirk, N. E., 1965, Studies of dietary intake and the expenditure of energy by New Guineans, *Technical Paper No. 147,* South Pacific Commission, Noumea (New Caledonia).

Hornabrook, R. W., 1968, Kuru, a subacute cerebellar degeneration: The natural history and clinical features, *Brain* **91**:53.

Hornabrook, R. W., 1976, Kuru: the disease, in: *Essays on Kuru,* (R. W. Hornabrook, ed.), Monograph Series No. 3, p. 53, Institute Human Biology, Papua, New Guinea.

Hornabrook, R. W., and Moir, D. J., 1970, Kuru—epidemiological trends, *Lancet* **2**:1175.

Hornabrook, R. W., and Wagner, F., 1975, Creutzfeldt-Jakob disease, *Papua New Guinea Med. J.* **18**:226.

Lawrence, G., 1979, The pathogenesis of Pigbel in Papua New Guinea, *Papua New Guinea Med. J.* **22**:39.

Mathews, J. D., 1967, The epidemiology of Kuru, *Papua New Guinea Med. J.* **10**:76.

Mathews, J. D., 1976, Kuru as an epidemic disease, in: *Essays on Kuru,* (E. W. Hornabrook, ed.), Monograph Series No. 3, p. 83. Institute Human Biology, Papua, New Guinea.

Mathews, J. D., Glass, R. N., and Lindenbaum, S., 1968, Kuru and cannibalism, *Lancet* **24**:449.

Murrell, T. G. C., 1979, A history of *Enteritis necroticans, Papua New Guinea Med. J.* **22**:5.

Norgan, N. G., Ferro-Luzzi, A., and Durnin, J. V. G. A., 1974, The energy and nutrient intake and energy expenditure of 204 New Guinean adults, *Philos. Trans. R. Soc. Lond. Ser. B.* **268**:309.

Oomen, H. A. P. C., 1961, The nutrition situation in Western New Guinea, *Trop. Geogr. Med.* **13**:321.

Oomen, H. A. P. C., 1970, Inter-relationship of the human intestinal flora and protein utilization, *Proc. Nutr. Soc.* **29**:197.

Shepherd, A., 1979, Clinical features and operative treatment of Pigbel—Enteritis necroticans, *Papua New Guinea Med. J.* **22**:18.

Sigurrdson, B., 1954, Rida, a chronic encephalitis of sheep with general remarks on infections which develop slowly and some of their special characteristics, *Br. Vet. J.* **110**:341.

Venkatachalam, P. S., 1962, *A Study of the Diet, Nutrition, and Health of the People of Chimbu Area,* Monograph No. 4, Department of Health, Port Moresby.

Walker, P. D., Batty, I., and Egerton, J. R., 1979, The typing of *C. perfringens* and the veterinary background, *Papua New Guinea Med. J.* **22**:50.

25

FISH AND SHELLFISH POISONING IN BRITAIN

P. C. B. Turnbull and R. J. Gilbert

"It's a wery remarkable circumstance, Sir," said Sam, "that poverty and oysters always seem to go together."
Charles Dickens—*Pickwick Papers*

Fish which cannot be sold must be declared as such and can then only be sold to foreigners.
Edict of the City of Basel, 13th Century

INTRODUCTION

Fish and shellfish have long been a popular part of the British diet. Mayhew observed in the 1840s that the poorer classes in London were consuming 875 million herrings and 124 million oysters annually (cited by Barker and Yudkin, 1971). Comparable figures are not available today principally because consumption is quoted in tonnage; also, they would have limited meaning since so much fish is now prepared and sold in a processed form. It can be said, however, that large quantities of fish and shellfish are consumed annually.

In a review of food poisoning caused by fish and fishery products Shewan and Liston (1955) stated that, in Britain at least, fish in all its forms was one of the safest articles of diet. A similar situation exists today although there is ample evidence that fish, when handled in an unhygienic manner, can be the vehicle of most of the well-established types of food poisoning. Typhoid fever and gastroenteritis following the consumption of molluscs, particularly oysters eaten raw, are well known in many countries, but in Britain routine examination of raw shellfish for *Escherichia coli* (the most sensitive bacterial indicator of pollution), has contributed greatly to minimizing such disease risks. Also in Britain, shellfish other than oysters are normally eaten only after cooking.

P. C. B. Turnbull and R. J. Gilbert • Food Hygiene Laboratory, Central Public Health Laboratory, London NW9 5HT, United Kingdom.

However, just since 1976, Britain has experienced an increase in the number of reported foodborne infections and intoxications, involving etiological agents rarely or not encountered before in connection with fish and shellfish in that country.

FISH AND SHELLFISH POISONING OF BACTERIAL ETIOLOGY

The fish and shellfish implicated in outbreaks of foodborne infections and intoxications in England and Wales from 1965 to 1980 and in Scotland from 1973 to 1980 are listed in Table I. Details of individual outbreaks are on record, but it is sufficient to make the generalization that the events leading to food poisoning with *Staphylococcus aureus, Salmonella* sp., *Clostridium perfringens* and *Vibrio parahaemolyticus* involved inadequate cooking, recontamination, or incorrect storage of the fish after cooking. Specific foods are frequently not identified in food-poisoning incidents but it should be emphasized that,

TABLE I
Fish and Shellfish Implicated in General and Family Outbreaks of Foodborne Infections and Intoxications in England and Wales (1965–1980)[a] and Scotland (1973–1980)[b]

Years	Etiology	Total no. of outbreaks	Food vehicle[c] (no. of outbreaks)
1965–1975	*S. aureus*	14	Prawns/scampi (6), salmon (1), salmon* (2), crab* (1), herring* (1), mackerel* (1), prawns* (1), sardines* (1)
	Salmonella sp.	2	Prawns (2)
	V. parahaemolyticus	4	Prawns (3), crab (1)
	C. perfringens	2	Salmon (2)
	Paralytic shellfish poisoning	1	Mussels (1)
	Unknown	6	Oysters (5), cockles and mussels (1)
1976–1980	*S. aureus*	4	Salmon* (2), crab* (1), sardines* (1)
	Salmonella	2	Salmon (2)
	V. parahaemolyticus	2	Prawns (2)
	C. botulinum	1	Salmon* (1)
	B. cereus	2	Lobster (1), prawns (1)
	A. hydrophila	2	Oysters (1), prawns (1)
	Scombroid	79	See Table III
	Hepatitis	3	Cockles (2), mussels (1)
	Small round viruses	9	Cockles[d] (7), oysters(2)
	Unknown	>20	Salmon* (8), cockles (>10), fish cakes (1), tuna* (1)

[a] Published and unpublished data from the Public Health Laboratory Service (Communicable Disease Surveillance Centre and the Food Hygiene Laboratory).
[b] Data supplied by Dr. J. C. M. Sharp, Communicable Diseases (Scotland) Unit, Glasgow.
[c] Asterisk (*) denotes canned food
[d] Includes one large outbreak associated with cockles (33 separate incidents affecting 797 persons)

where they have been, fish and shellfish are implicated in less than 3% of all the general and family outbreaks reported in Britain (Vernon and Tillett, 1974; Vernon, 1977; Gilbert and Roberts, 1979; Hepner, 1980).

Canned foods were incriminated on several occasions between 1965 and 1980; often the product was freshly opened, indicating a fault in the canning procedure, e.g., post-processing contamination with *S. aureus.* In a number of the canned-fish (salmon) outbreaks that are listed under unknown etiology (Table I), the incubation period and symptoms closely resembled staphylococcal food poisoning, but *S. aureus* was not isolated from remnants of fish or clinical specimens, and tests for enterotoxin were also negative.

Only one of the *V. parahaemolyticus* outbreaks (Hooper *et al.,* 1974) was associated with "home-produced" seafood (crab), and this was attributed to post-cooking recontamination followed by inadequate refrigeration of the crab during storage. The remaining five outbreaks have followed consumption of imported prawns. It is of interest that the organism is readily found in British coastal waters (Ayres and Barrow, 1978); presumably the fact that very little seafood is eaten uncooked in Britain accounts for the rarity of *V. parahaemolyticus* food poisoning there, despite increased awareness of the organism over the past decade.

The persistently poor bacteriological quality, over a number of years, of imported, frozen cooked prawns, particularly from Malaysia, led to various discussions between the public health authorities and the importing and frozen-food trade. Agreement was reached on guidelines of microbiological quality on which the acceptance of imported, frozen cooked prawns could be assessed. The guidelines (which had no statutory authority) were issued in October, 1975, and laid down (1) basic sampling plans, (2) recommended methods for the tests and (3) specific criteria with respect to the aerobic colony plate count and the presence of *E. coli, S.aureus,* and *Salmonella.* A criterion for *V. parahaemolyticus* was not included in the guidelines because knowledge of the organism in Britain was rather limited. The results of examination of frozen, cooked prawns in the Food Hygiene Laboratory during the period 1969–1980, with respect to isolations of *Salmonella* and *V. parahaemolyticus,* are summarized in Table II. The problems of undercooking or recontamination of the final product are apparent; the fall in the isolation rate of *V. parahaemolyticus* in the second of the two periods may have been a result of demands made by British authorities for more care during processing in Malaysia.

With the ever-increasing ease and speed of international travel, public health authorities in Britain have become increasingly aware of the need for monitoring diseases contracted overseas that manifest themselves during transit to or after arrival in this country. As far as fish and shellfish are concerned, crab (Peffers *et al.,* 1973), and prawns and shrimps (Barrow, 1974; Vernon, 1977) contaminated with *V. parahaemolyticus* have been implicated in this type of situation.

The most dramatic poisoning of bacterial etiology in Britain was the incident of botulism in Birmingham in the summer of 1978 affecting four

TABLE II
Isolation of Salmonella and Vibrio parahaemolyticus from Malaysian Frozen Cooked Prawns in the Food Hygiene Laboratory, 1969–1980[a]

	Years	No. of samples examined	No. positive	Percentage positive
Salmonella	1969–1975	2335	38	1.6
	1976–1980	595	9	1.5
V. parahaemolyticus	1972–1975	1310	122	9.3
	1976–1980	595	15	2.5

[a] Guidelines on microbiological quality for imported, frozen, cooked prawns were introduced in October, 1975, but these did not include a criterion for *V. parahaemolyticus.*

persons, two of whom died (Ball *et al.,* 1979). Neurological symptoms developed about 12 hr after eating salmon from a freshly opened can originating from False Pass, Alaska. Laboratory tests confirmed the presence of type E toxin from *C. botulinum* in sera from all four patients and the pathogen was isolated from remnants of the fish.

In Britain, botulism in man is rare and only 8 incidents involving 28 cases and 17 deaths have been recorded to date. Toxin types A, B, and E have been specifically implicated and it is of special interest to note that the two most recent episodes—in 1955 (Mackay-Scollay, 1958) and 1978—involved imported, preserved or canned fish.

Some 14,000 cans of salmon having the same batch number as that incriminated in the Birmingham outbreak were widely distributed throughout the country, and the implications were those of a potential disaster while the cans were being recalled. The contents of many cans had already been consumed, but some 539 cans from the precise batch were examined by laboratories of the Public Health Laboratory Service and were found to be sterile. Fortunately, it turned out to be a single-can problem, with all evidence indicating that the organism had entered through a small hole in the can during cooling. Economically, however, the effects were severe for the particular companies concerned with repercussions throughout the salmon-canning industry.

C. botulinum has been isolated on many occasions in Britain from both home-grown and imported farm trout (Huss *et al.,* 1974; Cann *et al.,* 1975; Jarvis and Patel, 1979); fortunately no cases of human intoxication have been reported. An advisory memorandum on the processing, handling and cooking of trout was issued by the Department of Health and Social Security in 1978.

SCOMBROID FISH POISONING

The precise biochemical events that occur between consumption of the implicated fish and appearance of symptoms in scombroid fish poisoning have yet to be elucidated. However, there is little doubt that the toxic factor(s) are produced as a result of the action on the flesh of fish, particularly those belonging to the families Scomberesocidae and Scombridae, of bacterial enzymes produced during spoilage.

Except for anecdotal cases, the first detailed account of scombroid poisoning in Britain was given by Cruickshank and Williams (1978). Since then the condition has become well recognized and by the end of 1980 a total of 79 incidents affecting 276 people had been brought to the attention of the Public Health Laboratory Service Communicable Disease Surveillance Centre in London. Features of the first 50 incidents to be reported have been described by Gilbert *et al.* (1980). The increase in incidents coincides with a dramatic increase in the annual landings of mackerel in Britain, from 8800 ton in 1972 to 320,900 ton in 1978 (although some 90% of mackerel catches are, in fact, exported). At the same time, landings of herring decreased from 151,000 MT in 1973 to 14,700 MT in 1978 (Central Statistics Office, 1980) because of a depleted herring stock from overfishing. Before this, mackerel was not a popular fish in recent years, partly because of its poor storage qualities. This characteristic has clearly been known for many years; in the mid-1800s mackerel was the only fish permitted to be sold on the Sabbath (Stern, 1971). In fact, mackerel keeps relatively well under refrigeration (G. Hobbs, personal communication).

Of the 79 incidents of scombroid fish poisoning reported since 1976, some 52 (66%) have been associated with smoked mackerel and a further 6 incidents have been from canned or soused mackerel or pâté (Table III). The number of reports of scombroid toxicity associated with canned food became particularly apparent in 1980. Canned tuna/bonito imported from a number

TABLE III
Scombroid Poisoning in Britain (1976–1980)

			Fish incriminated (no. incidents)			
Years	No. of incidents	No. of illnesses	Mackerel		Other[a]	
1976–1978	7	19	Smoked	(7)		—
1979	44	178	Smoked	(37)	Bonito	(1)
			Canned	(2)	Sprats	(1)
			Paste	(1)	Canned pilchards	(1)
					Canned sardines	(1)
1980	28	79	Smoked	(8)	Canned tuna/bonito	(10)
			Canned	(1)	Canned sardines	(4)
			Pâté	(1)	Canned pilchards	(1)
			Soused	(1)	Canned anchovies	(1)
					Raw tuna	(1)
Total	79[b]	276		58		21

[a] Includes: from U.K.—canned mackerel and pilchards; Morocco—sardines and pilchards; Spain—anchovies; Fiji, Japan, Malaysia, Peru, Taiwan, and Thailand—tuna/bonito.
[b] Mainly family outbreaks and sporadic cases, but in two outbreaks more than 25 persons were affected.

of countries was incriminated in 10 incidents, but the most interesting development has been the increasing involvement of nonscombroid fish such as sardines, pilchards, and anchovies (Murray *et al.*, in preparation).

Since scombrotoxin has not been identified with certainty, the only confirmatory laboratory test is the detection of high concentrations of histamine in the flesh of suspect fish. Histamine assays were carried out by Hobbs and Murray, at Torry Research Station, Aberdeen, on samples of fish (remnants from meals or samples from the same batch) from 63 of the 79 reported incidents in Britain. Concentrations of < 50 mg/100 g fish were found in 25 incidents, between 50 and 100 mg in 8 incidents, and > 100 mg in 30 incidents. The highest level recorded was > 1000 mg/100 g in the remnants of a sample of canned tuna from Taiwan, with up to 450 mg/100 g in other freshly opened cans from the same batch. Similar problems with canned tuna have occurred elsewhere in Europe, particularly in Sweden (Lonberg *et al.,* 1980).

As the illness is the result of bacterial spoilage, the most important preventive measure is to ensure that fish are kept properly refrigerated between the time of catching and consumption (or canning).

PARALYTIC SHELLFISH POISONING (PSP)

In reviewing the history of PSP in Britain, Ayres (1975) concluded that between 1827 and 1968 only ten incidents, all associated with mussels, were sufficiently well documented to be reliably classified as PSP. The largest and most recent outbreak occurred on the northeast coast of England in May, 1968, following the consumption of locally gathered mussels (McCollum *et al.,* 1968). Of the 78 persons affected, most had purchased cooked mussels from a retail outlet and a few had cooked them at home. Rapid publicity through the newspapers, radio, and television prevented subsequent intoxications, and the absence of fatalities among those afflicted was attributed in part to the British habit of boiling mussels thoroughly.

The responsible agent was *Gonyaulax tamarensis* a dinoflagellate, and two toxic fractions were extracted from the mussels. The first was apparently identical to "saxitoxin," earlier described by Schantz (1967) as the toxic principle of *G. catanella,* the dinoflagellate associated with PSP in North America. This was, however, the minor of the two fractions; the major fraction was similar to saxitoxin but had a different affinity in ion-exchange chromatography (Evans, 1970).

The monitoring of mussels from the coast of Britain has revealed the presence of dinoflagellate toxin every year since the program was initiated in 1968 (Ayres and Cullum, 1978). However, the levels recorded and the areas affected have not reached the proportions observed in 1968. The level of toxin in mussels that is safe for human consumption has been set at 400 mouse units/100 g; the maximum level recorded in 1968 was 50,000 mouse units/100 g.

Recently, our attention was drawn to the report of a case of curare-like poisoning traced to consumption of red whelks. The whelk, *Neptuna antiqua,* secretes the toxin tetramine in its salivary glands (Fleming, 1971).

CIGUATERA POISONING IN BRITAIN

Ciguatera poisoning, while recognized as a serious problem in tropical regions, is rarely encountered in temperate zones; in fact, the first cases in Britain have only recently been recorded (Tatnall *et al.,* 1980). In the most serious report, the patient had purchased what were subsequently identified as moray eel and amberjack on the quayside in Antigua. He had dried and salted them himself and then brought them back to Britain, where he ate them and developed the symptoms. Two other cases which are referred to following a personal communication to the authors occurred about the same time. The patients had eaten barracuda during a meal in the Caribbean just prior to flying home and symptoms had developed during the flight.

Importing potentially ciguatoxic fish such as moray eels, amberjack, and barracuda are strongly discouraged in Britain, so it is likely that ciguatera will continue to be encountered only in association with travelers returning from regions such as the Caribbean or certain zones in the Pacific.

FOOD-BORNE VIRUS INFECTIONS

During the period 1965–1975, at least six outbreaks of food poisoning of unknown etiology were attributed to the consumption of oysters or cockles and mussels (Table I). Two of these outbreaks are well documented (Preston, 1968; Gunn and Rowlands, 1969) and in both, British cultured oysters were incriminated. The report by Gunn and Rowlands gave a typical picture: 37 of 42 persons at risk, with nausea, vomiting, abdominal pain, and diarrhea as the principle symptoms, and an incubation period ranging from 10 to 85 hr, but averaging 35–50 hr. Laboratory examination failed to reveal a bacteriological cause, and thus a viral etiology is suspected.

Between December, 1976 and January, 1977 a very large single-source outbreak of food poisoning was reported to affect 797 persons in 33 separate incidents in several areas of the country (Appleton and Pereira, 1977). Similar but much smaller outbreaks occurred in 1978, 1979, and 1980. The incubation period in all of these outbreaks was between 24 and 72 hr and symptoms included nausea, vomiting, abdominal pain, and diarrhea.

Intensive efforts to identify a pathogen have been undertaken, but no conventional pathogen has been found either in specimens from patients or in samples of cockles or other foods. Viruses were not isolated by tissue culture methods or inoculation into suckling mice, but electron microscopy regularly revealed the presence of small, round, featureless, virus-like particles in the majority of fecal specimens tested (Appleton and Pereira, 1977; Appleton, 1979, 1980). The particles were 23–26 nm in diameter with a buoyant density of 1.40g/cc in cesium chloride, and were similar to viruses that have been found in outbreaks of winter-vomiting disease. Examination of serial specimens from a few patients indicated that maximum excretion of virus particles occurred 4–6 days after onset of symptoms and that excretion continued for at least a month after illness.

In the 1976–1977 outbreak, the fault was linked to underprocessing of the cockles in recently installed steaming equipment. While it was clear that the heat processing had been adequate to kill bacteria—total colony plate counts, and coliform and *E. coli* counts were low in all samples examined—it appeared that the virus was relatively heat stable and had survived. The cause of recurrent outbreaks has not been identified, but for some episodes in 1980 it was thought that a sudden increase in demand, associated with unusually good weather over the Easter holiday period, may have led to some processors increasing their turnover rate with subsequent undercooking.

It is likely that the small viruses observed in some of these outbreaks were acquired from cockles that had been harvested from polluted water, and inadequately heat treated. Several of the outbreaks occurred when gastroenteritis was prevalent in the population living near the area where the cockles were grown, and most outbreaks were associated with cockles harvested during cold periods of the year.

Hepatitis A has also recently been associated with consumption of shellfish in Britain. An epidemiological investigation of reports of hepatitis A infection in Leeds early in 1978 identified 41 cases associated with the eating of mussels from a common source (Bostock *et al.*, 1979). Nineteen cases were associated with 7 restaurants, and the remaining 22 cases with 11 fish shops and market stalls. Further investigation revealed that all of the suspected mussels had been purchased live from the local wholesale fish market and that they had been imported from Eire. A further 41 mussel-associated cases came to light as a result of a retrospective search through case histories in other towns receiving mussels from the same source. It was concluded that the amount of heat used in the preparation of the mussels was usually sufficient to open the shells, but was probably inadequate to cook them thoroughly.

Two further outbreaks of hepatitis A infection involving some 20 persons occurred in two areas of Britain in 1980, but in these, cockles appeared to be the epidemiological link. In both outbreaks the cockles were thought to have come from the same town in southeast England from which those causing the viral gastroenteritis had originated. It is possible, though not proven, that the two events—outbreaks of viral gastroenteritis and hepatitis A infection—were related and resulted from the same underprocessing.

CONTROL MEASURES

In many of the incidents discussed in the preceding sections, follow-up investigations identified the poisoning agents and control measures were taken to prevent subsequent intoxication. These measures have been initiated by public health authorities and include inspection of premises and processes or monitoring of the food itself. In general, this involves regular examination of bacteriological quality, e.g., the prawn guidelines referred to earlier, or specific tests such as the annual assay of PSP-toxin levels in mussels, when dinoflagellate levels are at their peak. The reports on fish and shellfish hygiene published by the World Health Organization (WHO, 1974; Wood, 1976) are considered to be useful guidelines.

Control over polluted shellfish layings can be enforced under the Public Health (Shellfish) Regulations of 1934, and shellfish from such layings must be subjected to purification treatment or heat before sale for human consumption. However, perhaps the most important ingredient of control practiced in Britain is communication—that is, informing the public, fishermen, and fish or shellfish processors and retailers of the principles of and reasons for the appropriate hygienic measures.

ACKNOWLEDGEMENT. We are grateful to Dr. J. C. M. Sharp, Communicable Diseases (Scotland) Unit, Glasgow; Dr. S. Palmer and Mrs. Enid Hepner, Communicable Disease Surveillance Centre, Colindale; Dr. G. Hobbs and Mr. C. K. Murray, Torry Research Station, Aberdeen; Dr. P. A. Ayres, Fisheries Laboratory, Burnham-on-Crouch; and Dr. Hazel Appleton, Central Public Health Laboratory, Colindale. Without their assistance and information this paper could not have been written.

REFERENCES

Appleton, H., 1979, Small round viruses in outbreaks of food poisoning, in: Colloque Entérites Virales (F. Bricout and R. Scherrer eds.), *Inserms Symposia* **90:** 237–240.

Appleton, H., 1980, Outbreaks of viral gastroenteritis associated with the consumption of shellfish, in: *Viruses and Wastewater Treatment* Proceedings of the International Symposium on Viruses and Wastewater Treatment, Guildford, Surrey, September 15–17 1980, (M. Goddard and M. Butler eds.), Pergamon Press, Oxford.

Appleton, H., and Pereira, M. S., 1977, A possible virus aetiology in outbreaks of food poisoning from cockles, *Lancet* **1:** 780–781.

Ayres, P. A., 1975, Mussel poisoning in Britain with special reference to paralytic shellfish poisoning, *Environ. Health* **83:**261–265.

Ayres, P. A., and Barrow, G. I., 1978, The distribution of *Vibrio parahaemolyticus* in British coastal waters: Report of a collaborative study 1975–6, *J. Hyg.* **80:**281–294.

Ayres, P. A., and Cullum, M., 1978, Paralytic shellfish poisoning: An account of investigations into mussel toxicity in England 1968–77, *Fisheries Research Technical Report, No. 40,* Ministry of Agriculture, Fisheries and Food, London.

Ball, A. P., Hopkinson, R. B., Farrell, I. D., Hutchinson, J. G. P., Paul, R., Watson, R. D. S., Page, A. J. F., Parker, R. G. F., Edwards, C. W., Snow, M., Scott, D. K., Leone-Ganado, A., Hastings, A., Ghosh, A. C., and Gilbert, R. J., 1979, Human botulism caused by *Clostridium botulinum* type E: The Birmingham outbreak, *Q. J. Med.* **48:**473–491.

Barker, T. C., and Yudkin, J. (eds.), 1971, *Fish in Britain. Trends in its supply, distribution and consumption during the past two centuries,* Department of Nutrition, Queen Elizabeth College, University of London.

Barrow, G. I., 1974, Microbiological and other hazards from seafoods with special reference to *Vibrio parahaemolyticus*, *Postgrad. Med. J.* **50:**612–619.

Bostock, A. D., Mepham, P., Phillips, S., Skidmore, S., and Hambling, M. H., 1979, Hepatitis A infection associated with the consumption of mussels, *J. Inf.* **1:**171–177.

Cann, D. C., Taylor, L. Y., and Hobbs, G., 1975, The incidence of *Clostridium botulinum* in farm trout raised in Great Britain, *J. Appl. Bacteriol.* **39:**331–336.

Central Statistical Office, 1980, *Annual Abstracts of Statistics, 1980.* Her Majesty's Stationery Office, London.

Cruickshank, J. G., and Williams, H. R., 1978, Scombrotoxic fish poisoning, *Br. Med. J.* **2:**739–740.

Evans, M. H., 1970, Two toxins from a poisonous sample of mussels *(Mytilus edulis)*, *Br. J. Pharmacol.* **40:**847 865.

Fleming, C., 1971, Case of poisoning from red whelk, *Br. Med. J.* **3:**520.

Gilbert, R. J., and Roberts, D., 1979, Food poisoning risks associated with foods other than meat and poultry—outbreaks and surveillance studies, *Health Hyg.* **3:**33–40.

Gilbert, R. J., Hobbs, G., Murray, C. K., Cruickshank, J. G., and Young, S. E. J., 1980, Scombrotoxic fish poisoning: Features of the first 50 incidents to be reported in Britain (1976–9), *Br. Med. J.* **281:**71-72.

Gunn, A. D. G., and Rowlands, D. F., 1969, A confined outbreak of food poisoning, *Med. Off.* **122:**75–79.

Hepner, E., 1980, Food poisoning and salmonella infections in England and Wales, 1976–1978, *Public Health* **94:**337–349.

Hooper, W. L., Barrow, G. I., and McNab, D. J. N., 1974, *Vibrio parahaemolyticus* food poisoning in Britain, *Lancet* **1:**1100–1102.

Huss, H. H., Pederson, A., and Cann, D. C., 1974, The incidence of *Clostridium botulinum* in Danish trout farms. I. Distribution in fish and their environment, *J. Food Technol.* **9:**445–450.

Jarvis, B., and Patel, M., 1979, The occurrence and control of *Clostridium botulinum* in foods, *Leatherhead Food R.A. Technical Circular No. 686.*

Lonberg, E., Movitz, J., and Slorach, S., 1980, Histamine in canned fish, *Var Foda* **32:**114–123.

Mackay-Scollay, E. M., 1958, Two cases of botulism, *J. Pathol. Bacteriol.* **75:**482–485.

McCollum, J. P. K., Pearson, R. C. M., Ingham, H. R., Wood, P. C., and Dewar, H. A., 1968, An Pepidemic of mussel poisoning in northeast England, *Lancet* **2:**767–770.

Peffers, A. S. R., Bailey, J., Barrow, G. I., and Hobbs, B. C., 1973, *Vibrio parahaemolyticus* gastroenteritis and international air travel, *Lancet* **1:**143–145.

Preston, F. S., 1968, An outbreak of gastroenteritis in aircrew, *Aerosp. Med.* **39:**519–521.

Schantz, E. J., 1967, Biochemical studies on purified *Gonyaulax catenella*, in: *Animal Toxins* (F. E. Russell and P. R. Saunders, eds.) pp. 91–95, Pergamon Press, Oxford.

Shewan, J. M., and Liston, J., 1955, A review of food poisoning caused by fish and fishery products, *J. Appl. Bacteriol.* **18:**522–534.

Stern, W. M., 1971, The fish supply to Billingsgate from the nineteenth century to the second world war, in: *Fish in Britain* (T. C. Barker and J. Yudkin, eds.), pp. 30–82, Department of Nutrition, Queen Elizabeth College, University of London.

Tatnall, F. M., Smith, H. G., Welsby, P. D., and Turnbull, P. C. B., 1980, Ciguatera poisoning, *Br. Med. J.* **281:**948–949.

Vernon, E., 1977, Food poisoning and *Salmonella* infections in England and Wales, 1973–1975, *Public Health* **91:**225–235.

Vernon, E., and Tillett, H. E., 1974, Food poisoning and *Salmonella* infections in England and Wales, 1969–1972, *Public Health* **88:**225–235.

Wood, P. C., 1976, Guide to shellfish hygiene, *WHO Offset Publication No. 31,* World Health Organization, Geneva.

World Health Organization, 1974, Report of a WHO Committee on Fish and Shellfish Hygiene, *W. H. O. Tech. Rep. Ser. No. 550.*

26

LIVER FLUKES
Relationship to Dietary Habits and Development Programs in Thailand

Panata Migasena

It is now recognized that *Opisthorchis viverrini* is the only human hepatic trematode that causes liver fluke infection and is highly prevalent in Thailand (Vajrasthira and Harinasuta, 1957; Harinasuta, 1969). The clinical presentation of *O. viverrini* infection depends upon the number of worms present in the liver and the duration of infection (Harinasuta and Vajrasthira, 1960). On the other hand, Wykoff *et al.* in 1966 failed to relate any of the specifically studied symptoms to the presence of infection with *O. viverrini.* Although many of the subjects did exhibit the symptoms reportedly associated with liver fluke infection, the noninfected subjects also exhibited the same symptoms with essentially the same frequency. Moreover, when noninfected persons from the same area were compared with infected persons, hematological and biochemical abnormalities were also found in nearly the same frequency. For that reason, detection of *O. viverrini* eggs in stool remains, in general, the most satisfactory method for diagnosis (Harinasuta, 1969). It is interesting to note that there is a significant increase of ceruloplasmin and hemopexin in the serum of persons with subclinical opisthorchiasis (Schelp *et al.,* 1974). These substances were also found in patients with primary liver cell carcinoma (Migasena *et al.,* 1979) and cholangiocarcinoma (Migasena *et al.,* 1980a).

The adult worm of *O. viverrini* usually inhabits the distal bile duct and the gall bladder of a man and animal. The eggs laid by the adult fluke in the biliary passages of man, cats, dogs, and some fish-eating mammals are passed out of the host's body in the feces. The eggs containing fully developed miracidia are then swallowed by suitable snails, i.e., *Bithynia siamensis, B. goniomphalos,* and *B. funiculata,* which are the first intermediate hosts in which

---

Panata Migasena • Department of Tropical Nutrition and Food Sciences, Faculty of Tropical Medicine, Mahidol University, Bangkok 4 Thailand.

miracidia hatch and develop into sporocysts, rediae, and cercariae. After leaving the snail, cercariae swim around in the water until they find an appropriate cyprinoid fish, which serves as the second intermediate host. A period of at least 21 days is required for the full maturation of metacercariae in the fish (Wykoff *et al.*, 1965). Man and animals acquire infection by eating raw fish that contain viable metacercariae (Harinasuta and Vajrasthira, 1960).

DIETARY HABITS

The following factors determine the distribution of liver fluke infection among people in northeastern Thailand: the habit of eating raw fish, in the form of *Goi-pla;* haphazard defecation; lack of latrines and basic sanitation, which leads to the pollution of soil and water; and the prevalence in this region of intermediate hosts. Migasena in 1978 studied the eating patterns of preschool children by giving prepared questionnaires about the frequency that certain foods were eaten to interview mothers in the Ubolratana Dam and Nam Pong areas. They found no difference between people who lived in nonirrigated and irrigated areas: 1% eat raw fish every day, 80% have it one or more times a week, and only 19% never eat raw fish. Furthermore, a total of 99% eat fermented fish every day (see Table I). The same studies were conducted in the southern part of Thailand at Pattani (Migasena, 1976a,b) and Surat Thani Province (Migasena, 1980). It was found that the people in the southern Thailand never eat either raw or fermented fish. These results were supported by reports on the incidence of *O. viverrini* infection throughout the country: 22.1% of the total population were found to be infected, with 29.8% incidents found in the northeast, 10.3% in the North, 0.3% in the central part, and none in the south (Vajrasthira and Harinasuta, 1957). Early reports indicated that about 2 million people in the northeastern region were infected (Sadun, 1955). Later, Wykoff *et al.* (1966) estimated that more than 3.5 million people in the northeast harbored this parasite, with

TABLE I
Food Patterns of Families Consuming Fish[a] (Percent of All Families)

	Nonirrigated area (N = 51)[b]			Irrigated area (N = 236[c])		
Fish	Daily	Moderately frequent[d]	Never	Daily	Moderately frequent[d]	Never
Fresh	62.0	38.0	—	67.9	32.1	—
Raw	1.0	82.0	17.0	1.0	80.1	18.8
Fermented	98.0	2.0	—	99.0	1.0	—

[a] From Migasena, 1978.
[b] Nonirrigated area in Nong Wai zone.
[c] Downstream villages, fishing villages, and Non Sang resettlement village in Nong Wai zone.
[d] Ate less frequently (2–4 meals/week).

infection rates varying from province to province and ranging from 3% to 88%. In that study, there was no difference in the incidence between the sexes. The ages of infected individuals ranged from 6 months to 87 years, and the incidence increased with age to a maximum at 15–40 years.

CARCINOGENIC EFFECT

In 1974, N-nitroso compounds were detected (Migasena and Changbumrung, 1974) in many locally preserved Thai foods containing protein, including *Pla ra* (fermented fish), which is consumed by nearly all people in the northeast daily (Migasena, 1978). Migasena and Changbumrung (1974) concluded that during the processing and preparation of *Pla ra,* which is sometimes stored for several months, secondary and tertiary amines are probably formed. These amines may react with nitrates or nitrites that are present as contaminants in salt or water, to form nitrosamines. Samples of *Pla ra* from the northeast and central regions of Thailand were found to contain nitrate concentrations of 83–803 and 30–1080 ppm, respectively, while nitrites were found in lower concentrations of 0–2 and 0–10 ppm, respectively. It is interesting to note that nitrate and nitrite concentrations in rock salt, which is commonly used in the northeast, were in the range of 28–248 and 0.1–0.5 ppm, respectively. In water supplies both from the northeast and from Bangkok (including tapwater) nitrite and nitrate concentrations ranged from 3–19 and 0–21 ppm, respectively (Migasena *et al.,* 1980b).

The interaction between *O. viverrini* infection and N-nitroso compounds was studied in Syrian gold hamsters by Thamavit *et al.* in 1978. It was found that animals that received both dimethylnitrosamine (DMN, 25 ppm) and parasites (100 metacercariae) developed cholangiocarcinoma (100%) and cholangiofibrosis (100%). The tumor was not observed in control animals that received either DMN or *O. viverrini* alone, although cholangiofibrosis was found in some animals in the DMN group. It was postulated that cholangiocarcinoma developed in animals because the DMN exerted a carcinogenic effect on the altered, proliferating epithelial cells of bile ducts that had been stimulated by the fluke. These findings suggest that the combination of DMN ingestion and liver fluke infestation may play an important role in the production of cholangiocarcinoma in human beings, which is found in high incidence in the northeast of Thailand.

In northeastern Thailand, uncooked *Pla ra* is commonly consumed in combination with vegetables (Migasena, 1978), which are also a source of nitrate. The preparation of *Pla ra* is carried out under primitive village conditions without any particular concern for modern standards of hygiene. Improper cleaning of fish will contaminate the flesh. Rock salt and sea salt have both been shown to be contaminated with bacteria (Bain *et al.,* 1957). The fish flesh in a preparation of *Pla ra* has a pH of 6–7 and the liquid portion has a pH of 5–6 (Sunthornvipart, 1970). These pH ranges are not sufficiently acidic for N-nitrosocompound formation by chemical reaction alone, but certain bacteria can stimulate N-nitroso compound formation under neutral conditions (Hawksworth and Hill, 1971).

The consumption of foodstuffs containing high concentrations of nitrates, nitrites, and secondary amines should be considered a health hazard because of the formation of carcinogenic N-nitroso compounds in the gastrointestinal tract. Furthermore, N-nitroso compounds are possible environmental carcinogens and could be formed in fermented fish rich in basic secondary amines and nitrates or nitrites through the action of bacteria at a neutral pH. Consumption of these foodstuffs may be linked to the high incidence of primary liver carcinoma in Thailand (Migasena *et al.*, 1979, 1980a).

DEVELOPMENT PROGRAM

The Lower Mekong Basin Development Program (LMBDP) is a multipurpose water resource development scheme, involving the lower part of the Mekong River and its tributaries. This international river is the major water resource of peoples of the Indo-China Peninsula. The riparian countries involved are Thailand, Laos, Vietman, and the Khmer Republic. The nutritional situation, health status, and impact of development in this area have been summarized (Migasena, 1972). The role of the biogeographic and medical geographic studies, especially concerning nutrition and public health policy, in planning the LMBDP has been discussed, with emphasis on the physical features that have changed after construction of the Ubol Ratana Dam (e.g., the manmade lakes and the irrigated areas) (Migasena, 1976b). The mode of life of people in each area has also changed. These changes will affect the epidemiological pattern of human endemic diseases and of zoonosis.

The increase in fishing in the manmade lakes and the poor sanitary practices of the lakeside dwellers promote the spread of liver fluke infection in this area (Harinasuta *et al.*, 1975; Migasena, 1976b; Migasena, 1978). Harinasuta *et al.* (1975) reported that 3.1% of the fish (carps) and 0.4% of the *B. goniomphalos* snails in the irrigated villages of Nam Pong area were positive for *O. viverrini* metacercariae and cercariae, respectively. The use of latrines is not a common practice among the villagers. In the irrigated villages of Nam Pong area, 71.5% of the households had no latrine, while in the nonirrigated villages, the percentage with no latrine reached 97.7%. Most of the villagers used small bushes near their houses or ricefields as defecating places. They usually defecated very early in the morning or at night.

At the present time, fish caught from Ubolratana Lake are distributed throughout the northeast region and as far as Bangkok. The need for laborers from the northeast for construction of dam projects in the South may contribute to the spread of liver fluke infection, because several types of intermediate fish hosts and suspected snail hosts of the disease exist in those areas already (Migasena, 1976a, 1980). Good environmental health and sanitation should be established and maintained, not only during the construction period but also in the postimpoundment period. Projects should provide adequate supplies of treated water and hygienic latrines, with no drains discharging untreated wastewater into public waterways. If sufficient preventive measures are undertaken at dams and resettlement villages, the hazard of importation of *O. viverrini* by migrating laborers can be greatly reduced (Migasena, 1976a,b, 1980).

REFERENCES

Bain, N., Hodgkin, W., and Shewan, J. M., 1957, The bacteriology of brines used in smoke curing of fish, in: *The Microbiology of Fish and Meat Curing Brines* (Proceedings of the Second International Symposium on Food Microbiology), p. 103, HMSO, London.

Hawksworth, G. and Hill, M. J., 1971, The formation of nitrosamines by human intestinal bacteria, *Biochem. J.* **122:**28.

Harinasuta, C., 1969, Opisthorchiasis in Thailand: A review, in: *Proceedings of Fourth Southeast Asian Seminar of Parasitology, Tropical Medicine, Schistosomiasis and Other Snail-Transmitted Helminthiasis,* Manila, Feb. 24–27, 1969 (C. Harinasuta, ed.), pp. 253–264, Thai Watana Press, Bangkok.

Harinasuta, C. and Vajrasthira, S., 1960, Opisthorchiasis in Thailand, *Ann. Trop. Med. Parisitol.* **54:**100.

Harinasuta, C., Sornmani, S. and Migasena, P., 1975, *Studies on Nutrition and Some Related Parasitic Diseases Arising from Water Resource Development in The Lower Mekong Basin,* Southeast Asian Ministers of Education Organization-Regional Tropical Medicine and Public Health Project./ PP-05, pp. 1–92.

Migasena, P., 1972, Nutrition, health status and the impact of development in the lower mekong basin, *SEADAG Papers on Problems of Development in Southeast Asia,* pp. 1–19. Asia Society, New York.

Migasena, P., 1976a, Public health condition, in: *Environmental and Ecological Investigation of Pattani Multipurpose Dam Project,* Southeast Asia Technology Co. Ltd., Bangkok, Report prepared for Electricity Generating Authority of Thailand, pp. 434–465.

Migasena, P., 1976b, *Biogeographic and Medical Geographic Studies in Mekong Development Program,* Proceeding XXIII International Geographical Congress, Moscow, August 1976, Vol. 4, p. 152.

Migasena, P., 1978, Public health and nutrition, in: *Study of Environmental Impact of Nam Pong Project, Northeast Thailand,* Southeast Asia Technology Co. Ltd., Bangkok, Report prepared for National Energy Administration, Office of The Prime Minister pp. 6.1–6.65.

Migasena, P. (1980). Public health and nutrition, *in: Environmental and Ecological Investigation of Chiew Larn Project,* TEAM Consulting Engineers, Bangkok, Report prepared for Electricity Generating Authority of Thailand, pp. VI.177–VI.213.

Migasena, P., Reansuwan, W., and Changbumrung, S., 1980b, Nitrates and nitrites in local Thai preserved protein food, *J. Med. Assoc. Thailand* **63:**500.

Migasena, P., Juttijudata, P., Changbumrung, S., Schelp, F. P., and Juttijudata, P., 1979, The serum protein pattern in primary hepatoma and amoebic liver absess, *Ann. Trop. Med. Parasitol.* **73:**355.

Migasena, P., Juttijudata, P., Changbumrung, S., Schelp, F. P., Harinsuta, C., and Juttijudata, P., 1980a, The serum protein pattern in primary hepatoma and other liver disease, in: *Proceedings of the Second Symposium of the Federation of Asian and Oceanian Biochemistry, Food and*

Nutritional Biochemistry (H.T. Khor, K. K. Ong, and K. C. Oo, eds.), p. 255, The Malaysian Biochemical Society.
Migasena, P., Reansuwan, W., and Changbumrung, S., 1980b, Nitrates and nitrites in local Thai preserved protein food, *J. Med. Assoc. Thailand* **63:**500.
Sadun E. H., 1955, Studies on *Opisthorchis viverrini* in Thailand, *Am. J. Hyg.* **62:**81.
Schelp, F. P., Migasena, P., Saovakontha, S., Pongpaew, P. and Harinasuta, C., 1974, Polyacrylamide gel electrophoresis of human serum in subclinical opisthorchiasis, *Southeast Asian J. Trop. Med. Public Health* **5:**435.
Sunthornvipart, U., 1970, personal communication, Fisheries Technology Section, Department of Fisheries, Ministry of Agriculture, Bangkok.
Thamavit, W., Bhamarapravati, N., Sahaphong, S., Vajrasthira, S., and Angsubhakorn, S., 1978, Effects of dimenthylnitrosamine on induction of cholangiocarcinoma in *Opisthorchis viverrini* infected syrian golden hamsters, *Cancer Res.* **38:**4634.
Vajrasthira, S., and Harinasuta, C., 1957, Incidence, distribution and epidemiology of seven common intestinal helminths in Thailand, *J. Med. Assoc. Thailand* **40:**309.
Wykoff, D. E., Harinasuta, C., Juttijudata, P., and Winn M. M., 1965, *Opisthorchis viverrini* in Thailand—The life cycle and comparison with *O. felineus, J. Parasitol.* **51:**207.
Wykoff, D. E., Chittysothorn, K., and Winn, M. M., 1966, Clinical manifestation of *Opisthorchis viverrini* infection in Thailand, *Am. J. Trop. Med. Hyg.* **15:**915.

27

ADVERSE EFFECTS OF NORMAL NUTRIENTS AND FOODS ON HOST RESISTANCE TO DISEASE

M. John Murray, Anne B. Murray, Nigel J. Murray, Megan B. Murray, and Christopher J. Murray

INTRODUCTION

Our studies show that under certain circumstances normal nutrients may have adverse effects on host resistance. Undernutrition and deficiencies of certain nutrients may protect the host against intracellular infections, while refeeding with normal foods and repletion of nutrient deficiencies seem to have the potential for activating quiescent disease. In some societies, long standing deficiencies of nutrients are part of an ecological compromise favoring optimum co-existence of host and pathogens. Once this equilibrium is disturbed by upgrading nutritional status, disease may erupt with unusual violence.

It is not widely appreciated that normal nutrients in the diet may lead to undue suscepibility to disease, especially during refeeding after starvation. Such adverse effects are hard to detect in technological societies that are moderated by preventive and therapeutic medicine. Many suspect, but cannot prove, that they may play a part in the genesis of malignancy and degenerative disease in Western societies.

In closed societies living intimately with their environment, the ecological balance between man and animals and their food supply is often critical in the defense of man against indigenous disease. It is not unreasonable to suggest that both positive and negative means toward preventing indigenous disease may exist within the local food resources of these societies, although the latter may be unaware of these preventive factors. Should this ecological

M. John Murray, Anne B. Murray, Nigel J. Murray, Megan B. Murray, and Christopher J. Murray • Department of Medicine, University of Minnesota Medical School, Minneapolis, Minnesota 55455.

balance be disrupted by the introduction of new and exotic crops, or by the deliberate repletion of long-standing dietary deficiencies, latent disease may erupt with unusual violence.

In the following paragraphs we will describe and discuss observations that support these concepts. Briefly, they include:

1. Refeeding activation of intracellular infection.
2. Activation of infection by iron repletion.
3. Increased susceptibility to infection following a radical change in diet.
4. Disruption of ecological balances between host, diet, and parasites.
5. Suppression of spontaneous malignant transformation by undernutrition or nutrient deficiency.

REFEEDING ACTIVATION OF INTRACELLULAR INFECTION

During the Sahel famine of 1973–1974 we observed that refeeding was quickly followed by activation of apparently quiescent *Plasmodium falciparum* malaria. The implication was that the disease was sufficiently suppressed by undernutrition to lie dormant until refeeding was begun. A prospective study revealed a peak rate of parasitemia and clinical malaria five days after institution of refeeding undernourished patients and their relatives who had arrived at our hospital (Murray *et al.,* 1975; Fig. 1). As early as 1953, in an extensive series of experiments, Ramakrishnan (1953, 1954) and Ramakrish-

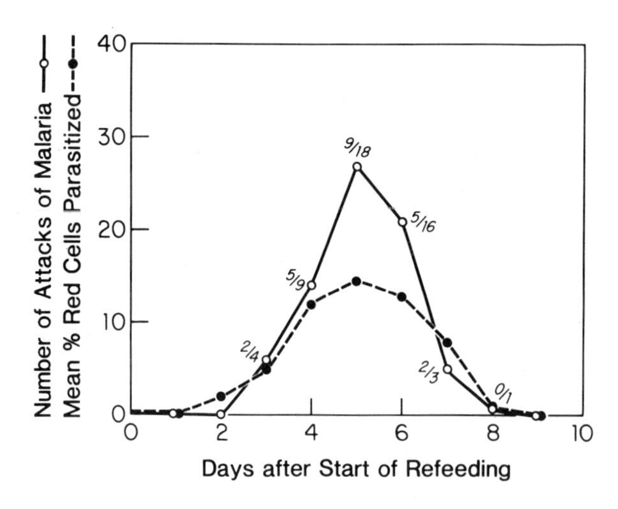

FIGURE 1. Mean percentage parasitemia of red cells and number of attacks of *P. falciparum* malaria in 72 patients and 109 relatives related to time in days after the start of refeeding with grain. The numerical ratios at each point (9/12) refer to the numbers of infected inpatients over numbers of infected relatives. From Murray (1980b). Published with permission and by courtesy of the *Yale Journal of Biology and Medicine.*

nan *et al.* (1953) confirmed and extended the observations that starvation could suppress malaria (Mackee, 1949). Ramakrishnan *et al.* (1953) also showed that suppressed infection could be activated simply by adding P-aminobenzoic acid (PABA) or methionine to the starvation diet. The mechanism of this phenomenon is probably as follows: plasmodia, unlike humans, often cannot use preformed folate but must have PABA for DNA synthesis and replication (Hawking, 1953). During starvation in the course of an all-milk diet (Hawking, 1953) PABA is in short supply so the organism must lie dormant in the tissues, most likely as the newly described hypnozoite form, until a supply of PABA or methionine is restored. Since PABA is nearly ubiquitous in most diets many refeeding diets have the potential to activate the disease.

We observed the same apparent refeeding activation of a variety of infections resulting from intracellular microorganisms in the Somali feeding camps of the Ogaden in 1975 (Murray *et al.*, 1976). They included brucellosis, salmonellosis, malaria, tuberculosis, and viral diseases, but we did not conduct a controlled prospective study. It is not clear why diseases other than malaria were suppressed, but it seems likely that starvation either reduces metabolism and replication of the host cell, which in turn discourages replication of the microorganisms, or deprives the microorganisms of a specific nutrient essential for their replication. One such nutrient could be zinc, which is essential for DNA synthesis and growth of all life forms (Falchuk *et al.*, 1975). Its biological unavailability to microorganisms during human undernutrition may have a greater inhibitory effect on the rate of replication of the microorganisms that is required for infection, than on the immune defense mechanisms of the host. We believe that this is an important principle in the interrelationships of infection and nutrition. Although many laboratory measures of host defense mechanisms are depressed during severe nutrient deficiency, affected animals are often more resistant to inoculations with intracellular, pathogenic microorganisms.

There are curious similarities between the types of infection encountered in refeeding and those occurring in the immuno-suppressed host. Physicians dealing with Vietnamese refugees in camps have reported cases of herpes zoster occurring in young children despite restoration to good nutritional status (Waldman, 1980).

ACTIVATION OF INFECTION BY IRON REPLETION

Simple observations by Armand Trousseau (1872) led him to suggest that iron deficiency mitigated the course of pulmonary tuberculosis, while subsequent iron repletion accelerated that disease. Our own observations of the Somali nomads of the Ogaden (Murray *et al.*, 1978a) confirmed his observations, and supported the reports by East African physicians of reduced infections in iron-deficient patients as compared with patients with other types of anemia (Masawe *et al.*, 1974). Somali nomads are often iron deficient, probably because of their predominantly camel-milk diet which contains little and poorly available iron (most mammalian milks contain less than 0.5 μg iron/ml). Our controlled prospective study on the course of infection in these nomads revealed a significant increase in intracellular infections during iron repletion (Table I). Unlike the peak incidence seen 5 days after refeeding following famine, the maximum incidence of infection occurred between 23 and 30 days after the start of treatment, as iron status was approaching normal. Similar findings have been observed in experimental models. The susceptibility of mice to fatal infection following inoculation with *Listeria monocytogenes* varies directly with their total iron body content, i.e., susceptibility increases when the iron content is increased above normal.

Iron is required by hosts and microorganisms alike (lactobacilli possibly excepted), for a variety of enzymes and factors concerned with intracellular metabolism and DNA synthesis. Once again, iron depletion and biological unavailability (iron being tightly bound to certain host proteins at the lower saturations of iron deficiency) have a greater short-term effect on the rapidly replicating microorganisms than on host immune mechanisms.

Another example of the effect of iron deficiency on infection is the relative freedom of milk-drinking nomads from infective diarrheas, especially amebiasis (Murray *et al.*, 1980a). We observed that even the addition of small amounts of iron (300 mg/week) to the all-milk diet of nomads in a controlled study over one year increased infection sharply as determined by detection of antibodies in the serum (Table II). The *Entomoeba histolytica* requires large amounts of iron (44 μM) for replication and invasiveness, while milk contains 6 or less μM of the element (Diamond *et al.*, 1978).

A final word on iron: normal iron status as currently defined in humans (pregnant women possibly excepted) probably does not increase susceptibility to most infections. However, rapid repletion of depleted iron stores is physiologically very different from the normal state, and probably allows an ex-

TABLE I

Observations during the Course of Treatment of Two Matched Groups of Iron-Deficient Somali Nomads Treated with Either Placebo (P) or Iron as Ferrous Sulfate (F)[a]

	Iron depleted (P)	Iron repleted (F)
Number of cases	66	71
Episodes of fever	6	29
Attacks of malaria	1	13
Malarial parasites seen	2	21
Clinical brucellosis	0	5
Tuberculosis	0	3
Enlarged liver	0	6
Enlarged spleen	0	11

[a] From Murray *et al.* (1980b). Published with permission and by courtesy of the *Yale Journal of Biology and Medicine.*

TABLE II
Serological Evidence of Infection with Entamoeba histolytica (EH) in Previously Uninfected (1977) Maasai with and without Iron Supplementation of Milk over One Year[a]

Year	Number	Serum Fe (μg/dl)	Transferrin saturation (%)	Sera positive for infection with EH
1977	70	33	15	0
1978	35 (without Fe)	37	14	3
	35 (with Fe)	60	29	29

[a] From Murray *et al.* (1978a).

cessive "catch-up" phase of DNA replication in host microorganisms. Since the latter benefit more quickly from repletion than the host, clinical infection may ensue.

INCREASED SUSCEPTIBILITY TO INFECTION FOLLOWING A RADICAL CHANGE IN DIET

In Western cultures, diets are rarely subject to radical change. In developing countries that are often the sites of recurrent famine, both refeeding with exotic foods and attempts to introduce more effective agricultural technologies may lead to drastic changes in the diet. We have recently reviewed these problems and for a more comprehensive treatment of the subject the reader is referred to our article (Murray *et al.*, 1980b). Such changes in the diet may increase the amount of certain nutrients that have long been deficient in indigenous foods. The classical example is the substitution of grains such as wheat or sorghum for milk.

Animal milk is an extraordinary secretion with a vast array of biological defense mechanisms against infection. Some of these mechanisms operate through reduced availability of specific nutrients such as iron, zinc, PABA, and folate to microorganisms. All have probably evolved to maintain the transfer of the mammalian gene pool through the suckling infant. Iron, zinc, and folate appear to be packaged in such a way as to make them optimally available to the host and minimally available to pathogens. In addition, milk contains antibacterial factors, e.g., oleic acid which inhibits streptococci (Speert *et al.*, 1979), and antiviral substances that are independent of antibody (Mathews *et al.*, 1976). The sudden supply of various nutrients that were not readily available in earlier diets, and the loss of the anti-infective qualities of milk, may reduce host resistance to infection.

We have observed this phenomenon in connection with cerebral malaria in children (Murray *et al.*, 1978b), with molluscum contagiosum in Maasai (Murray *et al.*, 1980c), and with planar warts in nomads (Murray *et al.*, 1980d). During a period of refeeding after famine when non-milk produce (usually grain) was substituted for milk we encountered an epidemic of molluscum contagiosum (a DNA pox virus infection) in Maasai that peaked at the height

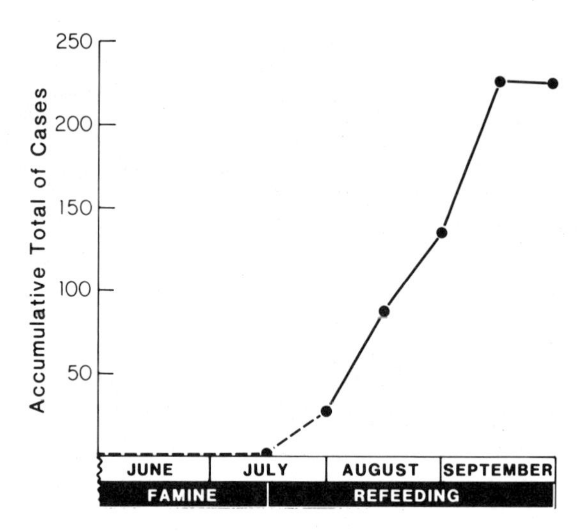

FIGURE 2. Clinical infection with *Molluscum contageosum* was not apparent during famine but became epidemic during refeeding of 1327 milk-drinking Maasai with grain.

of refeeding, despite a simultaneous improvement in nutrition (Fig. 2). Similarly, an outbreak of warts in Cushite nomads occurred in those refed with grain but not in those refed with milk. The sequence of events seems to involve disruption of a long-standing ecological compromise. A diet unchanged and partially or wholly deficient in a specific nutrient for millenia may allow pathogens to live harmoniously with the host at the price of their combined nutrient deficiency. The addition of a new or extra nutrient may swing the balance in favor of increased virulence of pathogens.

DISRUPTION OF ECOLOGICAL BALANCES BETWEEN HOST, DIET, AND PARASITES

Observations in the Comorro Islands (Murray *et al.*, 1977, 1978c) led us to suspect mutually advantageous ecological interrelationships between humans, malarial parasites, and the intestinal worm *Ascaris lumbricoides* on a nutritional basis. In the presence of moderate to heavy infestation of *A. lumbricoides* in Anjouan children, malaria was mild and infrequent. When the infestation was treated with piperazine, malaria became more frequent and severe. We speculated that the release of antitryptic and antichymotryptic factors by the worm to prevent its own dissolution, led to defective absorption of nutrients essential for the replication of the malarial parasites. Since then we have found that pigeon peas *(Cajanus cajan)*, which are a major staple in Anjouan, contained heat-stable inhibitors of trypsin and chymotrypsin. We believe these antienzymes interfere with the release from food and subsequent absorption of PABA or some other nutrient essential for replication of the plasmodia. When green peas *(Pisum sativum)* are substituted for pigeon peas, malaria increases in frequency and *A. lumbricoides* appears in the feces. During consumption of pigeon peas but not of green peas, the release of PABA from the synthetic peptide *N*-benzoyl-L-tyrosyl PABA, which is dependent on chymotrypsin in the gut, is largely blocked (Murray and Murray, 1980).

There appears to be an ecological relationship between the malarial parasite, the ascarid, and the host, in which coexistence of all three is favored by a diet of pigeon peas. When the relationship is disturbed by normally acceptable but not indigenous dietary items such as green peas, the ecological compromise is upset enough to allow one parasite to thrive at the expense of the host and the other parasite. It seems likely that such compromises may operate in many societies living closely with their environment.

SUPPRESSION OF SPONTANEOUS MALIGNANT TRANSFORMATION BY UNDERNUTRITION OR NUTRIENT DEFICIENCY

Experimental observations that undernutrition in animals favors reduced occurrence of spontaneous and inoculated malignancies are unequivocal. Similar relationships in man, although suspected, are unsubstantiated. There is, however, a surprising infrequency of malignancies in many undernourished societies, although "hot spots" of high frequencies of specific tumors occur in diverse locations. An interesting relationship occurs between diet, hepatitis B Virus (HBV) infection, and the malignant tumor of the liver, hepatoma. The incidence of hepatoma parallels the carrier rate for HBV in a population as well as the annual intake of aflatoxin in the diet (Lutwick 1979). Hepatomas are common in Africa, and reach peak frequency in Mozambique. However, in our experience the tumor is unusually rare in milk-drinking nomads, with none observed in over 20,000 examinations.

The known antiviral factors in milk and the almost certain low content of aflatoxin in milk of animals grazing in the wild may account for part of this infrequency. However HBV tends to persist unduly in livers and plasma containing more iron than normal (Felton *et al.,* 1979). Milk drinkers may also be protected from persistent HBV infection and from hepatomas through low levels and poor availability of specific nutrients such as iron in milk. Our observations on the rarity of warts in Cushite nomads and their appearance during grain refeeding may represent similar effects of nutrient deficiency on the oncogenic capacity of the Papova virus (Murray *et al.,* 1980d). In the same manner chronic malnutrition may favor decreased frequencies of malignancies.

CONCLUSION

We have shown that normal nutrients and food items may be responsible for the aggravation of preexisting quiescent or stable disease. These adverse effects are most readily seen in societies living closely with their environment. Every attempt should be made to avoid disrupting ecological compromises of a nutritional nature that favor optimum resistance against indigenous disease.

REFERENCES

Diamond, L. S., Harlow, D. R., and Phillips, B. P., 1978, *Entamoeba histolytica:* Iron and nutritional immunity, *Arch. Invest. Med. (Mex)* **9 (Suppl. 2):**329–338.

Falchuk, K. H., Fawcett, D. W., and Vallee B. L., 1975, Role of zinc in cell division of *Euglena gracilis, J. Cell Sci.* **17:**52–78.

Felton, C., Lustbader, E. D., Marten, C., and Blumberg, B. S., 1979, Serum iron levels and responses to hepatitis B virus, *Proc. Nat. Acad. Sci. U.S.A.* **76:**2438–2441.

Hawking, F., 1953, Milk diet, para-aminobenzoic acid and malaria *(P. berghei), Br. Med. J.* **1:**1201–1202.

Lutwick, L. I., 1979, Relation between aflatoxin, hepatitis B virus and hepatocellular carcinoma, *Lancet* **1:**755–757.

Mackee, R. W., 1949, Biochemistry and metabolism of malarial parasites, in: *Parasitic Infections in Man* (H. Most, ed.), p. 123, NY Acad. Med., Symposium on Microbiology No. 4.

Masawe, A. E. J., Muindi, J. M., and Swai, G. B. R., 1974, Infections in iron deficiency and other types of anaemia in the tropics, *Lancet* **2:**314–316.

Mathews, T. H. J., Nair, C. D. G., Lawrence, M. K., and Tyrrell, D. A. G., 1976, Antiviral activity in milk of possible clinical importance, *Lancet* **2:**1387–1389.

Murray, M. J., and Murray, A. B., 1981, Dietary trypsin-inhibitors: A clue to nutritional enlargement of the parotids, *Clin. Res.* **29:**627a.

Murray, M. J., Murray, A. B., Murray, N. J., and Murray, M. B., 1975, Refeeding malaria and hyperferremia, *Lancet* **1:**653–655.

Murray, M. J., Murray, A. B., Murray, M. B., and Murray, C. J., 1976, Somali food shelters in the Ogaden and their impact on health, *Lancet* **1:**1283–1285.

Murray, M. J., Murray, A. B., Murray, M. B., and Murray, C. J., 1977, Parotid enlargement, forehead edema and suppression of malaria as nutritional consequences of ascariasis, *Am. J. Clin. Nutr.* **30:**2117–2121.

Murray, M. J., Murray, A. B., Murray, M. B., and Murray, C. J., 1978a, The adverse effects of iron repletion on the course of certain infections, *Br. Med. J.* **2:**1113–1115.

Murray, M. J., Murray, A. B., Murray, N. J., and Murray, M. B., 1978b, Diet and cerebral malaria—the effect of famine and refeeding, *Am. J. Clin. Nutr.* **31:**57–61.

Murray, M. J., Murray, A. B., Murray M. B., and Murray, C. J., 1978c, The biological suppression of malaria: An ecological and nutritional interrelationship of a host and two parasites, *Am. J. Clin. Nutr.* **31:**1363–1366.

Murray, M. J., Murray, A. B., and Murray, C. J., 1980a, The salutary effect of milk on amebiasis and its reversal by iron, *Br. Med. J.* **1:**1351–1352.

Murray, M. J., Murray, A. B., and Murray, N. J., 1980b, The ecological interdependence of diet and disease in tribal societies, *Yale J. Biol. Med.* **53:**295–306.

Murray, M. J., Murray A. B., Murray, N. J., Murray, M. B., and Murray, C. J., 1980c, *Molluscum contagiosum* and herpes simplex in Maasai pastoralists: Refeeding activation of virus infection following famine? *Trans. R. Soc. Trop. Med. Hyg.* **74:**371–374.

Murray, M. J., Murray, A. B., Murray, M. B., and Murray, C. J., 1980d, Rarity of planar warts in Cushite nomads: Antiviral effect of milk? *Lancet* **2:**143–144.

Ramakrishnan, S. P., 1953, Studies on *Plasmodium berghei* N. SP. Vincke and lips 1948, VIII. The course of blood induced infections in starved albino rats. *Indian J. Malariol.* **7:**53–60.

Ramakrishnan, S. P., 1954, Studies on *Plasmodium berghei* Vincke and lips 1948, XVII. Effect of different quantities of the same diet on the course of blood induced infection in rats, *Indian J. Malariol.* **8:**89–96.

Ramakrishnan, S. P., Prabash, S., Krishnaswamie, S. P., and Sing, L. C., 1953, Studies on *Plasmodium berghei* Vincke and lips 1948, XIII. Effect of glucose, biotin, para-aminobenzoic acid and methionine on the course of blood induced infection in starving albino rats, *Indian J. Malariol.* **7**:225–231.

Speert, D. P., Wannamacher, L. W., Gray, E. D., and Clawson, C. C., 1979, Bactericidal effect of oleic acid on group A streptococci: Mechanism of action, *Infect. Immun.* **26**:1202–1210.

Trousseau, A., 1872, *Lectures on Clinical Medicine,* p. 96, New Sydenham Society, London.

Waldman, E., 1980, personal communication, Save The Children Fund, Tsim Sha Tsui, Hong Kong.

VI

GENETICS

28

ADVERSE EFFECTS OF FOODS IN GENETIC DISORDERS

Meinhard Robinow

INTRODUCTION

Most products of food metabolism have adverse effects on health if their concentrations in blood and extra- and intracellular fluids exceed narrow physiologic limits. Homeostasis, the maintenance of normal concentrations, is achieved by a vast number of genetically determined enzymatic processes that regulate digestion, absorption, transport, intermediate metabolism, and excretion.

Every inherited disorder of nutrient metabolism is a disturbance of homeostasis due to the absence or reduced function of a specific enzyme. Every enzyme deficiency produces a block in a metabolic sequence, usually resulting in an excess of the metabolite preceding the block and a deficiency of the metabolite following the block. In the majority of genetic metabolic disorders the clinical picture is dominated by the metabolites accumulating ahead of the block. They may cause direct chemical injury (as in galactosemia), extracellular deposition (as in gout), or intracellular storage with mechanical distension and eventual dysfunction (as in the lysosomal storage diseases). In addition, the accumulated metabolites can cause inhibition of other enzymatic processes (as in galactosemia and phenylketonuria). Downstream from the metabolic block occur deficiencies, and in some metabolic diseases the deficiencies are responsible for the clinical picture (as, for instance, the hypoglycemia of glycogen synthesase deficiency or of type I glycogen storage disease).

"NATURAL" AND "UNNATURAL" FOODS

A discussion of the adverse effects of foods must also consider the question of what are "natural" foods. Man is an omnivorous animal. In the course

Meinhard Robinow • Department of Pediatrics, School of Medicine, Wright State University, Dayton, Ohio 45404.

of evolution man has been able to adapt himself to a variety of raw unprocessed foods, to cooked, roasted, cured, and otherwise processed foods and, in more recent millenia, to many more or less "unnatural" foods as, for instance, the milk of other animals. Since the industrial revolution man has used more and more factory-processed nutrients (refined sugars, alcoholic beverages, and a variety of food additives and preservatives), i.e., nutrients to which he is not fully adapted by his genetic endowment. Some of these "seminatural" or "unnatural" foods have resulted in adverse effects. Individual tolerance to several of these foods (e.g., cow's milk)—just as tolerance to certain drugs—is partly determined by genetic factors.

Adverse effects of foods may occur soon after ingestion (e.g., bloating after ingestion of cow's milk), while in other cases they may not become evident until after many years of excess consumption (e.g., hepatic cirrhosis from heavy alcohol use).

What follows is not, nor can it be, a comprehensive review. It is only a sampling of the great variety of conditions in which adverse effects of foods are known or suspected to be due to genetic disorders. Nor will it be possible to describe the biochemical and genetic aspects of these anomalies in detail. The interested reader must consult the many excellent reviews and texts. Particularly recommended is *The Metabolic Basis of Inherited Disease* (Stanbury *et al.*, 1978), which will frequently be referred to below.

Food allergies, although they may be partly determined by genetic predisposition, will be considered outside the scope of this chapter.

ADVERSE EFFECTS OF FOODS IN GENETIC DISORDERS OF INTESTINAL DIGESTION AND TRANSPORT

Celiac Disease

Celiac disease is due to intolerance of the jejunal mucosa to the glutens of wheat and rye, the major proteins of these cereals. The biochemically similar glutens of oats and barley are tolerated. Ingestion of wheat or rye products by susceptible individuals leads to atrophy of the jejunal mucosa. Clinical symptoms do not occur on first contact with the offending foods but only after an appreciable time interval. The onset usually occurs during the first 2 years of life, occasionally in later childhood, rarely in adult life. Clinical manifestations are those of malabsorption: diarrhea, steatorrhea, abdominal distension, irritability and, eventually, wasting and growth failure. The clinical diagnosis must be confirmed by biopsy of the jejunal mucosa. Once wheat and rye are removed from the diet, all symptoms subside and the mucosa slowly regains its normal structure, but the intolerance persists. Attempts to reintroduce gluten will reactivate the disease at any age. Adults with celiac disease are at an increased risk for intestinal lymphoma (Barry and Read, 1973).

Patients with celiac disease have a high incidence of HL-A type 8 (Human Lymphocyte-Antigen type 8) (Falchuk *et al.*, 1972). Celiac disease is often familial but the data do not fit simple Mendelian inheritance (David and Adjukiewicz, 1975; Falchuk *et al.*, 1972). The incidence of celiac disease is variably estimated as 1 : 400 to 1 : 3000 (Ament, 1979).

Lactase Deficiency

Lactose is the only sugar of human and animal milk. It is not a constituent of any other natural food. Lactose is a disaccharide which must be split by intestinal lactase into its two absorbable monosaccharides, glucose and galactose. Lactase deficiency manifests itself by osmotic diarrhea, bloating, and abdominal discomfort. There are four types of lactose intolerance. First, congenital lactase deficiency, a very rare genetic defect of small infants, which has a tendency to subside spontaneously (Sunshine and Kretschmer, 1964). A second, more severe disorder is lactose intolerance without lactase deficiency (Holzel *et al.*, 1962). Patients may present with severe diarrhea, profound dehydration, and cachexia. It can be fatal unless milk is removed from the diet. Third, and far more common, is acquired postenteritic lactase deficiency, causing transient milk intolerance. The last type is the genetic lactose intolerance of older children and adults, relatively rare in whites but common in American blacks (Bayless and Rosenzweig, 1966; Gray, 1978). The condition is an inability of the intestinal mucosa to continue expression of the lactase gene past infancy, the "natural" period of milk drinking. This late lactase deficiency with its distressing symptoms is a trait that is maladaptive only in societies that drink cow's milk, and should probably not be called a "disease."

The domestication of cattle is a relatively recent event in many parts of the world. "You never outgrow your need for milk" has been a popular advertising slogan, but drinking the milk of another species is hardly natural and certainly does not constitute a universal need.

Two Other Defects of Intestinal Carbohydrate Metabolism

Only brief mention shall be made here of two other rare genetic disorders of intestinal carbohydrate metabolism in which ingestion of natural carbohydrates causes severe, life threatening diarrhea. In *glucose–galactose malabsorption* (Lindquist, *et al.*, 1962; Gray, 1978), these two monosaccharides cannot be transported through the intestinal epithelium. Symptoms arise with the first feeding, whether it is milk or glucose water. The other condition is *sucrase–isomaltase deficiency* (Ament *et al.*, 1979). In this disorder, sucrose (cane sugar) and starch are the offending nutrients. Diarrhea usually does not begin until the introduction of solid foods.

ADVERSE EFFECTS OF FOODS IN GENETIC DISORDERS OF INTERMEDIARY METABOLISM

Galactosemia

Virtually all dietary galactose is derived from the lactose of milk. After the galactose has been absorbed from the intestine it is carried to various body tissues where it is converted to galactose-1-phosphate by the enzyme galactokinase; (galactokinase deficiency is a very rare metabolic defect, manifested mainly by cataracts). The second step in intermediary galactose metabolism is the conversion of galactose-1-phosphate to glucose-1-phosphate by the enzyme galactose-1-phosphate uridyl transferase (Segal, 1978). "Galactosemia" is the traditional name for this defect, though *transferase deficiency* is the more precise term. The enzyme block results in elevated levels of galactose-1-phosphate that injure various organs, especially liver, brain, and kidney. Untreated galactosemia usually presents soon after birth as failure to thrive, vomiting, hepatomegaly, jaundice and acidosis, occasionally accompanied by hypoglycemia or severe hemolysis. Death may occur in the newborn period. In other untreated or partially treated infants the condition progresses more slowly but inexorably to cirrhosis of the liver and to irreversible mental retardation. Cataracts may be present at birth or develop soon afterwards. As in galactokinase deficiency, they are due to the accumulation of galacticol, another galactose metabolite, in the lens. Galactokinase and transferase deficiency are both autosomal recessive traits.

The only effective treatment is immediate, complete, and permanent elimination of galactose-containing foods, namely milk and milk products. Since substantial amounts of milk are contained in many common foods, strict adherence to a carefully prescribed food list is mandatory.

Hereditary Fructose Intolerance

In this disorder the deficient enzyme is fructose-1-phosphate aldolase. The offending foods are those containing fructose and sucrose (cane sugar). The latter is a fructose–glucose disaccharide which is split into its components in the intestine. Manifestations in infancy may be similar to those of classical galactosemia: jaundice, hypoglycemia, lethargy, and seizures (Froesch *et al.*, 1963). Milder cases may not be diagnosed until adult life. Fructose intolerance has approximately the same incidence as galactosemia. It is also transmitted as an autosomal recessive trait.

AMINOACIDOPATHIES

Phenylketonuria

Of all known heritable disorders of intermediary metabolism, by far the largest number are the aminoacidopathies. The best known and the most common amino acid disorder is phenylketonuria (PKU). As most other enzyme deficiencies, it is inherited as an autosomal recessive trait. The incidence is 1 : 6000 to 1 : 18,000 in western Europe and the U.S. (Tourien and Sidbury, 1978).

Protein foods are the offending agents. Phenylalanine constitutes approximately 5% of all food proteins. There is no natural low-phenylalanine protein and thus no natural low-phenylalanine diet. Since phenylalanine is an essential amino acid, a fraction of absorbed phenylalanine is used for the building of new proteins. The remainder is converted to tyrosine by the enzyme phenylalanine hydroxylase and its cofactor tetrahydrobiopterine. Classical PKU is caused by a deficiency of the hydroxylase. A rare form of PKU is due to deficiency of dihydropteridine reductase (Smith *et al.,* 1975; Danks *et al.,* 1979). Blockage of the normal phenylalanine → tyrosine pathway causes elevation of plasma phenylalanine and its diversion into an alternate degradative pathway, resulting in the appearance of phenylpyruvic, *o*-hydroxyphenyllactic, phenylactic acids, and phenylethylamine in plasma and urine. These metabolites are barely detectable in normal individuals. An additional effect of high phenylalanine levels is inhibition of other enzymes, particularly those involved in tryptophan metabolism.

Untreated PKU leads almost invariably to severe mental retardation by impairment of brain growth and myelinization. The pathogenetic mechanism is still not fully understood. In experimental animals, high serum phenylalanine levels, produced by feeding or injecting excess phenylalanine, result in defective mental development. To what extent the metabolites of the alternate pathway contribute to the retardation remains unknown.

The affected fetus is protected by the placental circulation. After birth the infant must be placed on a low-phenylalanine diet. This is a low-protein diet, supplemented by a protein hydrolysate from which most of the phenylalanine has been removed. Serum phenylalanine levels must be monitored. Optimal growth and development require serum phenylalanine levels appreciably higher than those of normal individuals. The diet must be started within 2–3 months after birth and must be maintained until brain growth and myelinization are completed. The exact age at which discontinuation of the diet is safe is still controversial. Most experts permit an unrestricted diet after 6–8 years of age.

The disorders of tetrahydrobiopterin metabolism do not respond to the low phenylalanine diet. Mental deterioration progresses in spite of reduced serum phenylalanine levels (Smith *et al.,* 1975; Danks *et al.,* 1979).

Other Aminoacidopathies

There are many other genetic disorders of amino acid metabolism in which enzyme blocks cause elevated levels of specific amino acids or some of their metabolites. In all these disorders proteins are the offending nutrients. Some of these disorders manifest themselves in the newborn period, clinically, by rapid deterioration, vomiting and seizures, and biochemically, by ketoacidosis or hyperammoniemia. They may rapidly lead to death unless protein intake is immediately reduced. In other instances the condition may initially be asymptomatic but later cause mental retardation, neurologic deterioration, or other dysfunctions. In histidinemia about half the affected individuals are retarded, while the remainder are normal (La Du, 1978).

ADVERSE EFFECTS OF METALS IN GENETIC METABOLIC DISORDERS

There are at least two genetic disorders in which metals contained in foods have serious adverse effects.

Familial Hemochromatosis

Familial hemochromatosis is the result of excessive absorption of iron from the small intestine. The excess iron is stored in the reticuloendothelial system and eventually causes cirrhosis of the liver. The mode of inheritance is probably autosomal dominant (Pollycove, 1978). Although ingested iron is responsible for the disorder, dietary iron restriction is not part of the medical management. Treatment consists of bloodletting and of chelation of the iron with deferoxamine.

Wilson Disease

In Wilson disease (hepatolenticular degeneration) the offending metal is copper. Copper accumulates in various organs, especially the brain, liver, and kidneys, because of a deficiency of ceruloplasmin, the copper transporting protein (Sass-Kortsak and Bearn, 1978). Symptoms have their onset in late childhood or early to mid-adult life. Initial manifestations are protean and are easily misinterpreted. Patients may present with liver disease, hemolytic anemia, change of personality, intellectual deterioration, or as a disorder of the basal ganglia. A constant and easily recognized clinical feature is the Kayser-Fleischer ring, a greenish ring at the corneal limbus of the eye. It is due to deposition of copper in Descemet's membrane. Left untreated, Wilson disease is relentlessly progressive and ends fatally. The course may range from a few months to 10–15 years. The only effective treatment is chelation of the body's excess copper with penicillamine. Early treatment will reverse all or most of the clinical abnormalities. In addition, copper intake should be reduced. The minute amounts of copper dissolved from copper cooking vessels and from high copper-containing foods (e.g., shellfish, liver, and chocolate) should be avoided. Treatment must be lifelong.

Wilson disease is an autosomal recessive disorder. Siblings of patients have a 25% chance of being affected. Since the disease can be diagnosed long before it becomes symptomatic, it is mandatory that all siblings of patients be tested appropriately, i.e., by monitoring serum cerulplasmin levels and urinary copper excretion after penicillamine injection.

FAVISM

The previous sections have dealt with adverse effects of common foods or their components in heritable metabolic disorders. Rarely, a minor constituent of a natural food (not a common nutrient) is injurious to the carrier of a mutant gene. The best known example is favism.

Favism is a tendency to develop acute hemolytic anemia after ingestion of the broad bean *(Vicia faba)*. Patients present with hemoglobinuria, anemia (which may be fatal if untreated), and mild jaundice. Symptoms begin 24 hr to 9 days after the eating of beans (Kattamis *et al.*, 1969). Even inhalation of the pollen can induce hemolysis (Sansone *et al.*, 1958). Young children are more susceptible than older children or adults (Kattamis *et al.*, 1969). Favism has been repeatedly observed in breast-fed infants 3–6 days after their mothers had eaten fava beans (Kattamis *et al.*, 1969). There is even a report of a fatal antenatal hemolytic crisis due to maternal ingestion (Mentzler and Collier, 1975).

The broad bean is the only native bean of Europe and the Near East. It is a popular food in the Mediterranean countries, where the young pods with the seeds are eaten as a cooked vegetable. Dried beans are used less often. Broad beans do not thrive in the hot summer climate of the U.S., except along the narrow, cool-weather strip of the Pacific coast. They are popular with some ethnic groups and are available as a frozen vegetable in many stores.

Favism occurs only in individuals with a deficiency of glucose-6-phosphate dehydrogenase (G6PD) (Szeinberg *et al.*, 1957), an enzyme that plays an important part in red blood cell metabolism. G6PD-deficient individuals are subject to hemolytic crises from a variety of substances, among them primaquine, an antimalarial drug (Tarlov *et al.*, 1962).

The gene for G6PD deficiency is located on the X chromosome, and deficiency is therefore found predominantly in males. But where the mutant gene is highly prevalent, as in Sardinia, parts of Greece, and among the Sephardic Jews of Israel, homozygous females occur who may experience favism. There is amazing genetic heterogeneity of G6PD. Over 100 allelic varients have been identified (McKusick, 1978). In Europe and the Near East, favism is found mainly in individuals carrying the "Mediterranean" G6PD variant. Favism is also common among Chinese carrying the "Canton" mutant but does not occur in blacks, many of whom carry the more benign A and A- "African" G6PD alleles. Since G6PD deficiency provides some protection against malaria, the defective variants have attained high population frequencies in areas where malaria was formerly endemic (Motulsky, 1964).

The constituent of the bean and the mechanism of action responsible for favism have not been identified with certainty. Pyrimidines derived from fava beans have been incriminated (Mager *et al.*, 1965). Other explanations have been offered (Beutler, 1970; Bottini *et al.*, 1971). Also, the relationship between G6PD deficiency and favism is still obscure. Even among individuals with the "Mediterranean" G6PD deficiency, not all develop favism after eating the beans; and even those subject to favism may, at times, eat the beans with impunity (Stammatoyannopoulis *et al.*, 1966). Family studies suggest that an additional gene is required to produce favism in G6PD-deficient individuals (Stammatoyannopoulis *et al.*, 1966; Kattamis *et al.*, 1969).

IDIOSYNCRASIES TO SOME PROCESSED FOODS AND FOOD ADDITIVES

Tyramine Headaches

Tyramine is a vasoactive amine, produced in the intestinal tract by the bacterial breakdown of tyrosine. It is also found in foods that have been fermented by bacteria and yeasts, as are some cheeses and wines. In most individuals tyramine is further metabolized by the action of monoamine oxidase (MAO). MAO inhibitors are widely used in psychiatry as antidepressive drugs. In patients taking MAO inhibitors substantial amounts of tyramine are absorbed and may produce hypertension and pounding headaches (Blackwell, 1963). But in some people who are not on any medication, cheese and wine produce the same symptoms (Hannington, 1967). Genetically determined MAO deficiency seems a likely explanation.

Hot Dog Headache

A phenomenon similar to the preceding description is the "hot dog headache," induced by eating hot dogs or luncheon meats preserved with nitrites. Only a rare number of individuals appear to be sensitive (Henderson and Raskin, 1972). There is nothing to suggest that allergy plays a role in the pathogenesis. The most plausible explanation is a genetic predisposition due to an enzyme deficiency.

The Chinese Restaurant Syndrome

The third of these well popularized food idiosyncrasies is the "Chinese Restaurant Syndrome" (Schaumberg *et al.*, 1969). The symptoms are headaches, vomiting, and a sensation of chest pressure. They are caused by sodium glutamate, a taste enhancer widely used in Chinese cookery and a constituent of soy sauce. Again, susceptibility is likely to have a genetic basis (McKusick, 1978).

Migraine

In many sufferers of classical migraine, attacks are initiated by ingestion of specific foods. Migraine is often familial and is transmitted as an autosomal dominant trait (McKusick, 1978). Food allergies are no longer considered to be responsible for attacks (Longdon and Forsythe, 1979). Migraine may be another genetic disorder in which foods or certain minor constituents of foods have adverse effects. One enzyme defect, phenylethylamine oxidase deficiency, has been proposed as the cause of chocolate-induced migraine (Sandler *et al.*, 1974).

Diet and the Hyperactive Child

Various foods and food additives (e.g., sucrose, salicylates, and food dyes) have been blamed for the hyperkinetic syndrome of childhood (Feingold, 1975). A genetically determined sensitivity to these substances has been postulated. Diets that avoid the alleged culprits are claimed to be beneficial (Feingold, 1975). All that can be stated at this time is that claims and counterclaims are difficult to substantiate.

CHRONIC DISORDERS WITH A GENETIC COMPONENT IN WHICH FOODS HAVE ADVERSE EFFECTS

There are several common chronic diseases in which genetic predisposition plays a part and that are adversely affected by some foods, either on a short-term or long-term basis. Table I provides a partial listing. A detailed discussion is beyond the scope of this chapter.

CANCER: ADVERSE EFFECTS OF FOOD AND GENETIC PREDISPOSITION

It is now thought that most cancers are caused by environmental agents, including foods (Wynder, 1979). Among foods linked to cancers are natural

TABLE I
Chronic Disorders with a Heritable Predisposition in Which Foods Can Have Adverse Effects

Disorder	Food
Atherosclerosis	Animal fats
Gout	Glandular meats (nucleoproteins)
Hypertension	Table salt
Hypercalcuria	Cow's milk
Diabetes mellitus	Sucrose

foods [high-fat diet linked to colon and breast cancers (Wynder, 1980)], processed foods (smoked and cured foods to stomach cancer, alcohol to upper gastrointestinal tract cancers), and food additives (saccharine to bladder cancer). On the other hand, cancer does not occur at random in individuals exposed to carcinogens. There is a genetic predisposition to many cancers. Close relatives of cancer patients have a much higher incidence than unrelated persons.

The pathogenetic mechanisms for most types of cancer are still obscure. One explanation has recently been proposed: many carcinogens seem to be inactive until they have been enzymatically converted into active carcinogens. One of the enzymes responsible for activating carcinogens is aryl hydrocarbon hydroxylase. Three alleles of this enzyme with different activities are known and these are inherited as simple Mendelian traits (Kellermann *et al.,* 1973a). It has been claimed that they play a role in the susceptibility to lung cancer (Kellermann, *et al.,* 1973b), but this claim has more recently been disputed (Paigen *et al.,* 1977).

REFERENCES

Ament, M. E., Perera, D. D., and Esther, L. J., 1973, Sucrase–isomaltase deficiency—a frequently misdiagnosed disease, *J. Pediat.* **83:**721–727.

Ament, M. E., 1979, Celiac Sprue in: *Nelson, Textbook of Pediatrics,* 11th ed. (V. Vaughn, J. McCay, and R. E. Behrman, eds.), p. 1083, W. B. Saunders, Philadelphia.

Barry, R. E., and Read, A. E., 1973, Celiac disease and malignancy, *Q. J. Med.* **42:**665–675.

Bayless, T. M., and Rosenzweig, N. S., 1966, A racial difference in incidence of lactase deficiency. *J. Am. Med. Assoc.* **197:**968–972.

Beutler, E., Baluda, M. C., Sturgeon, P., and Day, R. W., 1965, A new genetic abnormality resulting in galactose-1-phosphate uridyl transferase deficiency, *Lancet* **1:**353.

Beutler, E., 1970, L-dopa and favism, *Blood* **36:**523–525.

Blackwell, B., 1963, Hypertensive crisis due to monoamine oxidase inhibitors, *Lancet* **2:**849.

Bottini, E. Lucarelli, P., Agostino, R., Palmarino, R., Businco, L., and Antagnoni, G., 1971, Favism: Association with erythrocyte acid phosphatase phenotype, *Science* **171:**409–411.

Danks, D. M., Cotton, R. G. H., and Schlesinger, P., 1979, Diagnosis of malignant hyperphenylalanemia, *Arch. Dis. Child.* **54:**329.

David, T. J., and Adjukiewicz, 1975, A family study of celiac disease, *J. Med. Genet.* **12:**79–82.

Falchuk, Z. M., Rogentine, G. M., and Strober, W., 1972, Predominance of histocompatibility antigen HL-A 8 in patients with gluten sensitive enteropathy, *J. Clin. Invest.* **51:**1602–1605.

Feingold, B. F., 1975, Hyperkinesis and learning disabilities linked to artificial food flavors and colors, *Am. J. Nurs.* **75:**797–803.

Froesch, E. R., Wolf, H. P., Baitsch, H. H., Prader, A., and Labhart, A., 1963, Hereditary fructose intolerance: An inborn error of fructose metabolism, *Am. J. Med.* **34:**151.

Gray, G. M., 1978, Intestinal disaccharidase deficiencies and glucose–galactose malabsorption, in: *The Metabolic Basis of Inherited Disease,* 4th ed. (J. B. Stanbury, J. B. Wyngaarden, and D. S. Frederickson, eds.), pp. 1526–1536, McGraw-Hill, New York.

Hannington, E., 1967, Preliminary report of tyramine headache, *Br. Med. J.* **2:**250.

Henderson, W. R., and Raskin, N. H., 1972, "Hot dog" headache: Individual susceptibility to nitrate, *Lancet* **2:**250.

Holzel, A., Mereu, T., and Thomson, M. L., 1962, Severe lactose intolerance in infancy, *Lancet* **2:**1346.

Kattamis, A., Kyriazakou, M., and Chaidas, S., 1969, Favism, clinical and biochemical data, *J. Med. Genet.* **6**:34–41.

Kellermann, G., Luyten-Kellermann, M., and Shaw, C. R., 1973a, Genetic variation of aryl hydrocarbon hydroxylase in human lumphocytes, *Am. J. Hum. Genet.* **25**:327–331.

Kellermann, G., Shaw, C. R., and Luyten-Kellermann, M., 1973b, Aryl hydrocarbon hydroxylase inductibility and bronchogenic carcinoma, *N. Engl. J. Med.* **289**:934–937.

La Du, B. N., 1978, Histidinemia, in: *The Metabolic Basis of Inherited Disease,* 4th ed. (J. B. Stanbury, J. B. J. Wyngaarden, and D. S. Frederickson, eds.), pp. 317–327, McGraw-Hill, New York.

Lindquist, B., Meuwisse, G. W., Melin, K., 1962, Glucose–galactose malabsorption, *Lancet* **2**:666.

Longdon, P. J., and Forsythe, W. I., 1979, Migraine in childhood. A review, *Clin. Pediatr.* **18**:353–356.

Mager, J., Glaser, G., Razin, A., Isak, C., Bien, S., and Noam, M., 1965, Metabolic effects of pyrimidines derived from glycosides on human erythrocytes deficient in glucose-6-phosphate dehydrogenase, *Biochem. Biophys. Res. Comm.* **20**:235–240.

McKusick, V. A., 1978, *Mendelian Inheritance in Man,* 5th ed., Johns Hopkins University Press, Baltimore, Maryland.

Mentzler, W. C., and Collier, E., 1975, Hydrops fetalis associated with G-6-PD deficiency and maternal ingestion of fava beans and ascorbic acid, *J. Pediatr.* **86**:565–567.

Motulsky, A. G., 1964, Hereditary red cell traits and malaria, *Am. J. Trop. Med.* **13**:147–155.

Paigen, B., Gurtoo, H. L., Minowada, J., Houten, L., Vincent, R., Paigen, K., Parker, N. E., Ward, E., and Hayner, N. T., 1977, Questionable relation of aryl hydrocarbon hydroxylase to lung cancer, *N. Engl. J. Med.* **297**:346–350.

Pollycove, M., 1978, Hemochromatosis, in: *The Metabolic Basis of Inherited Disease,* 4th ed. (J. B. Stanbury, J. B. Wyngaarden, and D. S. Frederickson, eds.), pp. 1127–1164, McGraw-Hill, New York.

Sandler, M., Yondim, M. B., and Harrington, E., 1974, A phenylethylamine oxidizing defect in migraine, *Nature* **250**:337–339.

Sansone, G., Piga, A. M., and Segni, G., 1958, El Favismo, *Minerva Med.,* Torino, Italy.

Sass-Kortsak, A. and Bearn, A., 1978, Hereditary disorders of copper metabolism, in: *The Metabolic Basis of Inherited Disease,* 4th ed. (J. B. Stanbury, J. B. Wyngaarden, and D. S. Frederickson, eds.), pp. 1098–1126, McGraw-Hill, New York.

Segal, S., 1978, Disorders of galactose metabolism, in: *The Metabolic Basis of Inherited Disease,* 4th ed. (J. B. Stanbury, J. B. Wyngaarden, and D. S. Frederickson, eds.), pp. 160–181, McGraw-Hill, New York.

Schaumberg, H. H., Byck, R., Gerstl, R., and Mashman, J. H., 1969, Monosodium l-glutamate: Its pharmacology and role in the Chinese Restaurant syndrome, *Science* **163**:826–828.

Smith, I., Clayton, B. E., and Wolff, O. H., 1975, new variant of phenylketonuria with progressive neurologic illness, *Lancet* **1**:1108–1111.

Stammatoyannopoulis, G., Fraser, G. R., Motulsky, A. G., Fessas, P., Akrirakisn, A., and Papayannapoulou, T., 1966, On the familial predisposition of favism, *Am. J. Hum Genet.* **18**:253–263.

Stanbury, J. B., Wyngaarden, J. B., and Frederickson, D. S. (eds.), 1978, *The Metabolic Basis of Inherited Disease,* 4th edition, McGraw-Hill, New York.

Sunshine, P., and Kretschmer, N., 1964, Studies of small intestine during development III. Infantile diarrhea associated with intolerance to disaccharides, *Pediatrics* **34**:38–50.

Szeinberg, A., Sheba, C., Hirshorn, N., and Bodonyi, E., 1957, Studies on erythrocytes in cases with past history of favism and drug induced hemolytic anemia, *Blood* **12**:603–616.

Tarlov, A. R., Brewer, G. J., Carson, P. E., and Alving, A. S., 1962, Primaquine sensitivity. Glucose-6-Phosphate dehydrogenase deficiency and biological significance, *Arch. Int. Med.* **109**:209–234.

Tourian, A. Y., and Sidbury, J. B., 1978, Phenylketonuria, in: *The Metabolic Basis of Inherited Disease,* 4th ed. (J. B. Stanbury, J. B. Wyngaarden, and D. S. Frederickson, eds.), pp. 240–255, McGraw-Hill, New York.

Wynder, E. L., 1979, Dietary habits and cancer epidemiology, *Cancer* **43**:1955–1961.

Wynder, E. L., 1980, Dietary factors related to breast cancer, *Cancer* **46**:899–904.

VII

ALLERGY

29

FOOD ALLERGY IN CHILDREN AND ADULTS

Tatsuo Matsumura

DEFINITION

Two thousand years ago Lucretius (96–55 BC) said that "One man's food might be another's poison." Food allergy is defined in a broader sense as an altered allergic reaction to food. In a narrower sense, it means the development of an allergic reaction mediated by humoral and cellular factors involving immune mechanisms.

As a matter of fact, limited methods have been developed for proving the allergic-natured clinical features, and therefore many patients have little chance of finding relief from their illnesses, because of narrow immunological definitions by physicians. Although a broader concept has long been held by a small number of allergists specializing in food allergy, a marked tendency toward the wider definition has emerged only recently in certain societies of allergists.

DISEASES AND SYMPTOMS

Table I lists the diseases and clinical features that can be associated with allergy, described in the book *Food Alergy,* by Albert H. Rowe, a pioneer in food allergy, in collaboration with his son, Albert Rowe, Jr. (1972). Tables II and III concern the Lee–Miller provacative-neutralizing method. The former shows common subjective symptoms and objective signs stimulated by the injection of provocative doses of allergens; and the latter, syndromes sometimes relieved by the injection of neutralizing doses. These tables have been cited here because most of these symptoms, signs, and syndromes were observed by this author in food allergy patients (Miller, 1972). As a reference

---

Tatsuo Matsumura • Department of Pediatrics, School of Medicine, Gunma University, Maebashi, Gunma, Japan.

TABLE I
Food Allergy Diseases and Symptoms[a]

1. *Gastrointestinal:* Canker sores, coated tongue, heavy breath, distension, belching, epigastric heaviness, sour stomach, burning, pyrosis, nausea, vomiting, diarrhea, constipation, gas in bowels, pruritus ani, abdominal pain, gastric ulcer, duodenal ulcer, colitis, proctitis, irritable colon, hemorrhoid, gastrointestinal bleeding, ulcerative colitis, regional enteritis.
2. *Respiratory:* Bronchial asthma, chronic bronchitis, bronchiectasis, pulmonary emphysema, wheezing, recurrent colds, nasal congestion, nasal blocking, nasal discharge, nasal polyp, sneezing, hoarseness.
3. *Dermatological:* Atopic dermatitis (eczema), pruritus, acne, urticaria, angioneurotic edema, dermatographism.
4. *Hematological:* Purpura, eosinophilia, bleeding and hemorrhages in body tissues, hematuria, easy bruising of tissues.
5. *Cerebral and neural:* Headache, migraine, neuralgia, allergic toxemia and fatigue, allergic epilepsy.
6. *Urogenital:* Allergic cystitis, orthostatic albuminuria, nephrotic syndrome,[b] dysmenorrhea.
7. *Cardiovascular:* Arrhythmia, paroxysmal tachycardia.
8. *Aural:* Serous otitis media, dizziness, vertigo, Menier's syndrome, dermatitis of ear and ear canal, edema of Eustachian tubes, tinnitus.
9. *Ocular:* Blepharitis.
10. *Musculoskeletal:* Arthralgia, myosalgia, allergic synovitis, hydroarthrosis.
11. *Pediatric:* Colic, pylorospasm, celiac syndrome, cyclic recurrent vomiting, anorexia, food dislike, malnutrition, protein-losing enteropathy, croup, bronchiolitis, enuresis, diaper rash.
12. *Systemic:* Fever.

[a] From Rowe and Rowe, Jr., 1972.
[b] Matsumura *et al.*, 1971b.

to aid in the understanding of the great variety of manifestations of food allergy, which has been appropriately described as the "great masquerader" by Crook (1975), it would seem helpful to explain Miller's description of these symptoms.

Miller (1972) made the observation, in relation to Table II, "It is not suggested here that these symptoms are always due to allergy, but that they are sometimes due to allergy. This is supported by the fact that patients undergo testing because they are experiencing these symptoms." In connection with Table III, he commented that "this list of indications consists only of some of the conditions seen with sufficient frequency in an allergist's office to allow definite conclusions at this time."

Milk allergy symptoms in infants are vomiting, diarrhea, constipation, erythema, urticaria, papular urticaria, eczema (atopic dermatitis), diaper rash, perianal dermatitis, severe sudamina, nasal congestion, rhinorrhea, wheezing, night crying, and hiccup. The same symptoms have also been frequently observed in breast-fed infants, though usually milder in degree. By indirect elimination and ingestion of suspected allergenic foods by the mother, it was ascertained that these symptoms were caused not by breast milk itself, but by foods ingested by the mother (Matsumura, 1969).

As will be described, a portion of ingested foods is absorbed as large

TABLE II
Common Symptoms Stimulated by Testing[a]

1. *Skin:* Itching, burning, flushing, warmth, coldness, tingling, sweating behind neck, etc. Hives, blisters, blotches, red spots, "pimples."
2. *Ear, nose, throat:* Nasal congestion, sneezing, nasal itching, runny nose, postnasal drip. Sore, dry, or tickling throat, clearing throat, itching palate, hoarseness, hacking cough. Fullness, ringing, or popping of ears, earache, intermittent deafness, dizziness, imbalance.
3. *Eye:* Blurring of vision, pain in eyes, watery eyes, crossing of eyes, glare hurts eyes; eyelids twitching, itching, drooping, or swollen; redness and swelling of inner angle of lower lid.
4. *Respiratory:* Shortness of breath, wheeze, cough, mucus formation in bronchial tubes.
5. *Cardiovascular:* Pounding heart, increased pulse rate, skipped beats, flushing, pallor, warm, cold, tingling, redness or blueness of hands, faintness, precordial pain.
6. *Gastrointestinal:* Dryness of mouth, increased salivation, canker sores, stinging tongue, toothache, burping, retasting, heartburn, indigestion, nausea, vomiting, difficulty in swallowing, rumbling in abdomen, abdominal pain, cramps, diarrhea, itching or burning of rectum.
7. *Genitourinary:* Frequent, urgent, or painful urination; inability to control bladder; vaginal itching or discharge.
8. *Muscular:* Fatigue, generalized muscular weakness, muscle and joint pain, stiffness, soreness, chest pain, backache, neck muscle spasm, generalized spasticity.
9. *Nervous system:* Headache, migraine, compulsively sleepy, drowsy, groggy, slow, sluggish, dull, depressed, serious, crying; tense, anxious, stimulated, overactive, restless, jittery, convulsive, head feels full or enlarged, floating, silly, giggling, laughing, inebriated, unable to concentrate, feeling of separateness or apartness from others, amnesia for words or numbers or names, stammering or stuttering speech.

[a] From Miller, 1972.

TABLE III
Syndromes Sometimes Relieved by Food Injection Therapy[a]

A. Headache
   1. Migraine
   2. Vascular
   3. Histamine
   4. Tension
   5. "Emotional"
   6. Muscle spasm
B. Ophthalmic
   1. Eye pain
   2. Photophobia
   3. Episodic blurring of vision
   4. Transistory refractive changes
   5. Tearing
C. Otologic
   1. Serous otitis
   2. Tinnitus
   3. Meniere's syndrome
   4. Hearing loss
   5. Vertigo
D. Respiratory
   1. Laryngeal edema
   2. Asthma
   3. Postnasal discharge
   4. Allergic tracheitis
   5. Allergic laryngitis
   6. Allergic rhinitis
E. Cardiovascular
   1. Extrasystoles
   2. Tachycardia
   3. Palpitation
   4. Episodic syncope
   5. Generalized angio-edema
   6. Angio-edema of lungs, liver, etc.
   7. Flushing, chilling
F. Gastrointestinal
   1. Cheilitis
   2. Aphthous stomatitis
   3. Aerophagia
   4. Nausea
   5. Vomiting
   6. Heartburn
   7. Indigestion
   8. Gassiness
   9. Abdominal pain

(continued)

Table III
(Continued)

10. Cramps
11. Diarrhea
12. Pruritus ani
13. Irritable bowel
14. Spastic colon
15. Mucous colitis
16. Nervous stomach
17. Food intolerances

G. Dermatologic
1. Urticaria
2. Atopic dermatitis
3. Neurodermatitis
4. Adult acne
5. Erythema multiforme
6. Skin lesions of porphyria
7. Hand dermatitis
8. Nondescript syndromes

H. Muscular
1. Muscle spasm
2. Muscle pain
3. Muscle cramps
4. Muscle weakness
5. Nuchal pain or rigidity
6. Undue fatigue
7. Sluggishness

I. Cerebral Depression
1. Acute and chronic depression
2. Drowsiness approaching narcolepsy
3. Episodic dullness or dreaminess
4. Learning disorders
5. Tension-fatigue syndrome
6. Minimal brain dysfunction

J. Cerebral Stimulation
1. Restlessness
2. Nervousness
3. Jitteriness
4. Insomnia
5. Hyperactivity
6. Behavior problems

K. Psychiatric
1. Feelings of apartness
2. Floating sensation
3. Episodic amnesia
4. Pathologically poor memory
5. Inability to concentrate
6. Personality changes

L. Urological
1. Frequency
2. Dysuria
3. Nocturia
4. Enuresis

M. Hematological
1. Anemia
2. Neutropenia
3. Purpura

[a] From Miller, 1972.

molecules and retains its antigenicity. These molecules circulate through the body and are excreted into breast milk as well as urine. It has been demonstrated that food antigen, antibody, and antigen–antibody complexes are involved in inducing allergic symptoms in breast-fed infants (Berger and Bürgin-Wolff, 1962; Matsumura, 1969).

As food allergy can affect all the tissues and organs, many diseases and symptoms are now being reported to be due to allergies. Interestingly, the brain is not an exception; behavioral disorders such as hyperkinesis and learning disabilities in children, as well as schizophrenia in adults, have now begun to be investigated from a nutritional viewpoint (Randolph, 1962; Rowe and Rowe, Jr., 1972; Speer, 1978; Crook, 1980, and others).

TYPES OF ALLERGY

There are two classifications. The first is the obvious type (fixed, manifest, acute), and the second is the hidden type (masked, latent, chronic). The former occurs when the patient is highly sensitized to food. Symptoms are prompt and violent, and urticaria, wheezing, asthma, abdominal pain, vomiting, diarrhea, and coma are most common. This type is characterized by onset within two hours after consumption (immediate onset type), provocation by relatively uncommon foods such as seafood, and simple detection. The hidden type is manifested by mild and varied symptoms. Special characteristics are delayed onset 2 hr to 2 days and later after consumption (delayed onset type), and induction by foods ingested practically every day, including milk, corn, wheat, legumes, citrus, eggs, and sugar in the U.S.; and milk, eggs, soybeans, pork, and seafood in Japan. This type of allergy is rarely recognized by the patient. Worthy to note is that the hidden type occurs about a hundred times more often than the obvious type. In other words, the major part of food allergy remains hidden, just like an iceberg under the sea, unrecognized by patients and even most nonspecialized physicians (Crook, 1978).

IMMUNOLOGY

As described above, food allergy is supposed to be mostly caused by immunologic mechanisms. However, this assumption is rarely verified, and is indeed the fatal defect in food allergy research. Fortunately, over the past several years, interest in food allergy has rapidly increased, and this has somewhat relieved that drawback.

It has been shown conclusively that the obvious type is mediated by specific IgE (immunoglobulin E) and specific IgG_3 action against food allergens. Now, immunological mechanisms responsible for the hidden type, such as inhibitory or provocatory activities of immunoglobulins, including IgA, IgG, IgM, and IgD, as well as immune bodies and immunoregulants such as lymphocytes (T-cell and B-cell), mast cells, basophiles, eosinophiles, and complements are being seriously studied (Bahna, 1978, and others).

Foods, quite simply, are usually not absorbed through the digestive tract before they are broken down into small molecules—for example, proteins into amino acids. However, small quantities are absorbed as large molecules. This means that foods are foreign bodies that can act immunologically when ingested as antigens, often causing antigen–antibody reactions. In this way, foods not only provide nutrition, but are also responsible for the adverse effects of antigen–antibody reactions.

Fairly recently, an exogeny hypothesis has gained acceptance (Randolph, 1976). Also, heredity and the environment seem to determine the health and kind and severity of disease in each individual. Among the elements composing the environment, the most dreadful are microorganisms (infectants), followed by ingestants, inhalants, contactants, and, at times, injectants. The amount of ingestants absorbed through the digestive tract is assumed to be considerably larger than that of inhalants absorbed through the respiratory tract, and skin contactants, which suggests the extreme importance of ingestants in allergy.

Now, in order to minimize the allergic reaction that may be induced by foods every day, the following inhibitory factors must be more adequately understood: first, those involved in immune mechanisms such as IgA; second, those in the endocrine system such as steroid hormone and epinephrine; and third, those in the autonomic nervous system such as sympathetic nerves. Allergic reactions are initially characterized by local symptomatic changes, which seems, teleologically, to be an attempt to save the whole body by sacrificing a part.

FOOD ALLERGENS

Generally, foods high in protein and consistently ingested, often from infancy, tend to cause food allergy. Such allergenic foods vary somewhat among races, depending on different eating customs, although similar trends are observed. In the U.S., ten common food allergens are mentioned by Speer (1978). In patients with food allergy, common food allergens were milk (65% of total incidence), chocolate and cola (45%), corn (30%), legumes (peanuts, soybeans, common beans, peas, etc.) (26%), eggs (26%), citrus (25%), tomatoes (16%), wheat and rice (11%), pork (10%), and cinnamon (8%), followed by fish, potatoes, onions, apples, beef, crustaceans (especially shrimp), artificial food colorings, black pepper, and bananas. Other food allergens reported were mustard, alcohol, berries, buckwheat, cane sugar, coffee, coconut, and yeast.

In Japan, three major food allergens are milk, eggs, and soybeans, followed by pork, and fish such as mackerel and tuna. Rice is the staple food and is consumed in large quantities. It is one of the common allergens in Japan, just as wheat is in the U.S. (Matsumura 1969, 1973a).

Racial differences in food allergy are most markedly observed with milk. The incidence of late onset lactose intolerance (developmental lactase deficiency), which is a lactase deficiency in the brush border of the intestinal cells, is much higher in Blacks and Orientals, who have enjoyed milk for only one or two generations, than Caucasians who have been drinking milk since before Christ. This corresponds well to the patterns of incidence of milk allergy. The reason is not clear. It is common knowledge that man should differentiate lactose intolerance and milk allergy. However, from the positive results of provocation tests with lactose-free casein or lactose-free milk in six patients with late onset lactose intolerance, this writer holds the opinion that this illness may be the result of allergy to milk protein and that the demonstration of lactose intolerance does not exclude, but rather suggests, the copresence of milk protein allergy (Matsumura *et al.*, 1971a).

TRIGGER MECHANISMS

In inducing allergic reactions, i.e., altered reactions to foreign bodies, two factors are involved: inhibitory and stimulatory. Taking asthma as an example, there appear to be various stimuli that together produce the disease entity. A patient with asthma can be figuratively thought of as a loaded gun with numerous stimuli pulling the trigger and causing the gun to fire. Some trigger mechanisms that may result in asthmatic attack are known to be emotional stress, such as tension and anxiety, fatigue, climatic change (especially exposure to cold), and endocrine factors, such as menstruation.

DIAGNOSIS

Food allergy may be suspected in the following cases:

1. When symptoms and diseases described by Rowe are observed.
2. When symptoms and syndromes described by Miller persist or reappear repeatedly.
3. When stress diseases and psychosomatic diseases cannot be cured.
4. When autonomic dysfunction is not cured.
5. When menstruation difficulties and climacterium difficulties are not cured.
6. When diseases and symptoms appear or get worse in cold seasons, and in air-conditioned rooms.
7. When diseases and symptoms appear or become aggravated upon infection, such as the common cold.

Incidentally, five of the items (3–7) are also manifestations of diseases caused by infectious agents, e.g., those leading to focal infections.

Diagnosis of food allergy depends predominately on evident amelioration or disappearance of allergic manifestations on elimination of suspected food allergens and obvious aggravation or reappearance of manifestations on reintroduction of that food into the diet. In this way, allergens are clinically elucidated. This is not an easy technique to employ, however, because patients are sensitized to varying degrees to multiple food allergens and demonstrate diversified manifestations.

Rowe proposed his elimination diets omitting foods that most frequently cause allergy. Such foods as milk, wheat, eggs, fish, nuts, chocolate, coffee, condiments, and a few vegetables and fruits were eliminated during treatment. Rinkel discovered from his own experience that a hidden type of food allergy can be made obvious by avoiding specific foods for 4–11 days. Based on this discovery, suspected foods in all forms are now eliminated for 4 days and then upon trial ingestion, immediately occurring allergic symptoms are carefully observed for about an hour to detect food allergens (individual deliberate feeding test) (Rinkel *et al.*, 1951).

In vivo and *in vitro* methods are also used in the detection of food allergens as supportive measures. Among the latter, the skin test is most important. The prick or scratch test, using a droplet of food allergen extract, along with specific IgE tests of serum, is helpful in detecting food allergens that cause the obvious type of food allergy. However, because of its fallibility in the case of the hidden type of food allergy, the skin test was abandoned for a long period. While using various dilutions of food extract intradermally to examine a single food at a time, Lee discovered that food allergy symptoms are produced with certain dilutions and relieved with others, described by Dickey (1976). Miller (1972) refined this technique and described it in detail for general use. This Lee–Miller provocative-neutralizing method seems to have paved a new way to food allergy diagnosis and treatment and is slowly gaining acceptance. Interesting and suggestive is the fact that Miller extended this method to relieve patients with influenza and herpes. Sublingual antigen testing and therapy methods for food allergy developed by Hansel, Pfeiffer, and Dickey are also used. These methods employ sublingual absorption as a pathway to general circulation comparable to intravenous injection.

Other *in vivo* tests include passive transfer, accelerated pulse rate, leukopenia, mast cells in nasal secretion, skin window eosinophilia, rectal provocation, and intestinal biopsy in humans, and passive cutaneous anaphylaxis (PCA) in guinea pigs. *In vitro* studies include precipitation (gel diffusion), hemagglutination, complement fixation, cytotoxic test, Farr's technique, leukocyte histamine release, and leukocyte transformation. Although many methods have been developed, there is still no test that clearly and conclusively detects food allergens of the hidden type (Bahna 1978 and others).

Another method of diagnosing this type of allergy is the lordotic test. Orthostatic albuminuria (postural proteinuria) is a proteinuria which appears after standing or sitting and disappears after lying down. The diagnosis of orthostatic albuminuria is usually made by the lordotic test. This contributor showed that orthostatic albuminuria is caused by food allergy, and that the lordotic test is applicable for the determination of food allergens (Matsumura *et al.,* 1966, Matsumura, 1976).

SUSCEPTIBILITY TO CHEMICALLY CONTAMINATED FOOD

In recent years, as a result of industrial growth and the advent of agrochemical farming, chemicals have begun to rapidly and seriously pollute the environment. Unexpectedly, this has presented a new, intricate problem, as these chemicals are difficult for humans to metabolize and adapt to. It was found by Randolph that chemicals give rise not only to toxic symptoms, but usually manifestations that do not differ from those of allergy. These compounds are found in foods contaminated by pesticide residues, preservatives, colorings, flavorings, and sweetners, in addition to being ingredients of indoor and outdoor air pollution. In this way, a new problem emerged, that of differentiating symptoms due to food *per se* and those caused by chemicals contaminating foods (Randolph, 1962).

In order to diagnose and treat a patient susceptible to chemicals and all environmental elements (including food itself), Randolph, in 1956, developed a comprehensive environmental control program in an ecological unit in which all chemicals are avoided. Chemically less contaminated, organically grown foods including organic meats, and contaminated foods from commercial sources were administered as in the individual deliberate test after complete elimination of foods by fasting for 4 days (Randolph, 1976b).

THERAPY

The essential method of treating patients with food allergy is to continue avoiding allergenic foods. Allergic symptoms will disappear and the patient will gradually develop a tolerance for these foods. The time required to develop such tolerance depends on the grade of sensitization to foods and the age of the patient. In general, at least 10 years is required for adults and 3 years for children sensitized from infancy. When a patient becomes considerably less sensitive to a food after years of avoidance, it is advisable to begin consumption in small amounts and at gradually shorter intervals of 10, 7, and 5 days. This prevents resensitization. Incidentally, the Lee–Miller provocative-neutralizing method has shown us that a patient can develop tolerance earlier with controlled injection of a neutralizing dose of allergen extract, in spite of the continued consumption of allergenic foods. Recently, good substitutes for milk have become available in Japan (*MA-1,* made by Moringa Milk Industry, Minatoku, Tokyo) and Europe (*Berodiät,* made by Böhringer, Ingelheim, West Germany) for infants and patients with milk allergy. Their main contents are milk casein hydrolysate, sugars other than lactose, and vegetable oils rather than butter.

There is no drug that effectively and safely relieves food allergy symptoms. Antihistamines have some effect against symptoms of the obvious type. Corticosteroids work well, sometimes dramatically in both types. However, the side effects do not allow its continued use. Tranquilizers are somewhat effective in mild cases, which some researchers have taken as an indication that food allergy is a psychosomatic illness. This is not totally true, however. Many of these researchers have ignored the antiallergic effect of tranquilizers, which indicates that food allergy is actually a somatopsychic disease (Tajima *et al.,* 1970).

Another, more fundamental medication has also been developed: disodium cromoglycate (DSCG). This drug has been used with good results since 1977. It is inhaled for the prophylaxis of asthma attacks and atopic rhinitis and blocks the release of mediators from mast cells upon antigen–antibody interaction. It was shown recently that this drug also prevents food allergy, not only the obvious type, but also the hidden type, when given orally. DSCG is expected to become the drug of choice in food allergy treatment (Nizami *et al.,* 1977; Gerrard, 1979, and others).

PROPHYLAXIS

Food allergens are not only excreted into urine and breast milk, but also transfer to the fetus, suggesting the possibility of intrauterine (congenital, pre-natal) sensitization to foods. As to intrauterine sensitization, Ratner (1927) demonstrated such sensitization in animals, and this writer (Matsumura, 1969; Matsumura *et al.*, 1974/1975; Matsumura *et al.*, 1975) has observed this phenomenon in humans. Intrauterine sensitization to foods seems to have a significant role in the prophylaxis of food allergy.

Prophylaxis is the best rule in the treatment of all diseases, but this is especially true with food allergy, in which the methods of treatment remain primitive. The risk of sensitization is greatest during fetal life, and only gradually decreases during the child and adult stages of life. *Shizen Ikuji Ho,* which translates to *The Natural Way or Rearing Children,* a thesis by this author (Matsumura, 1973b, 1979) in which the prophylaxis of allergy including food allergy, is discussed, contains the following suggestions:

1. Make it a rule to breastfeed the baby.
2. During the period of breastfeeding, detect and avoid allergenic foods that give rise to allergic symptoms in the foods that are consumed by the mother.
3. During the weaning period, avoid allergenic foods that acted indirectly to cause allergic symptoms in the nursing period and directly incite allergic symptoms thereafter.
4. During childhood, give food always with a mind to prevent food allergy, and watch for symptoms of food allergy.
5. Eliminate allergenic foods from the mother's diet that caused food allergy in the sibling, during the next pregnancy.

A supplementary explanation follows. Food allergens that induce allergic symptoms in nursing infants also cause allergic symptoms in mothers, though always milder in degree. Accordingly, almost no mothers are aware that they suffer from food allergy. However, following elimination of food allergens afflicting their babies, mothers feel better, relieved from chronic illnesses such as constipation, stuffy noses, coughing, wheezing, headaches, drowsiness, and fatigue.

In this way, mothers can recover from associated discomfort, the cause of which can be accurately detected by carefully watching for allergic symptoms in their babies and eliminating food allergens from their diet. In this particular sense, the mother and child are one flesh. This can be easily understood by considering the intrauterine sensitization to food that basically determines postnatal food allergy, including milk allergy in bottle-fed infants and food allergy in breastfed babies.

Further information on food allergy prophylaxis can be found in the work of Glaser and Dreyfuss (1979).

REFERENCES

Bahna, S. L., 1978, Control of milk allergy: A challenge for physicians, mothers and industry, *Ann. Allergy* **41**:1.

Berger, E., und Bürgin-Wolff, A., 1962, Der übertritt von antikörpern gegen nahrungsmittel-antigene in die frauenmilch, *Experientia* **18:**429.

Crook, W. G., 1975, Food allergy, the great masquerader, *Pediatr. Clin. North Am.* **22:**227.

Crook, W. G., 1978, *Are You Allergic?* Professional Books, Jackson, Tennessee.

Crook, W. G., 1980. Can what a child eats make him dull, stupid, or hyperactive? *J. Learn. Disabil.* **13:**281.

Dickey, L. D., 1976, History and documentation of coseasonal antigen therapy, intracutaneous serial dilution titration, optimal dosage, and provocative testing, in: *Clinical Ecology* (L. D. Dickey, ed.), pp. 18–25, Charles C. Thomas, Springfield, Illinois.

Gerrard, J. W., 1979, Oral cromoglycate: Its value in the treatment of adverse reactions to foods, *Ann. Allergy* **42:**135.

Glaser, J., and Dreyfuss, E. M., 1979, Prenatal and postnatal prophylaxis of allergic disease, in: *Brennemann's Practice of Pediatrics* (V. C. Kelley, ed.), Chapter 72, Harper and Row, Maryland.

Matsumura, T., 1969, Clinic on food allergy, *Jpn. J. Allergy* **18:**169 (in Japanese with English abstract).

Matsumura, T., 1973a, Pediatric allergy in Japan, in: *Allergy and Immunology in Childhood* (F. Speer and R. J. Dockhorn, eds.), pp. 718–721, Charles C. Thomas, Springfield, Illinois.

Matsumura, T., 1973b, Shizen ikuji ho, *Jpn. Med. J.* **(2941):**7 (in Japanese).

Matsumura, T., 1976, Postural proteinuria, in: *Clinical Ecology* (L. D. Dickey, ed.), pp. 233–243, Charles C. Thomas, Springfield, Illinois.

Matsumura, T., 1979, Allergies and breastfeeding, *La Leche League News* **21:**87.

Matsumura, T., Kuroume, T., Matsui, A., Fukazawa, T., Kimura, S., and Tamura, H., 1971b, Therapy of the nephrotic syndrome by eradication of foci and elimination diets, *Separatum of the 13th International Congress of Pediatrics,* pp. 5–13, 41–56, Verlog der Wiener Medizinischen Akaedemie, Wien, Österreich.

Matsumura, T., Kuroume, T., Oguri, M., Kobayashi, K., Kanbe, Y., Yamada, T., and Matsumoto, S., 1974/1975. Demonstration of haemagglutinating antibody against food antigens in the amniotic fluid and cord blood, *Alleg. Immunol.* **20/21:**373.

Matsumura, T., Kuroume, T., Oguri, M., Iwasaki, I., Kanbe, Y., Yamada, T., Kawabe, S., and Negishi, K., 1975, Egg sensitivity and eczematous manifestations in breast-fed newborns with particular reference to intrauterine sensitization, *Ann. Allergy* **35:**221.

Miller, J. B., 1972, *Food Allergy: Provocative Testing and Injection Therapy,* Charles C. Thomas, Springfield, Illinois.

Nizami, R. M., Lewin, P. K., and Baboo, M. T., 1977, Oral cromolyn therapy in patients with food allergy: A preliminary report, *Ann. Allergy* **42:**135.

Randolph, T. G., 1962, *Human Ecology and Susceptibility to the Chemical Environment,* Charles C. Thomas, Springfield, Illinois.

Randolph, T. G., 1976a, Historical development of clinical ecology, in: *Clinical Ecology* (L. D. Dickey, ed.), pp. 9–17, Charles C. Thomas, Springfield, Illinois.

Randolph, T. G., 1976b, Hospital comprehensive environmental control program, in: *Clinical Ecology* (L. D. Dickey, ed.), pp. 70–85, Charles C. Thomas, Springfield, Illinois.

Ratner, B., Holmes, C., Jackson, X., and Gruehl, H. L., 1927, Transmission of protein hypersensitiveness from mother to offspring; active sensitization in utero, *J. Immunol.* **14:**303.

Rinkel, H. J., Randolph, T. G., and Zeller, M., 1951, *Food Allergy,* Charles C. Thomas, Springfield, Illinois.

Rowe, A. H., and Rowe, A., Jr., 1972, *Food Allergy: Its Manifestations and Control and the Elimination Diets,* Charles C. Thomas, Springfield, Illinois.

Speer, F., 1978, *Food Allergy,* PSG Publishing, Littleton, Massachusetts.

Tajima, S., Kuroume, T., and Matsumura, T., 1970, Antiallergic action of psychotropic drugs as observed in passive cutaneous anaphylaxis, *Jpn. J. Allergy* **19:**777.

VIII

FOOD ACCIDENTS

30

FOOD ACCIDENTS

Faith W. Winter and Roy E. Brown

INTRODUCTION

Among the many conditions related to the adverse effects of food that may fall into the category of food accidents, this chapter will discuss only those physical conditions that result from mechanical obstruction. Accidental poisonings, environmental contamination, and both excesses and deficiencies occurring accidentally will not be addressed.

Under consideration will be accidents that occur from the ingestion of foods, first causing problems in the gastrointestinal tract and then causing difficulties in the respiratory tract. The various conditions will be considered from the aspect of pathogenesis, diagnosis, and treatment, and finally, with consideration given to prevention.

The most likely victims of food accidents are the very young and the elderly, and unlike certain other conditions, the most conservative approach to therapy is early intervention rather than watchful waiting. When obstruction of the airway occurs there is usually no time for anything beyond the emergency application of the "Heimlich maneuver" by the person in the immediate area of the victim, details of which will be presented. Although each form of intervention carries with it a certain potential for morbidity, or even mortality, such rates of procedural complication must be balanced against the problems associated with delayed intervention.

INGESTION OF FOREIGN BODIES

In the course of consuming ordinary food, individuals may inadvertently swallow nondigestible portions, which may then be treated as foreign bodies by the gastrointestinal tract. Such commonly ingested foreign bodies associ-

---

Faith W. Winter and Roy E. Brown • Mount Sinai School of Medicine, New York, New York 10029.

ated with food include fish, chicken, and meat bones; poorly chewed food masses; and fruit pits.

In general, the most likely victim of foreign body ingestion is the denture-wearing adult. Approximately 85% of ingested foreign bodies are reported in edentulous patients (Berk and Reit, 1971; Bunker, 1962; Maleki and Evans, 1970). Dentures interfere with the tactile sensitivity of the palate and prevent the recognition of sharp or hard objects in ingested food. The problem may be compounded in the elderly, who may be less careful in the selection, preparation, and mastication of foods. Excessive alcohol ingestion will tend to decrease sensory acuity of the oral cavity, predisposing inebriated individuals to foreign body ingestion (Maleki and Evans, 1970; Roseman, 1978). This has been designated "the martini–olive pit" syndrome. Young children, often through parental carelessness or haste, and mental patients are other categories of likely victims.

Between 80 and 90% of swallowed foreign bodies pass through the gastrointestinal tract spontaneously, without causing significant damage or injury (Eldridge, 1961; Maleki and Evans, 1970; Schwartz and Polsky, 1976). The remaining 10–20% of foreign bodies, by failing to pass through the entire gastrointestinal tract, may result in serious complications which most notably include obstruction, hemorrhage, abscess and perforation. Bones are particularly associated with a very high incidence of perforation.

Physiologic sphincters and areas of acute angulation are the most susceptible to injury, although sites in all areas of the intestinal tract have been reported in the literature (Maleki and Evans, 1970; MacManus, 1941; Roseman, 1978). Specifically, ingested foreign bodies most frequently cause complications in the cricoharyngeus, pylorus, and ligament of Treitz. The ileocecal region has been documented as the most common site of perforation, perhaps associated with a decrease in bowel motility at this junction. Approximately three-quarters of all perforations occur at this site (MacManus, 1941; Schwartz and Polsky, 1976). In patients with a history of preceding gastrointestinal lesions and disease, the affected sites are more prone to injury by foreign bodies. Common complicating lesions include strictures, rings, and webs of the esophagus, stenosis of the pylorus, diverticuli, and tumors (Kassner *et al.*, 1975; Roseman, 1978).

Foreign body ingestion may manifest itself clinically by abdominal pain, nausea, vomiting, abdominal distention, bleeding, drooling associated with an overflow of saliva, and elevated temperature. Laboratory findings may include a leucocytosis with a left shift of immature polymorphonucleocytes, indicating a focal infection, or iron deficiency anemia associated with prolonged blood loss with chronically impacted objects. Diagnostic problems arise in the patient who has unknowingly ingested a foreign body, since the symptoms initially are nonspecific and may mimic several more common abdominal conditions. Initial misdiagnosis is very likely and the differential diagnosis may include appendicitis, peritonitis, acute cholecystitis, pancreatitis, acute diverticulitis and malignancy (Maleki and Evans, 1970; McPherson *et al.*, 1957).

Since the majority of ingested foreign bodies will pass through the digestive system without incident, the initial management tends to be conservative, watching and waiting (Roseman, 1978; Schwartz and Polsky, 1976). X-ray films may prove valuable in following the object's progress through the gastrointestinal tract and the patient should be closely observed for signs of obstruction, perforation, or hemorrhage. If the foreign body is not eliminated after 72 hours, removal is indicated.

A more aggressive approach is required in the 10–20% of instances where the swallowed object becomes lodged in the intestinal tract. Perforation, the most critical complication of foreign body ingestion, occurs in less than 1% of incidents (Moreira *et al.*, 1975; McPherson, 1957). Surgery is therefore rarely necessary.

Most foreign bodies can be located and treated endoscopically (Schwartz and Polsky, 1976). Removal of foreign bodies, in fact, was one of the primary concerns during the development of endoscopy. The original straight endoscope, however, with its large forceps and biopsy channel, had itself been a significant cause of perforation with associated morbidity and mortality (Katz, 1969; McCaffery and Lilly, 1975). The removal of foreign bodies was questioned as a primary use of endoscopy. Further development in the instrumentation has brought the endoscope back into the forefront as a treatment modality for removal of foreign bodies in the gastrointestinal tract. The flexible fiberoptic endoscope has been used successfully to extract a wide variety of foreign bodies (Dunkerley *et al.*, 1975; Roseman, 1978; Schwartz and Polsky, 1976; VanThiel and Stefan, 1975), and in skilled hands has proved generally acceptable. Its use has been limited by the small diameter of its biopsy channel and small caliber of forceps, which make removal of particularly large objects difficult. An alternative, in the case of a large food bolus, has been to tear the object piecemeal and let it pass distally (Dunkerley *et al.*, 1975). However, this is not amenable for application with sharp, irregular objects such as bones (Olsen *et al.*, 1975). For some of these cases, an umbilical tape may be employed along with the fiberendoscope, if the object has an opening through which the umbilical tape can be threaded (Dunkerley *et al.*, 1975). Another technique developed to adapt the flexible endoscope to removal of large foreign bodies has been the polypetomy snare. It can secure large objects, grasping and removing them under direct vision with the fiberendoscope (McCaffery and Lilly, 1975; Roseman, 1978). Due to the refinements in endoscopic techniques, surgery is currently rarely indicated in the retrieval of foreign bodies from the gastrointestinal tract. Signs of perforation, small bowel obstruction, or bleeding remain indications for immediate surgical intervention, as is the persistent delay of an object in a segment of the intestine for more than 72 hr (Roseman, 1978).

BEZOAkS

Bezoars are defined as insoluable aggregates of ingested materials which form in the gastrointestinal tract (DeGerome, 1973). The most commonly seen food-related bezoar is the lactobezoar, which is a protein coagulum, usually found in an infant's stomach as a result of mixing powdered milk formula with inadequate amounts of water. The overly concentrated formula may flow through the feeding nipple, but becomes inspissated when it reaches the stomach (Wolf and Bruce, 1959). Upon qualitative analysis the lactobezoar coagulum has been found to consist of calcium paracasinate (Levkoff *et al.*, 1970).

The pathogenesis of lactobezoar in infancy involves a vicious cycle as follows: a highly concentrated and incorrectly prepared milk formula precipitates diarrhea and vomiting, symptoms which lead to dehydration; in the body's efforts to replenish fluid volume, water is absorbed from the gastrointestinal tract, including the stomach; a lactobezoar forms and by mechanical obstruction causes the persistence of vomiting and dehydration.

Clinical manifestations of lactobezoar include the aforementioned diarrhea, vomiting, dehydration and elevated temperature (Wolf and Davis, 1963). There may be abdominal distension and often an abdominal mass is palpable (Levkoff *et al.*, 1970; Wolf and Davis, 1963). Radiographic examination usually reveals a distinctive mass in the stomach, characteristic of a lactobezoar. In addition to the presenting symptoms and X-ray, the work-up should include a careful feeding history as well as details related to the preparation of the formula (Majd and LoPresti, 1972).

Conservative therapy for lactobezoar initially consists of intravenous fluids for management of the rehydration. Rehydration alone can result in spontaneous resolution in approximately 60% of the patients (Majd and LoPresti, 1972). Surgery may be indicated in patients with severe obstruction or evidence of perforation. In addition, management must include careful instructions and reviews of the correct formula preparation with the parents. Instances of lactobezoar have been reported from England where the cited cause is the consumption of Jersey milk, which has far greater concentrations of calcium, protein, fat, and total solids than either human or other cows' milk (Bennett and Herman, 1975; Davis, 1975).

Phytobezoars, or bezoars of plant origin, are less commonly reported than lactobezoars. The composition may reveal a variety of plant materials including fibers, skin, seeds, leaves, roots, and stems, often moulded together to form a compact, insoluable mass (DeBakey and Ochsner, 1938). Phytobezoars tend to be small, but abrasive and are known to cause trauma frequently to the intestinal epithelium (DeGerome, 1973). Formation of a phytobezoar requires a particular combination of gastrointestinal stasis and ingestion of appropriately nondigestible plant material (McNamara and Paulson, 1967). Many reported phytobezoars are due to the ingestion of unripe and unpeeled persimmons (DeGerome, 1973).

Persimmons are berries containing as many as eight seeds and a nonfibrous pulp. Shiboul, a phlobatannin including phloroglucin and gallic acid is found in unripe persimmons, causing their astringent taste. When the fruit ripens, shiboul coagulates, rendering a sweet taste. Exposure of phloroglucin and gallic acid to gastric juices produces an insoluable coagulum. Shiboul remains in the ripe fruit, just under the skin. Thus a phytobezoar may be formed when unripe or unpeeled persimmons are eaten on an empty stomach or intestinal stasis is concurrently present (DeGerome, 1973).

Other vegetable materials reported to cause phytobezoars include coconut, celery, pumpkin, heather roots and twigs, grape skins and stems, prunes, and raisins (DeBakey and Ochsner, 1938). Phytobezoar is clinically manifested by a freely moving abdominal mass, generally located in the epigastrium and readily displaced beneath the left coastal margin. The patient may experience some pain or tenderness, nausea, vomiting, moderate weakness or weight loss. There is either constipation or diarrhea present. Fever is absent, but there may be a secondary anemia or mild leukocytosis (DeBakey and Ochsner, 1938).

Diagnosis is usually made radiographically, but should include a careful history as to recent eating habits. Possible complications common to phytobezoars include intestinal obstruction, gastroduodenal ulceration, perforation of ulcer, and peritonitis (DeBakey and Ochsner, 1938). The treatment is essentially surgical, although esophageal phytobezoars have been removed endoscopically (DeBakey and Ochsner, 1938; DeGerome, 1973).

ASPIRATION OF FOREIGN BODIES INTO RESPIRATORY TRACT

Aspiration of foreign bodies into the respiratory tract is a serious and potentially fatal occurrence, particularly in young children and infants. Although relatively rare in adults, aspiration represents a leading cause of mortality in children under 6 years of age. The National Safety Council of America has designated foreign body aspiration as the foremost cause of accidental death in the home for children in this age group (Aytac *et al.*, 1977).

Aspiration occurs most frequently in children between 1 and 3 years of age (Daniilidis *et al.*, 1977). The etiology of this event is attributed to the inability of toddlers to chew their food completely, associated with the tendency to put a wide variety of objects into their mouths. Parental neglect may be a contributing factor, either due to improper food preparation or selection, or the lack of supervision during mealtime. Foreign body aspiration often occurs when children have been playing, shouting, or crying while eating (Daniilidis *et al.*, 1977).

Nearly 93% of aspirated foreign bodies are organic in nature, with nuts and seeds predominating (Blazer *et al.*, 1980; Kim *et al.*, 1973). This fact is particularly significant because objects of vegetable origin may swell within the respiratory tract due to secretions, further complicating an already dangerous situation. The majority of foreign bodies are aspirated into the mainstem bronchus, with the remainder into the larynx and trachea (Blazer *et al.*, 1980; Daniilidis *et al.*, 1977).

Early diagnosis and removal of the foreign body is essential in order to prevent chronic complications and fatal outcomes. Diagnosis, however, may be difficult, since symptoms tend to mimic infections or allergies. The classical triad of symptoms, coughing, wheezing and decreased air entry, are commonly associated with many disorders of the respiratory tract (Blazer *et al.,* 1980; Kim *et al.,* 1973). A high index of suspicion should be present to consider foreign-body aspiration in any young child who presents these clinical signs, particularly during seasons when respiratory tract infections or allergic phenomena are uncommon (Kim *et al.,* 1973).

A complete diagnostic workup includes a careful history and physical examination, and possible radiographic evaluation. A comprehensive history of recent events, especially those that pertain to food and eating, is essential since as many as 85% of cases studied report a positive history of foreign body inhalation (Blazer *et al.,* 1980; Kim *et al.,* 1973; Meade, 1962). Adequate inspiratory and expiratory X-rays, appropriately supplemented by fluoroscopy when indicated, will provide an adequate diagnosis in over 95% of lower-airway foreign-body aspirations (Meade, 1962). In young children, however, such radiographic studies are often difficult to obtain. In the presence of negative X-ray findings accompanying a positive history and clinical picture, foreign-body aspiration should be strongly suspected in the differential diagnosis.

The clinical course generally varies depending on age, characteristics of the particular foreign body, and the duration of its stay in the respiratory tract (Banks and Potsic, 1977). There might be a lag time between the initial symptoms and therapy. Delays of up to 2 years have been reported, although 1–4 months are more commonly found (Banks and Potsic, 1972; Daniilidis *et al.,* 1977; Kim *et al.,* 1973). Factors cited in treatment delay include parental negligence, initial misdiagnosis, lack of symptoms, and the diversity of the clinical picture (Daniilidis *et al.,* 1977).

Prompt and adequate removal of the aspirated foreign body is the foremost aim of treatment. Bronchoscopic removal is the treatment of choice (Aytac *et al.,* 1977) and is successful in up to 98% of some reported series (Blazer *et al.,* 1980). The success of bronchoscopy is related to recent developments in flexible endoscopy and modern anesthesia (Daniilidis *et al.,* 1977). Before bronchoscopy was perfected as a lifesaving procedure, mortality rates for foreign body aspiration were as high as 56% (Aytac *et al.,* 1977). With modern techniques, the mortality figures have fallen to approximately 7% in 1968, and more recent studies show mortality to between 1.8 and 3.8% (Aytac *et al.,* 1977).

Good communication and teamwork between broncoscopist and anesthesiologist, as well as careful preparation of instruments, are essential for lower mortality figures (Blazer *et al.,* 1980). As part of postbronchoscopy therapy, steroids have been suggested by some authorities to avoid the edema of the larynx which often necessitates tracheotomy (Aytac *et al.,* 1977; Daniilidis *et al.,* 1977). Antibiotics may be necessary if accompanying infection is present (Blazer *et al.*, 1980). In addition, the patient should be kept under close observation for clinical signs, since the removal of one foreign body does not preclude the existence of another.

Bronchoscopic removal of a foreign body is usually sufficient to prevent or reverse chronic pulmonary changes. Surgical resection only becomes indicated when obvious and irreversible changes such as bronchiectasis have occurred (Banks and Potsic, 1972).

Waiting for a spontaneous expulsion of the foreign body is never justified. This occurs in as few as 1–2% of the cases, and the ensuing complications of a neglected foreign body are serious and potentially fatal (Aytac *et al.*, 1977). Untreated aspirated foreign bodies may lead to pneumonia or atelectasis and eventually chronic pulmonary problems, including bronchiectasis, lung abscess, or bronchopulmonary fistula (Banks and Postic, 1977). Development of these complications makes bronchoscopic removal of the foreign body more difficult and increases the likelihood of surgical intervention.

Aspiration accidents can be prevented or minimized by educating both the general public and the medical practitioner. Both groups should be aware of clinical manifestations of foreign body aspiration and the age group that is particularly vulnerable. Parents should be encouraged to make certain that food preparations and mealtimes of children under 6 years of age are closely supervised. There are advocates of withholding particularly dangerous foods, such as nuts or seeds, from children in this age group.

CHOKING

Food choking takes the lives of approximately 3900 individuals each year in the U.S. and represents the sixth leading cause of accidental death (Heimlich, 1976). Traditional methods have had little success in overcoming fatal outcome following food choking. Standard practices have included slapping the victim on the back, reaching into his mouth with a hand or instrument, and emergency tracheostomy (Heimlich, 1975, 1976). The food-choking phenomenon is often dubbed "the café coronary," since the untrained observer may mistake the rapid onset of a choking attack for a heart attack. In such an instance, one would not attempt to remove the obstructing piece of food since choking may not even be considered.

In 1974, a revolutionary procedure for overcoming food choking was introduced by Heimlich (Heimlich, 1974). The method was designated the "Heimlich maneuver". Heimlich had substantial experience in treating diseases of the esophagus and had a special concern for food swallowing. With an interest in the tragedy of food choking and in his search for a first aid technique available to laymen, Heimlich applied his knowledge of intrapleural and intrapulmonary pressures from past research (Heimlich, 1968, 1976). Rather than removing the food bolus through the mouth, a method was sought to force the bolus out from below. The basic concept is to compress the lungs and utilize the resultant increased intrapulmonary air pressure to force out a bolus of food blocking the airway.

Animal experiments established that pressure applied to the ribcage was not effective in expelling an obstruction, and could cause fractured ribs, along with internal injury. It was theorized and proven that pressing quickly upward just below the ribcage, thereby elevating the diaphragm, would consistently be successful in dislodging food from the airway and restoring normal respiration (Heimlich, 1976).

In employing the Heimlich maneuver, the first consideration is to make certain that the victim is indeed choking. The cardinal symptom is the victim's inability to speak. This differentiates him from the victim of a myocardial infarction who usually can both breathe and speak (Heimlich, 1976). The food-choking victim, in addition to his inability to speak, cannot breathe, turns first pale and then blue or black, evidences great distress and will die in 4–5 min unless the airway is cleared.

The execution of the Heimlich maneuver if both victim and rescuer are standing is as follows:

1. Stand behind the victim and wrap your arms around his waist.
2. Place your fist, thumb side against the victim's abdomen, slightly above the naval and below the rib cage.
3. Grasp your fist with your other hand and press into the victim's abdomen with a quick upward thrust.
4. Repeat several times if necessary.

If the victim is sitting, the rescuer stands behind the victim's chair and performs the maneuver in the same fashion as above. If the victim is supine, the maneuver is performed as follows:

1. Have the victim lying on his back, with the rescuer facing the victim, kneeling astride his hips.
2. With one of your hands on top of the other, place the heel of your lower hand on the abdomen, slightly above the navel and below the ribcage.
3. Press the victim's abdomen with a quick upward thrust.
4. Repeat several times if necessary.

If the victim is an infant:

1. Hold him in your lap in a sitting position.
2. Place your arms around him and put your index and middle fingers of both hands against the infant's abdomen, above the navel and below the ribcage.
3. Press the abdomen with a quick upward thrust.

An alternative, if the victim is an infant:

1. Place the baby on his back on a firm surface;
2. Face the baby, give a quick upward thrust, using the index and middle fingers of both hands.

With an infant, in either position, it is important to be gentle and remember not to press on the ribs or chest. In all positions, the Heimlich maneuver, through pressing upward into the abdomen, suddenly elevates the diaphragm, thereby compressing the lungs. The basic principle is that tracheobronchial air pressure is increased and suddenly expelled through the trachea, ejecting an obstructing food bolus (Heimlich *et al.*, 1975).

Risks of the Heimlich maneuver are minimal if the above directions are carefully followed. There have been reported instances of the victim being grasped too high and having his ribs fractured. A growing number of cases indicates that the Heimlich maneuver is a simple, effective method for saving the life of someone choking on a bolus of food. Food items reported to have been ejected from the airway of victims include beef, chicken, veal, clams, spaghetti, parts of rolls, popcorn, lettuce, apples, coughdrops, and candy (Heimlich, 1976, 1977).

Initially it was thought that the Heimlich maneuver would be successful only when the airway was totally obstructed, thereby permitting air pressure to build up to the level necessary to expel the bolus. Recent reports, however, establish the fact that even small objects causing partial obstruction of the airway, as evidenced by the victim's stridor, will be expelled (Heimlich *et al.*, 1975; Heimlich, 1976).

In order to avoid food choking altogether, individuals should keep in mind a few simple precautions. Food should always be cut into appropriately small pieces and chewed carefully. Talking, laughing, or moving about with food in one's mouth should be avoided. This is often problematic with small children. Special care must be taken while eating when large amounts of alcoholic beverages are consumed, since alcohol tends to numb the sensory nerves of the palate. The resultant accident has been referred to as the "martini–olive pit" syndrome of choking. Children should always be supervised during mealtimes, and walking or running while eating should never be permitted.

REFERENCES

Anonymous, 1974, Simple method relieves "cafe coronary," (editorial), *J. Am. Med. Assoc.* **229:**746.

Aytac, A., Yurdakul, Y., Ikizler, C., Olga, R., and Saylam, A., 1977, Inhalation of foreign bodies in children: Report of 500 cases, *J. Thorac. Cardiovasc. Surg.* **74:**145.

Banks, W., and Potsic, W. P., 1977, Elusive unsuspected foreign bodies in the tracheobronchial tree, *Clin. Pediatr.* **16:**31.

Bennett, P., and Herman, S., 1975, Curious curd, *Lancet* **1:**1430.
Berk, R. N., and Reit, R. J., 1971, Intra-abdominal chicken-bone abscess. *Diagn. Radiol.* **101:**311.
Blazer, S., Naveh, Y., Friedman, A., 1980, Foreign body in the airway. A review of 200 cases. *Am. J. Dis. Child.* **134:**68.
Bunker, P. G., 1962, The role of dentistry in problems of foreign bodies in the air and food passages, *J. Am. Dent. Assoc.* **64:**782.
Daniilidis, J., Symeonidis, B., Triaridis, K., and Kouloulas, A., 1977, Foreign body in the airways. A review of 90 cases, *Arch. Otolaryngol.* **103:**570.
Davis, J. C., 1975, Curious curd, *Lancet* **2:**276.
DeBakey, M., and Ochsner, A., 1932, Bezoars and concretions, *Surgery* **5:**132.
DeBakey, M., and Ochsner, A., 1938, Bezoars and concretions, *Surgery* **4:**934.
DeGerome, J. H., 1973, Snare extraction of a gastric foreign body, *Gastrointest. Endosc.* **20:**73.
Dunkerley, R. C., Shull, H. J., Avant, G. R., and Dunn, G. D., 1975, Fiberendoscopic removal of large foreign bodies from the stomach, *Gastrointest. Endosc.* **21:**170.
Eldridge, W. W., 1961, Foreign bodies in the gastrointestinal tract, *J. Am. Med. Assoc.* **178:**665.
Heimlich, H. J., 1968, Valve drainage of the pleural cavity, *Dis. Chest* **53:**282.
Heimlich, H. J., 1974, Pop goes the cafe coronary, *Emergency Med.* **6:**154.
Heimlich, H. J., 1975, A life saving maneuver to prevent food choking. *J. Am. Med. Assoc.* **234:**398.
Heimlich, H. J., 1976, Death from food-choking prevented by a new life-saving maneuver, *Heart and Lung* **5:**755.
Heimlich, H. J., 1972, The Heimlich Maneuver: Prevention of Death from Choking on Foreign Bodies, *J. Occup. Med.* **19:**208.
Heimlich, H. J., Hoffmann, K. A., and Canestri, F. R., 1975, Food-choking and drowning deaths prevented by external subdiaphragmatic compression, *Ann. Thorac. Surg.* **20:**188.
Kassner, E. G., Rose, J. S., Kottmeier, P. K., Schneider, M., and Gallow, G. M., 1975, Prevention of small foreign objects in the stomach and duodenum, *Radiology* **114:**683.
Katz, D., 1969, Morbidity and mortality in standard and flexible gastrointestinal endoscopy, *Gastrointest. Endosc.* **15:**134.
Kim, I. G., Brummitt, W. M., Humphry, A., Siomra, S. W., and Wallace, W. B., 1973, Foreign body in the airway: A review of 202 cases, *Laryngoscope* **83:**347.
Levkoff, A. H., Gadsden, R. H., Hennigar, G. R., and Webb, C. M., 1970, Lactobezoar and gastric perforation in a neonate, *J. Pediatr.* **77:**875.
Majd, M., and LoPresti, J. M., 1972, Lactobezoar, *Am. J. Roentgenol. Radium Ther. Nucl. Med.* **116:**575.
Maleki, M., and Evans, W. E., 1970, Foreign-body perforation of the intestinal tract, *Arch. Surg.* **101:**475.
McCaffery, T. D., and Lilly, J. O., 1975, The management of foreign affairs of the GI tract, *Digest. Dis.* **20:**121.
MacManus, J. E., 1941, Perforation of the intestine by ingested foreign body, *Am. J. Surg.* **53:**393.
McNamara, J. J., and Paulson, D. L., 1967, Bezoar of the esophagus, *Ann. Thorac. Surg.* **4:**171.
McPherson, R. C., Karlan, M., and Williams, R. D., 1957, Foreign-body perforation of the intestinal tract. *Am. J. Surg.* **94:**564.
Meade, R. H., 1962, Laryngeal obstruction in children, *Pediatr. Clin. North Am.* **9:**233.
Moreira, C. A., Wongpakdee, S., and Gennaro, A. R., 1975, A foreign body (chicken bone) in the rectum causing extensive perirectal and scrotal abscess: Report of a case, *Dis. Colon. and Rectum.* **18:**407.
Olsen, H., Lawrence, W., and Bernstein, R., 1974, Fiberendoscopic removal of foreign bodies from the upper gastrointestinal tract, *Gastrointest. Endosco.* **21:**58.
Roseman, D. M., 1978, Foreign bodies in the gut and upper airway, in: *Gastrointestinal Disease* (M. H. Sleisenger and J. S. Fordtran, eds.), J. B. Saunders, Philadelphia.
Schwartz, G. F., and Polsky, H. S., 1976, Ingested foreign bodies of the gastrointestinal tract, *Am. Surg.* **42:**236.
Steichen, F. M., Fellini, A., and Einhorn, A. H., 1971, Acute foreign body laryngo-tracheal obstruction: A cause for sudden and unexpected death in children, *Pediatrics* **48:**281.

VanThiel, D. H. V., and Stefan, N. J., 1974, Removal of soft foreign bodies from the esophagus using a flexible instrument, *Gastrointest. Endosc.* **20:**163.

Ward-McWuaid, J. N., 1952, Perforation of the intestine by swallowed foreign bodies, *Br. J. Surg.* **37:**349.

Watson, J. T., 1965, Airway foreign body fatalities in children, *Ann. Otol. Rhinol. Laryngol.* **74:**1144.

Wolf, R. S., and Bruce, J., 1959, Gastrotomy for lactobezoar in a newborn infant, *J. Pediatr.* **54:**811.

Wolf, R. S., and Davis, L. A., 1963, Lactobezoar: A foreign body formed by the use of undiluted powdered milk substance, *J. Am. Med. Assoc.* **184:**782.

IX

EXCESSIVE INTAKES

31

EXCESSIVE NUTRIENT INTAKES

Donald S. McLaren

INTRODUCTION

Food and drink are prepared and consumed as a matter of custom, to satisfy the pangs of hunger and thirst, and as one of the more delightful social activities of man. It is only as an afterthought that the ordinary person pays attention to the nutritional value of food and drink. For the masses in developing countries, knowing where tomorrow's meal is to come from presents a constantly recurring problem. It is only in recent years that those who live in technologically developed societies have been made increasingly aware of the possible hazards to health of excessive consumption. Attention in this regard has mainly been directed toward the major energy sources, fats and carbohydrates, but consideration also needs to be given to some vitamins and elements.

It is by no means easy to decide what is an adequate intake of any nutrient and when it become excessive. Recommended Dietary Allowances (RDAs) (Food and Nutrition Board, 1980) have a limited application and should not be misinterpreted. They are considered to be adequate to meet the needs of all but approximately 2.5% of the healthy population in the United States. Except for energy, these RDAs are equivalent to a supply of nutrients in amounts 30–40% above the average reported in biologic measurements of groups of normal healthy subjects (Alfin-Slater and Mirenda, 1980). The safety margin above the average requirement, varying from nutrient to nutrient, is designed to make allowance for individual variation and fluctuations from day to day. It is of considerable interest in the present context to note that the latest RDA revision (Food and Nutrition Board, 1980) includes ranges of estimated *safe* and adequate intakes for three vitamins (vitamin K, pantothenic acid, and biotin), six trace elements (copper, manganese, fluoride, chromium, selenium, and molybdenum), and three electrolytes (sodium, potassium, and chloride). This trend indicates an increased awareness of the

Donald S. McLaren • Department of Medicine, Royal Infirmary, University of Edinburgh, EH9 3YW Scotland.

possible harmful effects of prolonged, high intake of certain micronutrients. These ranges are intended to serve as guidelines for manufacturers of formulated foods, food analogs, and vitamin and mineral supplements.

Mention will also be made here of the dietary limitations that have to be imposed on patients with certain diseases that impair their ability in one way or another to utilize nutrients in amounts that are normal for the healthy, but are excessive for them.

TOTAL ENERGY

The presence of an excessive amount of adipose tissue, or obesity, can only arise if the total energy intake from food exceeds the total energy expenditure. The control of the balance between the two sides of the energy equation is complex and incompletely understood. In the present context, attention will be confined to consideration of the consequences of excessive intake.

The energy requirements of an individual are related not only to body size but also to activity that is difficult to estimate. The RDAs are intended to meet the average needs of a population group and do not include a margin of safety.

The relationship between obesity and overeating has been investigated on a number of occasions and it has been found in general that there is no difference between the average estimated intake of groups of fat people and thin people of similar age and sex, although at the upper limit of obesity there are often a few people in whose food intake is unusually high (Ries, 1973).

There is strong evidence that most obesity in adult life commences in childhood. Genetic predisposition appears to be less important than familial influences, and an increase in number of adipocytes from overfeeding in infancy as the cause of obesity is now considered to be less well established for man than in experimental animals (Hirsch, 1978). A recent study (Garrow *et al.*, 1980) suggests that predisposition to obesity is more likely to relate to energy intake, perhaps through maternal example, than to energy expenditure. In a group of women ranging from normal to grossly obese, resting metabolic rate was related to the degree of obesity, but the age of onset and family history of obesity had no effect on this relationship.

Large-scale investigations of mortality and weight have been conducted by life insurance companies in the U.S. A summarization of the data from three of these studies, Build and Blood Pressure Studies, 1957 (Society of Actuaries, 1959), Build Study, 1979 (Society of Actuaries and Association of Life Insurance of Medical Directors, unpublished) and American Cancer Society (Lew and Garfinkel, 1979) is presented in Table I. In all these studies, in both male and female, there is a steady increase in mortality from all causes with increasing weight.

The major contributions to this increased mortality are shown to be cor-

TABLE I
Actual to Expected Mortality Ratios According to Variation in Weight

Weight group	Build and Blood Pressure study, 1959[a] %		American Cancer Society study[b] %		Build study, 1979[c] %	
	Male	Female	Male	Female	Male	Female
20% Underweight	95	87	110	100	105	110
10% Underweight	90	89	100	95	94	97
10% Overweight	113	109	107	108	111	107
20% Overweight	125	121	121	123	120	110
30% Overweight	142	130	137	138	135	125
40% Overweight	167	—	162	163	153	136
50% Overweight	200	—	210	—	177	149
60% Overweight	250	—	—	—	210	167

[a] From Society of Actuaries (1959).
[b] From Lew and Garfinkel (1979).
[c] From Society of Actuaries and Association of Life Insurance Medical Directors (unpublished study, 1979).

onary heart disease, stroke, diabetes, and digestive diseases according to data in two of these studies (Table II).

Other studies have suggested an association of obesity with cancer of the endometrium (Blitzer *et al.*, 1976) and cancer of the breast (DeWaard and Baarders-van Halewun, 1969) in both of which increased conversion in adipose tissue of androstenedione to estrone may be a factor. Colorectal carcinoma, a condition especially prevalent among affluent societies, has been related to excess energy intake from a variety of sources (Williamson, 1980).

Almost every system is affected to some extent in obesity, and there is

TABLE II
Mortality Ratios (as a Function of Body Weight)

Cause of death	Men 20% above average weight: Build study, 1979[a] %	Men 20% above average weight: American Cancer Society study[b] %	Men 40% above average weight: Build study, 1979 %	Men 40% above average weight: American Cancer Society study %
Coronary heart disease	118	128	169	175
Cerebral "hemorrhage" (stroke)	110	116	164	191
Cancer	100	105	105	124
Diabetes	250	210	500+	300+
Digestive disease	125	168	220	340
All causes	120	121	150	162

[a] From Society of Actuaries and Association of Life Insurance Medical Directors (unpublished study, 1979).
[b] From Lew and Garfinkel (1979).

psychosocial impairment as well as physical. Van Itallie (1979) has provided a comprehensive summary.

ALCOHOL (ETHANOL)

Ethanol is a high energy source, providing 7 kcal/g. It has been estimated that, on the average, alcohol accounts for 210 kcal/day per person in the U.S., and in studies of heavy drinkers it was found to account for 36% of energy intake in men and 22% in women for sustained periods of intake (Neville *et al.*, 1968).

Chronic alcoholism is known to be associated with a wide range of disorders. These include hematologic abnormalities, cardiomyopathy, peripheral neuropathy, acute and chronic muscle disorders, central nervous system degeneration (particularly of the cerebellum and cortex), chronic pulmonary disease, certain malignancies (particularly of the head and neck), intestinal malabsorption, hypoglycemia, ketosis, hyperuricemia, gout, alterations in drug metabolism, increased susceptibility to infection, and is a major underlying cause of malnutrition. Of primary concern, however, is the role of alcohol in liver disease and in heart disease and conditions associated with it. Attention here will be focused on these.

The epidemiologic evidence linking alcohol consumption with cirrhosis of the liver is extremely strong, although this does not mean that cirrhosis may not occur in abstainers nor that there are not other causes of cirrhosis. In general, cirrhosis is seven times more common in alcoholics than nondrinkers. A striking decline in the occurrence of deaths from cirrhosis has been reported from the U.S., Sweden, and Denmark in relation to alcohol control programs. Cirrhosis tends to occur in 10–20% of heavy drinkers and the risk of developing the disease relates directly and linearly both to the duration and the magnitude of alcohol intake. An intake of 180 g/day for 25 years (4200 liters of 100 proof whisky) resulted in severe liver disease in 20% of the group of German alcoholics studied by Lelbach (1975). Abstinence from alcohol has little impact on the progression of cirrhosis once evidence of portal hypertension is present, but the other two major manifestations of alcohol-related liver disease, fatty liver and alcoholic hepatitis, can reverse completely to normal when alcohol is removed from the diet.

A link between alcohol intake and coronary heart disease has been investigated in a number of studies. Cigarette smoking and other risk factors have confounding influences, but when these are taken account of there is some evidence that alcohol may have a protective effect (Klatsky *et al.*, 1974; Yano *et al.*, 1977). Analysis of data from 18 developed countries showed a negative correlation between wine consumption and coronary heart disease (St. Leger *et al.*, 1979). Greater than average alcohol consumption has been specifically correlated with blood pressure greater than 160/95 mm Hg (Klatsky *et al.*, 1977; D'Alonzo and Pell, 1968). The rise in HDLP reported in alcoholics may not necessarily indicate a protective effect, according to Johnson and Nilsson-Ehle (1978). Alcohol consumption does not appear to be related to obesity, except at very high levels of intake (Klatsky *et al.*, 1974; D'Alonzo and Pell, 1968).

CARBOHYDRATE

Coronary Heart Disease

The effects of a high consumption of carbohydrate-rich foods can be looked at in two ways. Of necessity, a high carbohydrate diet is a diet that is low in fat.. There is epidemiologic evidence that high food energy intake from carbohydrates has a protective effect on coronary heart disease (Keys, 1971). The possible relationships of such a diet to atherosclerosis and coronary heart disease are discussed under Fat (and Chapter 42). The other way of considering high carbohydrate intake is the replacement of starches by naturally occurring sugars, mainly sucrose. Yudkin (1957) suggested a relationship between high sugar intake and coronary heart disease on epidemiologic grounds but subsequent studies have failed to substantiate this claim (Joint Working Party of the Royal College of Physicians of London and the British Cardiac Society, 1976; Morris *et al.*, 1977). The per capita sugar consumption in the U.S. has shown essentially no change during the past 60 years, but deaths from coronary heart disease increased dramatically between 1945 and the early 1960s and have since fallen steadily.

The influence of dietary carbohydrates on serum lipids is complex and has been reviewed recently (Little *et al.*, 1979). Both complex carbohydrates and sugars tend to produce postprandial hyperinsulinemia and hypertriglyceridemia in the basal state, but these effects are transient and are not considered to be important in relation to the development of susceptibility to coronary heart disease.

Diabetes Mellitus

Although it has frequently been suggested that excessive sugar consumption predisposes an individual to the development of diabetes mellitus, there is a wealth of epidemiologic evidence to the contrary (West, 1978). A high intake of complex carbohydrates has been shown to result in improved glucose tolerance and stabilization of insulin requirements in some diabetics (Stone and Connor, 1963). It has been suggested that a high complex carbohydrate diet in the treatment of diabetes mellitus may be beneficial by lowering saturated fat intake and reducing the threat of atherosclerosis and hyperlipidemia to which diabetics are prone, but this has never been put to the test. There is certainly a striking correlation between the fall in diabetes mortality and the rise in carbohydrate and fall in fat intake in Britain during the two world wars (Himsworth, 1949). Yudkin (1964) related sucrose consumption to the incidence of diabetes deaths in many countries, but strong contrary evidence emerged from a prospective study of 10,000 Israeli civil servants that showed that those destined to develop diabetes during the observation period actually consumed less sugar during their prediabetic period than those who remained nondiabetic (Kahn *et al.*, 1971). There is, thus, little evidence of a relation between diabetes and high carbohydrate consumption, but much to support the role of excessive total caloric intake.

Obesity

There is a widely prevalent notion that sugar is especially fattening. However, a considerable body of epidemiologic evidence shows a negative relationship between sugar intake and the occurrence of obesity (West, 1978). Carbohydrate content of the diet appears to be a major factor in stimulating increased active thyroid hormone production. When the proportion of carbohydrate is increased in isocaloric or hypocaloric diets, T3 (active thyroid hormone triiodothyronine) concentrations are decreased (Danforth *et al.*, 1975). It is not yet known whether corresponding changes in metabolic rate occur under these conditions. In conclusion, it is unlikely that an increased proportion of carbohydrates in the diet *per se* is a cause of obesity. Of primary importance is the balance between energy intake and energy expenditure.

Dental Caries

The role of carbohydrates in the development of dental caries has been reviewed by Bierman (1979). It is clearly not a nutritional disease but is an infection, the prime agent probably being *Streptococcus mutans* in man. This organism can ferment simple carbohydrates, resulting in rapid acid production which is the major force leading to cavitation of the teeth. There is convincing evidence that sucrose consumed in high amounts, especially as sticky sweet foods between meals, is closely associated with a high rate of dental caries. Substitution of sucrose in the diet by a nonfermentable and non-*S. mutans*-supporting sugar alcohol resulted in the decline of caries attack rate to virtually zero (Scheinin *et al.*, 1975). Although the importance of limiting the consumption of sucrose appears to have been clearly demonstrated, the part played by other factors such as low fluoride content of drinking water also has to be borne in mind.

Fiber

Fiber consists of the structural components of plants that are for the most part indigestible by man. The major components, such as cellulose, hemicellulose, pectins, gums, and mucilages, are complex carbohydrates but lignin is composed of polymers of substituted phenylpropanes.

Much attention has been devoted in recent years to the beneficial effects of the fiber component of the diet. However, there is also evidence that consumption of a diet with excessive fiber may have some deleterious consequences.

Some types of fiber, especially those with a high content of phytate, bind divalent cations and impair their absorption, resulting in the possibility of deficiency of iron, zinc, magnesium, calcium, and phosphorus in populations consuming a suboptimal diet. Negative balances for calcium, magnesium, phosphorus, and zinc were shown to occur in subjects fed wheat in wholemeal bread (Reinhold *et al.*, 1973, 1976). Fecal losses of these elements were correlated with fecal dry matter, which was, in turn, proportional to fecal fiber excretion. The same subjects utilized these elements more efficiently when fed white bread. Similar results have been observed in relation to the absorption of iron (Haghshenass *et al.*, 1972; Dobbs and Baird, 1977).

Other possible harmful effects of a high-fiber diet have been the common occurrence of volvulus of the sigmoid colon attributed to increased colonic content of volatile gases (Sutcliffe, 1968) and the presence of trypsin and chymotrypsin inhibitors in fiber, especially wheat bran, that may have detrimental effects in populations subsisting on a marginal diet (Mistuanaga, 1974; Schneeman, 1977).

Metabolic Disorders

For the management of patients with certain disease, sugars have to be greatly reduced in the diet or completely eliminated. These include disaccharidase deficiency disorders, galactosemia, galactokinase deficiency, and fructose intolerance. Detailed descriptions and dietary control may be found in Bondy and Rosenberg (1980).

FAT

This subject is discussed in detail elsewhere and several general comments only will be made here. Fat is the most concentrated source of dietary energy (9 kcal/g), but severe restriction of intake results in a diet of low palatability, resulting in impairment of intake of all nutrients. In comparison with a mixed diet containing normal proportions of carbohydrate, protein, and fat, it has been shown that excess energy provided solely from fat produces greater weight gain in long-term (3–6 months) overfeeding studies (Sims *et al.*, 1973).

Patients with a variety of intestinal disorders that result in malabsorption frequently are unable to tolerate customary amounts of dietary fat in the form of long-chain triglycerides. Medium-chain triglycerides, which are absorbed through the portal circulation, are frequently effective in providing energy and fat-soluble vitamins.

CHOLESTEROL

Cholesterol is an essential metabolite; it is an important constituent of membranes, and precursor of bile acids, steroid hormones, and vitamin D. Although it is not a dietary nutrient in the ordinary sense, as it is synthesized in the body at a rate sufficient to meet the normal needs (0.5–1.0 g/day in the adult), yet it forms an important part of the diet. The average intake in the U.S. is much the same today as it was in 1909, about 500 mg/person per day (Marston and Page, 1978). Cholesterol occurs in all animal tissues, but only in trace amounts in plant tissues. Principal dietary sources include meats (liver, 372 mg/3 oz serving; veal, lamb, and beef, 80–86 mg/3 oz cooked serving), eggs (1 large, 252 mg), shellfish (shrimp, 128 mg/3 oz serving), poultry (chicken, 74 mg/3 oz cooked serving), fish (raw, 43–60 mg/3 oz serving), and dairy products (whole milk, 34 mg/8 oz serving; ice cream, 27–49 mg/half cup; American cheese, 28 mg/oz; yogurt, 17 mg/cup; butter, 12 mg/pat; and skim milk, 5 mg/cup) (Adams, 1975). Human milk contains 26–52 mg/8 oz (Jensen *et al.*, 1978).

The major concern over cholesterol is the association between raised plasma cholesterol concentration and coronary heart disease. Many studies of large population groups have shown a strong association between cholesterol intake, plasma cholesterol concentration and mortality rates from atherosclerotic disease. However, cross-sectional studies of individuals *within* populations have failed to show a correlation between dietary cholesterol intake and plasma cholesterol levels, or between dietary cholesterol intake and incidence of atherosclerotic disease. On the other hand, many experiments have shown that in healthy adults, "normolipidemic" by U.S. standards, plasma cholesterol concentration increases as cholesterol intake increases within the range of 0–600 mg/day. For every additional 100 mg dietary cholesterol/1000 kcal, estimates of the average increase in plasma cholesterol vary from 3 to 12 mg/dl (McGill, 1979).

Dietary cholesterol affects cholesterol biosynthesis. In man, 40–60% of circulating cholesterol is accounted for by diet (Dietschy and Wilson, 1970a,b,c), but many other factors also have an influence. This complex subject has recently been reviewed by Kritchevsky and Czarnecki (1980). Beneficial effects from restriction of cholesterol intake have never been demonstrated experimentally, and it would be extremely difficult to do so as atherosclerosis begins in childhood and does not usually manifest as a clinical disease until middle age or later. Nevertheless, it would appear to be prudent to restrict intake to somewhat below the U.S. average of about 500 mg/day. For patients with hyperlipidemia, restriction to less than 300 mg cholesterol/day is indicated, together with weight control and limitation of fat content of the diet to 35% of total calories and saturated fat to less than 10% of total calories (Goodman, 1978).

ESSENTIAL FATTY ACIDS

It is generally accepted that the minimum dietary linoleate should be 1–2% of total energy (about 5% of total dietary fat) and that the optimal intake may be about 12–14% of total energy. The possible effects of ingestion of excessive amounts of essential fatty acids have recently been reviewed by Mead (1980). The known effects of polyunsaturated fatty acids (PUFA), including essential fatty acids, include depression of enzymes involved in fatty acid synthesis, increased requirement for vitamin E, enhancement of growth of tumor cells, increase of platelet aggregation and thrombus formation, and alteration in properties of cellular membranes. At present there is no evidence that customary diets provide essential fatty acids in excessive and harmful amounts, but there is clearly need for more research into this possibility.

PROTEIN

In technologically developed societies protein is habitually consumed in amounts considerably in excess of the RDA. For the 70 kg man and 55 kg woman, 56 g and 44 g of protein, respectively, are recommended (Food and Nutrition Board, 1980). To arrive at these figures, protein requirements have been estimated on the basis of data from nitrogen equilibrium, protein depletion-repletion, and growth promoting studies. Nitrogen losses have been calculated, converted to protein losses, and then to the minimum protein requirement. A margin of safety and an additional correction for 75% efficiency of utilization of mixed protein in the American diet are superimposed on the minimum requirement. From the estimates of food supplies moving into consumption, published annually in the U.K. the daily consumption of protein per person between 1909 and 1962 ranged between 79 and 89 g (Greaves and Hollingsworth, 1966). Under these circumstances there is an excess of amino acids that are deaminized, the nitrogen component being eliminated in the urine as urea and the carboxyl group being used as energy or converted to carbohydrate and fat; an expensive source of energy from the point of view of metabolism and cost. About half of the protein intake is from animal sources and this represents further inefficiency and economic loss in terms of energy cost.

Little attention has been paid to the possible harmful effects on health of this long-term overconsumption of protein. Although there is a positive correlation between protein intake and mortality from coronary heart disease in data from 25 countries, the association is weaker than for total energy, sucrose, total fat, and saturated fat (Masironi, 1970).

Patients suffering from certain genetic disorders have various enzyme defects that result in incomplete metabolism of one or more amino acids. As part of their treatment special diets have to be given in which the amino acids concerned are greatly restricted. Protein intake also has to be limited in patients with renal or hepatic failure. These conditions are reviewed by Bondy and Rosenberg (1980).

VITAMIN A (RETINOL)

The symptoms and signs of hypervitaminosis A vary considerably, depending on the age of the subject and the duration of the excessive intake. It is the most common form of any vitamin toxicity. The popular belief that vitamin A promotes good vision and protects against infection is partly responsible.

The young child is especially susceptible. The acute form has resulted from single large doses (30,000–90,000 μg retinyl palmitate) and manifests itself as raised intracranial pressure with vomiting, headache, stupor, and occasionally papilledema. Symptoms rapidly subside on cessation of the vitamin and complete recovery always results. The few instances reported appear to result from a special hypersensitivity. It is occasionally observed during massive dose prophylactic programs. In subacute or chonic toxicity, doses usually ranging from about 10,000–50,000 μg have been given daily for several months or one or two years. Symptoms and signs vary considerably but include anorexia; dry, itchy, desquamating skin; alopecia and coarsening of the hair; subperiosteal new bone growth and cortical thickening, especially of the small bones of the hands, feet, and long bones; enlargement of the liver and spleen; and sometimes papilledema, double vision, and symptoms suggestive of brain tumor (Lombaert and Carton, 1976). Plasma vitamin A is always elevated, usually well above 200 μg/100 ml. Cessation of the vitamin dosing leads to a slow recovery without reported after effects. Recently, chronic hypervitaminosis A has been reported in infants fed chicken liver (Mahoney *et al.*, 1980).

Acute symptomatology in the adult is mainly associated with the consumption of several hundred thousand micrograms of the vitamin in the liver of seal or polar bear in a single meal by Arctic explorers. Eskimos have long been aware of the danger. Headache, vomiting, vertigo, blurred vision, and peeling of the skin are the usual consequences and they rapidly subside. Manifestations of chronic poisoning in the adult are similar to those in the child, but the bone changes are mild or absent and menstrual disturbances occur in women (Jeghers and Marrato, 1958). The syndrome has bizarre features and often goes unrecognized unless a careful history is taken and plasma vitamin A is determined. Excessive intake often begins with prescribed dermatological treatment and continued long beyond medical supervision. Dosing is sometimes self-prescribed or a result of food faddism. Manifestations subside gradually following cessation of excessive intake, but there is some evidence of irreversible damage to the liver (Russell *et al.*, 1974).

CAROTENE

Excessive consumption of carotene-rich vegetables or fruits may result in hypercarotenosis with high levels of carotenoids in the plasma ($\geq$ 300 μg/100 ml). Yellowish staining of the skin, especially of the palms and soles, is the only manifestation and general health is not impaired. The sclerotica remain white and serve to differentiate this condition from jaundice. Large quantities of carrot juice may be the cause in infants and adults (Abrahamson and Abrahamson, 1962). Other cases have resulted from the consumption of several pounds of carrots weekly over many months (Almond and Logan, 1942) and oranges, squash, and spinach have also been implicated (Moore, 1957).

Carotenoids are converted to vitamin A only to a limited extent and consequently hypervitaminosis A does not result. One fatal case of self-medication in Britain (Leitner *et al.*, 1975) was given considerable prominence, as death was attributed by the coroner to excessive intake of carrot juice. This was not the case, as massive doses of retinol had also been consumed for a long period and plasma retinol was extremely high (13,000 μg/100 ml).

Hypercarotenemia has also been reported in diabetes mellitus, hypothyroidism and anorexia nervosa (Robboy *et al.*, 1974). It has been suggested that there is some defect in the conversion of carotene to vitamin A in these conditions, but this has never been demonstrated. The association with diabetes was especially common in the earlier years of this century when the consumption of vegetables was advocated as a measure to reduce carbohydrate consumption.

VITAMIN D (CALCIFEROL)

Most instances of hypervitaminosis D have occurred with large and prolonged therapeutic dosing of the vitamin in vitamin D deficiency states. These have become much less common with the availability of the active metabolite $1,25(OH)_2D$ and synthetic analogs.

Excessive dietary intake of vitamin D has been linked with the state of idiopathic hypercalcemia of infancy. It is the mild form that is so associated; the severe form appears to be a distinctly separate condition. The mild form was common in Britain in the 1950s and appeared to be due to excessive intake of vitamin D. When this was reduced, the incidence fell (British Paediatric Association, 1964). Whether it is due to a direct toxic effect of vitamin D or whether some individuals are hypersensitive to the vitamin has never been determined.

The patient is usually aged 3–6 months. He becomes irritable, anorexic, and constipated and develops vomiting, polyuria, and eventually wasting. Serum calcium and blood urea are raised but phosphorus,alkaline phosphatase, chloride, and bicarbonate are normal. Prolonged hypercalcemia from any cause may lead to nephrocalcinosis and renal failure, but this may be prevented by a low calcium diet with no added vitamin D.

A history of excessive vitamin D intake is important in differentiating this condition from hyperparathyroidism, where the triad of hypercalcemia, nephrolithiasis, and bone disease is also present. So-called hypercalcemia in infancy with failure to thrive has been seen with daily vitamin D intakes less than 50 μg, and as low as 25 μg, in which cases there may be a hypersensitivity to the vitamin. Vitamin D toxicity may occur during the treatment of hypoparathyroidism. This may be differentiated from hypercalcemia due to surreptitious ingestion of vitamin D by the presence of low serum iPTH together with elevated 25(OH)D.

The severe form is rare and hypercalcemia is part of a syndrome which includes osteosclerosis, hypercalciuria, nephrocalcinosis, mental retardation, elfin facies, supravalvular aortic stenosis, and multiple peripheral pulmonary stenoses (Black and Bonham Carter, 1963). Its etiology is unknown. By the time the child presents with cardiac anomalies, retardation, and elfin facies, the hypercalcemia has often disappeared. Familial cases have been reported and it has occurred in triplets who were thought to be identical (Manios *et al.*, 1966). The prognosis is poor but a low calcium diet, possibly combined with corticosteroids, may help. There is no evidence that the clinical findings are directly related to hypercalcemia.

Vitamin D in milk has a three- to tenfold greater activity than the same concentration in oil and thus those drinking milk freely are especially at risk to excessive intake. The Food and Nutrition Board (1980) draws attention to the dangers of high consumption and the American Academy of Pediatrics Committee on Nutrition (1974) has recommended the discontinuation of the enrichment of foods other than milk and infant foods. It has been shown that most of the vitamin D in breast milk is in the form of a water-soluble sulfur conjugate (Lakdawala and Widdowson, 1977). It has to be borne in mind that under normal conditions most of vitamin D in the body is obtained by synthesis from a precursor under the influence of ultraviolet light in the skin.

VITAMIN E (TOCOPHEROL)

The addition of large amounts of vitamin E to the feeds of animals has been shown to depress prothrombin levels and cause a coagulopathy (Mellette and Leone, 1960). No adverse effects were described in men habitually consuming 100–800 i.u. vitamin E per day (Farell and Willison, 1975). In a single patient reported by Corrigan and Marcus (1974), vitamin E might have interacted with warfarin and clofibrate in producing a coagulopathy. Thus, evidence for vitamin E toxicity in man is scanty.

VITAMIN K

Vitamin K occurs in several forms in nature, the K_1 group (phylloquinone) in plants, and the K_2 group (menaquinone) in bacteria that are derivatives of menadione (2-methyl-1, 4-napthoquinone). Large doses of vitamin K_1 and K_2 are well tolerated by children and adults, and little accumulates in the tissues. However, excessive doses of menadione and its derivatives have resulted in hyperbilirubinemia and kernicterus, especially in low birth weight infants with erythroblastosis fetalis. The mechanism of toxicity is not well understood but may be increased hemolysis and inhibition of glucuronide formation (Myer and Angus, 1956). The inclusion of menadione in over-the-counter supplements for the grown female has been prohibited in the U.S. and an upper, "safe," limit of 140 μg for adults has been set for the RDA (Food and Nutrition Board, 1980).

NIACIN (NICOTINIC ACID)

Occasional toxic effects have been reported in experimental animals and man from administration of large doses by the parenteral route. Nevertheless, care has been recommended in the prescription of niacin for women of child-bearing age and during pregnancy, and for patients with a history of peptic ulcer, gout, diabetes mellitus, or liver disease (Mosher, 1970).

FOLIC ACID (FOLACIN)

Folic acid is not known to be toxic to man, but if used alone in the treatment of pernicious anemia, subacute, combined degeneration of the cord may be precipitated or allowed to progress (Vilter *et al.*, 1950). Multivitamin capsules containing folic acid may allow neurologic degeneration to occur while the anemia is kept in check. Currently, the sale without prescription of vitamin preparations recommending doses of more than 0.1 mg folacin/day is prohibited (Food and Nutrition Board, 1980).

VITAMIN B_{12} (COBALAMIN)

This vitamin is not known to cause any toxic effects, but there are several facts that should give rise to some concern regarding customary dietary intakes and the present RDA (3 μg). Vitamin B_{12} accumulates in the body in amounts far in excess of daily requirements. It has been calculated that this "store" is sufficient to last the average person for 2–8years (Chanarin, 1969). This is almost certainly a consequence of the habitual lifelong dietary intake of vitamin B_{12} in amounts several times greater than the RDA, which is set at a level to maintain high liver concentration (about 900 ng/g) and plasma levels at the upper end of a wide range of normal values (200–900 pg/ml). As a consequence, vitamin B_{12} concentration in the liver rises steadily throughout life, from about 56 ng/g at birth to more than 20 times this value in adult life. Vitamin B_{12} is present in tissues mainly as two cobalamin-dependent enzymes, the levels of which are directly related to the concentration of the associated enzymes, which has suggested that vitamin B_{12} is not just inactively "stored." The possibility that these massive amounts of cofactor are engaged in unnecessary, and possibly harmful, enzyme induction has not been explored.

There is no doubt that vitamin B_{12} is a luxus nutrient in the customary diet, being consumed in amounts far beyond daily requirements (McLaren, 1981). It is therefore disturbing to note that the Food and Nutrition Board (1980) has recently recommended a substantial increase in the RDA of vitamin B_{12} for infants and children.

VITAMIN C (ASCORBIC ACID)

It is only in the past decade that concern has been expressed about the possible harmful effects of prolonged intake of many grams of vitamin C daily as the result of the claims that have been made that this is effective against the common cold, mental illness, and cancer. The case against these claims has recently been stated by Schrauzer (1979).

Damaging effects of overdosage in animals include reproductive defects and increased collagen catabolism. In man there is circumstantial evidence that acidosis, oxaluria, renal stones, gastrointestinal symptoms, and raised blood cholesterol may result. It is well substantiated that vitamin C enhances considerably the absorption of nonheme iron (Cook, 1977). The massive (up to 100 g daily) and prolonged doses now being consumed by large numbers of people in the U.S. and other countries may readily lead to damage from iron overload.

IRON

Excessive accumulation of iron in body tissues, especially the liver, bone marrow, spleen, skin, and heart, results in hemosiderosis and hemochromatosis. There are a number of causes of this, some of which are disorders of metabolism in which excessive dietary intake does not play a part. There are, however, other instances of iron overload that are associated with prolonged, excessive intake of iron. Hemochromatosis has on occasion resulted from extensive iron therapy (Fisher and Tisherman, 1960), and with the widespread consumption of iron preparations and tonics in countries like the U.S., where they are readily available without prescription, the condition may be more common than is generally appreciated.

Among the Bantu of South Africa iron overload is common, and has been attributed to iron pots used in cooking and to iron drums used in the preparation of Kafir beer (Bothwell and Finch, 1962). It is not known to what extent the cirrhosis of the liver that is frequently found is a result of iron overload or of chronic alcoholism or malnutrition (Moore, 1973).

In Ethiopia the highest iron intake anywhere reported (about 470 mg/day) is due to the very high iron content of the staple cereal, called teff. Even so, tissue siderosis is not reported to be so common, possibly because the iron is not in a readily absorbable form (Roe, 1966).

Acute iron poisoning is a common problem especially in children aged 1–5 years (Smith *et al.*, 1977). Iron preparations used in the correction of anemias are widely considered as dietary supplements, and insufficient care is taken in the home to keep them out of the hands of children, who have been known to take tablets or capsules as candy. The lethal dose is variable, as little as 2 g of ferrous sulfate has been reported to cause the death of a young child, while a dose of 14 g has been followed by recovery. Four stages have been described: vomiting, diarrhea, and bleeding from the bowel; a quiescent period; metabolic acidosis, convulsions, coma, and acute hepatic necrosis; and gastrointestinal scarring and strictures.

SODIUM

During recent years evidence has been accumulating that excessive salt consumption poses a real threat to health. There are several reasons why the infant is especially susceptible to develop the dangerous condition of hypernatremia. The kidney plays a central role in the maintenance of electrolyte homeostasis and is immature for some time after birth. There has been a tendency for the manufacturers of infant foods to add salt to their formulations, but this has been discouraged in recent years. The practice of mothers to heap the measures of powdered milk instead of leveling them off as instructed results in infants receiving excessive amounts of milk, including salt. If the infant is hot and thirsty, the tendency to give a feed instead of water results in salt in excess of requirements being ingested. Finally, a number of major disasters has been reported from infant nurseries where feeds have in error been made up with salt instead of sugar (Finberg *et al.*, 1963).

Hypernatremia (serum sodium > 145 meq/liter) usually results when water loss exceeds sodium loss in conjunction with inadequate water intake. Confusion, neuromuscular excitability, seizures, and coma may result.

The possible relationship between habitual salt consumption and the development of hypertension has been the subject of numerous animal experimental and human epidemiologic studies, and the evidence has recently been reviewed (Tobian, 1979). Most societies among whom hypertension of unknown etiology is common, consume salt far in excess of physiological requirements. The highest incidence of hypertension is in northern Japan where the sodium intake is above 400 meq (about 23 g salt) per day. The usual consumption in the U.S. is 100–200+ meq/day and the incidence of hypertension in middle life varies between 9 and 20%. On the other hand, some 20 primitive societies have been studied who ingest low-sodium diets (10–60 meq/day) and they all have virtually no hypertension. Moreover, the steady rise of blood pressure with age universally observed in industrialized societies does not occur among these primitive peoples. Potassium intake tends to be inversely related to sodium intake, and consequently to hypertension, but it is not known whether this is a causal relationship.

Where hypertension is common it tends to cluster in families, suggesting a genetic predisposition. Consequently, those not so disposed appear to be able to consume the customary daily intake of about 200 meq/day without the likelihood of developing elevated blood pressure. On the other hand, those predisposed would be advised to follow a lifelong modest restriction of less than 60 meq/day. There is no certain way of identifying susceptible individuals, but a greater than normal risk is carried by those who have a family history of hypertension, those with a blood pressure in the upper 20% of the population at any age, individuals whose resting heart rate is considerably greater than would be expected from their state of physical conditioning, and those who are more than 15% above optimal body weight.

A similar modest restriction of sodium intake enhances the effect of antihypertensive drugs and other measures in patients with established hypertension, but restriction to as little as 20 meq/day may be required to reduce severe hypertension levels. In applying sodium restriction, care has to be exercised to identify those with high sodium requirements, such as patients with salt-losing renal disease or adrenal insufficiency, and high losses from sweating.

FLUORINE

Fluorine intoxication or fluorosis is rare except in parts of the world where the fluorine content of water is high, as this is the main source of intake. Endemic areas tend to have semidesert conditions, such as parts of India, Iran, the Arabian Gulf states, Tanzania, and Texas.

The ideal water concentration is considered to be 1.0–1.5 ppm. In the range of 2.0–2.5 ppm the earliest sign of excessive intake is seen, consisting of a white mottling of the teeth in children who live in the area during the development period. At slightly higher levels brownish discoloration and increased brittleness of the enamel occurs, leading to marked pitting in some instances, especially affecting the upper central incisors. When the water concentration exceeds about 10 ppm the disabling condition of skeletal fluorosis results. Exostoses and increased density of spinal vertebrae produce compression of spinal cord and nerve roots with paraplegia (Singh *et al.*, 1963). In an endemic area in South India, genu valgum appears in a high proportion of those aged 10–25 years and this has been attributed to fluorosis (Krishnamachari and Krishnaswamy, 1973). In a comparison of two communities in Texas, where the water content was 0.4 and 8.0 ppm, no differences in health were found and only severely mottled teeth were observed in the latter (Shaw, 1954).

Acute fluoride toxicity, consisting of abdominal pain, nausea and vomiting with full recovery in 24 hr, was reported recently (Hoffman *et al.*, 1980) when a series of errors led to several hundred times the normal amount of fluoride being added to the drinking water in a school in New Mexico. Most children refused to drink it because of the salty taste.

COBALT

This trace element is required by the body only, so far as is known, for its role as an integral part of the vitamin B_{12} molecule. The only reported instance of excessive intake is its possible involvement in a syndrome of congestive heart failure with a distinctive cardiomyopathy, polycythemia, thyroid epithelial hyperplasia, and colloid degeneration. This highly fatal condition was described in very heavy beer drinkers (approximately 5 liters/day) consuming beer to which cobalt salts had been added (1.2–1.5 ppm Co) to improve the foaming properties (Grinvalsky and Fitch, 1969). The process is no longer in use. It is not clear that cobalt alone was responsible for the disease, and it has been suggested that the high alcohol intake and deficiencies of protein, thiamin, and perhaps other nutrients may have played a part.

COPPER

Most diets provide an amount of copper that lies within the range of 2–3 mg set as adequate and safe by the Food and Nutrition Board (1980) and there is no danger of excessive intake. Patients who suffer from the rare disease hepatolenticular degeneration (Wilson's disease) accumulate large amounts of copper in tissues of the liver, brain, kidney, and eye and as part of treatment their dietary intake of copper has to be strictly limited.

CALCIUM

One of the features of the aging process is the deposition of calcium in many tissues. There is no evidence to suggest that this is in any way related to excessive dietary intake. Populations vary considerably in their habitual consumption of calcium but homeostasis is maintained by the adjustment of intestinal absorption and urinary excretion to conform with the body's requirements.

REFERENCES

Abrahamson, I. A., and Abrahamson, I. A., Jr., 1962, Hypercarotenemia, *Arch. Ophthalmol.* **68:**4.

Adams, C. F., 1975, Nutritive value of American foods in common units, *Agriculture Handbook No. 456*, U.S. Department of Agriculture, Agricultural Research Service, Washington, D.C.

Alfin-Slater, R. B., and Mirenda, R., 1980, Nutrient requirements: What they are and bases for recommendations, in: *Human Nutrition a Comprehensive Treatise,* Vol. 3A, *Nutrition and the Adult: Macronutrients* (R. B. Alfin-Slater and D. Kritchevsky, eds.), pp. 1–48, Plenum, New York.

Almond, S., and Logan, R. F. L., 1942, Carotinaemia, *Br. Med. J.* **2:**239.

American Academy of Pediatrics Committee on Nutrition, 1974, Salt intake and eating patterns of infants and children in relation to blood pressure, *Pediatrics* **53:**115.

Bierman, E. L., 1979, Carbohydrates, sucrose, and human disease, *Am. J. Clin. Nutr.* **32:**2712.

Black, J. A., and Bonham Carter, R. E., 1963, Association between aortic stenosis and facies of severe infantile hypercalcaemia, *Lancet* **2:**745.

Blitzer, P. H., Blitzer, E. C., and Rimm, A. A., 1976, Association between teen-obesity and cancer in 56,111 women. All cancers and endometrial carcinoma, *Prev. Med.* **5:**20.

Bondy, P. K., and Rosenberg, L. E., 1980, *Metabolic Control and Disease,* 8th ed., Saunders, Philadelphia.

Bothwell, T. H., and Finch, C. A., 1962, *Iron Metabolism,* Little Brown, Boston.

British Paediatric Association, 1964, Infant hypercalcaemia, nutritional rickets and infantile scurvy in Great Britain, *Br. Med. J.* **1:**1659.

Chanarin, I., 1969, *The Megaloblastic Anaemias,* Blackwell, Oxford.

Cook, J. D., 1977, Absorption of food iron, *Fed. Proc. Fed. Am. Soc. Exp. Biol.* **36:**2028.

Corrigan, J. J., Jr., and Marcus, F. I., 1974, Coagulopathy associated with vitamin E ingestion, *J. Am. Med. Assoc.* **220:**1300.

D'Alonzo, C. A., and Pell, S., 1968, Cardiovascular disease among problem drinkers, *J. Med.* **10:**344.

Danforth, E., Jr., Desileto, E. S., Hortan, E. S., Sims, E. A. H., Burger, A. G., Braveman, L. E., Vagenakis, A. G., and Ingbar, S. H., 1975, Reciprocal changes in serum triiodothyronine (T3) and reverse triiodothyronine (rT3) induced by altering the carbohydrate content of the diet, *Clin. Res.* **23:**573A.

DeWaard, F., and Baanders-van Halewun, E. A., 1969, Cross sectional data on estrogenic smears in a postmenopausal population, *Acta Cytol.* **13:**675.

Dietschy, J. M., and Wilson, J. D., 1970a,b,c, Regulation of cholesterol metabolism, *N. Engl. J. Med.* **282:**1128,1179,1241.

Dobbs, R. J., and Baird, I. M., 1977, Effect of wholemeal and white bread on iron absorption in normal people, *Br. Med. J.* **1:**1641.

Farell, P. M., and Willison, J. W., 1975, Megavitamin E supplementation in man, *Fed. Proc. Fed. Am. Soc. Exp. Biol.* **34:**912.

Finberg, L., Kiley, J., and Luttrell, C. N., 1963, Mass accidental salt poisoning in infancy: A study of a hospital disaster, *J. Am. Med. Assoc.* **184:**187.

Fisher, E. R., and Tisherman, S., 1960, The association of idiopathic hemochromatosis and excessive iron overload. Report of a case with comment relative to the concept of exogenous hemochromatosis. *Arch. Pathol.* **69:**683.

Food and Nutrition Board, 1980, *Recommended Dietary Allowances,* 9th rev. ed., National Academy of Sciences, National Research Council, Washington, D.C.

Garrow, J. S., Blaza, S., Warwick, P., and Ashwell, M., 1980, Predisposition to obesity, *Lancet* **1:**1103.

Goodman, D. S., 1978, Diseases of lipid and lipoprotein metabolism, in: *The Year in Metabolism 1977* (N. Freinkel, ed.), pp. 183–218, Plenum, New York.

Greaves, J. P., and Hollingsworth, D. F., 1966, Trends in food consumption in the United Kingdom, *World Rev. Nutr. Diet.* **6:**34.

Grinvalsky, H. T., and Fitch, D. M., 1969, A distinctive cardiomyopathy occurring in Omaha, Nebraska: Pathological aspects, *Ann N.Y. Acad. Sci.* **156:**544.

Haghshenass, M., Mahlondji, M., Reinhold, J. G., and Mohammadi, N., 1972, Iron deficiency anemia in an Iranian population associated with high intakes of iron, *Am. J. Clin. Nutr.* **25:**1143.

Himsworth, H. P., 1949, Diet in the etiology of human diabetes, *Proc. R. Soc. Med.* **42:**323.

Hirsch, J., 1978, What's new in the treatment of obesity? in: *The Year in Metabolism 1977* (N. Freinkel, ed.), pp. 169–182, Plenum, New York.

Hoffman, R., Mann, J., Calderone, J., Trumbull, J., and Burkhart, M., 1980, Acute fluoride poisoning in a New Mexico elementary school, *Pediatrics* **65:**897.

Jeghers, H., and Marrato, H., 1958, Hypervitaminosis A. Its broadening spectrum, *Am. J. Clin. Nutr.* **6:**335.

Jensen, R. G., Hagerty, M. M., and MacMahon, K. E., 1978, Lipids of human milk and infant formulas: A review, *Am. J. Clin. Nutr.* **31:**990.

Johnson, B. G., and Nilsson-Ehle, P., 1978, Alcohol consumption and high density lipoprotein, *N. Engl. J. Med.* **298:**633.

Joint Working Party of the Royal College of Physicians of London and the British Cardiac Society, 1976, Prevention of coronary heart disease, *J. R. Coll. Physicians London* **10:** 213.

Kahn, H. A., Herman, J. B., Medalie, J. H., Neufeld, H. N., Riss, E., and Goldbourt, U., 1971, Factors related to diabetes incidence: A multivariate analysis of two years observations on 10,000 men, *J. Chronic Dis.* **23:**617.

Keys, A., 1971, Sucrose in the diet and coronary heart disease, *Atherosclerosis* **14:**193.

Klatsky, A. L., Friedman, C. D., and Siegelaut, A. B., 1974, Alcohol consumption before myocardial infarction. Results from the Kaiser-Permanente epidemiologic study of myocardial infarction, *Ann. Intern. Med.* **81:**294.

Klatsky, A. L., Friedman, G. D., Siegelaut, A. B., and Gerard, M. D., 1977, Alcohol consumption and blood pressure. Kaiser-Permanente multiphasic health examination. *N. Engl. J. Med.* **296:**1194.

Krishnamachari, K. A. V. R., and Krishnaswamy, K., 1973, Genu valgum and osteoporosis in an area of endemic fluorosis, *Lancet* **2:**877.

Kritchevsky, D. and Czarnecki, S. K., 1980, Nutrients with special functions: Cholesterol, in: *Human Nutrition: A Comprehensive Treatise,* Vol. 3A *Nutrition and the Adult: Macronutrients* (R. B. Alfin-Slater and D. Kritchevsky, eds.), pp. 239–258, Plenum, New York.

Lakdawala, D. R., and Widdowson, E. M., 1977, Vitamin-D in human milk, *Lancet* **1:**167.

Leitner, Z. A., Moore, T., and Sharman, I. M., 1975, Fatal self-medication with retinol and carrot juice, *Proc. Nutr. Soc.* **34:**44A.

Lelbach, W. K., 1975, Cirrhosis in the alcoholic and its relation to the volume of alcohol abuse, *Ann. N.Y. Acad. Sci.* **252:**85.

Lew, E. A., and Garfinkel, L., 1979, Variations in mortality by weight among 750,000 men and women, *J. Chronic Dis.* **32:**563.

Little, J. A., McGuire, V., and Derksen, A., 1979, Available carbohydrates, in: *Nutrition, Lipids, and Coronary Heart Disease* (R. Levy, B. Rifkind, B. Dennis and N. Erust, eds.), pp. 119–148, Raven, New York.

Lombaert, A., and Carton, J., 1976, Benign intracranial hypertension due to A hypervitaminosis in adults and adolescents: A review, *Eur. Neurol.* **14:**340.

Mahoney, C. P., Margolis, M. T., Knauss, T. A., and Labbe, R. F., 1980, Chronic vitamin A intoxication in infants fed chicken liver, *Pediatrics* **65:**893.

Manios, S., Panagou, M., Tsakalidis, D., and Kovatsis, A., 1966, Hypercalcemie idiopathique chez des triplees. *Arch. Fr. Pediatr.* **23:**63.

Marston, R., and Page, L., 1978, *Nat. Food Rev.* **NFR-5:**28.

Masironi, R., 1970, Dietary factors and coronary heart disease, *Bull. W.H.O.* **42:**103.

McGill, H. C., Jr., 1979, Appraisal of cholesterol as a causative factor in atherogenesis, *Am. J. Clin. Nutr.* **32:**2632.

McLaren, D. S., 1981, The luxis vitamins—A and B_{12}, *Am. J. Clin. Nutr.* **34:**1161.

Mead, J. F., 1980, Nutrients with special functions: Essential fatty acids, in: *Human Nutrition a Comprehensive Treatise*, Vol. 3A *Nutrition and the Adult: Macronutrients* (R.B. Alfin-Slater and D. Kritchevsky, eds.), pp. 213–238, Plenum, New York.

Mellette, S. J., and Leone, L. A., 1960, Influence of age, sex, strain of rat and fat soluble vitamins on hemorrhagic syndromes in rats fed irradiated beef, *Fed. Proc. Fed. Am. Soc. Exp. Biol.* **19:**1045.

Mistuanaga, T., 1974, Some properties of protease inhibitors in wheat grain, *J. Nutr. Sci. Vitaminol.* **20:**153.

Moore, C. V., 1973, Iron, in: *Modern Nutrition in Health and Disease*, 5th ed. (R. S. Goodhart and M. E. Shils, eds.), pp. 319–320, Lea and Febiger, Philadelphia.

Moore, T., 1957, *Vitamin A*, pp. 442–455, Elsevier, London.

Morris, J. N., Marr, J. W., and Clayton, D. G., 1977, Diet and heart: A postscript, *Br. Med. J.* **2:**1307.

Mosher, L. R., 1970, Nicotinic acid side effects and toxicity. A review, *Am. J. Psychiatry* **126:**1290.

Myer, T. C., and Angus, J., 1956, The effect of large doses of "synavit" in the newborn, *Arch. Dis. Child.* **31:**212.

Neville, J. N., Eagles, E. A., Samson, G., and Olson, R. E., 1968, Nutritional status of alcoholics, *Am. J. Clin. Nutr.* **21:**1329.

Reinhold, J. G., Nasr, K., Lahimgarzadeh, A., and Hedayati, H., 1973, Effects of purified phytate and phytate-rich bread upon metabolism of zinc, calcium phosphorus and nitrogen in man, *Lancet* **1:**28.

Reinhold, J. G., Faradji, B., Abadi, P., and Ismail-Beigi, F., 1976, Decreased absorption of calcium, magnesium, zinc and phosphorus by humans due to increased fiber and phosphorus consumption as wheat bread. *J. Nutr.* **106:**493.

Ries, W., 1973, Feeding behaviour in obesity, *Proc. Nutr. Soc.* **32:**187.

Robboy, M. S., Sato, A. S., and Schwabe, A. D., 1974, The hypercarotenemia in anorexia nervosa: A comparison of vitamin A and carotene levels in various forms of menstrual dysfunction and cachexia, *Am. J. Clin. Nutr.* **27:**362.

Roe, D. A., 1966, Nutrient toxicity with excessive intake. II. Mineral overload, *N.Y. State J. Med.* **66:**1233.

Russell, R. M., Boyer, J. L., and Bagheri, S. A., 1974, Hepatic injury from chronic hypervitaminosis A resulting in portal hypertension and ascites, *N. Engl. J. Med.* **291:**435.

St. Leger, A. S., Cochrane, A. L., and Moore, F., 1979, Factors associated with cardiac mortality in developed countries with particular reference to the consumption of wine, *Lancet* **1:**1017.

Scheinin, A., Makinen, K. K., and Ylitalo, K., 1975, Turku sugar studies. I. An intermediate report on the effect of sucrose, fructose and xylitol diets on the caries incidence of man, *Acta Odontol. Scand. Suppl.* **33(70):**5.

Schneeman, B. O., 1977, The effect of plant fiber on trypsin and chymotrypsin activity *in vitro*, *Fed. Proc. Fed. Am. Soc. Exp. Biol.* **36:**1118.

Schrauzer, G. N., 1979, Vitamin C: Conservative human requirements and aspects of overdosage, in: *International Review of Biochemistry, Biochemistry of Nutrition IA*, Vol. 27 (A. Neuberger and T. H. Jukes, eds.), pp. 168–188, University Park Press, Baltimore.

Shaw, J. H., 1954, *Fluoridation as a Public Health Measure*, p. 127, American Association for the Advancement of Science, Washington, D.C.

Sims, E. A. H., Danforth, E., Jr., Horton, E. S., Bray, G. A., Glennon, J. A., and Salaus, L. B., 1973, Endocrine and metabolic effects of experimental obesity in man, *Recent Prog. Horm. Res.* **29:**457.

Singh, A., Jolly, S. S., Bansal, D. C., and Mathur, C. C., 1963, Endemic fluorosis. Epidemiological, clinical, and biochemical study of chronic fluorine intoxication in Punjab, India, *Medicine Baltimore* **42**:229.

Smith, W. L., Franken, E. A., Jr., Grossfeld, J. L., and Ballantine, T. V. N., 1977, Pneumatosis of the bowel secondary to acute iron poisoning, *Radiology* **122**:192.

Society of Actuaries, 1959, *Build and Blood Pressure Studies, 1957*, Vol. 1, Society of Actuaries, Chicago.

Stone, D. B., and Connor, W. E., 1963, The prolonged effects of a low cholesterol, high carbohydrate diet upon the serum lipids in diabetic patients, *Diabetes* **12**:127.

Sutcliffe, M. M. L., 1968, Volvulus of the sigmoid colon, *Br. J. Surg.* **55**:903.

Tobian, L., Jr., 1979, Sodium: Dietary salt (sodium) and hypertension, *Am. J. Clin. Nutr.* **32**:2659, 2739.

Van Itallie, T. B., 1979, Obesity: Adverse effects on health and longevity, *Am. J. Clin. Nutr.* **32**:2723.

Vilter, R. W., Horrigan, D., Mueller, J. F., Jarrold, T., Vilter, C. F., Hawkins, V., and Seaman, A., 1950, Studies on the relationships of vitamin B_{12}, folic acid, thymine, uracil, and methyl group donors in persons with pernicious anemia and related megaloblastic anemias, *Blood* **5**:495.

West, K. M., 1978, *Epidemiology of Diabetes and its Vascular Lesions*, pp. 224–274, Elsevier, New York.

Williamson, R., 1980, Diet and bowel cancer, *Br. Med. J.* **2**:146.

Yano, K., Rhoads, G. G., and Kagan, A., 1977, Coffee, alcohol and risk of coronary heart disease among Japanese men living in Hawaii, *N. Engl. J. Med.* **297**:405.

Yudkin, J., 1957, Diet and coronary thrombosis: Hypothesis and fact, *Lancet* **2**:155.

Yudkin, J., 1964, Dietary fat and dietary sugar in relation to ischaemic heart disease, *Lancet* **2**:4.

32

AURANTIASIS

Teruo Honda

INTRODUCTION

The skin color is determined by the qualities and quantities of melanin, melanoid, carotenoid, blood circulation and hemoglobin, as well as the thickness and transparency of the stratum corneum.

Aurantiasis (or carotinemia) is a yellowness of the skin resulting from deposition of carotenoid in the skin, especially the stratum corneum of the epidermis. It sometimes accompanies diabetes mellitus, myxedema, nephrotic syndrome, hypercholesterolemia or xanthoma, but is usually of dietary origin. Dietary aurantiasis is generally considered to develop in persons who ingested excessive amounts of food containing carotenoids over a long period of time. Though aurantiasis itself is not harmful, it frequently signifies an unbalanced diet, besides being unsightly, and is not infrequently mistaken for jaundice.

CAROTENOIDS

Carotenoids are a group of pigments ranging in color from yellow to red, widely distributed in animals and plants, of which about 60 varieties are known. From a nutritional viewpoint, it is significant as provitamin A, though only a limited number of carotenoids such as α, β, and γ carotenes, cryptoxanthin, and myxoxanthin have a function as vitamin A, while lycopene and xanthophyll have no such function. In man, carotenoids are supplied through ingestion of vegetables, which are absorbed in the intestine in the presence of bile and lipase, while provitamin A is transformed to vitamin A (retinol) in the intestinal wall and liver to be stored chiefly in the liver in the form of retinol ester. The other carotenoids are considered to undergo metabolism in the liver.

Teruo Honda • Department of Pediatrics, Tokyo Medical College, Tokyo, Japan.

TABLE I
Rates of Absorption of Carotenoid in Food

Foods	Rate of absorption	References
Adults		
spinach, 150 g (boiled)	40%	Hara (1957)
carrot, 200 g (cooked and grated)	54%	
pumpkin, 150 g (cooked and mashed)	50%	
dried laver, 10 g (toasted)	9%	
natural β-carotene, 4–10 mg (90% solution in oil)	76%	
carrot (β-carotene, 2652 μg) (cooked)	21–25%	Fujita (1957)
Children		
tangerine juice, 50 ml (infants)	60–68%	Akai (1959)
carrot juice, raw, 20 ml (infants)	21–35%	
pumpkin, 60 g (cooked and mashed), (young children)	58%	

Daily foods that contain large amounts of carotenoids are tangerine, carrot, pumpkin, yolk, yellow sweet potato, spinach, tomato, laver, loquat, persimmon, etc. In some areas, papaya and red palm oil may be added to the above list.

Absorption Rate of Carotenoid in Food

Table I shows the results of an absorption test conducted in Japan, quoted from published data. While absorption rate exceeds 50% with tangerine juice, pumpkin, and carrot (cooked and grated), it seems to be lowered below one-third if the above materials are mixed with or given together with other foods. Absorption tends to be accelerated in the presence of fat.

Site of Carotenoid Deposition

Because carotenoids have a preference for the stratum corneum as the site of sedimentation, yellowness of the skin first becomes conspicuous at the sole, palm, then the alae nasi and nasolabialsulcus, and subsequently spreads all over the body. But we observed that yellowness first appeared in the alae nasi and the initial coloration at the sole was completely absent in infants before toddling.

CLINICAL OBSERVATIONS

Aurantiasis due to food consumption has occurred only under special conditions and has not been common in Japan. Since around 1972, however, a marked increase in the incidence of aurantiasis has been noted, chiefly in winter, among children in Tokyo and surrounding areas. In the absence of reports on large-scale occurrence of aurantiasis, we feel this deserves comment.

Causative Foods for Aurantiasis in Children

Table II shows the state of ingestion of causative foods in 84 infants and children (aged 2–10 years) who visited the hospital with aurantiasis in the period between January, 1973 and March, 1974. Our criteria for diagnosis of aurantiasis consisted of two points—generalized yellowness of the skin, and yellowness of the skin at the sole at Grade 4 or above (over a period of 1 year) according to icterometer (Gosset) readings. As the table shows, tangerine was the cause in 70 patients (83.3%). Various kinds of juice, including 100% tangerine juice, were responsible only in a small number of cases. It is interesting that as many as 20 children (23.8%) developed aurantiasis as a result of ingestion of 2–4 tangerines/day over more than 4 weeks. We think it noteworthy that aurantiasis resulted from ingestion of amounts of tangerines that could not be judged as excessive. The results of thyroid function test (Triosorb test, Tetrasorb test) and urinalysis (sugar, protein, acetone), performed in the above 20 children, were all in normal ranges, which ruled out possibilities of hypothyroidism, nephrotic syndrome and diabetes mellitus. In spite of the small intake of causative food, therefore, aurantiasis in these children seemed attributable to food.

Intake of Tangerines and Incidence of Aurantiasis

Table III shows the results of an investigation carried out in March (at the end of the tangerine season) in 233 first graders. The incidence of aurantiasis was 76% in the massive intake group, and decreased with reduction in intake to 44% and 18%, until it reached zero in children who disliked and did not eat tangerines. The data support a claim that foods other than tangerine are rarely responsible for aurantiasis in the Tokyo area (probably in

TABLE II
Intake of Foods Causing Aurantiasis to Children

Amount of food		No. of children
Tangerine	> 11 pc	14
	5–10 pc	36
	2–4 pc	20
100% Tangerine juice 100–500 ml		7
Homemade mixed juice (carrot, tangerine, apple)		3
Tomato juice 400–500 ml		2
Orange juice 500–600 ml		2

TABLE III
Intake of Tangerine and Frequency of Aurantiasis

Intake	Number of pupils	Aurantiasis
> 6 pc	50	38 (76.0%)
4–5 pc	100	44 (44.0%)
1–3 pc	72	13 (18.0%)
0 pc	11	0
Total	233	95 (40.7%)

all of Japan). The fact that one-fourth of the children with massive intake of tangerine failed to present aurantiasis suggests, when considered together with its development in children with small intake, that there are great individual variations involved, such as absorption, metabolism, capillary permeability, and skin transparency.

Serum Chemistry Test

Serum carotene level was determined by Bessey's modified method in 36 children with aurantiasis and ten control children. As Table IV shows, the level was 172 μg/100 ml at the highest and 142 μg/100 ml on the average in the control group (C), and as high as 819 μg/100 ml and 401 μg/100 ml respectively in 27 children with aurantiasis who were taking the causative food everyday (group A). In nine children with aurantiasis who had ceased to take the causative food for at least four weeks (group B), serum carotene level was below 230 μg/100 ml in all the cases but one and 180 μg/100 ml on the average (only slightly higher than in the control).

Table V shows laboratory data other than carotene level in 27 children with aurantiasis who were taking the causative food. The values were all normal except for the icterus index, which was slightly high, while hyperlipemia or impairment of liver function was not observed.

Vitamin A was simultaneously determined by Bessey's modified method in 12 children with aurantiasis (including 8 children who were taking causative food everyday) and 3 control children. It was between 65.3 and 278.0 i.u./100 ml, and M ± SD for it was 158.13 ± 66.94 vitamin i.u./100 ml.

TABLE IV
Level of Carotene in Serum[a]

Group (number)	Mean (μg/100ml)	SD	Range
A (27)	401.0	159.3	215.7–819.1
B (9)	180.4	39.2	110.3–324.0
C (10)	142.3	19.9	116.3–172.8

[a] Determined by Bessey's Modified Method (Bessey *et al.*, 1946).

TABLE V
Laboratory Findings of 27 Children with Aurantiasis[a]

Total protein	6.9 ± 0.37 g/100 ml
Albumin	54.2 ± 3.9%
β-Lipoprotein	1.5 ± 0.5 mm
Cholesterol	167 ± 45 mg/100 ml
Icterus index[b]	11.8 ± 3.7
GOT	24.4 ± 10.8 Karmen U
GPT	17.7 ± 8.0 Karmen U

[a] Values are means ± SD.
[b] Determined without treating with acetone.

The correlation coefficient between vitamin A and carotene was 0.14, and vitamin A level in 2 children with a serum carotene level over 740 μg/100 ml was below 100 i.u./100 ml.

Fluctuation in Serum arotene Level

Fluctuation in serum carotene level after cessation of the causative food was followed in eight children. Serum carotene level rapidly declined except in one case, but more than one month was needed for the carotene level to be normalized if it was abnormally high. It took a still longer time for yellowness of the skin to disappear, and 2–3 months were needed for its complete disappearance in the majority of cases.

AURANTIASIS IN INFANTS DUE TO MOTHER'S MILK

There are few publications on aurantiasis in infants. Kawakami *et al.* (1976) discovered aurantiasis in 20 breast-fed infants (2–5 months old) in winter, while finding it in none of the bottle-fed infants. Since none of the breast-fed infants were given juice or carrot soup, aurantiasis seemed attributable to mother's milk.

In the above infants, the skin was generally yellow, and conspicuously so at the alae and apex of the nose. Yellowness at the sole, however, was not marked; in this respect the pattern of coloration in the infants differed from that in young and school children.

The overwhelming majority of their mothers presented aurantiasis, whose cause was limited to tangerine. They were taking 11–20 tangerines/day over at least 4 weeks, though 2 mothers had 5–6 tangerines/day. Serum carotene level in the mothers was 201.0–462.0 μg/100 ml. Serum carotene level in five of their infants was 144.5–240 μg/100 ml. From this data we concluded that serum carotene level at the onset of aurantiasis was slightly lower in infants than in young and school children.

TABLE VI
Level of Carotene in Breast Milk[a]

Group (number)	Mean (μg/100 ml)	SD	Range
A (20)	425.63	284.16	226.5–1424.0
B (13)	137.96	63.67	41.0–223.5
C (24)	82.44	33.80	33.4–153.1

[a] Determined by McLaren's Method (McLaren *et al.*, 1967).

Carotene Content of Mother's Milk

Table VI shows the carotene content of mother's milk determined by McLaren's method. The carotene content of mother's milk in Group A, whose infants had developed aurantiasis, was 425.6 μg/100 ml on the average. The milk from two mothers with carotene level above 1000 μg/100 ml was colored deep yellow, like colostrum. In group C mother's milk that was collected while tangerines were out of season and aurantiasis was not found in mothers or infants, the mean carotene content was 82.4 μg/100 ml, with the value under 100 μg/100 ml in the majority of cases. This mean value was considerably high, as compared with 27 μg/100 ml obtained by the National Research Council (1953) or 20.9 μg/100 ml by Saito *et al.* (1963). In group B mother's milk that was collected while tangerines were in season and aurantiasis was not found in mothers or infants, the mean carotene level was 137.9 μg/100 ml, which ranked between group A and group C. To summarize, the carotene content of mother's milk that caused aurantiasis was high, as expected, but it was an interesting finding that the carotene content of mother's milk was higher in tangerine season than out of it even among subjects who did not develop aurantiasis.

Fluctuation in the Carotene Content of Mother's Milk

In three cases of mother's milk containing 350–500 μg/100 ml carotene, carotene content fell below 100 μg/100 ml two weeks after cessation of the causative food or below 150 μg/100 ml 8–9 weeks after that, which indicates that the pattern of decrease in carotene content was variable. In one case of mother's milk containing as much as over 1000 μg/100 ml carotene, carotene content fell below 600 μg/dl two weeks after cessation, then gradually decreased thereafter, and remained at 348 μg/100 ml even 12 weeks after cessation of the food.

Fluctuation in the Carotene Content of Mother's Milk after Ingestion of Tangerine

Three medium-sized tangerines (peeled, weighing 550–610 g) were given to five mothers during the first three postpartum months at early morning fasting time, and the carotene content of their milk was determined at certain fixed time intervals (Kawakami, 1979). No fluctuation occurred in two cases where the carotene content was low. In the other three cases, the carotene content rose 9 hr after ingestion of tangerines and the maximum range was 276 μg/100 ml. The above findings indicate that there is an extremely great individual variation in the mechanism by which an intake of a carotene-containing food is reflected in the carotene content of mother's milk. Milk samples in most of the tests, initial 10 ml portions, were obtained in the morning, (9–12 A.M.).

DIFFERENTIATION OF AURANTIASIS FROM JAUNDICE

The most characteristic findings in aurantiasis are conspicuousness of pigmentation at specific sites such as the sole, palm, alae nasi and nasolabial-sulcus; and noncoloration of the sclera.

Urine takes on a yellow color, but it is lemon-yellow and never yellowish brown or brown as in jaundice (direct bilirubin type). The color of feces is in the normal range or characterized by increased yellowness, but never white or yellowish white. There is a report (Aoyama, 1977) that diagnosis of congenital bile duct atresia was made difficult by coloration of feces due to carotene contained in mother's milk.

In chemical examination of blood, carotenoid level (carotene, lycopene, lutein, etc.) rises, but bilirubin level is not elevated. Serum turns yellow, and, without precipitation by acetone, icterus index is elevated.

CONCLUSION

Very high intakes of carotenoid-containing foods is correlated to a high incidence of aurantiasis. Some persons, however, do not develop aurantiasis in spite of massive ingestion of such foods, while others present yellowness of the skin after slight ingestion. What causes this discrepancy? I think the individual's response depends more on the clearance ability of the liver than on absorption or susceptibility of the skin, but cannot give a convincing proof.

Since a high serum carotene level does not always imply that the serum vitamin A level is in optimal range, it is advised to take carotenoid-containing foods only in moderate amounts.

REFERENCES

Akai, M., 1959, On the carotene absorption rate in the childhood, *Acta Paediatr. Jpn. Overseas Ed.* **63:**148.

Aoyama, K., 1977, Personal communication, Okayama National Hospital, Okayama, Japan.

Bessey, O. A., Lowry, O. H., Brock, M. J., and Lopez, J. A., 1946, The determination of vitamin A and carotene in small quantities of blood serum, *J. Biol. Chem.* **166:**177.

Costanza, D. J., 1968, Carotenemia associated with papaya ingestion, *Calif. Med.* **109:**319.

Fujita, A., 1957, Absorption of carotene in human body, *Vitamins* (*J. Vit. Soc. Jpn.*) **13:**1.

Hara, M., 1957, Studies on the absorbability of carotene in human body, *Vitamins* (*J. Vit. Soc. Jpn.*) **13:**545.

Kawakami, T., 1979, A study on carotenoid content in breast milk, *J. Tokyo Med. Col.* **37:**643.

Kawakami, T., Kohri, N. Ushiyama, M., Osakabe, T., Kohno, H., Morishima, N., Ohsumi, K., and Honda, T., 1976, Aurantiasis in infants due to mother's milk, *Pediatr. Jpn.* **17:**411.

McLaren, D. S., Read, W. C., Awdeh, Z. L, and Tchalian, M., 1967, Microdetermination of vitamin A and carotenoids in blood and tissue, in: *Method of Biochemical Analysis, XV* (G. David, ed.), pp. 1–23, Interscience Publishers, New York.

Saito, K., Furuichi, E., Noguchi, Takezaki, A., and Imamura, M., 1963, Study on human milk of the Japanese, *J. Jpn. Soc. Food Nutr.* **15:**408.

33

INFANTILE HYPERCALCEMIA

Brian A. Wharton and Sylvia J. Darke

INTRODUCTION

Hypercalcemia is defined as a plasma calcium concentration of more than 11 mg calcium/100 ml (2.75 mmole/liter) when the blood sample has been collected without venous stasis and after a minimum period of six hours since the introduction of any calcium compounds into the body, either by mouth or intravenously. Rigorous quality control between different laboratories is essential.

In infants, hypercalcemia may be due to hypervitaminosis D, hyperparathyroidism, subcutaneous fat necrosis and metastatic bone tumors, or may be regarded as idiopathic. Other causes of hypercalcemia are seen in older children and adults, for example sarcoidosis, immobilization, diuretic therapy, hypervitaminosis A, hyperthyroidism, phosphate depletion, milk alkali syndrome, and peptide or prostaglandin secreting tumors. Only occasionally do these conditions need to be considered in infancy.

CALCIUM HOMEOSTASIS AND VITAMIN D IN INFANCY

As far as is known, hypercalcemia is not caused by an intake of calcium that is in excess of the body's requirements. Vitamin D stimulates the synthesis of a specific calcium transport protein that facilitates the intestinal absorption of calcium and so ensures an adequate concentration of this mineral at the growing and active sites in bone. Vitamin D may also have a direct stimulating effect on bone formation. There are a number of excellent reviews of calcium metabolism in childhood (Harrison and Harrison, 1979; Tsang *et al.*, 1979).

The most important source of vitamin D is the synthesis of cholecalciferol (vitamin D_3) by the action of ultraviolet radiation from the sun on 7-dehy-

---

Brian A. Wharton • Sorrento and Birmingham Maternity Hospitals, Birmingham, United Kingdom. *Sylvia J. Darke* • Departments of Health and Social Security, London, United Kingdom.

drocholesterol in the skin. Cholecalciferol is rapidly metabolized in the liver to 25-hydroxy vitamin D, which has antirachitic properties. This compound is still further hydroxylated in the kidney to the much more active metabolite 1,25-dihydroxy vitamin D, which is responsible for the absorption of calcium and deposition of calcium in bone. Under normal physiological conditions a feedback mechanism operates in such a way that homeostasis is maintained and both hypo- and hypercalcemia are avoided.

Calcium homeostasis, however, is not achieved solely by regulation of the amount of vitamin D in the body. The secretion of the parathyroid glands controls the concentration of calcium and inorganic phosphorus in blood plasma by increasing the removal of both calcium and phosphorus from bone and by promoting the renal excretion of phosphorus. In addition, calcitonin, a secretion of the thyroid gland, may also affect plasma calcium since, if injected, the resorption of bone is inhibited and plasma-calcium concentration is decreased. Nevertheless, the importance of vitamin D in the maintenance of calcium homeostasis in infancy cannot be overstressed.

The diet is not the main source of vitamin D for the human body and comparatively few foods contain this vitamin. The chief natural food sources are fatty fish, such as herrings, mackerel, salmon, pilchards, sardines, and tuna, which contain amounts that vary, on average, from 22.5 μg/100 g in raw herring to 5.8 μg/100 g in tuna. Eggs, butter, and cheese contain smaller amounts, on average, 1.75, 0.76, and 0.26 μg/100 g respectively. Household milk is said to contain, on average, about 0.03 μg vitamin D/100 g in summer and 0.01 μg vitamin D/100 g in winter. The liver oils of some fish, notably tunney, halibut, and cod are very rich in vitamin D. Tunney liver oil contains 40,000–625,000 μg, halibut 500–10,000 μg, and cod 125–625 μg/100 g. However, these oils cannot be considered as foods, although they are used medicinally to provide supplementary vitamin D.

Infants initially receive dietary vitamin D either in breast milk or in infant formula. Breast milk contains about 0.4 μg vitamin D/liter in the fat soluble form but water soluble vitamin D sulphate is present in concentrations of 4–18 μg/liter (Sahashi *et al.*, 1967; Lakdawala and Widdowson, 1977; Department of Health and Social Security, 1977). The antirachitic activity of this water soluble fraction is not established, however. Leerbeck and Sondergaard (1980), using a biological assay, found that about three-quarters of the vitamin D activity in whole human milk was in the lipid fraction. Most infant formulas have vitamin D added to a concentration of 10 μg/liter.

Once weaning occurs, observed intakes of vitamin D in British infants have been found to decrease. Median daily intakes were 4 μg at 6–18 months of age and 2 μg in 1½–4½-year-old children (Darke *et al.*, 1980). Supplements by mouth and fortified foods were found to be important sources of the vitamin in the first 18 months of life. In older children these sources were replaced by natural foods such as fish and eggs.

The most important source of vitamin D is obtained by synthesis in the skin. Problems of vitamin D deficiency arise when too little ultraviolet radiation reaches the skin for the individual's requirements for vitamin D to be satisfied. This situation may occur in the winter months in countries like Britain. Only solar radiation of wavelength less than 313 nm is effective in forming vitamin D (Knudson and Benford, 1938; Kobayashi and Yasumura, 1973), and the average intensity of such radiation declines with distance from the equator (Johnson *et al.*, 1976). On the other hand, deficiencies of vitamin D may occur at or near the equator, where ultraviolet radiation from the sun is available all the year round if, for reasons of culture or religion, the body is almost completely covered by clothing or the day is spent indoors or in narrow streets where sunlight hardly penetrates. In the same way, cloud cover, pollution of the atmosphere, or living behind glass windows may give rise to symptoms of deficiency because the short wave ultraviolet radiation cannot penetrate glass and is scattered by a curtain of dust particles.

HYPERVITAMINOSIS D AND IDIOPATHIC HYPERCALCEMIA

The extent that hypervitaminosis D and "idiopathic" infantile hypercalcemia are distinct entities is debatable.

A commonly accepted concept is that idiopathic hypercalcemia represents an unusual susceptibility to comparatively normal intakes of vitamin D, and in these circumstances, the condition is often known as Williams syndrome. This syndrome, first described in New Zealand (Williams et al., 1961), may possibly be distinguished from hypervitaminosis D in that presentation is common in the early neonatal period; supra-aortic stenosis and peripheral pulmonary stenosis occur and there is a characteristic elfin face. Fraser *et al.* (1966) suggest that the severe form occurs when the disease starts in utero.

The British "Epidemic"

Rickets has long been known as a disease of children and was regarded in Europe as the "English disease," although it is uncertain whether this was because the disease was widespread in Britain or because two British authors, Daniel Whistler in 1645 and Francis Glisson in 1650, were among the first to describe the condition. The curative properties of cod liver oil for rickets were well known, but Mellanby (1921) and Chick *et al.* (1923) were responsible for establishing that rickets was associated with a deficiency of vitamin D and could be rectified by ultraviolet radiation from the sun or any other source, as well as by giving cod liver oil.

At the beginning of the Second World War, government in the U.K. was concerned with safeguarding the health of the nation and especially of pregnant women and young children. Margarine for domestic use was compulsorily fortified with vitamins A and D in 1940. Also in 1940, under the Welfare Foods Scheme, cheap household milk, as a source of protein and calcium, was made available to all pregnant and lactating women and to children under

5 years of age. National Dried Milk (dried cows' milk) was available for infants who were not breast-fed. Cod liver oil as a source of vitamins A and D and fruit juices as a source of vitamin C were also provided for pregnant women and for children up to 5 years of age. In 1942, the vitamin content of the cod liver oil was doubled in an attempt to compensate for the fact that many children received infrequent and irregular dosage. National Dried Milk was fortified with vitamin D in 1945 and manufacturers of all proprietary infant formulas were quick to follow suit. Some manufacturers had fortified infant formulas with vitamin D before 1945. At this time infant formulas in Britain were based on dried cows' milk that was reconstituted by the addition of varying amounts of water and sugar to achieve a sufficient energy content.

Lightwood (1952) and Payne (1952) from the Hospital for Sick Children, Great Ormond Street, London, first described the occurrence of a mild form of infantile hypercalcemia characterized by a "failure to thrive" and kidney damage. The disease was of relatively sudden onset between the ages of 3 and 9 months and had a good prognosis. Butler and Schlesinger (1951) described a more severe form which presented at an earlier age and had a poorer prognosis.

In 1953 and 1954, the British Paediatric Association (1956) reported about 100 new cases of infantile hypercalcemia per year with an incidence of 1 in 8,000 live births. At that time a conscientious mother might have taken cod liver oil during pregnancy and lactation and have weaned her infant on to fortified dried milk and fortified cereals and be giving cod liver oil daily. Calculations indicated that such an infant could be consuming as much as 100 μg vitamin D/day. But not all infants who developed hypercalcemia consumed as much vitamin D as this. Graham (1959) reported that of 38 patients, 26 (68%) has taken less than 50 μg and 8 (21%) less than 25 μg vitamin D daily.

A government committee that was set up by the Ministry of Health and the Department of Health for Scotland (1957) recommended that the amount of vitamin D in cod liver oil be halved and the vitamin D in National Dried Milk and in infant cereals be reduced. By 1960–1961, when there had been time for the recommendations to be implemented and for stocks of infant foods fortified with the larger amounts of vitamin D to be cleared, the incidence of infantile hypercalcemia had declined (British Paediatric Association, 1964). Stapleton *et al.* (1957) and the British Paediatric Association (1964) both made the point that the amount of vitamin D consumed from fortified products at different ages in the first year of life corresponded well with the reported age distribution of the disorder and provided some confirmation of a causal relationship between the disease and consumption of 25–100 μg vitamin daily.

Clinical Presentation

The major symptoms are vomiting with failure to thrive. Constipation is a common problem and occasionally it is possible to obtain a history of polyuria. A typical "elfin" or Pekinese face may be recognized in the severe variety. A systolic murmur and hypertension are occasionally found, and in severely affected infants, microcephaly with craniosynostosis may develop. Rarely, there is a subacute encephalopathy with drowsiness, stupor, or convulsions. A history of hypercalcemia in a sibling is sometimes obtained (Forfar *et al.*, 1956).

Investigation

Apart from hypercalcemia, further abnormalities are revealed on investigation. Radiology may show dense bones, particularly at the base of the skull and around the orbits giving a "spectacle" appearance. Occasionally nephrocalcinosis and, more rarely, metastatic calcification elsewhere in the body may occur. Biochemically there is evidence of glomerular failure with raised plasma urea and creatinine, hypercalciuria of more than 4mg (100 μmol/kg) per 24 hours, or more than 0.25 mg of calcium/mg creatinine and a raised plasma cholesterol. Plasma phosphorus remains normal and alkaline phosphatase is variable.

Differential Diagnosis

Vomiting plus constipation should always arouse suspicion of hypercalcemia. Pyloric stenosis is the most common cause of this combination of symptoms but metabolic disorders should always be considered.

Once the biochemical abnormality is detected the most likely diagnosis becomes that of hypercalcemia. Hyperparathyroidism is uncommon in infancy, but has been described even in the newborn (Hillman *et al.*, 1964; Spiegel *et al.*, 1977) and is usually associated with a low plasma phosphorus, a high alkaline phosphatase activity, and eventually demineralization of the skeleton with bone resorption. Hypercalcemia may complicate many malignancies, and occasionally these occur in infancy. The tumors may produce parathormone, prostaglandin PGE2, or osteoclast stimulating factor (Tsang *et al.*, 1979).

Treatment

The most important treatment of hypercalcemia is by means of a "low calcium, low vitamin D" diet. A preparation called *Locasol* made by Cow and Gate was used extensively in the British "epidemic" of infantile hypercalcemia. If the plasma calcium concentration does not fall rapidly on this regimen or if the child is severely ill, corticosteroids (e.g., Prednisone, 2 mg/kg body weight daily) should be given in addition. The treatment is continued until plasma and urinary calcium have been normal for at least one month. Calcium may then be slowly introduced into the diet, followed later by the introduction of vitamin D. An occasional child may remain sensitive to vitamin D for many months. Sudden death in the acute stage has been described (Rhaney and Mitchell, 1956).

Two additional treatments have been used since the British "epidemic." Either is indicated when encephalopathy is present and a reduction in plasma calcium concentration becomes urgent. The first treatment uses frusemide at a dosage of 1 mg/kg body weight. This is given every 6 hr. A thiazide diuretic is contraindicated. In addition 0.5 normal saline in dextrose that contains potassium in amounts that provide 30 mmole potassium/liter is administered intravenously in a volume equal to that of the urinary output together with the insensible fluid loss. The latter can be approximately calculated in infancy as 10–20 ml/kg body weight per 6 hr. The treatment is contraindicated if renal failure is already present. The second treatment is to give intramuscular salmon calcitonin in a dosage of 5–8 MRC units/kg body weight at intervals of 6–12 hr. If either of these two treatments is used, corticosteroids should also be given and the intake of both calcium and vitamin D should be restricted.

Prognosis

In nearly all mildly affected children recovery is usually complete (Mitchell, 1960). In the more severe variety of hypercalcemia the prognosis is poorer because problems may occur in relation to the heart (supravalvular aortic stenosis, peripheral pulmonary stenosis, and myocardial calcification), the kidney (nephrocalcinosis) and the brain (mental retardation sometimes associated with craniostenosis).

CURRENT VIEWS ON VITAMIN D SUPPLEMENTATION

More recently in Britain there has been a recrudescence in the incidence of rickets. This reappearance of vitamin D deficiency is almost entirely confined to Asian children, Asian adolescents, and young Asian women. There is evidence that the incidence of the disease mirrors immigration. Also, as the newcomers adapt to conditions here, and health professionals become more aware of the importance of such summer sunshine as the British weather permits and of the need for supplementary vitamin D during pregnancy, lactation, and early childhood, the incidence of deficiency at least in young children is declining. A program of health education in the use of facilities which already exist is considered to be more prudent than the risk of hypercalcemia by increased food fortification (Department of Health and Social Security, 1980a).

In some countries, fortification of certain foods with vitamin D is mandatory. In Britain, margarine for domestic use must contain 7.05–8.82 μg vitamin D/100g and in Canada the fortification of household milk and margarine (but of no other foods except infant formula) is compulsory. Many countries, including Britain and the U.S. permit voluntary fortification of various foods, and regulations ensure that the amount claimed to be added to the food is declared on the label.

In Britain, under the Welfare Foods Scheme, Childrens' Vitamin Drops are available at present at low price to all children up to 5 years of age and are free of charge to children from poor families. These drops contain vitamins A, C and D and provide about 7 μg vitamin D as cholecalciferol in the daily dose. A recent report from the Department of Health and Social Security (1980b) recommends that all infants and young children should receive adequate amounts of vitamins, and that a vitamin supplement is particularly necessary for low birth weight infants, those who receive household milk, Asian children and those for whom other sources of vitamin D (notably from the action of ultraviolet radiation on the skin) is in doubt. Household milk in Britain is not fortified with vitamin D. The report states that either Childrens' Vitamin Drops or a suitable proprietary alternative should be given, and that only one preparation of supplementary vitamin D should be taken daily.

The report emphasises that only when the mother's professional adviser is sure that the infant or child is receiving an adequate intake of vitamin D from other sources such as human milk (if the mother's vitamin D status is good), an approved infant formula, fortified baby foods, or a proprietary multivitamin preparation is there no need to advise the use of a supplement of vitamin D.

REFERENCES

British Paediatric Association, 1956, Hypercalcaemia in infants and vitamin D, *Br. Med. J.* **2**:149.

British Paediatric Association, 1964, Infantile hypercalcaemia, nutritional rickets and infantile scurvy in Great Britain, *Br. Med. J.* **1**:1659–1661.

Butler, N. R., and Schlesinger, B., 1951, Generalised retardation with renal impairment, hypercalcaemia and osteosclerosis of the skull, *Proc. R. Soc. Med.* **44**:296–297.

Chick, H., Dalyell, E. J., Hume, E. M., MacKay, H. M. M., Henderson Smith, H. and Wimberger, H., 1923, Studies of rickets in Vienna, 1919–1922: Report to the Accessory Food Factors Committee, *Medical Research Council Special Report Series No. 77,* H.M.S.O., London.

Darke, S. J., Disselduff, M. M., and Try, G. P., 1980, Frequency distribution of mean daily intakes of food energy and selected nutrients obtained during nutrition surveys of different groups of people in Great Britain between 1968 and 1971. *Br. J. Nutr.* **44**:243–252.

Department of Health and Social Security, 1977, The composition of mature human milk, *Report of a Working Party of the Committee on Medical Aspects of Food Policy,* p. 27, H.M.S.O., London.

Department of Health and Social Security, 1980a.- Rickets and osteomalacia, *Report on Health and Social Subjects No. 19,* H.M.S.O., London.

Department of Health and Social Security, 1980b, Present day practice in infant feeding 1980, *Report on Health and Social Subjects No. 20,* H.M.S.O., London (in press).

Forfar, J. O., Balf, C. L., Maxwell, G. M., and Tompsett, S. L., 1956, Idiopathic hypercalcaemia of infancy: Clinical and metabolic studies with special reference to the aetiological role of vitamin D, *Lancet* **1**:981–988.

Fraser, D., Kidd, B. S. L., Kooh, S. W., and Paunier, L., 1966, A new look at infantile hypercalcaemia. *Pediatr. Clin. North Am.* **13**:503–525.

Graham, S., 1959, Idiopathic hypercalcaemia, *Lancet* **2**:1261–1264.

Harrison, H. E. and Harrison, H. C., 1979, Disorders of calcium and phosphate metabolism in childhood and adolescence, in: *Major Problems in Clinical Paediatrics,* Vol. 19, pp. 15–47, Saunders, Philadelphia.

Hillman, D. A., Scriver, C. R., Pedvis, S., and Shragovitch, I., 1964, Neonatal familial primary hyperparathyroidism, *N. Engl. J. Med.* **270:**483.

Johnson, F. S., Mo. T., and Green, A. E. S., 1976, Average latitudinal variation in ultraviolet radiation at the earth's surface, *Photochem. Photobiol.* **23:**179–188.

Knudson, A. and Benford, F., 1938, Quantitative studies on the effectiveness of ultraviolet radiation of various wavelengths in rickets, *J. Biol. Chem.* **124:**287–299.

Kobayashi, T. and Yasumura, M., 1973, Studies on the ultraviolet irradiation of provitamin D and its related compounds, *J. Nutr. Sci. Vitaminol.* **19:**123–128.

Lakdawala, D. R. and Widdowson, E. M., 1977, Vitamin D in human milk, *Lancet* **i:**167–170.

Leerbeck, E., Sondergaard, H., 1980, The total content of vitamin D in human and cows milk, *Br. J. Nutr.* **44:**7.

Lightwood, R. C., 1952, Idiopathic hypercalcaemia in infants with failure to thrive, *Arch. Dis. Child.* **27:**302.

Mellanby, E., 1921, Experimental rickets, *Medical Research Council Special Report Series No.61,* H.M.S.O., London.

Ministry of Health and Department of Health for Scotland, 1957, *Report of the Joint Subcommittee on Welfare Foods,* H.M.S.O., London.

Mitchell, R. G., 1960. The prognosis in idiopathic hypercalcaemia of infants, *Arch. Dis. Child.* **35:**383–388.

Payne, W. W., 1952. The blood chemistry in idiopathic hypercalcaemia, *Arch. Dis. Child.* **27:**302.

Rhaney, K. and Mitchell, R. G., 1956, Idiopathic hypercalcaemia of infants, *Lancet* **i:**1028–1033.

Sahashi, Y., Suzuki, T., Higaki, M. and Ssano, T., 1967, Metabolism of vitamin D in animals V. Isolation of vitamin D sulphate from mammalian milk, *J. Vitaminol.* **13:**33.

Spiegel, A. M., Harrison, H. E., Marx, S. J., Brown, E. M., and Anbach, G. D., 1977, Neonatal primary hyperparathyroidism with autosomal dominant inheritance, *J. Pediatr.* **90:**269–272.

Stapleton, T., Macdonald, W. B., and Lightwood, R., 1957, The pathogenesis of idiopathic hypercalcaemia in infancy, *Am. J. Clin. Nutr.* **5:**533–542.

Tsang, R. C., Noguchi, A., and Steichen, J. J., 1979, Pediatric parathyroid disorders, *Pediatr. Clin. North Am.* **26:**223–249.

Williams, J. C. P., Barratt-Boyes, B. G., and Lowe, J. B., 1961, Supravalvular aortic stenosis, *Circulation* **24:**1311–1318.

X

DEFICIENT INTAKES: NATURALLY OCCURRING

34

NUTRIENT DEFICIENCIES IN NATURALLY OCCURRING FOODS

Daphne A. Roe

Nutrient deficiencies in foods ingested by animals and human beings may be inherent or acquired. Causal factors, responsible for nutrient deficiency may be single or multiple. A tenable argument is that all foods are nutrient deficient if they do not *per se* provide all nutrients necessary to physiological maintenance without addition of other foods or nutrients. However, given that different foods are consumed that complement one another in providing for nutritional needs, then nutrient-deficient foods are those that have inherent or acquired characteristics that render them inferior in comparison with other foods of similar origin. In the context of human nutrition, a nutrient-deficient food is one that cannot support life, growth, tissue repair, or optimal health if used as a staple food. The limitations of this definition will be obvious to the purist but may serve the purposes of this discussion.

Nutrient-deficient staple foods fall into two major groups: those that have constant and inherent limitations in nutrient content and those that have inconstant or acquired deficiencies.

Two classifications of nutrient-deficient foods that reflect this grouping are given in Table I and Table II.

FOODS OF LOW NUTRIENT-TO-CALORIE RATIO

Certain traditional staples as well as newer energy sources, now extensively consumed because of cheapness and availability, provide foods of low nutrient quality. Nutrient-poor traditional food sources include the edible parts of the starchy root cassava (*Manihot utilissima* and *sago*), from the sago palm *(Metroxylen sagu)*. Food based on these products is extremely low in

Daphne A. Roe • Division of Nutritional Sciences, Cornell University, Ithaca, New York 14853.

TABLE I
Classification of Nutrient-Deficient Foods

A. Foods having inherent deficiencies in nutrient content:
   1. Edible substances of low nutrient density.
   2. Animal or plant derived foods providing protein of biological value.
B. Foods having acquired or inconstant deficiencies in nutrient content[a]:
   1. Foods in which nutrients have been destroyed by post-harvest conditions, storage, processing, or preparation (nutrient lability owing to physical causes).
   2. Foods in which nutrient destruction has occurred due to the presence of a naturally occurring toxicant, a contaminant, or an intentional food additive.
   3. Foods in which nutrient availability is low due to the physical or chemical form of the nutrient or associated with the presence of a natural food toxicant or chemical additive.
   4. Foods that have a reduced content of minerals or trace elements, which they normally provide, owing to production of impoverished soils (environmental depletion).

[a] "... a wide variety in nutrient content is observed among different samples of food." (Watt, 1980.)

protein content. Banana and plantains provide attractive sweet foods that, however, are also low in protein. Foods that have achieved the status of staples in modern society, which are also of very low nutrient-to-calorie ratio, include candy, sodas, syrups, and other sugar-based products. Corn starch is another example of a modern product which supplies only food energy (Davidson *et al.,* 1975). In Africa and South America, where diets are based on these foods, protein malnutrition and avitaminoses are prevalent. Millions of children die in the post weaning periods, or if they survive, suffer permanent physical damage both because of shortage of food and because they subsist on these foods (den Hartog, 1975).

The nutrient-to-calorie ratio of alcoholic beverages is in general low. It is particularly important to appreciate that the vitamin content of alcoholic beverages is not only initially low, but also may further be reduced by various factors which diminish nutrient stability, including the ethanol itself, prolonged storage, and light exposure (Roe, 1979).

PROTEIN SOURCES OF LOW BIOLOGICAL VALUE

Estimates of protein requirements vary for infants, children, and adults in relation to protein quality. The quality of a protein denotes the efficiency of its utilization as a nitrogen source, which is related to the sufficiency or otherwise of energy intake, to the digestibility of the protein, and to the

TABLE II

Factors Responsible for Nutrient Deficiencies in Specific Foods and Effects on Nutrient Content

Factors	Foods of low nutrient density	Low biological value protein sources[a]	Nutrient lability in foods[b]	Foods containing antinutritional factors[c]	Environmental mineral depletion of foods[d]
Natural food composition	Cassava	Maize meal (tryptophan)		Goitrogen-rich cassava (I_2) Protease inhibitor in soybean (protein)	Keshan rice/corn (Se) Andean tubers and cereal grain (I_2)
Manufacture	Candy		Milled rice (thiamin)		Fiber-rich Iranian bread (Fe)
Storage			Irradiated milk (riboflavin)		
Processing			Autoclaved milk (vit. B_6)		
Cooking			Boiled peas (vit. C)		

[a] (limiting amino acid).
[b] (nutrient lost).
[c] (nutrient not available).
[d] (mineral/trace element "missing").

abundance of essential amino acids present in relation to need. Protein quality used to be expressed in terms of *biological value* which, in estimation of nitrogen balance, refers to (Retained Nitrogen/Absorbed Nitrogen) × 100 (Block and Mitchell, 1946). Biological value makes no correction for nitrogen losses in digestion. The more generally accepted term, *net protein utilization* (NPU), is equal to biological value × availability (Allison, 1955). Net protein utilization that actually expresses the percentage of dietary nitrogen retained in the body, when allowance is made for endogenous losses of nitrogen via the urine and feces, is greater for high quality animal protein sources than for vegetable protein sources (Food and Nutrition Board, 1974). The use of the term *biological value* in the subtitle is intended not in its limited definition, but to describe the overall nutrient quality of protein sources.

Whether the nutritive value of the protein in the diet is or is not adequate depends on the NPU of the admixture of protein sources consumed. The first limiting amino acid, that is, the essential amino acid in least supply, varies with different vegetable protein sources (Table III). However, the amino acid sources of diets, based on vegetable protein, are not always dependent on the first limiting amino acid of the staple foodstuff because of the contribution of the amino acid composition of supplementary foods.

Digestibility is an important determinant of protein quality. The low NPU of certain vegetable proteins such as sorghum, is due to low digestibility rather than amino acid imbalance (Nicol and Phillips, 1961). Sorghum products differ in NPU according to fiber content with higher fiber milled whole-meal sorghum having lower digestibility and NPU than lower fiber home-pounded sorghum. However, it has been shown that Nigerian men, adapted to a low protein diet, could achieve nitrogen equilibrium on a diet in which the protein was provided by sorghum flours, cassava, and rice, when the food energy intake was 43 kcal (180 kJ)/kg body weight per day and protein intake equal to 0.41–0.44 g/kg body weight per day (Nicol and Phillips, 1978).

Other determinants of the protein quality of practical diets include the treatment of vegetable protein sources by heat to destroy trypsin inhibitors or by fermentation to alter the level and availability of amino acids (Campbell-Platt, 1980). For example, in the case of locust beans, a small decrease in sulfur-containing amino acids and a larger decrease in aspartic and glutamic acids occurs with production of the fermented product.

TABLE III
Limiting Amino Acids in Vegetable Protein Sources

Protein source	Limiting amino acids
Maize	Tryptophan
Wheat	Lysine
Soy	Methionine

FOODS CONTAINING ANTINUTRITIONAL FACTORS

High Phytate Foods

Phytic acid is the synonym for the plant constituent, inositol hexaphosphate. Phytates constitute 1–2% by weight of many cereals and oilseeds and amounts as high as 6% have been found in certain plant foods. Phytic acid and its salts constitute one of the major food sources of phosphorus in many plant seeds.

Minerals are generally less available from foods of plant origin as compared to animal foods. This has been attributed largely to the presence of phytates.

Mineral–phytate chelates are soluble at low pH but are insoluble at the physiological pH of the human small intestine. Phytic acid complexes with minerals in food which can reduce their availability. It has been shown that there is generally an inverse relationship between intake of high phytic acid in food and mineral absorption but that this relationship varies for different minerals. Whereas earlier workers showed that phytates in food had an adverse effect on the absorption of calcium, magnesium, zinc and iron, recent work has shown minerals added to high phytate foods may be quite well absorbed and that effects on mineral availability, previously attributed to the presence of phytate, may be due to the high fiber content of these foods (Cheryan, 1980).

However, availability of zinc from cereal grains is decreased by the presence of phytate. Reinhold (1971) showed that various types of unleavened bread consumed by Iranian villagers who have been shown to be zinc deficient contains significantly more phytate than commercial leavened breads.

High-Fiber Foods

Evidence has been presented by Haghshenass *et al.* (1971) that iron deficiency anemia may be associated with high intake of dietary fiber in Iranian wheat breads. Further studies by Reinhold (1976) have indicated that the availability of zinc may be reduced when high-fiber diets are ingested, particularly these Iranian bread diets containing wheat bran.

Proponents of high fiber diets, however, emphasize that adverse effects of Iranian wheat breads consumption on mineral availability could be due to high-phytate content rather than high-fiber content (Macdonald, 1976).

In support of this viewpoint, Morris and Ellis (1980) showed that there was no indication in the rat that the fiber in low-phytate wheat brans affected the bioavailability of zinc. Iron in low-phytate bran was highly bioavailable to rats, but the bioavailability of iron from this source was not significantly different from the bioavailability of iron from raw bran. It was suggested by these investigators that the depressing effect of wheat bran on iron absorption in people may not be due to phytate but rather to interaction of fiber with phytate or possibly some other bran component.

TRYPSIN INHIBITORS IN LEGUMES AND CEREALS

In 1966 Mickelson and Yang wrote "Probably as many papers have appeared about the deleterious properties of the raw [soy] beans as about the nutritional advantages of the heated beans." While it has been shown experimentally that chicks, rats, and mice show markedly impaired feed efficiency when given raw soybean meal as a protein source, the adverse effects of soybeans on protein utilization are species specific, both among nonruminants and ruminants (Mickelsen and Yang, 1966). Within species that are sensitive to the adverse nutritional effects of raw soybeans and other raw legumes, younger animals show more severe effects with growth inhibition. Growth inhibition in these animals has been attributed to the presence of a trypsin inhibitor (antitrypsin factor) present in the beans which renders them less available for digestion. Trypsin inhibitors have been found in soybeans, in other legumes, and in wheat flour (Ambrose, 1966).

Barnes and Kwong (1964) reported that whereas unheated soybeans and certain other legumes reduced overall nitrogen absorbability, compared to the heated legumes, it was not known whether this was due to decreased intestinal hydrolysis of the legume protein, or to a decreased rate of amino acid absorption, or whether there was increased fecal nitrogen excretion associated with ingestion of these foods.

Feeding purified trypsin inhibitor as well as raw beans induces pancreatic hypertrophy in rats. It has been suggested that growth depression could be the result of endogenous loss of essential amino acids in pancreatic exocrine secretion (Abbey *et al.*, 1979). However, growth depression in rats fed beans is not entirely due to the presence of the trypsin inhibitor.

When raw soybeans or other legumes are properly heated, many of their adverse nutritional effects disappear. However, feeding heated soybean meal as a single protein source can still have adverse nutritional effects in rats with growth inhibition (particularly sulfur-containing amino acids) (Newberne, 1980).

Secondary effects of trypsin inhibitors in legumes may be to allow undigested protein residues to reach the large intestine where they undergo microbial digestion. Release of tryptophan from undigested protein residue and further microbial metabolism of this amino acid yields indole, which is further metabolized to yield indoxyl sulfate and indoxyl glucuronide. In the formation of indoxyl sulfate, the sulfur is provided by sulfur-containing amino acids in the diet. This may lead to inefficient utilization of both sulfur-containing amino acids in methionine as well as tryptophan (Roe, 1971).

Liener (1979) has further discussed the physiological significance of trypsin inhibitors in human diets and has pointed out that most studies of these protease inhibitors have been carried out in rats or chicks. Trypsin inhibitors are abundant in certain unprocessed, raw foods, such as soybeans used to make foods for infants, children and adults. Trypsin inhibitor activity is high in raw soybeans and 30% of the original activity remains in soybean protein isolates from which textured meat analogs are manufactured. However, the final product contains less than 10% of the antitryptic activity of the raw soybean flour.

The only subgroup of the human population who are likely to obtain their total protein intake from soybean protein are infants in the early months of life who have hypersensitivities to milk proteins. These infants are commonly fed processed, sterilized soy protein isolates. It is estimated that in the U.S. approximately 10% of infants are fed soy isolate-based formulas (Fomon, 1975).

Processing and sterilization of infant soybean formulas also reduces antitryptic activity to less than 10% of the original soy isolate. Potential risks of trypsin inhibitors in soybean products is further reduced because the predominant form of human trypsin (a cationic species in human pancreatic juice) is resistant to the antitryptic activity in the soybean. Whether trypsin inhibitors in other legumes, such as lima and navy beans, reduce nutritive value for people has not been adequately evaluated (Liener, 1979).

TOXICITY OF LECTINS IN FOODS

Plant substances, having the property of agglutinating erythrocytes have been collectively designated as phytohemagglutinins or lectins (Newberne, 1980). High concentrations of lectins have been found in kidney beans *(Phaseolus vulgaris)*. Lectins have also been isolated from navy beans and white kidney beans. It has been proposed that lectins from these sources may impair nutrient digestion and absorption. Pusztai *et al.* (1979) showed that in rats fed *P. vulgaris* lectins from this source, disruption of brush border enterocytes in the duodenum and jejunum occurred but absorption was not markedly compromised. Toxicity and antinutritional effects in the rat may be due to absorption through the damaged intestinal mucosa with resultant increase in protein catabolism leading to higher nitrogen excretion in the urine.

Another possible effect is increased absorption of microbiological toxins secondary to lectin-induced enterotoxicity. Whether or not lectins can actually (as previously suggested) cause tissue invasion by normally saprophytic bacteria has not been clarified. Further, the relevance of these studies to human populations consuming high legume diets is unknown.

GOITROGENS

Goitrogens inhibit uptake of iodine by the thyroid gland. Food sources of goitrogens are mainly from the brassica genus, e.g., cabbage, turnips, rutabagas, mustard greens, horseradish, radish, mustard seed, and cassava. All these foods contain thioglucosides. Thioglucosidases in the plants convert a progoitrin to a goitrin. Hydrolysis of the progoitrin can also occur due to microbiological activity in the intestine. Alternate explanations have been given for the course of events which leads to the adverse effect of the goitrogen on iodine uptake by the thyroid gland. It has been suggested that ingestion of a raw vegetable containing a progoitrin as well as a plant-derived thioglucosidase may occur, or that a cooked vegetable containing a progoitrin may be ingested and that the goitrin may be liberated by the effect of intestinal bacteria. A third suggestion is that progoitrins may be ingested in cows' milk, e.g. cows fed from rape seed, and other progoitrin-containing plants.

Cabbage seed, rape seed, mustard seed, turnip, and Swede roots have been shown to contain the precursor goitrogenic agent, which is a cyclized thiocyanate, which is 1-5-vinyl-2-thiooxazolidine, known as goitrin. Cabbage contains a thioglucoside named glucobrassicin, which on enzymatic cleavage yields thiocyanate, indole acetonitrial, and 3-hydroxymethylindol. The thiocyanate ion is a goitrogen, whereas a thioglucosidase exists in the same plants as the precursor goitrogens, hydrolysis of the progoitrogens to yield the active compound can also be due to bacterial hydrolysis in the large intestine. Plant thioglucosidase is destroyed by cooking. Whereas the antinutrient effect of goitrogens is through inhibition of iodine uptake into the thyroid gland and hence impairment in the synthesis of thyroid hormones, triiodothyronine and thyroxine, foods containing goitrogens may have positive effects.

The indolic compounds which are released when progoitrogens are hydrolyzed speed up drug metabolism in the gastrointestinal tract and may have a protective influence against colon cancer because of an increased rate of metabolism of potential carcinogens (Greer, 1957; Sapeika, 1969; Wills, 1966; Pantuck *et al.*, 1976).

Reports from endemic goiter areas have indicated that goiter may persist despite adequate iodine intake associated with iodine supplementation of a food or food substance (Gibson *et al.*, 1960; Vought *et al.*, 1967).

A high incidence of endemic goiter, apparently resistant to iodine prophylaxis has been found in Colombia and Ecuador. In both of these countries, the incidence of goiter is largely limited to the Andean regions. It has been reported by Meyer *et al* (1978) that in the Cauca–Patia Valley located in the Andean region of Colombia, 21 years of continuous iodine prophylaxis has not wiped out goiter. In the survey of 1948, prior to use of iodine prophylaxis, it was shown that 53% of Colombian school children in that area were affected by goiter; that still an average of about 15% had goiters in 1968 with a range from 1–42% in different districts (Gaitan *et al.*, 1968).

In order to explain persistent goiter incidence in this region it was suggested that persons with "iodine resistant goiters" were consuming goitrogens. Goitrogens are believed by some to be present in the drinking water. It has been proposed that the goitrogens in drinking water are organic, sulfur-containing compounds, probably disulfides of saturated and unsaturated aliphatic hydrocarbons. The hypothesis has been put forward that sedimentary rocks, rich in organic material, are the main source of the water-borne goitrogens in Colombia (Servicio Geologico Nacional, 1962).

Although it is claimed that persistent iodine deficiency in Colombia cannot be explained on the basis of availability to indigent people of foods of higher iodine content, or to breakdown in the distribution of the iodization program, these alternate explanations for the persistence of goiter have not been fully explored. It would be of interest to investigate the amounts of "goitrogenic waters" which are consumed directly and which might also be consumed in foods.

Goitrogens in the food supply, particularly in cassava, have also been considered to explain the differential distribution and severity of endemic goiter on Idjwi Island in the Congo. Whereas iodine deficiency of soils on Idjwi Island on Lake Kivu in the Congo was general, at the time of studies there in 1970–71, goiter was ten times more prevalent in the northern parts of the island than in the other areas. Cassava is eaten in large amounts on this island, and it has been shown that consumption of cassava reduces the radioiodine uptake by the thyroid gland, especially consumption of the cassava grown in the northern part of Idjwi Island. Consumption of cassava also increases iodine excretion. It has been further demonstrated that cassava contains linamarin, another goitrogen that is a cyanogenic glucoside. In Idjwi Island goiter it has therefore been suggested that iodine deficiency is a result both of geographically determined iodine deficiency of plant foods and also ingestion of goitrogens (DeLange and Ermans, 1971).

OXALATES AND CALCIUM ABSORPTION

Oxalates occur in foods in water soluble and insoluble forms. Studies conducted over a period of 40 years indicate that ingestion of oxalates interferes with calcium absorption on laboratory animals.

It has been shown that calcium present in Amaranthus spp. leaves is not available for absorption by human subjects, and furthermore, that eating these leaves with milk makes the milk calcium less available. A working hypothesis is that the adverse effects of Amaranthus leaf consumption on calcium absorption is due to the oxalate content of the plant.

It has been shown that if the water soluble oxalates are removed by discarding water in which the Amaranthus leaves are cooked, Amaranthus consumption no longer has an inhibitory effect on calcium absorption (Pringle and Ramasastri, 1978). From a practical standpoint, this approach to the oxalate problem confers nutritional disadvantages in that water soluble vitamins including folacin would also be thrown away with the cooking water. In India, and elsewhere, Amaranthus is eaten by people whose intake of folacin is marginal, thus cooking losses could impose a real nutritional risk.

THIAMINASES IN EDIBLE PLANTS AND FISH

Thiaminases have been found in fish, in bracken fern, and in many plants used as food by humans. Monogastric farm animals that consume large amounts of bracken fern develop thiamin deficiency (Kingsbury, 1964). Japanese investigators have shown that ferns, including bracken fern, may be eaten by Oriental people. It is further noted that some persons consuming these ferns develop visual impairment and it has been suggested that this could be due to a thiamin deficiency.

The Haff disease (acute alimentary myositis) that has been reported occurring in people living on the shores of the Baltic Sea is associated with eating fresh water fish such as bream, turbot, eel, pike, perch, and roach. Clinical features include muscle pains, paralysis and myoglobinuria that develop 18–24 hr after the fish is eaten, and recovery is within 24 hours. It has been suggested that this disease is due to the presence of a thiaminase in the fish. A similar disorder has occurred in silver foxes that have been fed on carp (Green *et al.*, 1942).

Doubt has been cast on the attribution of Haff disease and the fox counterpart to effects of fish thiaminase, both because the signs do not resemble thiamin deficiency, and because of rapid recovery. An alternate explanation of the Haff disease has been that of chemical intoxication due to the fish being poisoned by industrial waste (Ham, 1963).

UNUSUAL AMINO ACIDS IN PLANT FOODS

Mimosine

Mimosine or leucenine is an unusual amino acid (β-[N]-(3-hydroxypyridone-3)-α-aminoproprionic acid). It is found in the wild tamarind *(Leucaena glauca)*. The plant has a wide distribution in various parts of the world including South and Central America, Florida, and the Far East. In Indonesia where *L. glauca* is consumed, it has been found that, from time to time, groups of children and adults have developed acute alopecia which has involved the scalp and eyebrows. Localized edema of the scalp has also been observed (van Veen, 1966).

Mimosine forms complexes with pyridoxal-5-phosphate, and has been shown to be a vitamin B_6 antagonist (Lin and Ling, 1961, 1962).

Mimosine forms an iron complex that causes inactivation. If *L. glauca*-containing soups and stews are cooked in iron pots, then the amino acid no longer has any adverse effects (van Veen, 1966).

Selenocystathionine

In 1964, Kerdel-Vegas described a generalized scalp hair loss following ingestion of a Venezuelan seed, locally called *coco de mono (Lecythis ollaria)*. A man, aged 54, on a hunting trip in the state of Portuguesa in Venezuela had eaten 70–80 *coco de mono* almonds. That same day, he developed anxiety followed by violent chills, anorexia, and diarrhea followed by arthralgia and sudden loss of the scalp and body hair. Hair loss apparently occurred 8 days after the ingestion of the almonds. A violet streak was noted across his nails. In experimental mice, administration of *coco de mono* almonds inhibited normal growth of hair after partial depilation. The animals also lost weight and died if the *coco de mono* treatment was continued.

Aronow and Kerdel-Vegas (1964) extracted coco de mono and from the extracts obtained a crystal which on purification was shown to be selenocystathionine or L-2-amino-4-(L-2-amino-2-carboxyethyl) selenylbutyric acid. The toxicity of selenocystathione to fibroblast cells was demonstrated, and it was further shown that when cysteine was added to the culture medium the toxic effect was reversed. It was then proposed that the selenocystathionine interfered with the utilization of the sulfur-containing amino acid, L-cysteine, required in the synthesis of hair.

IODINE-DEFICIENT FOODS AND GOITER INCIDENCE

The iodine content of food crops depends on the soil content of iodine and on the type of fertilizer applied. Cereal grains and other vegetable foods are iodine deficient when they are grown on iodine-deficient soils. Iodine-deficient soils occur in certain mountainous regions of the world and on the sites of old glacial ice fields. Secondary factors which are determinants of iodine deficiency in the soil include irrigation by water, which does not pass through marine rock deposits containing iodine, little decaying vegetation, heavy cropping and farming that leach iodine out of the soil, and erosion (Norris, 1975). In iodine-deficient areas, foods of vegetable origin are low in iodine when they are locally grown, prepared with use of low-iodine water supply, and without use of iodized salt or other iodide fortified materials.

Although in iodine-deficient areas farm animals are iodine deficient and their milk reflects local iodine deficiency, nevertheless, animal protein foods are higher in iodine in all areas than foods of vegetable origin (Kidd, 1974). In contrast, foods in iodine-deficient areas which are sufficient in this element are foods imported from iodine-rich areas, foods prepared with iodized salt, iodine-containing food additives, foods containing iodine as a contaminant from iodophors, foods of marine origin, and foods prepared with water containing significant amounts of iodine.

In Egypt, it was determined by Coble *et al.* (1968) that water may significantly contribute to local iodine intakes. Goiter incidence in Egyptian oases is inversely related to the iodine content of well waters. It appears that in Egypt the iodine content of local water is increased when the water passes through marine deposits containing iodine. In other countries goitrous areas are generally distinguished by water supplies containing very little iodine (Karmarkar *et al.*, 1974).

Platzer *et al.* (1977) investigated factors influencing goiter incidence in the Alto Adige in northern Italy. The target population was that of Certosa, a village at an altitude of 1327 m that has long been known for goiter incidence. The region is poor in iodine; the mean iodine content of 55 samples of local drinking water was found to be 0.81 ± 0.96 μg/liter. The iodine content of several foods was lower than those from the markets of nearby Turin. Among foods from the valley of Certosa, for which iodine content was determined, lowest values were found in rye flour, potato, and onion. Higher food content of iodine was found in bacon, which was explained by addition of iodine to animal fodder by breeders. A combination of economic necessity and habit may explain intake of low iodine foods by goitrous villagers.

Endemic goiter incidence is influenced by socioeconomic factors that define access to particular foods. Koutras *et al.* (1970), working in mountainous areas of Greece, confirmed that goitrous areas were those where there had been a recent and intense glaciation where iodine had been washed out of the superficial layers of soil, where superficial drinking water sources had a low pH, and where there was bacterial pollution of drinking waters. Nongoitrous areas were those where there were calcareous rock formations, where the water had a higher pH, and where there was "no pollution." Differential levels of iodine intake were demonstrated between the goitrous and nongoitrous population which were related to socioeconomic status. Koutras showed that in the goitrous areas, consumption of animal protein fod, which is a better source of iodine, was lower than in the nongoitrous areas.

Koutras *et al.* (1973) also showed that in Greece endemic goiter in children is associated with malnutrition. Alternate explanations are that iodine deficiency impairs growth or that low energy-low protein diets that cause nutritional dwarfing are also low in iodine. In adults, Koutras found that goiter was associated with short stature.

SELENIUM-DEFICIENT FOODS

Staple foods are selenium deficient in areas of the world where there are selenium-poor soils. Selenium-impoverished soils occur widely in New Zealand, particularly in the central volcanic plateau of the Northern Island, but also over most of the Southern Island. Other areas of the world where there are selenium-deficient soils include large parts of China. Plants, including cereal grains, grow well on selenium-poor soils, apparently because selenium is not essential for most plant species. Milk, eggs, and meat derived from animals grazed in selenium-deficient areas show lowered selenium content. Biochemical evidence of selenium deficiency has been found in healthy New Zealanders living in the areas of selenium deficiency. Markers of selenium deficiency have included low red blood cell levels of glutathione peroxidase (GSHPx), as well as decreases in the selenium concentration of whole blood, erythrocytes, and plasma. Estimates have been made of the selenium intake of New Zealanders in the endemic selenium deficiency areas and it has been shown that if food is limited to that which is locally produced, selenium intake would be much lower than that of New Zealanders living in areas not selenium deficient and also lower than that of people living in other parts of the world where there is no regional selenium deficiency. However, there are many imported foods in New Zealand and consumption of these foods coming from selenium rich areas may to some extent compensate for low intakes of selenium from locally produced foods. Access to selenium-rich foods may well depend on socioeconomic factors.

Thomson and Robinson (1980) who have reported extensively on selenium deficiency in humans in New Zealand populations have not found a selenium responsive disease in healthy residents consuming well-balanced diets. They therefore do not think there is any justification for selenium supplementation of the general population there.

In selenium-deficient areas of New Zealand, selenium intakes only exceed 30 μg/day when diets include fish, liver, kidney, or several eggs. A total of 28 μg selenium/day represents a mean of values obtained for men and women of 33 and 23 μg selenium/day, respectively.

Studies of the selenium status of New Zealand infants and children with phenylketonuria and maple syrup urine disease who had to consume synthetic protein diets containing less than 5 μg selenium/day, show biochemical evidence of selenium deficiency but no clinical evidence to indicate an adverse effect to these low selenium intakes (McKenzie *et al.*, 1978).

Van Rij *et al.* (1979) reported on a patient receiving total parenteral nutrition who developed muscle tenderness that responded to selenium supplementation. Claims that various muscular pains are related to selenium deficiency have been questioned.

Until recently, Burke's statement that "no human disease has been vigorously proved to be causally related to a deficiency of this element: selenium", was unchallenged, but newer information indicates that this statement must be qualified (Burke, 1976). Not only is there some evidence that clinical signs of selenium deficiency may arise in patients on total parenteral nutrition, but also in China a fatal cardiomyopathy (Keshan disease) in children has been linked to a severe dietary lack of selenium. Children with this disease are usually between the ages of one and nine, and they present with cardiac failure. The areas of China where Keshan disease has been described include a belt stretching across the country from northeast to southwest. Dietary staples in this area include corn and rice. It seems that the children who have been reported with the cardiomyopathy did not have access to foods which contain significant quantities of selenium. Whether or not this disease represents a true dietary selenium deficiency is not as yet clear, particularly as the established disease does not respond to administration of sodium selenite in doses of 1 mg/week (Keshan Disease Research Group, 1979a,b; Anonymous, 1980).

ZINC-DEFICIENT FOODS

Fruits and vegetables are poor sources of zinc. Rich sources of zinc in the food supply include animal protein foods such as meat, eggs, milk products, and shellfish, particularly oysters. A number of factors may influence the zinc content of foods, including processing and food preparation. Zinc is lost during the milling of wheat for flour, and bread made from white flour has a lower zinc content than whole wheat bread. Similarly, cornstarch contains less zinc than whole corn kernels (Halstead *et al.*, 1974).

Preparation of food can also influence zinc content. Zinc content of foods is increased when acid foods are cooked in contact with galvanized metal containing zinc, but this type of "contamination" is unusual today in most countries because of an increased use of stainless steel and plastic-coated cooking utensils.

Diets which are zinc deficient are composed entirely of foods of low zinc content. Infant diets may be zinc deficient when cows' milk, low in zinc (3–4 μg/ml), and a cereal such as rice with a zinc content of 11–12 μg/g comprise the total food intake. It is considered that zinc deficiency is most likely to occur in infants, children, or adults who consume a cereal-based diet that is not only low in zinc but in which the availability of zinc is reduced by one or more antinutritional factors (Oberleas and Prasad, 1976).

However, recent studies indicate that the total zinc content of meals is the most important factor influencing the amount of zinc absorbed (Sandstrom and Cedarblad, 1980).

NUTRIENT LABILITY IN FOODS

During the storage, processing and preparation of foods for human consumption, vitamins may be destroyed and proteins may undergo chemical changes that diminish their nutritional value. Physical or physicochemical factors that destroy vitamins include irradiation, freezing and thawing, heat, changes in pH, milling, and leaching (Borenstein, 1980).

Thiamin is one of the least stable water-soluble vitamins. It readily loses activity during food storage, processing and cooking. Cereals, stored as whole grains, will lose thiamin if the moisture content is high. Canned meats, such as pork, lose thiamin when stored. Unenriched wheat flour is a poor source of thiamin because most of the vitamin is lost in milling. Thiamin is also lost into the water when cereal grains, such as rice, are washed. Cooking losses vary with the food product and the method employed. Baking is particularly destructive to thiamin in cereal products. Low temperature cooking of food products, having a $pH > 5.5$ leads to considerable losses. In wet processes, thiamin is leached out of food. The bleaching of vegetables is associated with loss of the vitamin. Thiamin losses from meat occur when frozen cuts are thawed and the vitamin passes into the thaw juice. Thiamin is also leached out into the meat juice when roasts are cut at the table (Aurand and Woods, 1973; Klaui, 1979).

FOLACIN

The folacin content of vegetable foods varies with plant variety, processing, and storage, as well as with cooking method. Destruction or change of folacin in foods may be due to heat, light, oxygen, or pH change.

Folacin is photodegradable. Storage losses of folacin in foods may be due to effects of light exposure or activity may be lost because of the natural absence or failure to add a reducing agent such as ascorbic acid.

Heat treatment of commercially available pasteurized cows' milk results in marked lowering of folacin concentration. These losses of folacin during heat treatment can be prevented by addition of sodium ascorbate before the boiling procedure (Ek and Magnus, 1980). Microwave cooking destroys folacin in convenience foods (Cooper *et al.,* 1978).

Food folacins are converted from one chemical form to another by storage and preparations. Freezing and thawing causes deconjugation and alters the state of oxidation of constituent folacins. Food additives may reduce folacin stability and availability.

The availability of dietary folacins varies not only from food to food but also from one sample of a food to another, according to source, manufacture, and cooking method (Perloff and Butrum, 1977).

RIBOFLAVIN

Major losses of riboflavin occur by photodegradation. Exposure of milk to direct sunlight for 4 hr causes loss of 71% of the riboflavin present. If eggs are exposed to light during the cooking process (as in open pan scrambling), there is loss of riboflavin. The sun drying of fruits and vegetables also destroys the vitamin, as does sun drying of meats as practiced in certain tropical countries.

Riboflavin is destroyed by alkali as when soda is added to soften dried beans and peas. Vegetables held in warm cooking water containing baking soda lose riboflavin. Riboflavin may also be leached out of meats or lost in thaw juices.

Milling of grain reduces the riboflavin content of cereal products because the riboflavin is contained in the germ and bran that are removed. Polished rice contains 59% of the riboflavin originally found in brown rice (Hunt, 1975).

Improper storage and food processing are major causes of riboflavin deficiency in foods that, in the U.S., are potentially offset by the use of riboflavin in the enrichment of bread and cereals.

VITAMIN A (RETINOL)

Vitamin A is susceptible to destruction by light and oxidation. The oxidation is catalyzed by traces of iron and copper. Vitamin A-deficient foods are those that are naturally poor sources of the vitamin, e.g., cereal grains or sources of the vitamin that have been exposed to light and air. Addition of antioxidants, such as vitamin E, to foods protects against vitamin A losses (DeRitter *et al.,* 1974).

VITAMIN B_{12}

Vitamin B_{12} deficient foods are vegetables, fruits, or cereals prepared without protein sources of animal origin. Plant foods *per se* are devoid of vitamin B_{12}. Whereas it has been reputed that vitamin B_{12} in animal protein foods is destroyed by concurrent intake of vitamin C supplements, further research has indicated that these findings were due to an artifact in analysis (Herbert and Jacob, 1974; Newmark and Scheiner, 1977). However, vitamin B_{12} is slowly decomposed by reducing agents including vitamin C.

VITAMIN C (ASCORBIC ACID)

Storage and food processing as well as cooking can cause major losses of vitamin C. Prolonged storage at room temperature can result in considerable loss. After processing there is less vitamin C content due to cutting, bruising, washing, blanching, and canning. Further losses occur in cooking. Highest losses from vegetables occur when they are cooked in large volumes of water for a prolonged period and exposed to the air (Aurand and Woods, 1973). Vitamin C is readily oxidized when exposed to the air and is also leached into water. Greater cooking losses of vitamin C occur with preparation of leafy rather than root vegetables (Paul and Southgate, 1978).

VITAMIN B_6 (PYRIDOXINE)

Major losses of vitamers of vitamin B_6 do not occur with usual methods of food storage, processing, and preparation. Destruction of the vitamin occurs when cows' milk is autoclaved as was formerly carried out in the preparation of infant formula, which led to one reported outbreak of vitamin B_6 deficiency in babies (Coursin, 1954).

PROTEINS

Excessive heating of food proteins may result in loss of certain amino acids or may make them unavailable for digestion and hence absorption. The most important changes in protein constituents due to overheating, consist of reaction of amino acid residues of the protein with sugars. The amino acid most commonly involved in this reaction is lysine, but arginine, tryptophan, and histidine may be involved in the process (National Research Council, 1950).

Strong alkali treatment of certain food proteins including soy protein may reduce protein quality. However, alkali-heated soy protein isolate, fed with a well-balanced diet, does not produce adverse nutritional effects in rats (DeGroot and Slump, 1969; Van Beek *et al.,* 1974).

COMMENT AND CONCLUSIONS

A review of nutrient deficiencies in naturally-occurring foods is incomplete without comment on the nutritional risk of these foods to human populations. Nutritional risk is significant under the following circumstances:

1. Foods that are naturally deficient in essential nutrients are consumed in a monotonous diet without other foods that supply the nutrient in short supply.
2. Foods that contain antinutritional factors are improperly processed or prepared, in order to destroy the toxic factor, before being eaten.
3. Foods comprising the bulk of the diet are stored, processed, and prepared in a manner that causes destruction of micro- or macronutrients to an extent in which intake of these essential nutrients is less than requirements.

REFERENCES

Abbey, B. W., Neale, R. J., and Norton, G., 1979, Nutritional effects of field bean (*Vicia faba* L.) proteinase inhibitors fed to rats, *Br. J. Nutr.* **41:**31.

Allison, J. B., 1955, Biological evaluation of proteins, *Physiol. Rev.* **35:**664–700.

Ambrose, A. M., 1966, Naturally occurring antienzymes (inhibitors), in: *Toxicants Occurring Naturally in Foods,* NAS/NRC Food Protection Committee, Washington, D.C.

Anonymous, 1980, Prevention of Keshan cardiomyopathy by sodium selanite, (editorial), *Nutr. Rev.* **38:**278–279.

Aronow, L., and Kerdel-Vegas, F., 1964, Cytoxic factor in Lecythis ollaria ("coco de mono"), *Rev. Dermat. Venezol.* **4:**186–208.

Aurand, L. W., and Woods, A. E., 1973, *Food Chemistry,* pp. 211–214, AVI, Westport, Connecticut.

Barnes, R. H., and Kwong, E., 1964, Methionine absorption and utilization from soybean protein and the effect of soybean trypsin inhibitor—a study of amino-acid availability, in: *The Role of the Gastrointestinal Tract in Protein Metabolism* (H. N. Munro, ed.), pp. 41–59, F. A. Davis, Philadelphia.

Block, R. J., and Mitchell, H. H., 1946, The correlation of the amino acid composition of proteins with their nutritive value, *Nutr. Abstr. Rev.* **16:**249–278.

Borenstein, B., 1980, Effects of processing on the nutritional value of foods, in: *Modern Nutrition in Health and Disease,* 6th ed. (R. S. Goodhart and M. E. Shils, eds.), pp. 497–505, Lea and Febiger, Philadelphia.

Burk, R. F., 1976, Selenium in man, in: *Trace Elements in Human Health and Disease* (A. S. Prasad, ed.), pp. 105–133, Academic Press, New York.

Campbell-Platt, G., 1980, African locust bean (*parkia* species) and its West African fermented product, dawadawa, *Ecol. Food Nutr.* **9:**123–132.

Cheryan, M., 1980, Phytic acid interactions in food systems, in: *Critical Reviews of Food Science and Nutrition* (T. E. Furia, ed.), pp. 197–328, CRC Press, Boca Raton, Florida.

Coble, Y., Davis, J., Schulert, A., Heta, F., and Awad, A. Y., 1968, Goiter and iodine deficiency in Egyptian oases, *Am. J. Clin. Nutr.* **1:**277–283.

Cooper, R. G., Chen, T. -S., and King, M. A., 1978, Thermal destruction of folacin in microwave and conventional heating, *J. Am. Diet. Assoc.* **73:**406–410.

Coursin, D. B., 1954, Convulsive seizures in infants with pyridoxine-deficient diets, *J. Am. Med. Assoc.* **154:**406–408.

Davidson, S., Passmore, R., Brock, J. F., and Truswell, A. S., 1975, *Human Nutrition and Dietetics,* pp. 212–226, Churchill Livingstone, Edinburgh, London, New York.

DeGroot, A. P., and Slump, P., 1969, Effects of severe alkali treatment on proteins on amino acid composition and nutritive value, *J. Nutr.* **98:**45–56.

DeLange, F., and Ermans, A. M., 1971, Role of dietary goitrogens in the etiology of endemic goiter in Idjwi Island, *Am. J. Clin. Nutr.* **24:**1354.

den Hartog, C., 1975, *World Food Needs and Resources—The Current Crisis and Future Prospect,* Proc. X International Congress of Nutrition, Kyoto, Japan, 1975, Victory-sha Press, Kyoto, Japan.

DeRitter, E., Osadca, M. Scheiner, J., and Keating, J., 1974, Vitamins in frozen convenience dinners and pot pies, *J. Am. Diet. Assoc.* **64:**391–397.

Ek, J., and Magnus, E., 1980, Plasma and red cell folacin in cows' milk-fed infants and children during the first 2 years of life: The significance of boiling pasteurized cows' milk, *Am. J. Clin. Nutr.* **33:**1220–1221.

Food and Nutrition Board, 1974, *Recommended Dietary Allowances,* 8th rev. ed., pp. 45–46, National Academy of Sciences, National Research Council, Washington, D.C.

Fomon, S. J., 1975, What are infants fed in the United States, *Pediatrics* **56:**350.

Gaitan, E., Wahner, H. W., Correa, P., Bernal, R., Jubiz, W., Gaiton, J. E., and Llanos, G., 1968, Endemic goiter in the Cauca Valley. 1: Results of limitations of 12 years of ildine prophylaxis, *J. Clin. Endocrinol. Metab.* **28:**1730.

Gibson, H. B., Howeler, J. F., and Traments, F. W., 1960, Seasonal epidemics of endemic goiter in Tasmania, *Med. J. Aust.* **1:**875.

Green, R. G., Carlson, W. E., and Evans, C. A., 1942, The inactivation of vitamin B_1 in diets containing whole fish, *J. Nutr.* **23:**165.

Greer, M. A., 1957, Goitrogen substances in food, *Am. J. Clin. Nutr.* **5:**440.

Haghshenass, M., Mahloweth, M., Reinhold, J. G., and Mohammadi, N., 1971, Iron deficiency anaemia in an Iranian population associated with high intakes of iron, *Am. J. Clin. Nutr.* **25:**1143.

Halstead, J. A., Smith, J. C., and Irwin, M. I., 1974, A conspectus of research on zinc requirements of man, *J. Nutr.* **104:**345–378.

Ham, T. H., 1963, The myoglobinurias, in: *Cecil-Loeb Textbook of Medicine,* 11th ed. (P. B. Beeson and W. McDermott, eds.), p. 1464, W. B. Saunders, Philadelphia, London.

Herbert, V., and Jacob, E., 1974, Destruction of vitamin B_{12} by ascorbic acid, *J. Am. Med. Assoc.* **230:**241.

Hunt, S. M., 1975, Nutritional intake of riboflavin, in: *Riboflavin* (R. S. Rivlin, ed.), pp. 199–219, Plenum Press, New York, London.

Irwin, M. I., and Hegsted, D. M., 1971, A conspectus of research on protein requirements of man, *J. Nutr.* **101:**385–430.

Karmarkar, M. G., Deo, M. G., Kochupillai, N., and Ramalingaswami, V., 1974, Pathophysiology of Himalayan endemic goiter, *Am. J. Clin. Nutr.* **27:**97–103.

Kerdel-Vegas, F., 1964, Generalized hair loss due to ingestion of "coco de mono" *(Lecythis ollaria), J. Invest. Dermatol.* **42:**91–94.

Keshan Disease Research Group of the Chinese Academy of Medical Sciences, Beijing, 1979a, Epidemiologic studies on the etiological relationship of selenium and Keshan disease, *Chinese Med. J.* **92:**477–482.

Keshan Disease Research Group of the Chinese Academy of Medical Sciences, Beijing, 1979b, Observations on the effect of sodium selenite in prevention of Keshan disease, *Chinese Med. J.* **92:**471–476.

Kidd, P. S., Trowbridge, F. J., Goldsby, J. B., and Nichaman, M. Z., 1974, Sources of dietary iodine, *J. Am. Diet. Assoc.* **65:**420–422.

Kingsbury, J. M., 1964, *Poisonous Plants in the United States and Canada,* Prentice Hall, Englewood Cliffs, New Jersey.

Klaui, H., 1979, Inactivation of vitamins, *Proc. Nutr. Soc.* **38:**135–141.

Koutras, D. A., Christakis, G., Trichopoules, D. et al., 1973, Endemic goiter in Greece: Nutritional status, growth, and skeletal development of goitrous and non-goitrous populations, *Am. J. Clin. Nutr.* **26:**1360–1368.

Koutras, D. A., Papapetrou, P. D., Yataganes, X., and Malamos, B., 1970, Dietary sources of iodine in areas with and without iodine-deficiency goiter, *Am. J. Clin. Nutr.* **23:**870–847.

Liener, I. E., 1979, The nutritional significance of plant protease inhibitors, *Proc. Nutr. Soc.* **38:**109–113.

Lin, Y.-Y., and Ling, K.-H., 1961, Isolation and identification of mimosine, *J. Formosan Med. Assoc.* **60:**657.

Lin, Y.-Y., and Ling, K.-H., 1962, Studies on the mechanism of toxicity of mimosine, *J. Formosan Med. Assoc.* **61:**997.

Macdonald, I., 1976, The effects of dietary fiber: Are they good? in: *Fiber in Human Nutrition* (G. A. Spiller and R. J. Amen, eds.), pp. 263–269, Plenum Press, New York, London.

McKenzie, R. L., Rea, H. M., Thomson, C. D., and Robinson, M. F., 1978, Selenium concentration and glutathione peroxidase activity in blood of New Zealand infants and children, *Am. J. Clin. Nutr.* **31:**1413.

Meyer, J. D., Gaitan, E., Merino, H., and DeRouen, T., 1978, Geological implications in the distribution of endemic goiter in Colombia, South America, *Int. J. Epidemiol.* **7:**25–31.

Mickelsen, O., and Yang, M. G., 1966, Naturally occurring toxicants in foods, *Fed. Proc. Fed. Am. Soc. Exp. Biol.* **25:**104–123.

Morris, E. R., and Ellis, R., 1980, Bioavailability to rats of iron and zinc in wheat bran: Response to low phytate bran and effect of the phytate/zinc molar ration, *J. Nutr.* **110:** 2000–2010.

National Research Council, 1950, *The Problem of Heat Injury to Dietary Protein,* Reprint and Circular Series No. 131, National Academy of Sciences, Washington, D.C.

Newberne, P. M., 1980, Naturally occurring food-borne toxicants, in: *Modern Nutrition in Health and Disease,* 6th ed. (R. S. Goodhart and M. E. Shils, eds.), pp. 463–496, Lea and Febiger, Philadelphia.

Newmark, H. L., and Scheiner, J., 1977, Destruction of vitamin B_{12} by vitamin C, *Am. J. Clin. Nutr. Letter* **30:**299.

Nicol, B. M., and Phillips, P. G., 1961, Reference groundnut flour (GNF) and reference dried skim milk (DSM) as supplements to the diets of Nigerian men and children, pp. 151–168, Publication Number 843, National Research Council, National Academy of Sciences, Washington, D.C.

Nicol, B. M., and Phillips, P. G., 1978, The utilization of proteins and amino acids in diets based on cassava *(Manihot utilissima)*, rice, or sorghum *(Sorghum sativa)* by young Nigerian men of low income, *Br. J. Nutr.* **39:**271–287.

Norris, H., 1975, Geochemistry and minerology of trace elements, in: *Trace Elements in Soil–Plant–Animal Systems* (D. J. G. Nicholas and A. R. Egan, eds.), pp. 75–76, Academic Press, New York, San Francisco, London.

Oberleas, D., and Prasad, A. S., 1976, Factors affecting zinc homeostasis, in: *Trace Elements in Human Health and Disease, Vol. 1, Zinc and Copper* (A. S. Prasad and D. Oberleas, eds.), pp. 155–162, Academic Press, New York, San Francisco, London.

Pantuck, E. G., Hsiao, K. C., Loub, W. D. et al., 1976, Stimulatory effects of vegetables on intestinal drug metabolism in the rat, *J. Pharmacol Exp. Ther.* **198:**278–283.

Paul, A. A., and Southgate, D. A. T., 1978, *McCance and Widdowson's The Composition of Foods,* 4th rev. ed. of MRC Special Report No. 297, p. 25, HMSO, London and Elsevier/North Holland Biomedical Press.

Perloff, B. P., and Butrum, R. R., 1977, Folacin in selected foods, *J. Am. Diet. Assoc.* **70:**161–172.

Pingle, U., and Ramasastri, B. V., 1978, Effect of water-soluble oxalates in *Amaranthus* spp. leaves on the absorption of milk calcium, *Br. J. Nutr.* **40:**591–594.

Platzer, S., Fill, H., Reccabona, G., et al. 1977, Endemic goiter in Alto Adige (Italy), *Acta Endocrinol.* **85:**325–334.

Pusztai, A., Clarke, M. W., and King, T. P., 1979, The nutritional toxicity of *Phaseolus vulgaris* lectins, *Proc. Nutr. Soc.* **38:**115–120.

Reinhold, J. G., 1971, High phytate content of rural Iranian bread: A possible cause of human zinc deficiency, *Am. J. Clin. Nutr.* **24:**1204–1206.

Reinhold, J. G., Faradji, B., Abadi, P., and Ismail-Bergi, F., 1976, Decreased absorption of calcium, magnesium, zinc and phosphorus by humans, due to increased fiber and phosphorus consumption as wheat bread, *J. Nutr.* **106:**493–503.

Roe, D. A., 1971, Effects of methionine and inorganic sulfate on indole toxicity and indican excretion in rats, *J. Nutr.* **101:**645–654.

Roe, D. A., 1979, *Alcohol and the Diet,* pp. 208–210, AVI, Westport, Connecticut.

Sandstrom, B., and Cedarblad, A., 1980, Zinc absorption from composite meals. II. Influence of the main protein source, *Am. J. Clin. Nutr.* **33:**1778–1783.

Sapeika, N., 1969, *Food Pharmacology,* Charles Thomas, Springfield, Illinois.

Servicio Geologico Nacional, Ministerio de Minas y Petroleos, 1962, *Mapa Geologico de Colombia* (Compiled by H. Hubach, L. Radelli, and H. Burgl).

Thomson, D. D., and Robinson, M. F., 1980, Selenium in human health and disease with emphasis on those aspects peculiar to New Zealand, *Am. J. Clin. Nutr.* **33:**303–323.

Van Beek, L., Feron, V. J., and DeGroot, A. P., 1974, Nutritional effects of alkali-treated soy protein in rats, *J. Nutr.* **104:**1630–1636.

Van Rij, A. M., Thomson, C. D., McKenzie, J. M., and Robinson, M. F., 1979, Selenium deficiency in total parenteral nutrition, *Am. J. Clin. Nutr.* **32:**2076.

van Veen, A. G., 1966, Toxic properties of some unusual foods, in: *Toxicants Occurring Naturally in Foods,* pp. 174–182, Food Protection Committee, Food and Nutrition Board, NAS/NRC, Washington, D.C.

Vought, R. L., London, W. T., and Stebbing, G. E. T., 1967, Endemic goiter in northern Virginia, *J. Clin. Endocrinol. Metab.* **27:**1381.

Watt, B. W., 1980, Tables of food composition: Uses and limitations, *N.Y. State J. Med.* **80:**1912–1913.

Wills, J. H., Jr., 1966, Goitrogens in foods, in: *Toxicants Occurring Naturally in Foods,* pp. 3–17, Food Protection Committee, NAS/NRC, Washington, D.C.

35

KE-SHAN DISEASE

Kung-lai Zhang

A unique heart disease called Ke-Shan disease (named for the county in northeast China where it was first noted) was reported in China in the 1930s (Jilin Medical University, 1977a). Up to the present time, Ke-Shan disease, thought to be related to selenium deficiency, has not been reported elsewhere in the world.

PATHOLOGY

The primary lesion of Ke-Shan disease is found in the parenchyma of the myocardium, characterized by multiple scattered small foci of necrosis. There are large areas of relatively normal myocardium between these necrotic foci. Inflammatory responses of mesenchyma in the myocardium include infiltration of inflamed cells, the majority of which are macrophages and phagocytes. Resorption and resolution are secondary to development of the necrotic foci, leading to formation of scar tissue; old and fresh lesions are therefore present simultaneously in most cases.

CLINICAL CHARACTERISTICS

The major clinical characteristics of Ke-Shan disease concern the extent of myocardial infarction and the state of cardiac function of the patients. There are four clinical types according to onset of the disease, the seriousness of myocardial lesion, and the general condition of the patients (Si-chuan Provincial Anti-Epidemic Center, 1977).

Kung-lai Zhang • Beijing Medical College, Beijing, China.

Acute Type

Patients with abrupt onset usually develop symptoms rapidly and suffer from acute circulatory failure, weakening of myocardial contraction, and inadequacy of cardiac outflow. Thus, heart, brain, and other organs suffer from acute hypoxia. This is manifested by wheezing and pulmonary râles in most patients.

Subacute Type

The common clinical manifestations of this category of Ke-Shan disease are capillary circulatory failure and congestive heart failure. This subacute type of Ke-Shan disease affects primarily children under ten. The majority of these patients belong to the 2–6 year age group. Patients suffer from general malaise; facies appear dull, pale, and swollen; there is cardiac dyspnea; moderate or significant heart enlargement on both sides; weakness of cardiac systolic sound; and tachycardia (often 130–180/min). Diastolic gallop rhythm and a significant amount of râles in both lungs can often be found.

Chronic Type

This type is characterized by chronic congestive heart failure. In most cases, the chronic type of Ke-Shan disease is developed from the latent type or from acute or subacute attacks which have not been thoroughly treated.

Latent Type

Heart function is usually normal with good recovery rate after stressful activity. Patients with the latent type of Ke-Shan disease are not affected in their routine activities. The onset of the disease is gradual, and it is often difficult to ascertain the exact date of onset. Manifestation of the disease may vary from the absence of symptoms to shortness of breath and general malaise. Some of the latent cases are regressions from acute or chronic types.

DIAGNOSIS

Relevant diagnostic data are the following:

1. Exact residence location at onset, season at onset, presence of Ke-Shan disease in area, age, sex, and family history. Typically, patients have had at least 3 months of residency in the endemic areas.
2. Heart condition: acute or chronic dysfunction, cardiac enlargement, gallop rhythm, arrythmia, multiple ventricular extrasystoles, atrial fibrillation, ventricular or epiventricular paroxysmal tachycardia; ECG: A–V block, bundle-branch block (complete or incomplete), ST–T ab-

normality, prolonged Q–T period, and ventricular extrasystoles of multiple origin.
3. X-ray findings: heart enlargement and pulmonary stasis with heart distortion.

The preceding descriptions provide general diagnostic data. Individual patients, however, often vary in their combination of symptoms and diagnostic characteristics.

EPIDEMIOLOGICAL CHARACTERISTICS

Formerly, most of the characteristic epidemiological data have been collected from acute and subacute cases because of the obvious nature of epidemics which necessitates control and prevention. Since acute and subacute cases are less likely to be misdiagnosed, it is easier to compare data on acute and subacute cases collected from various locations. It has been suggested, however, that the latent type may reveal the pathogenesis of Ke-Shan disease (Jilin Research Institute of Endemic Diseases, 1976); thus, more thorough investigation of latent Ke-Shan might be of significant importance epidemiologically. The epidemiological characteristics stated below are based on acute and subacute type cases.

Geographic Distribution

Within China, about half of the provinces have been affected. The affected areas are within latitude 21°–53°, longitude 97°–135° (Si-chuan Provincial Anti-Epidemic Center, 1977). The affected areas and the number of cases reported in each epidemic vary widely. The number of outbreaks is higher in mountainous and hillside areas, and lower in the plains and flatlands. The altitude of these mountainous and hillside areas ranges from 200 to 2000 m above sea level. If a mountainous area contains plateaus and valleys, the acute type of the disease is found primarily in the valleys.

In certain instances, the epidemic seems to spread from one population group to the next geographically contiguous population group. The acuteness of disease in the affected group lessens with increasing distance from the center of the epidemic (Helongjiang Research Institute of Endemic Diseases, 1976). Within a large affected area, there are often unaffected spots. This is called "foci distribution." There is a report relative to Jilin Province showing the foci distributed within an area of one-third of a square mile.

The path of an epidemic, however, may be random and unpredictable. In Jilin and Si-chuan Provinces, for instance, the affected spots change from year to year (Si-chuan Provincial Anti-Epidemic Center, 1977; Helong-jiang Research Institute of Endemic Diseases, 1976). In Helongjiang Province, by contrast, the areas affected have not only remained the same, but the number of cases occurring in each area is basically constant according to statistical criteria compiled over a period of 15 years (Harbin Medical University, 1977).

Originally, Ke-Shan was endemic in the northern part of China. As of the 1960s, the outbreaks began to occur in provinces of southwest China such as Si-chuan and Yunnan. In recent years, the number of cases in these newly affected areas has been considerably higher than that in northern China (Research Institute of Ke-shan Disease of Yunnan Province, 1976).

Season

In some provinces such as Helongjiang and Shan-xi, the epidemic peaks every 3–5 years in terms of number of people affected. In other provinces in northeast China, the peak occurs every 10 years. (See Fig. 1 and Table I for illustrations of this pattern.)

We should emphasize that the statistical number of people affected in a province is determined by the number of cases in a small but heavily stricken area. Using the same statistical approach, the detected rate of latent Ke-Shan in a northern province also fluctuates from one year to another, as shown in Table II.

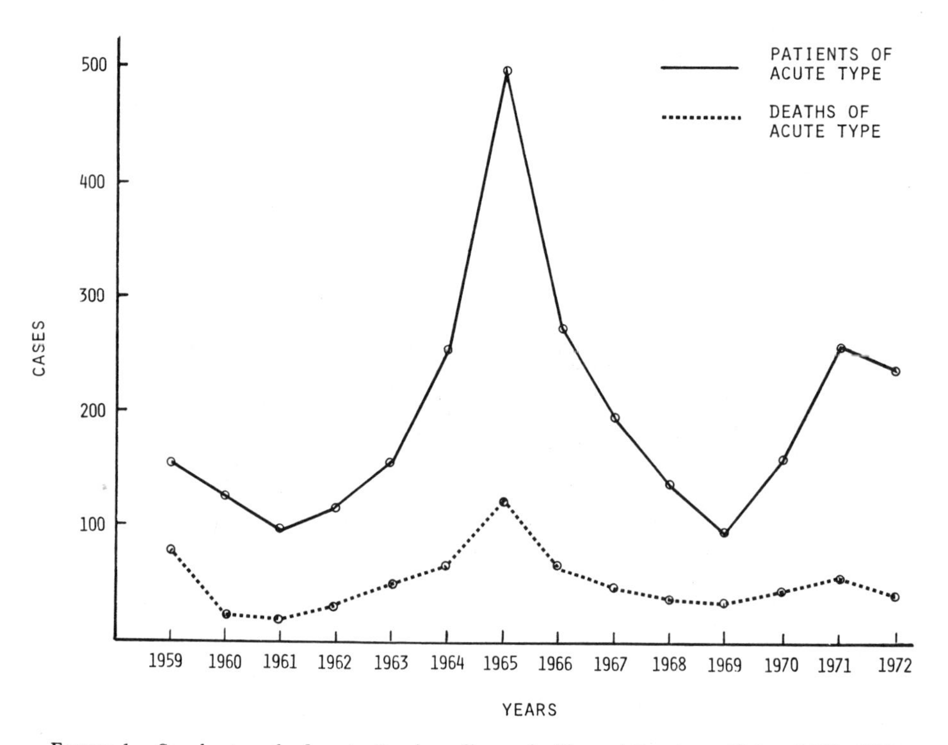

FIGURE 1. Secular trend of acute Ke-shan disease in Shan-xi Province, China, 1959–1972.

TABLE I
Incidence Rate of Ke-Shan Disease from 1950 to 1970 in a County of Helongjiang Province, China

Year	Incidence rate per 100,000	Year	Incidence rate per 100,000	Year	Incidence rate per 100,000
1950	17.0	1957	35.6	1964	37.2
1951	55.0	1958	132.8	1965	12.2
1952	121.0	1959	112.5	1966	21.4
1953	13.0	1960	0	1967	48.3
1954	4.0	1961	0	1968	35.0
1955	113.5	1962	0	1969	14.3
1956	32.5	1963	23.3	1970	15.0

The disease occurs all year round, but the peak season in the north is usually from December to February, accounting for 80–90% of the annual total (Si-chuan Provincial Anti-epidemic Center, 1977). In the southwest area of China, the peak season is from June to August, accounting for 60–70% of the annual total. The unique topography of Si-chuan—one of the southwest provinces—affords an interesting observation. In the mountainous area of Si-chuan, where the altitude is in excess of 2000 m above sea level and the average yearly temperature is around 10°C, the peak season coincides with that of the north (December to February) (Helongjiang Research Institute of Endemic Diseases, 1976). In the semitropical areas of that province, where the altitude is below 1000 m above sea level, and the average yearly temperature is around 10°C, the peak season falls in June to August.

Ke-shan has appeared abruptly in places where it has never before been reported. These sudden outbreaks are more violent and shorter in duration, usually lasting only 1–2 years and then disappearing (Helongjiang Research Institute of Endemic Disease, 1976).

TABLE II
Detected Rate of Latent-Type Ke-Shan Disease from 1950 to 1959 in a Province in Northern China

Year	Detected rate (%)
1950–1951	15.7
1953–1954	9.6
1954–1955	10.5
1955–1956	24.5
1956–1957	7.0
1957–1958	5.1
1958–1959	7.6

Demographic Distribution

The disease affects primarily women and children, especially new arrivals from nonepidemic areas. Twenty years ago, in the northeast and northwest areas, the people affected were mostly women in their fertile years. Since the late 1950s, an increasing number of children have been affected (about 60% of the total cases). In places where there is no influx of population, most of the cases are children aged 3–5. The youngest known case is a 6-month-old infant (Cheng Yun-yu, 1977). There is no significant difference in sex ratio among the aged (over 50 years) and children, whereas the sex ratio is one male to 2.6–8.7 females among those patients aged 20–50 years (Si-chuan Provincial Anti-Epidemic Center, 1977). In one study, almost half of the reported cases are children in the year after weaning, who are, interestingly, next to the youngest in their respective families (Harbin Medical University, 1977).

In both northern and southern epidemic areas, family clustering has been reported, with two or more cases in each family involved. Within a family, the occurrence of a second case may follow the occurrence of the first case in a few days, one month, or several years. One-third of the total cases occur in families previously stricken. One family had six members affected. In families with two or more cases, housing conditions are poor, crowded, and unsanitary.

Most acute, subacute, and chronic cases occur in farming families. Nonfarming families who live in an agricultural community for at least one year may also contract Ke-Shan, but statistics indicate that members of nonfarming families are less likely to be stricken. A retrospective study of a mixed community consisting of agricultural and industrial inhabitants in Helongjiang Province showed significant differences in incidence rate, death rate, and detected rate of latent Ke-Shan between the two groups of inhabitants. The agricultural group consisted of 180 families with 837 people; the industrial group, of 188 families with 1076 people. Of the 180 agricultural families, 33 (18.3%) contracted Ke-Shan. They include 48 acute or subacute patients, 22 of which were terminal cases. The cumulative incidence rate and death rate are 5.6% and 2.6%, respectively. In the industrial group, on the other hand, not even one acute or subacute case was found. Clinical screening revealed 25 latent and 8 chronic patients out of 592 inhabitants of the agricultural group, a rate of 5.6%. Among 734 industrial inhabitants, on the other hand, only 3 latent cases were detected, showing a rate of 0.4%. Of the two groups, sex ratio, age-specific distribution, time of residency, place of origin, immediate environment, and water supply are all identical. The only significant difference is the source and composition of their diet. The agricultural group grew and stored their own food, composed primarily of corn and sorghum, while the diet of the industrial group was wheat and rice, provided chiefly by the government (Harbin Medical University, 1977).

As to ethnic differences, in the northeast there is a significant difference in a number of cases between the Han people and the Korean minority in the same village. The majority of acute cases are concentrated in the Han people, but both ethnic groups have approximately the same number of latent cases. One fact worth noting is that the main food of the Koreans is rice, while the main staple of the Hans is maize (Harbin Medical University, 1977).

Birthplace and length of residency also may influence the number of affected people. New arrivals who have been in residence for only 2–3 years are twice as susceptible as the natives. The shortest residency is 6 months before the appearance of symptoms (Harbin Medical University, 1977). According to data collected from 3 villages in Helongjiang, the incidence rate for natives is 33.5/100,000, compared to 13.6/100,00 for non-natives.

ETIOLOGY

After 30 years of study, no definite conclusion has been reached as to etiology of Ke-Shan disease. Selenium deficiency has been proposed as one etiologic factor (Sino Academy of Medical Sciences, Institute of Health, 1978). Wherever Ke-Shan disease is found, there is noted also the appearance of "hepatosis diatetica" among domestic animals; this is known to be caused by selenium deficiency. Also, there is a similarity in clinical and pathological characteristics of these two diseases. Epidemiological studies showed a significant difference in selenium content of hair between the children in endemic areas and nonendemic areas. The hair of children in the nonendemic areas has a much higher content of selenium than that of children in endemic areas.

Since 1970, a medical corps of the Chinese Academy of Medical Science has been administering the oral medication sodium selenite in Si-chuan Province in an attempt to prevent acute Ke-Shan disease (Sino Academy of Medical Sciences, Institute of Health, 1978). Results are encouraging. In an agricultural population of 100,000, medication was administered 3 months prior to the epidemic season. Daily dosages were: ages 1–5, 0.5 mg; ages 6–9, 1.0 mg; adults, 2.0 mg; the medication was administered every 5–7 days during the epidemic season. Data collected in 1975 in four experimental communes in Si-chuan Province showed that within the group given sodium selenite only 3 (0.04%) subacute cases occurred among 6767 persons, with zero mortality, while 55 (1.01%) subacute cases and 19 deaths occurred among 5445 persons in the control group. In a follow-up investigation in 1976, among those 4510 who received sodium selenite in 1974, 10 persons (0.22%) contracted Ke-Shan disease, with a zero mortality rate; while among the control group, 54 patients (1.36%) developed Ke-Shan, with 23 deaths. These data confirm the effectiveness of selenium administration in proper dosage as a means of prevention of Ke-Shan disease.

In endemic areas, the selenium content in blood, hair, and grain were all lower than those in nonendemic areas. Blood glutathione peroxidase content in endemic areas is lower than in nonendemic areas. Thus, one can conclude that the selenium content is lower in both the population and the environment of the endemic areas.

Although the direct relationship of selenium deficiency to the incidence of acute Ke-Shan disease has been preliminarily established, it is still impossible to account for all of the epidemiological characteristics.

Many other hypotheses exist, including excessive amounts of nitrosamines in drinking water (Hebei Research Institute of Endemic Disease, 1975), molybdenum deficiency (Jilin Medical University, 1977b), or magnesium deficiency (Hebei Research Institute of Endemic Disease, 1975) in endemic areas.

According to the epidemiological and clinical characteristics and the data gathered from dietary investigations, it was suggested that lack of certain necessary trace elements, amino acids, or vitamins in myocardial metabolism play an important role in causing Ke-Shan disease (Research Institute of Ke-Shan Disease, 1975; Harbin Medical University, 1975). An experiment was set up in Helongjiang Province to evaluate the preventive results of supplementing soybean products in daily diets and comparing the results to a control group with ordinary food supply. After 1.5 years, the group with soybean supplement had no new cases of Ke-Shan disease and diagnosed patients of this group showed improvement, whereas new acute and latent cases occurred in the control group.

Although biological etiology still appears eligible in explaining the epidemiological characteristics of Ke-Shan disease, there is to date no conclusive evidence pointing to virus infections, fungi, toxins, or bacterial infections as being the etiological entity.

PREVENTION

Because of the etiological uncertainty of Ke-Shan disease, multifaceted preventive measures have been put into practice. In endemic areas, these measures include raising sanitation standards, improving water supplies, and controlling those factors which might lead to outbreaks of acute symptoms, i.e., avoidance of excessive cold or heat, avoidance of any kind of infection, the keeping of a regular lifestyle, and a balanced diet.

With regard to acute and subacute patients, the emphasis is on local and immediate treatment. In addition, screening among inhabitants in endemic areas to locate the latent cases is recommended.

ACKNOWLEDGEMENT. The author is grateful to Tamba McIntire Morgan for her assistance in typing the manuscript.

REFERENCES

Cheng, Yun-yu, 1977, Ke-Shan Disease, Harbin Medical University.
*Harbin Medical University, 1977, Ke-Shan Disease.
Hebei Research Institute of Endemic Diseases, 1975, Etiology of Ke-Shan Disease.
Helongjiang Research Institute of Endemic Diseases, 1976, Ke-Shan Disease.
Jilin Research Institute of Endemic Diseases, 1976, The Prevention and Treatment of Ke-Shan Disease.
Jilin Medical University, 1977a, Ke-Shan Disease.
Jilin Medical University, 1977b, Bulletin–Molybdenum Deficiency–A Possible Cause of Ke-Shan Disease.
Research Institute of Ke-Shan Disease, Harbin Medical University, 1975, Nutrition and Ke-Shan Disease.
Research Institute of Ke-Shan Disease of Yunnan Province, 1976, Ke-Shan Disease in Yunnan.
Si-chuan Provincial Anti-Epidemic Center, 1977, The Prevention and Treatment of Ke-Shan Disease.
Sino Academy of Medical Sciences, Institute of Health, 1978 (article to be published).

* Attribution to specific authors is usually impossible at this epoch in China's history. Listing is alphabetical by name of institution.

36

ZINC-DEFICIENT DWARFING

Hossain A. Ronaghy

BIOCHEMICAL CONSIDERATIONS

The role of zinc as an essential micronutrient in animal life was not well established until recently. For the first time Todd et al. (1934) produced zinc deficiency in rats. The essential role of zinc in animal nutrition is through its effect as a metaloenzyme in the production of DNA-dependent RNA polymerase (Wacker and Vallee, 1959).

Certain tissues of some animals are more sensitive to zinc deficiency. Prasad *et al.* (1967) demonstrated that the zinc content of liver, kidney, testis, pancreas, bone, and thymus was significantly lower in zinc-deficient rats than in pair-fed control animals. The metaloenzymes affected in these animals were alcohol dehydrogenase, alkaline phosphatase, and carboxypeptidase. It is therefore concluded from these animal experiments that zinc does play a significant role in cellular function.

While Swenerton and Hurley (1968) relate lactic dehydrogenase (LDH) activity of the liver to zinc, Chvapil (1973) links the zinc effect to an increased stability of lysosomal membrane. Despite this controversy, it is obvious that zinc does play a significant role in cellular function and protein synthesis. Therefore, its deficiency produces growth retardation in animal and man.

ZINC REQUIREMENT AND INTAKE

The requirement of zinc for adults is estimated to be about 6 mg/day (Scoular, 1939). However, actual intake is almost twice that amount; a number of metabolic studies reveal intake to be 9.8–14.4 mg/day (Tribble and Scoular, 1954). Despite this fact the National Academy of Science has recommended 15–20 mg zinc as the minimal daily requirement for adults (National Academy of Sciences, 1970).

Hossain A. Ronaghy • Formerly Department of Community Medicine, Medical School, Shiraz University, Shiraz, Iran. Present address: Trauma Medical Clinic, San Diego, California; and School of Public Health, University of California, Los Angeles, California 90024.

Oysters and herring are among the foods richest in zinc concentration. Animal protein, whole grains, lima beans, and peas also contain high amounts of zinc. Fruit and vegetables contain small amounts of this trace element. Those populations consuming refined carbohydrates and fats are most likely to suffer from a negative zinc balance.

In some disease states such as alcoholism, liver cirrhosis, malabsorption, and certain types of cancer, zinc concentrations in serum fall below normal levels. The significance of this biochemical abnormality is not clear.

GROWTH RETARDATION AND DWARFISM DUE TO ZINC DEFICIENCY IN MAN

For the first time, in 1960, a new syndrome of nutritional dwarfism was described in Iran (Halsted and Prasad, 1960). Growth retardation had been shown to be a feature of zinc deficiency in animals and this led Prasad *et al.* (1963) to suggest zinc deficiency as a cause of growth retardation in Iranian village children. The idea of zinc deficiency as a cause of dwarfism was further demonstrated in Egypt in 1963 (Prasad *et al.,* 1963).

The classic zinc-deficient dwarfs were described as young adults aged 18–22 with hepatosplenomegaly, severe anemia, hypogonadism, and dwarfism. Almost all of them gave a history of geophagia. Iron therapy improved the anemia to some extent but not completely. Several groups of the affected young individuals were studied in full detail. The following is the result of one double-blind trial carried out in Shiraz, Iran.

PATIENTS AND METHODS

Nutritional dwarfism is common in Iran's villages. With the cooperation of Iranian Induction Centers an excellent source for case studies became available. Since all males were called up for the draft at ages 19–20, and since the nutritional dwarfs were always rejected, it was possible to identify 186 male cases from the Induction Center in Shiraz. Chronic organic disease was excluded from consideration for the study. Such diseases included liver cirrhosis, nephritis, heart disease, urolithiasis with pyelonephritis, and malabsorption associated with diarrhea and steatorrhea. From the rest, those willing to undergo a long period of observation were admitted to the University Hospital research ward where they received a well-balanced diet with sufficient animal protein. After a two- or three-week stay for clinical and laboratory examination, they were transferred to a nearby house where 12 could be accommodated at a time, receiving the same hospital diet. Although diets were not analyzed, similar diets have been reported by Prasad *et al.* (1963) to contain from 27 to 31 mg zinc, of which about 10% is reported to be absorbed (Osis *et al.,* 1969). The patients remained in this house for 6 months to a year.

Twenty cases studied; of these, 8 were observed for 12 months, 3 for 9 months, and 9 for 6 months.

Two of the 20 dwarfs were females, one of whom was found in a village, the other having been admitted to the hospital for malnutrition. They were aged 19 and 20, respectively, and were transferred after hospital studies to the home of the author, where they lived during the study (one for 6 months, the other for a year).

The dwarfs were divided into three treatment groups. Group I received a well-balanced hospital diet plus a placebo capsule from the start. Group II received the hospital diet plus one capsule daily of 120 mg zinc sulfate from the start ($ZnSo_4 \cdot 7\ H_2O$, containing 27 mg of elemental zinc). Group III received the hospital diet plus placebo capsules for 6 months and then zinc sulfate was substituted for the placebo. Selection of subjects for Groups I and II was by the method of lottery with replacements. The 3 members of Group III were drawn from Group I at the end of 6 months. The capsules were given 7 days a week between the noon and evening meals by an attendant who assured that they were swallowed. The entire program was personally supervised by the author.

Iron doses of 100 mg of ferrous fumarate were given once daily only to those dwarfs whose hemoglobin levels were 7.0 g% or less (four in all). The iron used was found to contain a negligible amount of contaminating zinc. Every 3 months the subjects were transferred back to the hospital for a few days for clinical and laboratory tests.

RESULTS

The mean height at the time of admission was 141 cm and the mean weight was 35 kg. This is compared to average persons of the region to be 174 cm and 61 kg, respectively. Their mean hemoglobin was 10.3, with total protein of 7g and albumin of 3.5g%. Serum iron was 45 mcg%, while serum zinc was 48 mcg as compared to normal medical students with serum zinc of 95 ± 12 mcg%.

The patients' plasma zinc levels in all groups reached 75–77 mcg-% after 6 months stay. Group I, which was receiving zinc, grew an average of 11.8 cm. Group II grew an average of 3.6 cm ($P < 0.01$), showing a statistically significant difference between the two groups.

During the second 6 months, 4 cases continued tc receive diet alone and they grew only 2.4 cm, while the other three cases, who received the diet plus zinc, grew 8.7 cm. The difference in these two groups was highly significant ($P < 0.001$).

Using menstruation and ejaculation as the first sign of puberty, it was noted that there was a striking difference between the zinc and placebo groups. The 12 cases receiving placebo reached puberty from 24 to 52 weeks (mean 33 weeks), while 8 persons in the zinc group showed signs of puberty between 8 and 24 weeks (mean 10.5 weeks, $P < 0.001$).

CONCLUSION

Evidence from this study makes it clear that zinc may be an essential nutrient in determining both growth and onset of puberty in man. In planning food supplementation programs in underdeveloped countries, this factor might be considered.

A study that was reported from Iran (Ronaghy *et al.*, 1969) indicated that zinc supplementation was effective in hastening sexual development in growth-retarded but outwardly healthy schoolboys with delayed puberty and low plasma-zinc levels. These observations indicate that a spectrum of zinc deficiency may occur, with "full-blown" dwarfism as defined by the features of this syndrome at one end, and moderate growth retardation with late puberty at the other. This less marked end of the spectrum may be very widespread in underdeveloped areas of the world, where the chief food supplies are of cereal origin with high phytate content and thus probably lesser amounts of available zinc.

REFERENCES

Chvapil, M. 1973, New aspects in the bioloical role of zinc: A stabilizer of macromolecules and biological membranes, *Life Sci.* **13:**1014.

Halsted, J. A., and Prasad, A. S., 1960, Syndrome of iron deficiency anemia—hepato-splenomegaly, hypogonadism, dwarfism, and geophagia, *Trans. Am. Clin. Climatol. Assoc.* **72:**130.

National Academy of Sciences, 1970, *Zinc in Human Nutrition,* summary of Proceedings of a workshop, December, 1970, Food and Nutrition Board, NAS, Washington, D.C.

Osis, D., Royston, K., Samachson, J., and Spencer, H., 1969, Atomic absorption spectrophotometry in mineral and trace-element studies in man *Dev. Appl. Spectrosc.* **7a:**227.

Prasad, A. S., Miale, A., Farid, Z., Sandstead, H. H., Schulert, A. R., and Darby, W. J., 1963, Biochemical studies on dwarfism hypogonadism and anemia. *Arch. Intern. Med.* **111:**407.

Prasad, A. S., Oberleas, D., Wolf, P., and Horwitz, J. P., 1967, Studies on zinc deficiency: Changes in trace elements and enzyme activities in tissues of zinc-deficient rats, *J. Clin. Invest.* **46:**549.

Ronaghy, H. A., Fox, M. R. S., Garn, S. M., Israel, H., Harp, A., Moe, P. G., and Halsted, J. A., 1969, Controlled zinc supplementation for malnourished school boys: A pilot experiment, *Am. J. Clin. Nutr.* **22:**1279.

Scoular, F. I., 1939, A quantitative study, by means of spectrographic analysis, of zinc in nutrition, *J. Nutr.* **17:**103.

Swenerton, H., and Hurley, L. S., 1968, Severe zinc deficiency in male and female rats, *J. Nutr.* **95:**8.

Todd, W. R., Elvejhem, C. A., and Hart, E. B., 1934, Zinc in the nutrition of the rat, *Am. J. Physiol.* **107:**146.

Tribble, H. M., and Scoular, F. I., 1954, Zinc metabolism of young college women on self-selected diets, *J. Nutr.* **52:**209.

Wacker, W. E. C., and Vallee, B. L., 1959, Nucleic acids and metals. I. Chromium, manganese, nickel, iron and other metals in ribonucleic acid from diverse biological sources, *J. Biol. Chem.* **234:**3257.

XI

DEFICIENT INTAKES: CULTURALLY INDUCED

37

CULTURE-MADE NUTRITIONAL INADEQUACIES

Christine S. Wilson

INTRODUCTION

The diets of all peoples are determined by the culture of their particular group, and by impinging ecologic aspects (kinds of food regionally available, seasonality) and economic factors (ability to obtain food by own efforts, cash, or barter). No society that has been studied has eaten every potential foodstuff in its surroundings, although a few have nearly done so. The Otomi Indians of the Mezquital Valley of Mexico, geographically separated from outside contact and obliged to exist on edible items in their environment, used most to attain a varied diet and good nutritional status (Anderson *et al.*, 1946). Most groups resemble the !Kung San Bushmen of the Kalahari (Lee, 1973), who selected 23 plant and 17 animal species from a potential of 85 plant and 223 animal foods. Although selections may be conscious, choice is influenced in a variety of ways by the culture in which the individual was reared.

Culture is the sum of a people's approach to life, their artifacts, objects of daily use, material technology, and the manner or style in which they manipulate them. Much of it is learned so early in life that it becomes unconscious, and it persists because it has purposes that are adaptive for its practitioners. This chapter considers diets of a number of populations or subgroups, arrived at by cultural determinations of what is suitable, fit, or proper for that population, or certain of its members, that result in inadequate nutrient intakes.

It should be noted that although these diets may be disadvantageous from health and nutritional points of views, they are seldom *culturally* deleterious. Indeed, they persist because they fit successfully into other aspects of the society that are functional to its continued maintenance. As McLaren

Christine S. Wilson • Department of Epidemiology and International Health, University of California, San Francisco, California 94143.

(1976) has noted, many rituals central to a culture are closely involved with acquiring food and eating it. Thus, periods of the life cycle identified as unusually dangerous are associated with ritual proscriptions of foods. Population segments most affected are those most vulnerable to nutritional deprivation and least able to exert political pressure—young children and childbearing women.

The nutritional disadvantages are most often due to restrictions on choice of food a particular type of individual may consume. Restrictions may be applied to one class of individual only, such as pregnant women or priests. They may be observed once in a lifetime such as fasts during adolescent rites, or they may be recurrent but limited, such as dietary taboos imposed on a woman for the month after child delivery. Nutritional disadvantages may be due to cultural overemphasis on reputed, sometimes magical, worth of a staple, leading to overreliance and overconsumption, to the exclusion of other diet items that help assure a varied nutrient intake.

Though nutritionally inadequate, diets dictated by a culture are *culturally* sound, reinforcing by repetition the assurance of long-standing traditional behavior. Observation of rituals and prescribed activities gives a feeling of security and assurance (Jelliffe and Bennett, 1972).

CULTURE-BASED DIET MODIFICATIONS

Restrictions on "usual" diets most often result from *beliefs* about what should be eaten in particular physiologic states, *taboos* (proscriptions) against a specific type of food during conditions that depart from normal, and a wish of the society to *set individuals apart* into a category different from the one generally held by persons of that class. A woman just past childbirth is an example.

Reasons for culturally altering diets are often interrelated. The widespread *belief* that small children—a *set-aside* physiologic group—are not "strong" enough to eat animal protein results in a *taboo* on the child eating eggs, fish, or meat.

Beliefs

Beliefs about foods that result in dietary restrictions are often parts of larger beliefs systems that attempt to order the universe as the society perceives it. Typically these beliefs concern qualities assumed to inhere in the foods and to be capable of altering the physiologic responses of the consumer. The most widespread food beliefs are those that impute heating and cooling properties to foods.

Other attributes that have been assigned to food may stem from hot-and-cold beliefs. Malays term certain foods "toxic" due to properties that seem to be related to their "hotness" or "coldness." Mexican-Americans and urban blacks categorize some foods as "heavy" or "light."

Hot and Cold Food Beliefs

The origin of the physiologic theory that imputes heating and cooling properties to food is attributed to Hippocrates, the Greek "father of medicine" (4th century B.C.). Systematized in the 2nd century A.D. by the Greek physician Galen, the philosophy spread through Arab influence around the Mediterranean and eastward to the Malay Peninsula and Indonesia, northward through Europe and thence to the New World by way of the Spaniards in the 16th century. It was the base of "scientific medicine" in Europe until the 17th century (Foster, 1979).

Man and animals were viewed as consisting of four fluids or humors that determine health and temperament, blood (hot and wet), phlegm (cold and wet), choler or yellow bile (hot and dry), and melancholy or black bile (cold and dry). Hot, cold, wet, or dry qualities were assigned to food, drink, illness, and alterations of body state. Good health required a balance among all (Foster, 1978), with dietary adjustment employed to alter imbalance and return one who was ill to a state of good health.

The system at present has lost the humoral or temperamental aspect and the wet–dry component, but the term *humoral pathology* is still used. Its role in food-related beliefs is most widespread in Latin America, where it has been extensively studied (Foster, 1967; Gonzalez, 1964; Mazess, 1968; Wiese, 1976; Harwood, 1971), and where it is incorporated into preexisting folk beliefs about emotionally "hot" disorders such as evil eye, fright, or anger (Foster, 1967). A closely associated concept is that contact with an object of cold temperature (or wind itself) will let "air" into the body, as may eating a food defined as "cold."

Today the hot-and-cold beliefs are primarily folk wisdom, passed on by oral communication. There is lack of agreement even within a community as to which category many foods should be assigned. Eggs are "cooling" in Thailand, "hottest" in Bangladesh (Lindenbaum, 1977). Generally, however, fruits and vegetables are "cold," animal protein foods are "hot." Since more diseases are "cold," "cold" foods are more often prohibited. There are foods intermediate between "hot" and "cold." Starchy staples are neutral and therefore minimally noxious in local views.

Hot-and-cold beliefs are of particular concern for vulnerable beings—infants, pregnant women, the aged—contributing to food restrictions for these groups.

Some cultures (Wilson, 1970) make no attempts to balance "hot" and "cold" foods in ordinary meals. In Latin America, however, certain foods may be "hot" or "cold" depending upon how they are prepared (Currier, 1966), and housewives are careful to mediate hot and cold in cooking, and to alternate qualities in menu items (Ingham, 1970; Reichel-Dolmatoff and Reichel-Dolmatoff, 1961; Messer, 1972). In many cultures both the indisposed and seriously ill should avoid "hot" and "cold" foods. Latin Americans also think that too much "hot" or "cold" food may be deleterious to the healthy.

Two points should be noted about hot-and-cold beliefs:

1. While serious illness requiring diet alterations is infrequent, indisposition ("hot" fever, "cold" respiratory disturbance) is common, particularly in regions of endemic parasitism and rudimentary sanitation. Frequent episodes of minor disturbances mean frequent avoidance of "hot" animal foods or "cold" fruits and vegetables, the principal carriers of minerals and vitamins.
2. The belief may have physiologic bases. Specific dynamic action of high protein foods may be felt as heat, while enzymes or complex compounds in plant foods may cause bodily reactions. Papaya and pineapple figure in many food beliefs—each contains proteolytic enzymes. Some foods, legumes, for example, *do* produce "wind" (flatulence) in the digestive tract.

Yin–Yang and Related Beliefs

Akin and possibly of similar origins to humoral pathology are the Chinese *yin–yang* system and the indigenous Indian Ayurvedic theory of three humors or *doshas.* The opposites represented by *yin* and *yang* encompass hot and cold, respectively. Chinese classify foods as "hot" or "cold," as do Vietnamese, Burmese and Thais (Hart, 1969). Ayurvedic medicine, like Hippocratic, qualifies foods as heating or cooling and requires proper combinations to restore equilibrium among the body humors wind (breath), fire (bile), and water (phlegm) (Ferro-Luzzi, 1975). Disease caused by "cold" or "hot" requires opposite-quality foods (Hart, 1969).

Another folk food classification associated with hot-and-cold is that of "heavy" or "light." In Latin America "heavy" foods are starchy, dense (potato, rice), or difficult to digest (Gonzalez, 1964; Wellin, 1955). To Afro-Americans "heavy" foods (staples) "stay with you," and are preferred to "light" foods (Jerome, 1975), which the Anglo majority considers food for invalids.

Changeable "Toxicity" of Foods

Whereas this book considers foods naturally toxic under usual conditions, Malays believe that certain foods, ordinarily harmless, become toxic (*bisa,* literally "poisonous") in illness or indisposition (Hart, 1969; McKay, 1971). Not all foods are potentially *bisa,* but those that are are many, and the health aberrations calling for the avoidance are minor, self-diagnosed, and frequent. For example, peanuts, chicken, and pineapple are *bisa* to open sores. Cashew nuts, eggs, and mackerel are *bisa* to skin infections. Eggplant, frying banana, and chicken are *bisa* to a sick stomach. Soursop is *bisa* to influenza (a "cold" food and a "cold" disease). People with boils should not eat mutton, beef, or mackerel (Wilson, 1970).

"Hot" foods become *bisa* to people taking medicine from the doctor, which is also "hot"—two "hot" things taken together will war inside, and the individual will become *mabok* (dizzy, intoxicated) (McKay, 1971). Papaya, a ubiquitous, inexpensive, generous source of carotene, is the most feared of these "hot" foods. It is *bisa* and forbidden to the child with "eye worm," symptoms medically diagnosed as xerophthalmia, that Malays believe are caused by intestinal worms rising to the eyes. Green or yellow vegetables and fruits are also kept from other children because they irritate the worms to ascend from the gut (McKay, 1971), restricting carotene and vitamin A sources to those with already equivocal intakes.

Taboos

The term *taboo* comes from a Polynesian word (*tapu* or *kapu*) meaning sacred or prohibited. By extension it has come to mean forbidden by tradition or convention, an interdiction against certain acts, things, or words. Taboos on eating and drinking may have arisen from primitive views that the soul could escape from the mouth at these times. Folk who adopted an animal as totem or "god" that they thereafter avoided eating created the first taboos (Frazer, 1963).

Present-day taboos against eating certain foods are outgrowths of their being forbidden by tradition or convention, plus religious "laws" interdicting some foods or types of foods. Prohibitions against "hot," "cold," or *bisa* foods are also termed "taboos."

Perhaps the most common prohibition encountered today is that of animal protein to women and young children. It may be considered too "hot" or "strong" for such delicate people, or simply deemed more suited to more prestigious adult males, particularly when supply of this expensive food is scarce. Tabooing these foods to women, and more especially to small children who do not contribute to food production, may be a cultural means of population control when resources are limited (Townsend, 1971).

Stated reasons for taboos may differ from the above. Meat or eggs may be said to cause reproductive failure in women (O'Laughlin, 1974), or thought to cause sterility (Begrie, 1966). In Burma, children were not given eggs until they could speak, and tell if they were a reincarnation (Foll, 1959). Indonesians think fish, beef, and eggs will give babies stomach-ache (Intengan, 1976). In Mexico milk is too "strong" for sick children (Mead, 1955). Malay aborigines taboo animal protein due to beliefs that animals with strong spirits will cause convulsions, fever, and rigors in the susceptible. Meat of larger animals is tabooed to young children, pregnant women, their husbands, and lactating women (Bolton, 1972).

In East Africa scarce foods were forbidden women so men might have a greater variety (Trant, 1954). Among the South African Zulu only kin of the household head could take milk of his cattle. A pregnant or menstruating woman could not take milk or go near the cattle enclosure for fear she would exert an evil influence. Only allowed milk from her father's cattle, and on marriage going to live with her husband's family in another locale, an adult woman was effectively forbidden milk (Cassel, 1955).

Fowl eggs, nature's sanitary package of high quality protein, a universal fertility symbol, are widely forbidden. Among the Zulu, eating eggs is a sign of greed (Cassel, 1955). Eggs are tabooed to many Africans, particularly mature women (especially in pregnancy or lactation) and children (Simoons, 1961). Fear of their effects ranges from barrenness to licentiousness, from mental retardation to becoming a thief (Ogbeide, 1974). This food's potential to become a living creature may be the ultimate reason for the prohibitions. Economic causes are sometimes cited—the chicken may be more useful than the egg, providing meat and more eggs. Eggs also represent items salable for cash.

Religious proscriptions against eating items abhorred or unfit for the members of that sect nominally have less detrimental consequences than the above taboos, for alternative protein sources are usually substituted. Prohibitions against pork for Muslims and Jews have few adverse health effects, since other meats, fish and dairy products may be eaten. Only the religious requirement of draining the blood from slaughtered animals before further preparation for eating may be deleterious, since it lowers iron content.

In India, the land of the sacred cow, the situation is otherwise. Avoidance of eating beef is not itself harmful. Other taboos superimposed place a nutritional strain on the population. Some of these are vertical as well as horizontal, in accord with the caste system. To Hindus, foods are inferior or "impure," and superior or "pure." Cooked foods are inferior to "pure" raw ones, nonvegetarian foods are inferior to vegetarian foods. Milk and its products, ghee and curds (clarified butter, yogurt) are pure. The caste hierarchy accords with types of foods permitted. High-caste vegetarians may consume "pure" milk or curds, although they avoid eggs (Srinivas, 1955). Eating flesh other than beef is the mark of a lower caste. Those who wish to rise in caste eliminate meat from their diet, to emulate high-caste Brahmans. Pervasive concepts of pollution from below to above add further complication. Not only is eating together by different castes forbidden, but one of a higher caste may not eat food cooked or touched by a lower caste person (Marriott, 1968).

Potential interveners should recognize that breaking or flouting taboos can cause emotional stress. People may experience physiologic rejections when presented with culturally disapproved food. And following usual ways is reassuring, particularly in times of stress.

Life Cycle Crises—Times Set Apart

Traditional communities and technologically-developed cultures alike classify individuals according to their biologic states. In most cases the categories are biologically appropriate. One means of setting persons apart is to make them more visible, through regulation of what they may wear, where they may go, and what they may eat. Physiologic statuses most frequently felt to require such separation are pregnancy, the puerperium, lactation, weaning, and puberty. Less universal are childhood, widowhood and old age. These population segments are the chief subjects of beliefs and taboos, the sometimes docile captives of their societies' "pressure groups" that dictate continuance of traditions for eating, drinking and avoidances.

Explanations for diet restrictions or alterations are that the individuals are in supernatural, if not natural, danger. Diet changes are prescribed to ward off harm and ensure safe passage through a time of recognized stress. When the danger period is passed, reverse rituals reinstate the person among the group. Among the Gurage of Ethiopia, a woman after childbirth reincorporated herself into the community by going from house to house with the child, receiving a gift of food from each (Shack, 1969).

Pregnancy

Among low-income women in the U.S. beliefs about pregnancy were recited to a woman pregnant for the first time by an older one as a means of verbally separating her from her previous state (Newman, 1969).

Sympathetic magic and beliefs about "marking" the child are common causes of food avoidances in pregnancy. In Nigeria snails are avoided to prevent salivation in the newborn (Ogbeide, 1974). Hares and rabbits eaten in pregnancy are believed to be potential causes of harelip or large ears in the unborn. Eggs may cause the child to become too large, or to "cry like a fowl" at birth (Hope, 1975). In India, where light skin is valued, it is thought that eating leafy vegetables will result in a dark baby (Katona-Apte, 1977). A high proportion of women attending prenatal clinics in the U.S. believed the pregnant mother could cause stigmata in the child resembling certain foods craved but denied, or eaten in too great quantity (cherries, strawberries, chocolate) (Newman, 1969; Snow and Johnson, 1978). [Cravings may get around some pregnancy taboos: In Ceylon denying foods culturally defined to be craved would prevent rebirth in a culture that believes in reincarnation (Obeyesekere, 1963).]

Desire for a small baby and ease of delivery were stated causes for a restricted diet in pregnancy in Burma (Sharma, 1955; Foll, 1959), Thailand (Chandrapanond *et al.*, 1972), the Philippines (Valdecañas, 1971) and Latin America (Wellin, 1955). South Indian women restrict fluids in pregnancy for fear of "washing away" the baby (Ferro-Luzzi, 1973).

Some people consider pregnancy a disease; the woman is "ill" (Foster and Anderson, 1978). Where hot-and-cold beliefs prevail, a pregnant woman is deemed "hot." In Peru, "heavy" and "hot" foods are avoided in pregnancy, and in the last trimester most "cold" ones too (Wellin, 1955). "Heavy" eggs, meat, and certain fruits cause a "dirty" stomach," "hot" foods are potential abortifacients. Papaya, a "hot" food in South India, and pineapple are greatly feared in pregnancy, because they could cause abortion (Ferro-Luzzi, 1973). Other "hot" foods avoided by pregnant South Indian women are sesame seeds, Italian millet, and horsegram. These otherwise common diet items are thought capable of damaging the fetus (Katona-Apte, 1977). Mexican–Americans avoid hot-and-cold foods in pregnancy, especially "hot" chilies and their products; they would give the unborn child a rash (Clark, 1959).

In East Africa most food taboos fell on women, especially in pregnancy (Trant, 1954). No food avoidance by pregnant Indian women had a clear adaptive value (Ferro-Luzzi, 1978). In Thailand there was no recognition pregnant women have special diet needs (Chandrapanond *et al.*, 1972). The examples illustrate widespread prohibitions on high quality protein foods and sources of minerals and vitamins, particularly vitamins A and C, in pregnancy.

There are peoples who treat pregnant women kindly so far as diet is concerned. The Ngoni of southern Tanzania reserved millet for pregnant women, which is a valuable source of calcium (Robson, 1974). Throughout pregnancy in Greece, a "joyous expectancy," a woman was expected to eat everything she craved (Mead, 1955). Pregnant Senegalese women, while abstaining from spicy foods, were encouraged to eat curdled milk, palm oil, meat and butter (Garine, 1972). Chinese (Pillsbury, 1978) and most Malays (Wilson, 1973) regard pregnancy as nothing unusual; the woman may eat or act as she will.

Puerperium and Lactation

The culturally defined period after child delivery may be as short as 4 days, as in Burma (Mead, 1955), or 2 months or longer. A common set-apart time is 40 days. Some cultures gradually bring the woman back into her group within this time, adding on originally forbidden foods and activities bit by bit (Wilson, unpublished information). For Chinese the period is one moon, and the ritual observances are termed *doing the month* (Pillsbury, 1978).

In Haiti a female, always "warmer" than a male, is "hottest" during the first three months postpartum. She is forbidden so many "cold" foods that she may eat only 10 out of 74 potential (locally available) foodstuffs. She is at a point where "an extreme life-state classification intersects an extreme food classification system," both based on the hot–cold beliefs system. The ten available foods allowed are neutral starch sources, a little meat or beans, and some "very hot" spices and coffee (Wiese, 1976).

Even stricter dietary rules affect Indian women after childbirth. The first few days, only liquids may be taken (Katona-Apte, 1975). The first "foods" are bread, milk, rice-water, soup and coffee; the first meal is rice, sesame oil, and perhaps a "hot" vegetable (Ferro-Luzzi, 1974). Coffee, tea, and bread are not limited because they are proper foods for the sick (Katona-Apte, 1975). Fat and fried foods are avoided. Quantities and variety are gradually increased, but nonvegetarians become pure vegetarians for a month. Meat, fish and eggs are "impure," and dangerous to mother and child. Meat and fish may be avoided from 3 months to the end of lactation; fish might cause diarrhea and fits in children (Katona-Apte, 1977). Milk and its products are too "pure" for the first polluted after-birth month. Most women avoided fruits throughout lactation, and "cold" dangerous vegetables for a month, to protect the child (Ferro-Luzzi, 1974). When the nursling is sick, the mother must limit her diet (Katona-Apte, 1975). Abstentions may continue until the child can walk or talk.

Malays do not allow "cold" fruits and vegetables to enter the "cold" body of the postpartum woman for 40 days, to prevent bleeding and aid the uterus to return to its former size. A variety of "toxic" *(bisa)* fish that could harm the woman, and the child through her milk, are also forbidden. The diet is limited to a few "hot" foods, neutral rice, bread, and biscuits. Fats and fried foods are avoided. Only dry roast fish is eaten. Other protein sources are an occasional "hot" egg. There appear to be no avoidances in lactation. Anemia and folic acid deficiency were found in two women recently completing their confinement (Wilson, 1973). Mothers, grandmothers and midwives perpetuate these traditional practices.

Mexican postpartum diets are also restricted by hot-and-cold beliefs, to what the mother-in-law or local custom allows (Sanjur *et al.*, 1970). It is believed the umbilicus will not heal if the mother eats beans (Kelly, 1956). Mexican–American practices demonstrate the use of food to set an individual apart. The usual, basic foods—maize corn tortillas and beans—are avoided for *la cuarentena* (the 40-day lying-in period, now about 2 weeks). So are chilies and foods that contain them, and the favorite meat, pork. These people too abjure fat and fried foods. The foods mother or grandmother allows are more beneficial than the diets of the aforementioned groups. Chicken, soups, stews, bread, wheat tortillas, milk, cereals and other breadstuffs are permitted. However, most vegetables and fruits are "too cold." Vegetables allowed are "the kind that go in the stew"—carrots and potatoes. The restrictions are to prevent unwanted substances reaching the child through the mother's milk (Wilson, unpublished information).

Chinese avoid cold foods postpartum, advocating hot ones to "prevent anemia" (Yeung *et al.*, 1973). Most Chinese women "do the month" following child delivery to rid themselves of dirty blood and cure the unbalanced *yin–yang* of pregnancy. All raw or "cold" foods, including animals that go in the water and plants that grow in it, leafy green and root vegetables (which grow in damp earth), and most fruits are avoided (Pillsbury, 1978). With what she is allowed, the woman is stuffed, five to six meals a day. One chicken a day is recommended, cooked in sesame oil and rice wine, all "hot." Pig livers, chicken livers, and kidneys are prescribed, as are sweet foods and brown sugar, certain "hot" herbs, chicken soup, ginger, and rice wine. Eggs and rice are allowed. "Doing the month" is a cultural invention assuring the parturient woman sufficient rest to return to normal duties. Accompanying diet and behavioral rituals mark the "month's" importance and help insure its observance (Pillsbury, 1978).

All these cultures restrict movement and activities of the woman for the ritual withdrawal period, including exposure to cold. Being housebound lowers contact with carriers of communicable disease (Chen, 1973). In Southeast Asia keeping the body warm extends to a literal roasting procedure, the woman lying above glowing embers. The practice has been associated with sudden cardiac failure owing to electrolyte disturbances from prolonged sweating (Chen, 1975).

The Child at Weaning

The child just removed from the breast is even more subject to nutritional risk than its mother. Many traditional cultures have little awareness that the food needs of these small, rapidly growing individuals differ from those of an adult (Foster, 1966). In Guatemala there was no concept of foods being good for children (Gonzalez and Scrimshaw, 1957). The extreme situation was the Zande belief that a child grows like a fingernail, so needs no food (Culwick 1954). In Burma, old people were privileged prescribers of diets. Both mothers and babies were severely restricted, and infants died of beriberi from thiamin-deficient milk (Sharma, 1955). In Mexico, infants were treated by a different set of rules until they were gradually incorporated into the family (Sanjur *et al.*, 1970). In the Caribbean it was thought that food should not be given to babies until they have teeth (Hope, 1975). Introduction of a varied, supplemental diet to Bengali children at 6 months of age by means of a traditional rice-feeding ceremony required a propitious day and funds to pay for the ritual. The child's illness also delayed the ceremony (Jelliffe, 1957).

Related cultural practices exacerbating child-feeding problems are traditional beliefs about sanitation and lack of special foods and food preparation for small children. Some nutritious foods offered may be too bulky to provide sufficient nourishment. People may feed children well *because* they are healthy, withhold food when they are ill (Foster, 1966). Weak, "cool" herb teas may be substituted for breast or other "strong" milk for a sick child.

In Senegal, thin millet gruel, to which pulp of the baobab fruit (rich in ascorbic acid) is added, is given to a sick child, replacing staple millet balls and foods rich in fat and protein. This limited diet continues until fever abates, because a weak person should not be stressed by need to digest more solid food. Baobab pulp is white, resembling mother's milk (Garine, 1972). Supplementary foods for small children are often white, neutral, like human milk, but of lower nutritive value.

The satiating value of animal protein foods may have led to their classification as too "hot" or "strong" for young people. In the Caribbean it is thought eggs can make a child become pregnant (Hope, 1975). In India they are forbidden to children under 2 years for fear they will cause fits (Katona-Apte, 1977). In Guatemala small children can be given small bits of "strong" meat, depending on the number of teeth they have (Gonzalez, 1964).

Animal protein foods—milk, meat, and fish—are also forbidden small children in several cultures for fear they will excite intestinal parasites common in developing countries to rise to the eyes and cause blindness, or encourage worms to grow in the gut (Gonzalez and Scrimshaw, 1957; McKay, 1971).

Besides prohibitions, the weanling experiences reduction in maternal attention when a new sibling appears. He may be abruptly denied the breast, the chief sanitary protein source hitherto available, when a new pregnancy begins, for it is improper for the child without to drink the milk of the child within (Wilson, unpublished information), folk recognition of the double drain on the woman of lactation during pregnancy.

Childhood

Once a youngster has passed the initial purges inflicted on him neonatally to correct his "dirty stomach" (Wellin, 1955), and survived the rigors of weaning to adult foods, his society may take few pains to assure his nourishment. There are, to be sure, foods considered "children's foods." In developing countries these often are tree or bush fruits.

Some cultures consider boy children able to find their own food. The Alorese, an Indonesian people, socialized boys to expect no food from home until evening, obliging them to catch insects or small game, or to eat raw fruits and vegetables. Girls, being closer to the cook, were likely to nibble bits from the family food supply (DuBois, 1961). Foraging for food may allay hunger pangs without balancing the diet. Preadolescent Malay males roasted tapioca (manioc) roots, fish cadged from catches, or cashew nuts outdoors, or picked or scavenged fruit. These snacks supplemented at least two home meals (Wilson, 1970). Youngsters elsewhere may derive a third of their intake from sources outside the home. Fruits plucked from the tree may be important sources of essential vitamins and minerals for children in rural areas.

The prepubertal period appears less a time of being "set apart" than one to be passed through on the way to defined maturity.

Adolescence

For both sexes the transition to adulthood is marked by rituals in many cultures, sometimes accompanied by surgical procedures, scarification, circumcision, or ear- or nose-piercing. These again rapidly growing individuals are also subject to diet changes that demonstrate their approaching adulthood, and may include abstention from various foods—partial fasting—and other purificatory rites. This period of trial often terminates in ceremonial feasting, although the celebrant may not partake.

Such ordeals are more frequently visited upon males, but food restrictions on newly nubile females are far more widespread than other "marking" ordeals. Not only are there food taboos for this pubertal period; on appearance of the menses young girls become ritually polluting, or cause fear among males, and may be literally shut away on limited foodstuffs for the length of the effluvium, or until it first appears (Mead, 1950). In India a menstruating woman is barred from the kitchen, and eats only what she is given, because she is not expending energy (Katona-Apte, 1975).

Taboos against "cold" ("acid") foods for pubertal and menstruating females are said to prevent cramps or heavy flow (Snow and Johnson, 1978). Mexican-Americans add "sodas" (soft drinks) to foods these women are forbidden, surely a beneficial prohibition (Wilson, unpublished information). Brazilians believe that citrus fruit, watermelon, peppers, papaya, alcohol, or fish without scales will make the menstruum rise to the brain, causing insanity.

Some foods are forbidden young women because they are feared to cause lasciviousness or infertility. Mexican-Americans recommend young girls not to eat avocado or peanuts, for assumed sexual stimulating properties (Wilson, unpublished information). For a young girl in East Africa, eating chicken placed her in a fertile adult category to which she was not yet entitled (Trant, 1954).

Widowhood

Aside from temporary abstinence from certain foods as a sign of mourning, women in some cultures must observe loss of a spouse by long periods of food restriction. In India, a widow is allowed one meal a day. To eat more in that state would be unhealthy (Katona-Apte, 1975). Among some groups a widow abstains from previous customary foods for one or more years. To eat them would remind her of their former provider. The Pacific island Tanga expressed grief by radical alteration of the diet to foods which do not give pleasure, in the case of a widow for 5–10 years after her husband's death (Bell, 1948).

Religious Dietary Restrictions

Many religions require partial fasts before holy days, ranging from a day to a month. These "fasts" are commonly substitutions of one kind of food for another, such as the lentils eaten instead of beans by Mexican–Americans in Lent, or a shift in time of eating, observed by Muslims during the 29-day fasting month of *Ramadan*. They do not eat, drink, smoke, or swallow during daylight hours, eating usual foods in normal amounts between nightfall and sunup (Wilson, unpublished information).

Such fasting has few adverse nutritional effects, unless substituted foods are inferior nutritionally. Yom Kippur is a day of complete fasting for Jews (except children under 13 years), but it occurs only once a year. There are, however, religions which require more stringent restrictions. In Burma, a Buddhist "Lent" permitted only one meal a day to the devout for three months (Mead, 1955). Modern-day Hindus fast for a number of interrelated reasons of caste, pollution, age, and sex as well as religion and degree of orthodoxy (Katona-Apte, 1976). For some individuals there could be more fast than nonfast days during the week. Some fasts are observed by all Hindus, others mark holidays, while still others are for atonement and forgiveness of sins, to bring about an event, or receive a reward. Like Christian Lent, most fasting substitutes other foods for a particular type of food. Fasting for a Hindu symbolizes purification of the body, thus he will not eat the common diet on fast days. Fasting is another way of setting oneself apart. The chief nutritional disadvantages of Brahmanist fasting fall on the more religious, who are more restrictive in what they allow themselves to eat, and where these beliefs crosscut interdictions imposed on childbearing women (Matter and Wakefield, 1971).

The most severe fasting occurs in Ethiopia, where the Orthodox Church and Islam both apply restrictions. The Ethiopian Church has unrelenting prohibitions based on seasonal and annual events, additional to life-cycle fasts. Only pregnant and lactating women and the ill are excepted (Knutsson and Selinus, 1970). The markets are geared to the Christian diet, which allows only fish for fasts. Muslims too are affected, because meat may be unavailable, and Islam forbids consuming animals killed by one of another faith. Excused women must fast later, when their pregnancy or lactation is over. Fasts are more frequent for Orthodox priests than for the laity, but the latter must observe 110–150 fast days a year.

UNEQUAL DISTRIBUTION OF FOOD WITHIN THE FAMILY

Food is allotted to family members by cultural rules related to food taboos and beliefs, especially protein foods suitable for "weak" individuals. Other aspects include prestige. FAO surveys in Africa, Asia and Latin America indicated the most prestigious person eats first, usually adult males (Hartog, 1972; Taha, 1978). Often men control food of animal origin, women the vegetable foods, and sex differences in foods eaten may result. Children or females do not always come last or least (Mackenzie, 1976). An Egyptian father apportioned meat from the common dish to each child, a larger part to older boys, but girls under 12 years of age shared his meal (Ammar, 1954). Among the polygynous Tallensi of Ghana the man ate with the last of his children to stop suckling (Fortes and Fortes, 1936). They said a child must eat properly to grow, and must be taught how. Each wife kept food for an absent child of hers, but the man, after eating something from each wife's offering, called in all his children to partake, redistributing the grain allotted to each wife.

Between-meal eating or snacking, common in developing regions, may help counteract nutritional inadequacies arising from the "pecking order" in which family members obtain their meals and foods within meals.

OVEREMPHASIS ON PRIZED FOODS

Populations which for generations rely on a staple as their principal source of energy and chief component of their cuisine tend to regard it as having supernatural strength-giving powers, what Jelliffe (1967) terms a "cultural superfood." It is the centerpiece of rituals and ceremonies, prepared in a multitude of elaborate confections. It is venerated, and believed to contain a god or spirit which must be propitiated before mortals may partake (Frazer, 1963). Such a staple is rice in Asia. If rice is part of a meal, it doesn't matter what else is served. To an Asian a meal lacking rice is not a meal, for rice alone is life-sustaining. In a number of Asian tongues "rice" is synonymous with "food" or "eating." Thais consider rice becomes the tissues of man when eaten (Hanks, 1972). In rice-growing regions life revolves around the annual cycles of planting and harvest, marked by religious and secular rituals. When double-cropping was instituted, some of the rituals were doubled, too.

It is no wonder that its consumption is overemphasized. In eating it, one is consuming power and strength. Rice is highly regarded as an infant feeding supplement in Nepal (Brown *et al.,* 1968), the only food urged on children in Thailand (Chandrapanond *et al.,* 1972) and in Malaysia cooked rice pounded into a paste with a little sugar is pressed upon the neonate (Wilson, 1970; Fig. 1)—a child cannot learn too early what is the proper thing to eat.

FIGURE 1. Malay grandmother feeding pounded cooked rice to four-week-old grandchild.

Jelliffe and Bennett (1961) noted similar overfeeding of plantain to Bagandese children. A congee or gruel of soft-boiled rice is given to sick individuals of all ages in the rice-eating sections of Asia. This simple carbohydrate food is quite easily digested.

Maize eaters of North and Central America use *atole,* maize corn gruel, similarly. Maize consumers prefer a whiter grain and meal than yellower varieties which are more nutritious (Berg, 1973). Whiteness and blandness are prized in staple foods. When to this is added the belief that all needs will be met if this cultural superfood, *the* food, is consumed, nutritional disadvantages will accrue to the entire population. Whiteness and blandness require a degree of milling that removes much of the essential nutrients in the germ. Staple roots and tubers (with some exceptions, e.g., potatoes) are primarily carbohydrate sources. For peoples for whom snack foods are also primarily refined carbohydrates—biscuits, bread, and sweets—serious malnutrition can ensue.

CONCLUSION

Adverse effects of food may as often be due to cultural as to physical attributes of the food. Examples of traditional food behaviors in this chapter represent observations recorded over several decades. Technologic development is altering beliefs and practices. Some of the cited situations no longer pertain. However, new behaviors undergo filtration through the cultural screen of present belief before incorporation into practice. Understanding the emotional values attached to food may diminish attempts to introduce culturally unacceptable items, and mitigate adverse reactions to worthwhile substitutions.

REFERENCES

Ammar, H., 1954, *Growing Up In An Egyptian Village. Silwa, Province of Aswan,* Routledge and Kegan Paul, London.

Anderson, R. K., Calvo, J., Serrano, G., and Payne, G. C., 1946, A study of the nutritional status and food habits of Otomi Indians in the Mezquital Valley of Mexico, *Am J. Pub. Health* **36:**883–903.

Begrie, I. C., 1966, Food and food habits in Uganda, *Rev. Nutr. Food Sci.* **No 3 (April):**8–11.

Bell, F. L. S., 1948, The place of food in the social life of the Tanga, *Oceania* **19:**51–74.

Berg, A., 1973, *The Nutrition Factor. Its Role in National Development,* The Brookings Institution, Washington, D.C.

Bolton, J. M., 1972, Food taboos among the Orang Asli in West Malaysia: A potential nutritional hazard, *Am. J. Clin. Nutr.* **25:**789–799.

Brown, M. L., Worth, R. M., and Shah, N. K., 1968, Food habits and food intake in Nepal, *Trop, Geogr. Med.* **20:**217–224.

Cassel, J., 1955, A comprehensive health program among South African Zulus, in: *Health, Culture and Community. Case Studies of Public Reactions to Health Programs,* (B. D. Paul, ed.), pp. 15–41, Russell Sage Foundation, New York.

Chandrapanond, A., Ratchatsilpin, A., and Tansuphasiri, S., 1972, Dietary survey of preschool children and expectant women in Soongnern District, Nakorn Rajsima Province, Thailand, *Am. J. Clin. Nutr.* **25:**730–735.

Chen, P. C. Y., 1973, An analysis of customs related to childbirth in rural Malay culture, *Trop. Geogr. Med.***25:**197–204.

Chen, P. C. Y., 1975, Food taboos of childbirth: The Malay example (letter to the editor), *Ecol. Food Nutr.* **4:**57.

Clark, M., 1959, *Health in the Mexican–American Culture,* University of California Press, Berkeley.

Culwick, G. M., 1954, *Social Factors Affecting Diet,* Philosophical Society of the Sudan, Khartoum.

Currier, R. L., 1966, The hot–cold syndrome and symbolic balance in Mexican and Spanish–American folk medicine, *Ethnology* **5:**251–263.

DuBois, C., 1961, *The People of Alor: A Social–Psychological Study of an East Indian Island,* Harper & Row, New York.

Ferro-Luzzi, G. E., 1973, Food avoidances of pregnant women in Tamilnad, *Ecol. Food Nutr.* **2:**259–266.

Ferro-Luzzi, G. E., 1974, Food avoidances during the puerperium and lactation in Tamilnad, *Ecol. Food Nutr.* **3:**7–15.

Ferro-Luzzi, G. E., 1975, Temporary female food avoidances in Tamilnad. Interpretations and parallels, *East and West* (new series) **25(3–4):**471–485.

Ferro-Luzzi, G. E., 1978, More on salt taboos, *Curr. Anthropol.* **19:**412–415.

Foll, C. V., 1959, An account of some of the beliefs and superstitions about pregnancy, parturition and infant health in Burma, *J. Trop. Pediatr.* **5:**51–59.

Fortes, M., and Fortes, S. L., 1936, Food in the domestic economy of the Tallensi, *Africa* **9:**237–276.

Foster, G. M., 1966, Social anthropology and nutrition of the pre-school child, in: *Pre-School Child Malnutrition. Primary Deterrent to Human Progress,* pp. 258–271, Publication 1282, National Research Council, National Academy of Sciences, Washington, D.C.

Foster, G. M., 1967, *Tzintzuntzan. Mexican Peasants in a Changing World,* Little, Brown, Boston.

Foster, G. M., 1978, Hippocrates' Latin American legacy: "Hot" and "cold" in contemporary folk medicine, in: *Colloquia in Anthropology,* (R. K. Wetherington, ed.), pp. 3–19, Southern Methodist University, Dallas.

Foster, G. M., 1979, Humoral traces in United States folk medicine, *Med. Anthropol. Newsletter* **10(2):**17–20.

Foster, G. M., and Anderson, B. G., 1978, *Medical Anthropology,* John Wiley & Sons, New York.

Frazer, J. G., 1963, *The Golden Bough: A Study in Magic and Religion,* Macmillan, New York.

Garine, I. de, 1972, The socio-cultural aspects of nutrition, *Ecol. Food Nutr.* **1:**143–163.

Gonzalez, N. L. S. de, 1964, Beliefs and practices concerning medicine and nutrition among lower-class urban Guatemalans, *Am. J. Pub. Health* **54:**1726–1734.

Gonzalez, N. L. S. de, and Scrimshaw, N. S., 1957, Public health significance of child feeding practices observed in a Guatemalan village, *J. Trop. Pediatr.* **3:**99–104.

Hanks, L. M., 1972, *Rice and Man: Agricultural Ecology in Southeast Asia,* Aldine-Atherton, Chicago.

Hart, D. V., 1969, Bisayan Filipino and Malayan humoral pathologies: Folk medicine and ethnohistory in Southeast Asia, *Southeast Asia Program Data Paper No. 76,* Cornell University, Ithaca.

Hartog, A. P. den, 1972, Unequal distribution of food within the household, *FAO Nutr. Newsletter* **10(4):**8–17.

Harwood, A., 1971, The hot-cold theory of disease. Implications for treatment of Puerto Rican patients, *J. Am. Med. Assoc.* **216:**1153–1158.

Hope, M., 1975, Food taboos and nutrition in the Caribbean, *Cajanus* **8:**190–193.

Ingham, J. M., 1970, On Mexican folk medicine, *Am. Anthropol.* **72:**76–87.

Intengan, C. L., 1976, Nutritional evaluation of breast-feeding practices in some countries in the Far East, *J. Trop. Pediatr. Environ. Child Health* **22:**63–67.

Jelliffe, D. B., 1957, Social culture and nutrition: Cultural blocks and protein malnutrition in early childhood in rural West Bengal, *Pediatrics* **20:**128–138.

Jelliffe, D. B., 1967, Parallel food classification in developing and industrialized countries, *Am. J. Clin. Nutr.* **20:**279–281.

Jelliffe, D. B., and Bennett, F. J., 1961, Cultural and anthropological factors in infant and maternal nutrition, *Fed. Proc.* (Supp. 7), **20 (No. 1, Part 3):**185–187.

Jelliffe, D. B., and Bennett, F. J., 1972, Aspects of child rearing in Africa, *J. Trop. Pediatr. Environ. Child Health* **18:**25–43.

Jerome, N. W., 1975, Flavor preferences and food patterns of selected U.S. and Caribbean blacks, *Food Technol.* **29(6):**46–51.

Katona-Apte, J., 1975, The relevance of nourishment to the reproductive cycle of the female in India, in: *Being Female: Reproduction, Power and Change* (D. Raphael, ed.), pp. 43–48, Mouton Publishers, The Hague.

Katona-Apte, J., 1976, Dietary aspects of acculturation: Meals, feasts, and fasts in a minority community in South Asia, in: *Gastronomy. The Anthropology of Food and Food Habits* (M. L. Arnott, ed.), pp. 315–326, Mouton, The Hague.

Katona-Apte, J., 1977, The socio-cultural aspects of food avoidance in a low-income population in Tamilnad, South India, *J. Trop. Pediatr. Environ. Child Health* **23:**83–90.

Kelly, I., 1956, The anthropological approach to midwifery training in Mexico, *J. Trop. Pediatr.* **1:**200–205.

Knutsson, K. E., and Selinus, R., 1970, Fasting in Ethiopia: An anthropological and nutritional study, *Am. J. Clin. Nutr.* **23:**956–960.

Lee, R. B., 1973, Mongongo: The ethnography of a major wild food resource, *Ecol. Food Nutr.* **2:**307–328.

Lindenbaum, S., 1977, The "last course": Nutrition and anthropology in Asia, in: *Nutrition and Anthropology in Action* (T. K. Fitzgerald, ed.), pp. 141–155, van Gorcum, Assen, The Netherlands.

Mackenzie, M., 1976, Who is a good mother? *Ethnomedizin* **4(1/2):**7–22.

Marriott, M., 1968, Caste ranking and food transactions: A matrix analysis, in: *Structure and Change in Indian Society* (M. Singer and B. S. Cohn, ed.), pp. 133–171, Aldine, Chicago.

Matter, S. L., and Wakefield, L. M., 1971, Religious influence on dietary intake and physical condition of indigent pregnant Indian women, *Am. J. Clin. Nutr.* **24:**1097–1106.

Mazess, R. B., 1968, Hot–cold food beliefs among Andean peasants, *J. Am. Dietet. Assoc.* **53:**109–113.

McKay, D. A., 1971, Food, illness, and folk medicine: Insights from Ulu Trengganu, West Malaysia, *Ecol. Food Nutr.* **1:**67–72.

McLaren, D. S. (ed.), 1976, *Nutrition in the Community,* John Wiley & Sons, New York.

Mead, M., 1950, *Sex and Temperament in Three Primitive Societies,* Mentor Books, New York.

Mead, M. (ed.), 1955, *Cultural Patterns and Technical Change,* Mentor Books, New York.

Messer, E., 1972, Patterns of "wild" plant consumption in Oaxaca, Mexico, *Ecol. Food Nutr.* **1:**325–332.

Newman, L. F., 1969, Folklore of pregnancy: Wives' tales in Contra Costa County, California, *Western Folklore* **28:**112–135.

Obeyesekere, G., 1963, Pregnancy cravings (dola-duka) in relation to social structure and personality in a Sinhalese village, *Am. Anthropol.* **65:**323–342.

Ogbeide, O., 1974, Nutritional hazards of food taboos and preferences in Midwest Nigeria, *Am. J. Clin. Nutr.* **27:**213–216.

O'Laughlin, B., 1974, Mediation of contradiction: Why Mbum women do not eat chicken, in: *Women, Culture, and Society* (M. Z. Rosaldo and L. Lamphere, eds.), pp. 301–318, Stanford University Press, Stanford.

Pillsbury, B. L. K., 1978, "Doing the month": Confinement and convalescence of Chinese women after childbirth, *Social Sci. Med.* **12B:**11–22.

Reichel-Dolmatoff, G., and Reichel-Dolmatoff, A., 1961, *The People of Aritama: The Cultural Personality of a Colombian Mestizo Village,* University of Chicago Press, Chicago.

Robson, J. R. K., 1974, The ecology of malnutrition in a rural community in Tanzania, *Ecol. Food Nutr.* **3:**61–72.

Sanjur, D., Cravioto, J., Rosales, L., and Veen, A. van, 1970, Infant feeding and weaning practices in a rural preindustrial setting, *Acta Pediatr. Scand.* **Supp. 200.**
Shack, D., 1969, Nutritional processes and personality among the Gurage of Ethiopia, *Ethnology* **8:**292–300.
Sharma, D. C., 1955, Mother, child and nutrition. *J. Trop. Pediatr.* **1:**47–53.
Simoons, F. J., 1961, *Eat Not This Flesh,* University of Wisconsin Press, Madison.
Snow, L. F., and Johnson, S. M., 1978, Folklore, food, female reproductive cycle, *Ecol. Food Nutr.* **7:**41–49.
Srinivas, M. N., 1955, The social system of a Mysore village, in: *Village India. Studies in the Little Community* (M. Marriott, ed.), pp. 1–35, University of Chicago Press, Chicago.
Taha, S. A., 1978, Household food consumption in five villages in the Sudan, *Ecol. Food. Nutr.* **7:**137–142.
Townsend, P. K., 1971, New Guinea sago gatherers: A study of demography in relation to subsistence, *Ecol. Food Nutr.* **1:**19–24.
Trant, H., 1954, Food taboos in East Africa, *Lancet* **2:**703–705.
Valdecañas, O. C., 1971, Barrio central: A study of some social and cultural factors in malnutrition, *Philipp. J. Nutr.* **24:**223–237.
Wellin, E., 1955, Maternal and infant feeding practices in a Peruvian village, *J. Am. Dietet. Assoc.* **31:**889–894.
Wiese, H. J. C., 1976, Maternal nutrition and traditional food behavior in Haiti, *Human Org.* **35:**193–200.
Wilson, C. S., 1970, *Food Beliefs and Practices of Malay Fishermen: An Ethnographic Study of Diet on the East Coast of Malaya,* Ph.D. Thesis, University of California, Berkeley.
Wilson, C. S., 1973, Food taboos of childbirth: The Malay example, *Ecol. Food Nutr.* **2:**267–274.
Yeung, D. L., Cheung, L. W. Y., and Sabry, J. H., 1973, The hot–cold food concept in Chinese culture and its application in a Canadian–Chinese community, *J. Can. Dietet. Assoc.* **34:**1–8.

38

A SYNDROME OF TREMORS

Malati A. Jadhav

INTRODUCTION

Human breast milk is a unique gift a mother bestows upon her infant; it promotes general health and brain growth and plays an overall protective roll. The balance of vitamins and minerals in human milk is such that there is optimum nourishment. Under exceptional circumstances, however, when mother is deficient in vitamin B_{12}, either because she is a vegetarian or suffers from malabsorption of vitamin B_{12}, her breast milk will be deficient in vitamin B_{12} and her exclusively breast-fed infant may manifest adverse effects typified in the syndrome of tremors.

A fully developed form of the syndrome is illustrated in the following case.

CASE 1

This baby was first seen at the age of 9½ months. The pregnancy in the mother, the delivery of the infant, his neonatal period and early development up to the age of 7 months had been normal. He had turned over at 3 months, crawled at 7 months and was able to respond to a call with a ready smile and recognize his parents.

He was exclusively breast-fed. No solid food was offered at any time and no vitamin supplements were added. When 7 months old, he had a slight fever and constipation, for which his parents had administered castor oil. Following a few loose stools, his whole body had turned cold suddenly, followed by a generalized seizure—probably due to aspiration. After this febrile illness he had stopped smiling, crawling, supporting his head, and failed to recognize his parents. However, he had continued to feed well at the breast. He weighed 6.03 kg; his nutrition was adequate.

Malati A. Jadhav • Department of Child Health, Medical School, Christian Medical College and Hospital, Vellore 632004, Tamil Nadu, India.

Involuntary Movements

Rhythmic tremors, jerks, tossing movements, and frog-like licking movements were the most salient clinical findings. When he was awake, rhythmic involuntary movements involving the whole body were noted, that disappeared completely when he was asleep. The tongue, which was very pale, was kept protruded through the open mouth. Fine rhythmic tremors of tongue alternated with frog-like to-and-fro licking movements. Involuntary side-to-side tossing movements of the head were pronounced when he lay in bed. Recurring jerky tremors involved the trunk and upper limbs every 3–4 minutes. These were violent enough to exhaust the boy and he would then go to sleep. Deep tendon reflexes were brisk.

Developmental Regression

Central nervous system examination showed marked apathy amounting to disorientation. He was drowsy and failed to respond to calls. He never smiled or showed any interest in toys. At times, when he appeared to be wide awake, he completely disregarded a dangling or colored toy. He would look at his mother or focus on an object, but not intelligently so. These responses simulated a state of profound mental retardation. From time to time, he sank into a comatose state.

Apathy

He was apathetic, listless, and inactive most of the time. Muscles were flabby. The anterior fontanelle was open. Respiratory and cardiovascular system examination was normal. The liver was palpable 2 cm below the right costal margin; it was smooth.

Dark Pigmentation over the Skin

His natural color was wheat color. In contrast to his pale face and trunk the extremities showed striking dark pigmentation. It was homogenous and symmetrical over the palms, dorsa of hands, the soles and dorsa of feet, and around the ankles. The nail beds and skin over the terminal phalanges of the fingers appeared intensely dark in contrast to the pale fingernails. Pigmentation was also intensely dark over the great toes. Similar homogenous pigmentation was present over the knees and buttocks. Skin over the axillae and medial and antero-medial aspects of the arms and thighs showed black reticular mottling. In spite of the pigmentation the skin was smooth and healthy. The hair on the scalp was sparse but of normal texture.

Severe Anemia

The fingernails and toenails appeared white when compared to the black pigmentation over the skin of terminal phalanges. Hemoglobin was 7.4 g%,

TABLE I

Hematological Findings and Vitamin B_{12} Absorption Studies of Baby and Mother

	Case 1		Case 2	
	Child	Mother	Child	Mother
Hemoglobin (g/100 ml)	7.4	7.2	6.5	12.0
Packed cell volume (%)	20.0	23.0	21.0	42.5
Reticulocyte (%)	2.0	0.7	3.0	3.0
White cells/mm^3	4200	7100	3400	5400
Bone marrow	Megaloblastic, grade III	Megaloblastic, grade II	Megaloblastic, grade III	Megaloblastic, grade III
Serum B_{12} (μμg/ml)	56	48	64	50
Breast milk - B_{12} (μμg/ml)	—	48	—	32
B_{12} absorption	Normal[a]	Defective[a] (5% excretion)	Normal[b]	Normal[a]
Serum proteins (g/100 ml)				
total	7.0	7.2	5.9	6.7
albumin	3.8	4.1	4.2	3.6
globulin	3.2	3.1	1.7	3.1
Serum iron (μg%)	280	—	—	—

[a] Schilling test.
[b] Oral therapeutic response.

PCV was 20%, reticulocyte count was 2%. The peripheral blood picture showed moderate anisocytosis and hypochromia. There were few macrocytes and basophilic stippling was present. Bone marrow showed grade III megaloblastic changes. The results of other hematological investigations are given in Table I. Fasting blood sugar was 95 mg%. Cerebrospinal fluid (CSF) was normal in cells and biochemistry. CSF serum glutamic oxaloacetic transaminase was 13 units.

The mother was young, and although she was thin and pale she appeared healthy. She was a vegetarian by choice. Her staple diet consisted of rice and porridge made from ragi (millet) flour, with vegetables, mango pickle and a small quantity of dal. She never took eggs, fish, or ghee; she took some meat and fruit once a year. Once a week she took a glass of buttermilk.

Mother's Health and Past Obstetrical History

Her previous health was reported to be good. Her first child, born normally, probably suffered from intracranial birth injury and died on the 15th postnatal day. Her second pregnancy had terminated in an abortion at 3 months. Her second child had died when 2½ years old due to diarrhea.

Clinical Diagnosis Was a Syndrome of Tremors

The results of hematological investigations on the child and the mother established deficiency of vitamin B_{12} in the mother and child. Vitamin B_{12} deficiency in the mother was a combination of dietary deficiency and malabsorption. The child had normal absorption of B_{12} but suffered from primary dietary deficiency of vitamin B_{12}.

Treatment

Exclusive breast-feeding was continued on admission, with no added cow's milk, solids, vitamins or anticonvulsants. Three injections of vitamin B_{12} of 50 μg each were given intramuscularly, the second injection was given 3 days after the first and the third 8 days later. With this vitamin B_{12} supplementation alone, tremors disappeared, megaloblastic bone marrow was reverted to normal, and pigmentation decreased almost completely. At the end of 41 days after commencement of treatment he had become normal in orientation, was playful, responsive, and could sit up with support. An electroencephalogram taken 40 days after treatment showed increased voltage and occasional short spikes. The improvement noted was probably due to correction of vitamin B_{12} deficiency.

Response to Treatment

Response to treatment was prompt and complete:

Day 4 Reticulocyte count rose to 11%. Bone marrow reverted to normoblastic.
Day 5 He showed alertness for the first time and tremors decreased in frequency and intensity.
Day 7 He responded to his mother's call and recognized her. He seemed to respond to sounds. Tossing movements of the head became less frequent. Rhythmic flexor movements of fingers and hands became less rapid. The jerks and twitching movements involving the arms totally disappeared. Fibrillary movements of the tongue persisted.
Day 9 There was a remarkable improvement. He appeared almost normal except for the dark pigmentation that persisted. He smiled with recognition, responded with cooing sounds, recognized his mother, attending nurse, and doctor. He started playing with his hands and normally kicked his legs. At this stage, however, he still did not reach for a toy or a dangling object. Rhythmic tossing movements of the head had stopped and he supported his head steadily when lifted up by his arms. Rhythmic tremors involving the trunk were totally absent. However, frog-like licking movements of the tongue persisted. He was not yet able to turn over, although muscle tone had become almost normal.

During the second week the response was slower.

Day 15 He was well oriented and could catch a toy. Pigmentation was less evident.
Day 26 He could turn over. Pigmentation had completely disappeared.
Day 33 He could sit up with support.
Day 35 Hemoglobin was 9 g%, PCV 33%, reticulocyte count 8%, peripheral blood picture showed double population of cells.
Day 41 Remarkably improved motor and psychological development.

A follow-up at age 14 months showed him to be healthy and developing normally.

CASE 2

This boy was seen at 9 months. Striking involuntary movements involved the head, trunk, and limbs. In frequency and character, these movements and fine fibrillary twitchings of the tongue resembled those in Case 1. He had hyperpigmentation over the extremities. There was developmental regression, pallor, and apathy. Hemoglobin was 6.5 g/100 ml and the bone marrow was megaloblastic. Mother, aged 21 years, was vegetarian but very occasionally took buttermilk. Her hemoglobin was 12 g/100 ml and her bone marrow was megaloblastic. Other hematological results are given in Table I.

Comment

Case 2 had all the clinical features of syndrome of tremors.

Treatment

The baby was given breast milk and the mother's diet was unchanged. He was given a daily dose of 0.1 μg of vitamin B_{12} by mouth. On day 5, the reticulocyte count was 25.5% and red cell development in the bone marrow had reverted to a normoblastic pattern. By day 18, when a total of 1.8 μg of vitamin B_{12} had been given orally, the infant could smile intelligently, recognize his mother, follow objects, and watch and play with toys. The pigmentation had almost cleared, and tremors had become less frequent and disappeared within the next week (Day 25). The vitamin B_{12} deficiency in the mother was presumably associated with a defective intake of vitamin B_{12} in her diet.

An incomplete form of the syndrome without abnormal movements has also been noted.

Discussion

A clinically recognizable syndrome of tremors in South Indian infants has been studied and shown to be due to deficiency of vitamin B_{12} in the breast milk, occurring in exclusively breast-fed babies of vegetarian mothers, themselves deficient in vitamin B_{12} in serum and breast milk. The syndrome, when fully developed, is characterized by generalized tremors; dark pigmentation of skin; marked pallor of mucosal membranes, finger, and toenails; apathy amounting to disorientation; and marked developmental regression. Full hematological investigation shows both mother and infant to be suffering from megaloblastic anemia due to vitamin B_{12} deficiency. In the mother the megaloblastic anemia is associated with defective or normal vitamin B_{12} absorption. The syndrome is completely reversed in all its aspects by oral or intramuscular treatment with vitamin B_{12}.

A similar tremor syndrome in infancy has been reported from all parts of India during the past 3 decades. In a majority of these cases, etiology has not been established, although cases from Hyderabad, Bombay, Jabalpur, and others had some evidence of megaloblastic anemia and improved after treatment with vitamin B_{12}. Full hematological studies were not done. Viral encephalitis, acute cerebellar ataxia, tuberculous meningoencephalitis, nutritional recovery syndrome, drug toxicity, and magnesium deficiency are associated with similar tremors. In tremor syndrome associated with magnesium deficiency, mental retardation is absent.

BIBLIOGRAPHY

Dikshit, A. K., 1957, Nutritional dystrophy and anaemia, *Ind. J. Child Health* **6:**132.

Jadhav, M., Webb, J. K. G., Vaishnava, S., and Baker, S. J., 1962, Vitamin B_{12} deficiency in India infants, A clinical syndrome, *Lancet* **2:**903–907.

Kaul, K. K., Belapurkar, K. M., and Parekh, P., 1973, Syndrome of mental retardation, tremors, pigmentation and anemia in infants—some biochemical observations, *Ind. J. Med. Research* **61:**86–92.

39

ANORENIA NERVOSA

Hilde Bruch

In this condition the adverse effects of food are due to the severe restriction of intake and the imbalance of the diet. Anorexia nervosa is an illness with severe weight loss due to *self-inflicted* starvation in the absence of organic disease. It occurs mainly in young women, rarely in men. There have been endless efforts to explain the condition as due to some psychological or physiological aberration since its description more than 100 years ago (Gull, 1874). A psychological orientation was replaced by endocrine theories that in turn led to a differentiation between a picture of hypopituitary cachexia and a psychological anorexia nervosa syndrome. The search for organic factors continues. The difficulty is that it is often not clear whether observed abnormalities are the cause or the consequence of the severe undernutrition (Vigersky, 1977). Weight loss for psychological reasons is also not a uniform condition. It may occur in association with conversion hysteria, schizophrenia, and depression, and these atypical forms need to be differentiated from true anorexia nervosa (Bruch, 1973).

PRIMARY ANOREXIA NERVOSA

The characteristic feature of primary anorexia nervosa is the relentless pursuit of thinness, with intense fear of fatness that does not diminish as the weight loss progresses. The starvation is associated with severe physiological and psychological changes. Menses cease, the skin becomes dry with a yellowish tint, the hair is stringy, every bone shows, and the abdomen is hollow. Other physiological changes due to food deprivation are low blood pressure, slow pulse, low basal metabolism rate, anemia, and disturbances in sleep and neuro-endocrine functions. These symptoms occur also in other starving people and do not differentiate primary anorexia nervosa from the atypical forms. The differences lie in the psychological attitude.

Hilde Bruch • Department of Psychiatry, Baylor College of Medicine, Texas Medical Center, Houston, Texas 77030.

Body Image Disturbance

Ordinary starving people will complain about the weight loss, but the anorexic takes excessive pride in it and actively maintains it. She will defend with vigor and stubbornness her gruesome emaciation as "not too thin" and claim that she does not "see" it as abnormal or ugly. This disturbance in body image or self-perception is characteristic of primary anorexia nervosa and differentiates it from atypical forms.

Disturbed Awareness of Bodily State

Other signs of psychological disturbance include a disturbance in the recognition of bodily states and needs. Awareness of hunger and appetite in the ordinary sense seems to be absent, or is vigorously denied. Absence or denial of desire for food often alternates with uncontrolled impulses to gorge oneself, followed by self-induced vomiting, or the use of laxatives, enemas, and diuretics.

Another characteristic manifestation of falsified awareness of bodily states is overactivity and denial of fatigue, which often appears before the noneating phase. Sometimes there is an intensified interest in athletics and sports; in other cases the activities appear to be aimless, such as walking by the mile, chinning and bending exercises, or literally running around in circles. Drive for activity continues until the emaciation is far advanced. The anorexic's feeling of not being tired and of wanting to do things stands in contrast to the lassitude, fatigue, and avoidance of any effort that usually accompany chronic food deprivation.

One may also consider the failure of sexual functioning and the absence of sexual feelings as falling within this area of perceptual disturbance. They also precede the starvation phase and often continue after nutritional restitution. These patients are limited in their ability to recognize emotional reactions, anxiety, and other feelings; even severe depressive feelings may remain masked for a long time.

Ineffectiveness

Underlying this broad range of symptoms, they suffer from a paralyzing sense of ineffectiveness. They experience themselves as acting only in response to demands from others, as not doing anything because they want to. While the other features are readily recognized, this defect is camouflaged by enormous negativism and stubborn defiance. The indiscriminate way in which they reject ordinary demands of living indicates a desperate cover-up for an undifferentiated sense of helplessness, a generalized parallel to the fear of eating one bite lest control be lost completely. The paramount importance of this characteristic was recognized in the course of intensive psychotherapy. Once defined, it can be readily identified early in treatment.

This deep sense of ineffectiveness seems to stand in contrast to the reports of normal early development that supposedly has been free of difficulties and problems. It seems that overconforming behavior camouflaged the underlying serious self-doubt and left the anorexic-to-be unprepared to grow beyond the immediate family and to engage in relationships with members of her own age group.

Psychological Antecedents and Family Transactions

Basically, anorexia nervosa is not a disorder of food intake and weight regulation but it involves extensive disturbances in personality development and self-concept. The manifest illness is a late step in an individual's desperate struggle to acquire a sense of control and identity. In retrospect many indications of a child's psychological plight can be recognized, but they had been overlooked until the desperate symptoms forced attention and action. As one 16-year-old girl expressed it, "I have been trying to find out what I am supposed to be. Maybe they should recognize my true self. They never paid attention to me until I starved myself."

The families of these patients give the impression of being stable, with very few broken marriages. The social and financial success is often considerable, with great preoccupation with outer appearances; often they are exceedingly weight-conscious. The parents will emphasize the happiness of their homes. Underlying this apparent harmony, without dramatic signs of discord, deep conflicts can be recognized with widely varying individual features. The mothers appear to be conscientious in their concept of motherhood but superimpose their ideas of a perfect life on their children. Fathers, themselves successful, hold up high standards of performance. The parents fail to stimulate or permit the development of an adequate sense of self-reliance and effectiveness in the child.

Analysis of the transactional patterns reveals this deficit to be the result of subtle disregard of clues about his needs emanating from the child while superimposing adequate care which guarantees seemingly normal development. If responses to child-initiated clues are continually inappropriate or contradictory, the child will fail to develop a sense of ownership of his body, does not feel in control of its functions, and is convinced of the ineffectiveness of all efforts and strivings. Anorexia nervosa patients approach life with a self-defeating psychic orientation and are convinced that basically they are inadequate, low, mediocre, inferior, and despised by others. All their efforts are directed towards hiding the fatal flaw of their fundamental inadequacy (Bruch, 1978; Bruch, 1979).

There seems to be a delay in conceptual development. They cling rigidly to early childhood concepts when interpreting human relationships. Frequently they become socially isolated during the year preceding the illness. The new ways of acting and thinking that are characteristic of other adolescents are strange and frightening to them, and they withdraw or feel excluded.

They interpret society's demand for slenderness in an overrigid way and embark on a weight-losing program in the expectation of earning respect. Then a self-perpetuating cycle develops. The undernutrition in itself has severe psychological effects, leads to greater rigidity, irritability, and depression that in turn is reacted to by more severe starvation.

ATYPICAL ANOREXIA NERVOSA

In this group the loss of weight is incidental to some other problem and is often complained of or valued only secondarily for its coercive effect. These nonspecific pictures vary considerably in the underlying motivation and problems and in the severity of illness. In some patients the conflicts of conversion hysteria and other psychoneuroses can be recognized, and in others there are signs of schizophrenic reaction or of depression. An atypical patient who has been cachetic for any length of time may look deceptively like she is suffering from true anorexia nervosa. Clarification is possible only through detailed reconstruction of the earlier situation, with focus on the individual basic issues.

ANOREXIA NERVOSA IN THE MALE

Anorexia nervosa occurs conspicuously less often in males than in females, the ratio being approximately 1 : 10. Both the primary and atypical syndromes are observed. The atypical picture may occur in adults who become anorexic, with true loss of appetite and exhaustion, in response to overdemanding life situations. It occurs also in pubertal boys, and the fasting may have a symbolic meaning or may be related to delusional concepts about food.

In males, too, pursuit of thinness is the leading motive in the primary form. There are also similarities in the family constellation, with overcompliance and exceptionally good performance during childhood. The illness becomes manifest when their assured status of superiority is threatened by new demands or changes in the environment. Rigid food restriction or eating binges followed by self-induced vomiting bring about the severe weight loss. Hyperactivity and drive for achievement persist, often with remarkable athletic accomplishments despite the severe emaciation.

TREATMENT

Anorexia nervosa is a complex condition, and treatment involves several distinct tasks that need to be integrated. Restitution of normal nutrition, or at least correction of the severe starvation, is essential. Nonattention to the nutrition in the unrealistic expectation that the weight would correct itself with psychological understanding, may unnecessarily prolong the course of the illness. If the therapist communicates an awareness of the patient's sense of helplessness, without insult to her fragile self esteem, she will begin to relax the overrigid discipline and meaningful therapeutic involvement becomes possible.

Weight gain alone is not a cure. It needs to be combined with resolution of the disturbed patterns of family interaction and correction of the underlying psychological disorder. Family therapy is particularly effective with young patients soon after the onset of the illness (Minuchin et al., 1978). When the condition develops in late adolescence or even later, or has existed for any length of time, disengagement and redirection of malfunctioning processes in the family are also essential but not sufficient.

The deficits in the psychic development require individual psychotherapeutic help. Therapy must encourage the patient to become an active participant in the treatment process by evoking awareness of impulses, feelings and needs originating within herself. With correction of these cognitive and emotional distortions she can come to the point of relying on her own thinking and feeling, become more realistic in her self-appraisal and capable of leading a more competent life.

REFERENCES

Bruch, H., 1973, *Eating Disorders: Obesity, Anorexia Nervosa and the Person Within*, Basic Books, New York.

Bruch, H., 1978, *The Golden Cage: The Enigma of Anorexia Nervosa*, Harvard University Press, Cambridge.

Bruch, H., 1979, Island in the river: The anorexic adolescent in treatment, in: *Adolescent Psychiatry, Vol. VII* (S. C. Feinstein and P. L. Giovacchini, eds.), pp. 26–40, Chicago University Press, Chicago.

Gull, W. W., 1874, Anorexia Nervosa, pp. 22–28, *Trans. Clin. Soc.*, London.

Minuchin, S. Rosman, L. R., and Baker, L. 1978, *Psychosomatic Families: Anorexia Nervosa in Context*, Harvard University Press, Cambridge.

Vigersky, R. (ed.), 1977, *Anorexia Nervosa*, Raven Press, New York.

40

ZEN MACROBIOTIC DIETS

John R. K. Robson

For several years it has been recognized that the extreme diversity of health practices and options existing in the United States include not only bizarre but sometimes dangerous customs (Rynearson, 1974). One consequence of the recognition of dietary risks to health is that relatively harmless diet practices, associated with religious beliefs, may be unjustifiably condemned, while other dangerous cults may be extolled. Vegetarianism provides one example of the need for the differentiation of those practices which are undoubtedly or potentially harmful from those that present few risks to health. It is widely believed that meat is necessary for strength with the result that vegetarian diets are frequently ridiculed or condemned. While very large numbers of people are adequately fed on vegetarian diets, there can be problems related to the quantity and quality of vegetarian diets. Studies indicate, however, that this concern is less important than the likelihood of energy deficits and deficiencies of vitamin B_{12} (Bender, 1979; Robson, 1977; Dickerson and Fehily, 1979). Vegans who avoid all animal foods have been identified as being particularly at risk, yet these risks are negligible compared with those who follow the Zen Macrobiotic cult. It is important, therefore, to identify the risks of Zen Macrobiotic diets, to make the risks known to the public, and to differentiate the harmful nutritional practices of this cult from that of vegetarianism as practiced by large numbers of surprisingly well-informed persons (Turner, 1979).

Zen itself implies meditation and it aims at the awakening of the mind to the total personality of human existence. Training in Zen aims to overcome psychological defenses which interfere with existential aspects of human life, such as solitude, separation, and helplessness. It is to be expected that the characteristics of Zen might well attract susceptible persons to a dietary cult bearing the word *Zen* as part of its title. Whatever the meditation component of the cult may be, the dietary rules of Zen Macrobiotics have health impli-

John R. K. Robson • Department of Family Practice, Medical University of South Carolina, Charleston, South Carolina 29403.

cations far beyond "awakening to the truth of human nature" (Ikemi *et al.*, 1978.

Zen Macrobiotics owes its existence to George Ohsawa (Yukikazu Sakurazawa), born in 1893 in Kyoto, Japan. Ohsawa wrote over 300 books before his death in 1966. In one book Ohsawa (1966) describes in detail the concepts of Zen Macrobiotics and the philosophy of Oriental medicine. The combined effect of a diet that is basically unsound with unorthodox concepts of medicine has obvious dangers, but there appears to be no shortage of people willing to face these risks. One of the current manifestations of interest in Macrobiotics is the East West Foundation, which has recently published *Cancer and Diet* (Kushi, 1980).

In the *Book of Judgment* Ohsawa (1966) gives a brief historical review of macrobiotics that can be traced to the early Yin and Yang theories of 4000 years ago through the metaphysical era to the present day. For example, the early concept of Yin was the sky, which was considered centrifugal; the antagonistic property of Yang was the earth, which was considered centripetal. In the metaphysical era, the sky was recognized as being the generator of all phenomena and, therefore, Yin; at the same time, phenomena were also recognized as being passive and empty, and, therefore, Yang. Such paradoxes led to the concept of phenomena being relative and having more of one property than the other.

The scientific update uses this concept and designates all hollow organs that are considered to be passive and receptive as Yin. Such organs include the stomach, intestines, bladder, and lungs. On the other hand solid organs such as the kidneys, heart, and pancreas are Yang. The *Unique Principle* is that there are two antagonistic categories, Yin and Yang, which represent two complementary forces such as man and woman and night and day. Yin can be distinguished from Yang by virtue of heat (Yang) and cold (Yin), form (vertical is Yin, horizontal is Yang), weight (heavy is Yang), color, taste, moisture content, chemical composition, and geographical region (living things in cold regions are Yang). Dietary items thus may be Yin or Yang depending on the above properties. Diets and compounds of food items may contain more Yang items than Yin and therefore should be classed as Yang. Conversely, a trend to Yang can be corrected by Yin influences whether they be cold items, items living or grown in a cold climate, items with a vertical form, or light in weight. Chemical compounds are included in the classification; thus substances with hydrogen, carbon, lithium, and arsenic are more Yang than compounds which contain less of these items but more of potassium, sulfur, phosphorus, oxygen, and nitrogen. As in the metaphysical era, the situation is always relative with a tendency for the results of any equation to be more Yin, or more Yang. This concept allows a detailed analysis of the various influences to be made and undesirable trends towards more Yin or Yang can be corrected.

In Zen Macrobiotics, food plays a prominent part in setting trends towards Yin or Yang and countering the resultant trend. This concept is not only applied to the promotion of health but also the treatment of disease. For example, "... cancer and in fact all diseases result from a growing separation and disharmony between ourselves and our natural environment ..." (Kushi, 1980).

The dietary approach in disease and cancer causation has been defined by Kushi (1980) and deserves detailed consideration. Two types of cancer can be classified according to cause by applying macrobiotic theory. The first type results from an excess of animal foods such as eggs, meat, and fish and is, therefore, Yang. The second type is caused by excessive intakes of soft drinks, sugar, citrus, stimulants, chemicals, and spices, all of which are Yin. In general, Yang cancers are in the deeper parts of the body and the Yin cancers at the periphery of the body. This rule is not inviolable, however, for it is explained that cancers of the stomach (which are classed as Yin by virtue of the stomach being Yin) may actually be Yang if the growth develops in the contracted regions of the stomach. Followers of these principles are advised to look at the overall condition to see if Yin is greater than Yang or *vice versa*. A balance can be restored that then frees the body from any obligation to form a fortress against the disease process. As may be expected, balance is achieved by dietary manipulations. The basic Zen Macrobiotic dietary system involves ten different stages (Table I). Ohsawa (1966) claims that diet no. 7 can cure cataract, epilepsy, detachment of the retina, and various other diseases. Standard dietary recommendations have also been offered for cancer and for use "in nearly any other type of sickness" (Kushi, 1980). Verbatim details of these recommendations are:

> At least 50% by volume cooked of every meal should be whole cereal grains, prepared with a variety of cooking methods. Whole cereal grains include brown rice, whole wheat, whole wheat bread, whole wheat chapatis, whole wheat noodles, barley,

TABLE I
Zen Macrobiotic Diets

		Percentage of total consumption					
Diet no.	Cereals	Vegetable, nituke	Soup	Animal	Salads, fruits	Dessert	Beverages
7	100	—	—	—	—	—	
6	90	10	—	—	—	—	
5	80	20	—	—	—	—	
4	70	20	10	—	—	—	
3	60	30	10	—	—	—	As little as possible
2	50	30	10	10	—	—	
1	40	30	10	20	—	—	
−1	30	30	10	20	10	—	
−2	20	30	10	25	10	5	
−3	10	30	10	30	15	5	

millet, oats, oatmeal, corn, corn on the cob, corn grits, buckwheat groats, buckwheat noodles, rye, rye bread, etc.

Approximately 5% of daily food intake by volume should include miso soup or tamari broth soup (one or two small bowls). The taste should not be too salty. The ingredients should include various vegetables, seaweeds, beans and grains; alter the recipe often.

About 20–30% of each meal may include vegetables: two-thirds of them cooked in various styles, including sautéing, steaming, boiling, baking; up to one-third of them as raw salad. Mayonnaise and commercial dressings should be avoided. Potatoes, including sweet potatoes and yams, tomatoes, eggplants, asparagus, spinach, beets, zucchini squash, avocado, and any other tropical vegetables should be avoided, unless you live in a tropical region.

From 10 to 15% of daily intake should include cooked beans and seaweed. Beans for daily use are azuki beans, chickpeas, lentils, black beans. Other beans are for occasional use only. Seaweeds such as hiziki, kombu, wakame, nori, dulse, agar agar, and Irish moss can be prepared with a variety of cooking methods. These dishes should be flavored with a moderate amount of tamari soy sauce or sea salt.

Once or twice a week, a small volume of whitemeat fish may be eaten. The method of cooking should vary every week. A fruit dessert may also be eaten two or three times a week, provided the fruits grow in the local climatic zone. Thus, if you live in a temperate zone, avoid tropical and semitropical fruits. Fruit juice is not advisable, although occasional consumption in very hot weather may be moderately indulged. Roasted seeds and roasted nuts, with a slight salty taste, may be enjoyed as a snack or supplement as well as dried fruits and roasted beans.

Beverages recommended include bancha twig tea (roasted), Mu tea, dandelion tea, cereal grain coffee or tea, all for daily use, as well as any traditional tea which does not have an aromatic fragrance and a stimulant effect.

Foods to be avoided for the betterment of health include the following. Meat, animal fat, poultry, dairy food, butter. Tropical, semitropical fruits and fruit juice; soda, artificial drinks and beverages; coffee, colored tea, and all aromatic stimulant teas such as mint or peppermint tea.

Sugar, honey, all syrups, saccharine and other artificial sweeteners (rice honey or barley malt may be used in very small quantities occasionally to add a sweet taste if necessary.)

All chemicalized food such as colored, preserved, sprayed, and chemically treated foods; all refined, polished grains, flours and their derivatives; mass-produced industrialized food, including all canned and frozen food.

Hot spices, any aromatic, stimulant food, food accessories, and artificial beverages; also artificial vinegar.

Additional suggestions: Cooking oil should be of vegetable origin; if you wish to improve your health, limit oil to good quality sesame oil and corn oil, in moderate volume.

Salt should be unrefined sea salt. Tamari soy sauce and miso, prepared in the traditional way, may be used as salty seasoning.

The following condiments are recommended:

1. Gomasio (10–12 parts roasted sesame seeds to one part sea salt)
2. Roasted kelp powder, roasted wakame powder
3. Umeboshi plums
4. Tekka
5. Tamari soy sauce (moderated use, only for mild taste).

You may eat two to three times per day regularly, as much as you want, provided the proportion is correct and chewing is thorough. Please avoid eating for approximately three hours before sleeping. For thirst, you may drink a small amount of water, but not iced.

Other advice on variations according to whether the cancer is more Yin than Yang are also given. Clearly not all is known concerning the therapeutic effects of components of foods or foods and food mixtures, and an open mind must be kept on the possibility that dietary manipulations may reduce cancer risks and, as is known, play a useful part in cancer therapy as an adjuvant to other therapeutic modules such as chemotherapy or radiation therapy. Scientific evaluations of Zen Macrobiotics in cancer are not available but the documented hazards of Zen Macrobiotic diets as advocated by Ohsawa strongly suggest that the Zen Macrobiotic cancer diets may also be hazardous.

In the Passaic Grand Jury (1966) enquiry into deaths from Zen Macrobiotic diets six cases of malnutrition were described in adults. Four of the patients died, three of whom had used diet no. 7. Over periods of up to 9 months there had been weight losses of up to 40 lb. In one case, scurvy and iron deficiency anemia were documented (Sherlock and Rothchild, 1967). This 36-year-old female eliminated meat and milk from her diet after following Zen Macrobiotics for approximately 4 months. After another 3 months her diet was limited to brown rice and sesame seeds, ground oatmeal, cornmeal, buckwheat, and bread made from cooked rice. Her fluid intake was restricted to 12 oz/day in the form of soup or tea but never water. After 8 months on diet no. 7 she was admitted to the hospital. Prior to admission she had complained of weakness, fatigue, weight loss of 34 lb, a large purple spot on her left leg, swelling, pain, and tenderness of joints, and bleeding gums. On examination, she had clinical signs of scurvy, as well as anemia, hypoalbuminemia, and low circulating vitamin A levels. The patient responded to ascorbic acid and iron therapy, multivitamins, and a diet of 4500–5000 cal/day. These cases are particularly distressing because the dangers of the diet regimes were not recognized by the followers of the diet, and ill effects were not associated with the diet and were even ignored in the belief that Macrobiotics would overcome the problems. For example, in four of the cases cited medical attention had been refused.

The additional dangers for the young whose diet intake is completely controlled by parents adopting macrobiotics is obvious. This is especially true for infants fed the Macrobiotic baby food *Kokoh*. A typical formula for *Kokoh* is given in Table II. Characteristically, infants fed *Kokoh* exclusively present a clinical picture of failure to thrive; usually after a period of breast feeding

TABLE II
Ingredients of Kokoh

Ingredient	Proportion (%)
Sesame seeds	30
Brown rice	30
Sweet brown rice	20
Aduki beans	10
Wheat, oats and soybeans, equal parts	10

TABLE III
Food Energy and Protein Intakes of Three Infants Fed the Macrobiotic Food Kokoh Compared with Recommendations

Daily intake	Case 1	Case 2	Case 3	Recommendations
Cal/kg body wt	58	37	52	100–130
Protein/kg body wt	1.9	1.2	1.3	2.0–3.0

(Robson *et al.*, 1974; Roberts *et al.*, 1979) there is severe stunting of growth. The actual intakes of protein and food energy were determined in three of the five cases reported above (Table III). It can be seen that the intakes were inadequate, yet from a superficial appraisal of *Kokoh* it might be believed that the food was capable of providing adequate nutrition. For example, based on two parameters used to evaluate the value of protein in a diet, *Kokoh* scores well. The first parameter is concerned with the proportion of calories derived from protein. Approximately 7–8% of calories are derived from the protein in breast milk, while commercial formulas on an average provide about 12% of calories from protein. By comparison, *Kokoh* provides 11%. On this criterion *Kokoh* is similar to commercial formulas. The second parameter measures nutritional value of the protein by the amino acid content. It is well known that proteins vary in their ability to meet requirements for growth and this ability depends on the composition of the different amino acids in the protein. For example, on one hand, gelatin neither supports growth nor maintains normal body processes. On the other hand, it is well-known that milk, meat, and other animal proteins will maintain vigorous growth.

Mixtures of vegetable proteins are also capable of supporting growth if they contain the right amino acid pattern. The amino pattern of *Kokoh* has been evaluated and found to be good, and quite capable of supporting growth (Robson *et al.*, 1974). The problem is caused by the need to dilute *Kokoh* to a consistency similar to that of human milk or infant formulas, so that it might be fed by bottle. In one of the cases cited by Robson *et al.* (1974) the dilution was tenfold and the bulk of the feed prevented the infant achieving an adequate intake. In all of the cases cited above, the infants responded to proper feeding. In one case, rickets was present and it is becoming increasingly obvious that this bone disease is also a risk to older children being fed macrobiotically.

Dwyer *et al.* (1979) have studied the dietary intakes of preschool children. Sixteen of the 52 children in the sample were being fed macrobiotic diets, although the exact details of the diets were not given. Serum alkaline phosphate levels were elevated in 8 of the 16 children. Four cases are described in detail and they presented with signs and symptoms which included a slow awkward gait and flaring of the wrists. In three of the cases the diagnosis was confirmed by radiology that revealed findings consistent with active rickets. All cases responded to therapy. Stunting of growth was present in three of the cases (data was not available for the fourth).

The macrobiotic diets were being followed "by various yogic groups, or Seventh Day Adventists." In this publication no distinction is made between Zen Macrobiotics and the "microbiotic" (vegetarian) diets followed by religious groups and it is not intended to imply that the Seventh Day Adventists are followers of Zen Macrobiotics. It is apparent, however, that the term macrobiotic is being increasingly used in a context possibly beyond its original taxonomy. Additional evidence is appearing in the literature that suggests that other religious groups are following a macrobiotic diet regime which may have hazards similar to Zen Macrobiotics. Zmora *et al.* (1979) describe four infants suffering from protein calorie malnutrition, rickets, osteoporosis, pathological fractures, and nutritional anemia in a community known as Black Hebrews in Israel. This community owes its origins to conversion to a mixture of Christianity and Judaism in Chicago, Detroit, Cleveland, and Los Angeles. It is possible, therefore, that other followers of this faith may be facing similar hazards in the U.S. The infants, after being breast-fed for the first 3 months of life, were given almond "milk" and soya "milk" made from grinding up almonds or soya beans, mixing them with water, and straining the mixture. The filtrate constituted the milk that was supplemented with vegetables and oats. The resemblance to *Kokoh* is obvious. As in the case of the infants fed *Kokoh*, stunting of growth had taken place and hypoproteinemia and anemia were the result in the four cases described. Radiological confirmation of rachitic changes were documented in one case, delayed bone age in another. One of the infants died 10 hours after admission to the hospital.

From the preceding discussion it can be seen that there are undoubted dangers for both adults and infants receiving Zen Macrobiotic diets. The implications of Zen Macrobiotics as a therapeutic agent or adjuvant in cancer therapy are also obvious. It is also apparent that the basics of Zen Macrobiotic dietary practices are being used in religions which practice vegetarianism. While vegetarianism can be a satisfactory way of attaining nutritional adequacy, new religious cults may inadvertently adopt the harmful practices of Zen Macrobiotics. For the infant who has no choice in his diet this can have tragic results. It is hoped, therefore, that this discussion will help to carry out the recommendations of the Passaic Grand Jury in making the public aware of the dangers of Zen Macrobiotics.

REFERENCES

Bender, A., 1979, Health foods, *Proc. Nutr. Soc.* **38:**163.

Dickerson, J. W. T., and Fehily, A. M., 1979, Bizarre and unusual diets, *Practioner* **222:**643.

Dwyer, J. T., Dietz, W. H., Hass, G., and Suskind, R., 1979, Risk of nutritional rickets among vegetarian children, *Am. J. Dis. Child.* **133:**134.

Ikemi, Y., Ishikawa, H., Goyeche, J. R. M., and Lasaki, Y., 1978, "Positive" and "negative" aspects of the "altered states of consciousness" induced by antogenic training, Zen and yoga, *Psychother. Psychosom.* **30:**170.

Kushi, M., 1980, A dietary approach to cancer, in: *Cancer and Diet,* pp. 6–12, East West Foundation, Brookline, Massachusetts.

Ohsawa, G., 1966, *Book of Judgment*, Vol. II, *The Philosphy of Oriental Medicine*, Ohsawa Foundation, New York.
Passaic County Grand Jury, 1966, Zen Macrobiotic diet hazardous: Presentment of Passaic Grand Jury, *Public Health News* (**June, 1966**): 132.
Roberts, I. F., West, R. J., Ogilvie, D., and Dillon, M. J., 1979, Risk malnutrition in infants receiving cult diets: A form of child abuse *Br. Med. J.* **1**:296.
Robson, J. R. K., 1977, Food faddism, *Pediatr. Clin. North Am.* **24(1)**:189.
Robson, J. R. K., Konlande, J. E., Larkin, F. A., O'Connor, P. A., and Liw, H.-Y., 1974, Zen Macrobiotic dietary problems in infancy, *Pediatrics* **53(3)**:326.
Rynearson, Edward H., 1974, Americans love hogwash, *Nutr. Rev.* **32(7)**:1.
Sherlock, P., and Rothchild, E. O., 1967, Scurvy Produced by a Zen Macrobiotic diet, *J. Am. Med. Assoc.* **199(11)**:130.
Turner, R. W. D. 1979, Vegan diet and health (letter), *Br. Med. J.* **1(6163)**:613.
Zmora, E., Gorodischer, R., and Bar-Ziv, J., 1979, Multiple nutritional deficiencies in infants from a strict vegetarian community, *Am. J. Dis. Child* **133**:141.

XII

NUTRITIONAL IMBALANCE

41

DIETARY FIBER AS A PROTECTION AGAINST DISEASE

Denis P. Burkitt

CAUSE AND PREVENTION

Disease is generally viewed as being a result of various causative factors ranging between bacteria and radiation, poisons and trauma. Although this is true, it must not be forgotten that disease is the product of imbalance between causative and protective factors. The latter include not only host defense mechanisms but also all that mitigates against excessive exposure to noxious elements in the environment.

Commendable—though some might think excessive—precautions are taken to protect the public from the addition of even minute quantities of potential poison to processed foods. Yet the removal of elements now widely believed to be strongly protective is allowed with impunity. This applies to dietary fiber, a hitherto largely neglected component of food, but one which has received enormous attention during the past decade.

Lack of fiber has generally been considered to play a causative role in a wide variety of diseases that are characteristic of modern Western culture. But it seems more reasonable, and has proved more acceptable, to view its inclusion in diet in adequate amounts as being protective against disease, rather than its removal as causing disease.

Certain diseases now recognized as characteristic of modern western culture are commonly referred to as "Western diseases." Those specifically but not exclusively related to dietary factors include coronary heart disease (CHD), diverticular disease of the colon, appendicitis, gallstones, colon cancer, hiatus hernia, diabetes, obesity, varicose veins, hemorrhoids, and possibly Crohn's disease, venous thrombosis, and pelvic phleboliths. Others in which diet has not as yet been incriminated as a causative factor include ulcerative colitis,

Denis P. Burkitt • Unit of Geographical Pathology, St. Thomas's Hospital, London SE1 United Kingdom.

multiple sclerosis, thyrotoxicosis, Coeliac disease, and certain skeletal defects such as Paget's disease and Perthe's disease.

Only the diseases for which there is reason to incriminate dietary changes will be considered here. The possible relationship of these diseases to diet was in most instances first suspected as a result of epidemiological studies of their geographical, chronological, and socioeconomic distribution, which will be summarized.

EPIDEMIOLOGICAL FEATURES

Geographical

All diseases have their maximum prevalence in the more economically developed areas, typified by North America, Western Europe and Australia, and minimal occurrence in rural communities in the Third World.

Chronological

All the Western diseases for which evidence is available were much less common, and some were distinctly rare, even in Western countries, before the second quarter of the present century. Others, like appendicitis, diabetes, and obesity, had been increasing in prevalence during the 19th century.

Changes in disease patterns that occurred slowly over a prolonged period in Western countries have been taking place more rapidly during the past 30 years in Japan. Prior to the Second World War all the diseases listed, with the possible exception of appendicitis, were much less prevalent in Japan than in Western countries. Gallstones containing cholesterol, which are far the most common type in Western countries, were rare. Diabetes, obesity, CHD, colon cancer, and polyps were all uncommon. All these diseases have been increasing in prevalence. Hiatus hernia, diverticular disease, varicose veins, and venous thrombosis are still uncommon and Crohn's disease remains almost unknown.

Changes in Emigration

Black Americans

It can be assumed that when the ancestors of present black Americans arrived in that country, they must have had prevalences of these diseases no greater than those observed in rural Africa today. Thirty and more years ago there was still a marked discrepancy between the prevalences of these diseases in the black and white communities, the latter being more affected. Such discrepancies have now largely disappeared and both racial groups have comparable risks of developing any of these diseases (Burkitt and Trowell, 1975; Cleave, 1974).

Japanese in Hawaii

Although these diseases have been uncommon in Japan, their prevalence in second and subsequent generations of Japanese emigrants to Hawaii is comparable to other Americans. The diseases most extensively studied in this respect include both benign and malignant tumors of the colon, diverticular disease of the colon, and CHD (Stemmermann, 1970; Stemmermann and Yatani, 1973), and their prevalence is now comparable to that found in the Japanese and Caucasians in Hawaii. Coronary heart disease is still less prevalent in the Japanese in Hawaii than in Caucasians, but much more prevalent than in Japanese in Japan. There is good anecdotal evidence that the other diseases listed have also increased in prevalence.

It would appear that a rapid change of environment, and in this case a dietetic environment, has more severe ill effects than a gradual change.

Polynesians

Although Western diseases are rare or uncommon among those living in a traditional manner in Polynesian islands, they increase in frequency following emigration to New Zealand and the adoption of its Western way of life. In fact, several of these diseases, including diabetes, obesity, varicose veins, and CHD are now more common in Maoris than they are in the white population. A similar situation exists among Pima Indians in the Southern part of the U.S.A., who now have among the highest rates of diabetes and gallstones in the world.

It would appear that a rapid change of environment, and in this case a dietetic environment, has more severe ill effects than a gradual change.

Effects of Westernization and Modernization in Third World Communities

These diseases emerge or increase in frequency in urban or otherwise Westernized communities and in upper socioeconomic groups before they do so in peasant communities. Medical records in India reveal that half a century ago appendicitis occurred occasionally among army officers, but not among less affluent ranks (Hallilay, 1924). In Africa and India, Western diseases have been emerging more rapidly in upper socioeconomic groups and urban communities than in rural communities.

Order of Emergence

The order in which these diseases tend to emerge or increase in prevalence in Third World countries corresponds to that in which they appear with increasing age in the West. This suggests that each must be in part dependent on different periods of exposure to some common causative factors (Burkitt,

1977). In Western countries, appendicitis, diabetes, hemorrhoids, and obesity begin to occur at a relatively low age, whereas CHD and gallstones are uncommon before middle age and diverticular disease of the colon, hiatus hernia, and bowel cancer are primarily diseases of the elderly.

As communities change from a traditional style of life to that associated with modern Western culture these diseases tend to emerge in the same order. Coronary heart disease, for example, doesn't emerge as a medical problem until 20 years or more after diabetes has become common, and diverticular disease is very rare until more than a generation after appendicitis has become relatively common.

THE INCRIMINATION OF DIETARY FACTORS

The epidemiological features of these diseases indicate that they are primarily the result of environmental rather than of genetic factors. The fact that all these diseases are associated with one another, being common or rare together in different geographical locations or socioeconomic communities, and increasing in prevalence at approximately the same period in history, suggests that they may share some common causative or protective factors (Burkitt, 1970).

There are, of course, many changes in the environment associated with changes in prevalence of these diseases in Third World communities today, including the availability of transistor radios, plastic utensils, and ballpoint pens. These are associated with the diseases in question in Africa, but would not have been so associated with their emergence in North America. To be considered causative, associations must provide grounds for hypotheses of causation that make biological sense. Moreover, since all these diseases can be shown to be related either directly or indirectly to alimentary tract it would seem reasonable to consider changes in diet that invariably precede increases in incidence of these diseases, before incriminating other factors in the environment.

DIETARY CONTRASTS BETWEEN WESTERN AND THIRD WORLD COMMUNITIES

With few exceptions the fundamental differences between the diets of affluent and of less-developed countries are similar to the contrasts between Western diets today and those consumed a century and more ago. Economic development is associated with reduced carbohydrate, increased sugar, reduced fiber, and increased fat consumption.

The contrasts in dietary patterns can be summarized as follows:

Protein	This provides a comparable proportion of energy throughout most of the world, 8–12%.
Carbohydrate	Peasant communities in Africa and Asia derive about 70% of their energy from starchy plant foods retaining their full complement of fiber. Refined sugar was until recently unavailable. In the West only about 40% of energy is derived from carbohydrate and half of this is comprised of sugar.
Fat	This provides 10–15% of energy in poor communities (the fat is largely of plant origin); in the West fat (largely derived from animals) provides 40% of energy.
Fiber	This has been the neglected Cinderella of nutrition. Because it is of little nutritional value it has been considered disposable, and has consequently been removed and discarded almost as a contaminant. Much recent work has, however, shown that it is by no means the inert product it was believed to be, but plays a major role in virtually all the chemical, metabolic, and physiological processes that occur in the gut. This escalation of interest is reflected in the fact that whereas a decade ago only about ten scientific publications on the role of fiber in human nutrition appeared annually, the current figure exceeds 400. Whereas in Western countries dietary fiber intake is only in the region of 20 g, (except in vegetarians who consume about 40 g), in Third World countries it exceeds 60 g and may even exceed 100 g/day.

WHAT IS FIBER?

One of the reasons for the neglect of fiber has a been a misunderstanding of its structure and nature. It was wrongly assumed to be almost synonymous with cellulose, and the generally used term *crude fiber* was, from a physiological point of view, totally misleading. The test used to determine crude fiber values was devised in the early part of the 19th century and measured only the lignin and part of the cellulose. The all-important noncellulose polysaccharides were entirely omitted. These include pentoses, hexoses, pectins, gums, and other components. The different components of fiber are now known to have distinct physiological and other functions.

Generally speaking, fiber consists of the structure of cell walls as distinct from the contained nutrients. It can be viewed as that part of the plant food that resists digestion by the endogenous secretions in the small intestine. Most of it is, however, degraded by bacteria in the large bowel. In all seeds, which include all cereals, much of the fiber is on the outside, and is removed during milling.

Although the greatest single contrast between the diets of communities with high and low prevalences of Western diseases is in the fiber content, it must constantly be remembered that a high-fat diet is also a low-carbohydrate diet as well as being a low-fiber diet, and *vice versa*. Consequently it is wise to consider high-fat/low-fiber diets in contrast to low-fat/high-fiber diets.

Although fiber may be viewed as being to some degree protective against all the Western diseases that have been shown to be diet-related, fat is probably causative of several of them, including CHD, colon cancer, obesity, and diabetes.

RELATIONSHIP BETWEEN CHANGES IN DIET AND DISEASE PREVALENCE

In all the situations referred to above, geographical, chronological, socioeconomic, and following emigration from low-prevalence to high-prevalence countries, the dietary changes associated with increased disease frequency have been similar. This in itself does not prove a cause and effect relationship, but it provides grounds for postulating hypotheses of causation.

Since the protein content of diet changes relatively little between communities other than in its nature, changes in carbohydrate fat and fiber deserve most attention, and as mentioned above, these are interrelated. Because carbohydrate and fat have received much attention, emphasis will be placed on the neglected value of fiber.

No one component of food should be considered in isolation from others. As mentioned above high fat diets are almost invariably low-fiber diets and vice versa. It is, however, possible to suffer from the effects of overconsumption of fiber-depleted carbohydrate foods and yet to derive enough fiber from other foods to afford protection from diseases such as diverticular disease and hiatus hernia. The Pima Indians, eating many "junk" foods, but also consuming an abundance of beans, are a good example.

INFLUENCE OF FIBER ON BOWEL CONTENT AND BEHAVIOR

The best demonstrated property of fiber is its ability to increase the bulk and reduce the viscosity of feces and to maintain adequate speed of transit of fecal content through the large bowel. Fiber, however, only reduces transit time when it is prolonged, and has been shown to have the opposite effect in patients suffering from diarrhea (Payler et al., 1975). This bulking effect is mainly due to its pentose content, and is consequently manifested most by the foods rich in this component, which are mainly the cereals. The mechanism whereby fiber increases stool volume is, at the time of writing, not fully understood. Both the holding of water within the gut and bacterial proliferation would appear to be implicated. Cereal fiber seems to have the greater influence on the former, and vegetables on the latter (Cummings, 1980).

Daily stool output in Third World communities is usually between 250 and 500 g daily, but may exceed 700 g. In contrast it is usually between 80 and 120 g in Western countries, and considerably less in elderly patients.

The time taken for ingested markers to traverse the alimentary tract from mouth to anus is usually under 40 hr in Third World countries in contrast to 72 hr in the West and often over 300 hr in the elderly.

A deficiency of fiber in the diet, and of cereal fiber in particular, is the fundamental cause of the constipation that can be considered almost universal in Western countries. Cereal foods have been shown to be most effective in combatting constipation, due to their high pentose fraction.

DISEASES RELATED TO THE MECHANICAL EFFECTS OF CONSTIPATION

Raised Intra-luminal Pressures

In the Colon—Diverticular Disease

It is now generally accepted that the increased muscular activity required to propel forward fecal content of small volume and increased viscosity raises the pressures within the lumen of the colon. This forces protrusions of the mucosal lining out through weak spots in the overlying muscle wall to form diverticula (Painter, 1975). Vegetarians in Britain consume approximately twice as much fiber as do nonvegetarians and suffer only one-third as often from diverticular disease (Gear, 1979). Even their fiber intake is much less than that in Third World communities.

In the Appendix—Appendicitis

Appendicitis is usually an inflammation of the whole of the mucosa of the appendix distal to a line of demarcation between healthy and diseased tissue. It is now generally agreed that the initial lesion is obstructive. One of the major cases of obstruction is believed to be the presence of firm fecal particles within the appendix (Burkitt, 1975a).

Fiber, through its property of maintaining soft bowel content, is considered to be protective against this disease.

Raised Intra-abdominal Pressures

Straining of abdominal muscles, as if attempting to pass a stool, has been shown to raise intra-abdominal pressures, in a sitting position to a mean of 195 cm H_2O, while those above the diaphragm only reached 67 cm (Fedail *et al.* 1979). The differential was thus 128 cm. When squatting the pressures generated below and above the diaphragm were only 132 and 54 cm respectively, with a differential of 78.

Hiatus Hernia

It seems reasonable to suppose that these raised intra-abdominal pressures are the major cause of hiatus hernia, a protrusion of the upper end of the stomach upwards through the hiatus or opening in the diaphragm that transmits the esophagus (Burkitt, 1975b). Consistent with this is the fact that when a radiologist desires to demonstrate a hiatus hernia, he raises intra-abdominal pressure in order to do so.

A simple analogy of this proposed mechanism is a ball with a hole in its wall. When filled with water and squeezed, the water is extruded through the hole. The esophageal hiatus in the diaphragm corresponds to the hole in the bowel wall. A prospective radiological study of black and white Americans revealed no significant differences in the incidence of either diverticular disease or hiatus hernia, whereas both diseases, so common in the West, are very rare in Africa (Ward *et al.*, in press). This indicates that their causative factors must be environmental rather than genetic.

Varicose Veins

The etiology of varicose veins is almost certainly multifactorial. Certainly some of the traditional explanations of the maladaptation of man to his present erect posture, or to pregnancy, are at total variance with epidemiological evidence. One factor that appears to play an important role is the transmission of raised intra-abdominal pressures to the venous trunks draining the leg veins. These force blood retrogradely down the veins of the saphenous system causing their dilation, with consequent separation of the valve cusps and resultant incompetence. This process affects the valves from above downwards until the whole saphenous system becomes incompetent and varices ensue (Burkitt, 1976). Fiber, by reducing the necessity for abdominal straining, avoids the pressures claimed to be factors causative of both hiatus hernia and varicose veins.

Reduced Volume and Prolonged Transit of Colon Content

The reduction in volume and reduced speed of transit of fecal content that are results of fiber deficiency, are believed to contribute to the cause of both large bowel cancer and hemorrhoids.

Colo-rectal Cancer

It is now generally agreed that colo–rectal cancer is predominantly the result of dietetic factors. It is currently believed that excessive fat in the diet contributes to the causation of this disease and that fiber is protective on several counts. Because it increases fecal volume it dilutes any carcinogens or precarcinogens in the feces. By shortening intestinal transit time it reduces duration of contact between fecal carcinogens and bowel mucosa (Walker and Burkitt, 1976). Moreover, it has been shown to bind various toxins, rendering them less active (Ershoff, 1972) and to reduce fecal pH. Bowel cancer risk in different communities has been shown to relate directly to fecal pH (MacDonald *et al.*, 1978). It also reduces bacterial degradation of primary bile acids into potentially carcinogenic metabolites (Pomare and Heaton, 1973).

Hemorrhoids

Among the factors responsible for hemorrhoids appear to be the forcing of blood backwards down the superior mesenteric veins during abdominal straining (Graham-Stewart, 1962), and the shearing stress exerted by the passage of firm fecal masses through the anal canal. The venous engorgement of the vascular submucosal cushions that constitute hemorrhoids renders them more susceptible to the pressures exerted by the firm feces passing through the anal canal (Thomson, 1975). This hypothesis is supported by increasing evidence that fiber-rich diets are highly effective in the treatment of hemorrhoids, and many would emphasize that they should always be recommended no matter what surgical measures are found to be necessary.

FIBER AS A CONTROL ON ENERGY INTAKE AND OBESITY

Communities subsisting mainly on plant food, while retaining their natural complement of fiber, seldom have an obesity problem even when there is no scarcity of food.

Haber and his colleagues (Haber *et al.*, 1977) fed volunteers the same quantity of apples in three different forms. Apples in their natural state, retaining both their normal structure and their fiber; apple puree, with the structure destroyed but the fiber retained; and apple juice with the fiber removed. All had the same energy value. The first took longest to eat and gave the greatest sense of satiety. The juice was swallowed quickly and was least satisfying of hunger, and the puree was intermediate between the other two.

This, together with epidemiological evidence and the weight-reducing effect of fiber-rich carbohydrate foods, suggest that foods with a low energy/satiety ratio are protective against the development of obesity. This does not imply that addition of fiber to fiber-depleted diets has the same effect. The reduction in fat-rich foods, which is the almost inevitable result of consuming a diet based predominantly on fiber-rich starchy foods, and a reduction of sugar, which is totally fiber-depleted, are important components of a weight-reducing diet.

FIBER AS A PROTECTION AGAINST DIABETES

As with obesity, diabetes is rare among communities consuming diets consisting largely of fiber-rich starchy plant foods, whether based on cereals, tubers, or plaintains, with little or no refined carbohydrate added. In the experiment described above (when discussing obesity) the rise in blood-sugar levels was maximum after the rapid consumption of apple juice and minimum after the slower intake of apples in their natural state.

In the case of diabetes, probably more important than the speed of eating is the rate at which energy is absorbed from the intestinal content. Fiber-rich foods provide a more viscid intestinal content that slows down the rate of absorption of energy from the gut. This reduces demand on insulin secretion.

The components of fiber that have been shown to have the greatest effect in retarding the escape of energy from the bowel content are the guar gums derived mainly from the Indian cluster bean (Anderson, 1976; Jenkins *et al.*, 1978).

The geographical distribution of diabetes indicates that prescribing diets low in carbohydrate content for patients suffering from diabetes is fundamentally a faulty approach. All communities deriving most of their energy from carbohydrates, but from starches and not from sugars, have low prevalences of diabetes. It is consequently not surprising that diets rich in high-fiber carbohydrate foods but reduced in sugar and fat content have proved highly effective in treating the adult-onset type of diabetes.

INFLUENCE OF FIBER ON CHOLESTEROL AND BILE ACID METABOLISM

The nature of the diet influences cholesterol synthesis in the liver, its conversion to bile acids and the absorption of fecal excretion of the bile acids and their bacterial metabolites. Fiber-rich diets have been shown to be associated epidemiologically with bile which is less lithogenic, i.e., showing less tendency to form stones, than are fiber-depleted diets. There is experimental evidence that fiber reduces the lithogenicity of bile both by increasing the solvents—bile acid and lecithin—and by reducing the cholesterol content. In addition, the reduction in obesity associated with fiber-rich diets is also associated with reduced cholesterol synthesis. This and other evidence suggest that whereas there are likely to be several factors contributing to the formation of cholesterol gallstones, fiber may exert a protective influence (Heaton, 1978).

Both epidemiological (Trowell, 1975) and prospective population studies (Morris, 1977) suggest that fiber-rich diets may also exert a protective influence against CHD. The mechanisms involved are, however, less well understood. It would seem that diets high in fiber-rich carbohydrate foods and low in fat reduce risk factors, such as raised blood lipid levels, and are thus protective against CHD. The mere addition of fiber, as in the form of bran, is not effective in lowering blood levels of cholesterol and other lipids.

VENOUS THROMBOSIS AND PELVIC PHLEBOLITHS

Venous thrombosis, which can result in pulmonary embolism, is a major postoperative hazard in Western communities. It is universally rare in Third World countries. Pelvic phleboliths—little stones in veins which are readily detected by X-ray— are calcified blood clots in pelvic veins. These are also much more common in Western than less economically developed communities (Burkitt *et al.*, 1977). They are equally common in black and white Americans (W. J. Ward and T. J. Reinhart, unpublished data), and both diseases are related to the formation and dissolution of blood clots in the veins. Their occurrence in animals is almost unknown. A recent study has shown that the number of phleboliths detectable at X-ray is a measure of the risk of the patient developing D.V.T. (deep vein thrombosis) after surgery (Janvrin and Latto, 1978).

There is some evidence that the fiber content of diet may relate to the occurrence of venous thrombosis. In several British hospitals prescribing wheat fiber in the form of bran pre- and postoperatively, there is a marked reduction in the prevalence of venous thrombosis (Janvrin and Latto, 1978).

SOURCES OF FIBER

Southgate *et al.* (1976) have estimated the dietary fiber content of the different categories of plant foods. Fruits and leaf vegetables have a low content, probably never exceeding 3%, largely because they are mostly composed of water. Root vegetables or tubers, such as potatoes, carrots, and parsnips, have more fiber per weight, but this seldom exceeds 7%. Legumes or pulses, the families of peas and beans, have rather more fiber, but the richest source is found in cereals, wholemeal bread and breakfast cereals having the highest fiber content. Miller's bran, which cannot be classified as a food, is the richest source of all, containing 26% by weight dietary fiber.

Not only do cereals have a higher content of fiber than do other plant foods, but they are particularly rich in the pentose fraction which is the most effective in combating constipation.

Although wholemeal bread has only about 2.5 times as much dietary fiber as does white bread, it has 8 times the effect in increasing stool volume because of its higher pentose content.

TOTAL EFFECT GREATER THAN COMBINED EFFECT OF COMPONENTS

It must be emphasized that the total function of fiber is much more than the sum of the functions of its individual components. It is involved in a multitude of interrelations and activities, both physical and chemical. If a clock is dismantled, various wheels can be shown to engage with one another, but the sum total of this knowledge doesn't enable one to tell the time.

The addition of fiber to a fiber-depleted diet is beneficial in counteracting constipation and thus avoiding its various consequences. It is, however, no more than a compromise and is not an adequate substitute for foods from which the fiber has not been removed in the first place.

PRACTICAL CONSIDERATIONS

As has been emphasized above, high-fat diets are almost invariably low in fiber and *vice versa*. Excessive fat consumption has been implicated in the causation of several diseases. Fiber appears to be protective against a great number of diseases in which fat may play a causative role. In the light of current knowledge it would therefore seem prudent to advise an increased consumption of fiber-rich starchy carbohydrate foods, such as minimally processed cereals, potatoes, and legumes, and a reduced consumption of sugar and fats. In practical terms this would mean eating much more bread (but where possible avoiding white bread), less sugar, sweets, sweetened drinks, and confectionery, more potatoes (but neither cooked nor eaten with fat), and much less meat, since about 40% of lean red meat as currently produced in Western countries is composed of fat, even after the visible fat has been removed.

In view of the general fiber-deficiency of Western diets, it would be prudent for most people to add one or two tablespoonsful of miller's bran to their daily diet, but introducing it slowly—perhaps a teaspoonful at first—to minimize the intestinal flatus which it initially promotes.

REFERENCES

Anderson, J. W., 1976, Beneficial effects of high carbohydrate high fiber diet on hyperglycemic diabetic men, *Am. J. Clin. Nutr.* **29:**895–899.

Burkitt, D. P., 1970, Relationship as a clue to causation, *Lancet* **2:**1237.

Burkitt, D. P., 1975a, Appendicitis, in: *Refined Carbohydrate Foods and Disease* (D. P. Burkitt and H. C. Trowell, eds.) pp. 87–97, Academic Press, London and New York.

Burkitt, D. P., 1975b, Hiatus hernia, in: *Refined Carbohydrate Foods and Disease* (D. P. Burkitt and H. C. Trowell, eds.) pp. 161–169, Academic Press, London and New York.

Burkitt, D. P., 1976, Varicose veins: Fact or fantasy, *Arch. Surg.* **111:**1327–1332.

Burkitt, D. P., 1977, Relationship between diseases and their etiological significance, *Am. J. Clin. Nutr.* **30:**262–267.

Burkitt, D. P., Latto, C., Janvrin, S. B., and Mayo, B., 1977, Pelvic phleboliths. Epidemiology and postulated etiology, *N. Engl. J. Med.* **296:**1387–1389.

Burkitt, D. P., and Trowell, H. C., 1975, *Refined Carbohydrate Foods and Disease,* Academic Press, London and New York.

Cleave, T. L., 1974, *The Saccharine Disease,* John Wright & Sons, Bristol.

Cummings, J. H., 1980, Some aspects of dietary fiber metabolism in the human gut, in: *Food, Health, Science and Technology* (G. G. Birch and K. J. Parker, eds.), pp. 444–458, Applied Science Publishers, London.

Ershoff, B. H., 1972, Comparative effects of a purified diet and stock ration on sodium cyclamate toxicity in rats, *Proc. Soc. Exp. Biol. Med.* **141:**857–862.

Fedail, S. S., Harvey, R. F. and Burns-Cox, C. S., 1976, Abdominal and thoracic pressures during defecation, *Br. Med. J.* **1:**1.

Gear, J. S., 1979, Diverticulitis in vegetarians, *Lancet* **1:**511–514.

Graham-Stewart, C. W., 1962, Hemorrhoids and carcinoma of the rectum. *Surg. Gynecol. Obstet.* **115:**89–90.

Haber, G. B., Heaton, K. W., Murphy, D., and Burrough, D. F., 1977, Depletion and disruption of dietary fiber. Effects of satiate, plasma-glucose and cerium insulin, *Lancet* **2:**679–682.

Hallilay, H., 1924, Intestinal stasis and cancer in Indians, *Indian Med. Gaz.* **59:**403–405.

Heaton, K. W., 1978, Are gallstones preventable? *World Med. J.* **14:**21–23.

Janvrin, S. B., and Latto, C., 1978, Practical experience in fiber, in: *Third Kellogg Nutrition Symposium—Dietary Fiber*, (K. W. Heaton, ed.), p. 151.

Jenkins, D. J. A., Wolever, T. M. S., Leeds, A. R. et al. (1978) Dietary fibers, fiber analogues and glucose tolerance: Importance of viscosity, *Br. Med. J.* **1:**1392–1394.

MacDonald, I. A., Webb, G. R., and Mahoney, D. E., 1978, Fecal hydroxysteroid dehydrogenase activities in vegetarians, Seventh Day Adventists, control subjects and bowel cancer patients, *Am. J. Clin. Nutr.* **Suppl. 31.**

Morris, J. N., 1977, Diet and heart: A postscript, *Br. Med. J.* **2:**1307–1314.

Painter, N. S., 1975, *Diverticular Disease of the Colon*, Wm. Heinemann Medical Books, London.

Payler, D. K., Pomare, E. W., Heaton, K. W., and Harvey, R. F. 1975, The effect of wheat bran on intestinal transit, *Gut* **16:**209–213.

Pomare, E. W., and Heaton, K. W., 1973, Alteration of bile salt metabolism by dietary fiber (bran), *Br. Med. J.* **4:**262–264.

Southgate, D. A. T., Bailey, B., Collinson, E., and Walker, A. R. P., 1976, Fiber, a guide to calculated intakes of dietary fiber, *J. Hum. Nutr.* **30:**303–313.

Stemmermann, G. N., 1970, Patterns of disease among Japanese living in Hawaii, *Arch. Environ. Health* **20:**266–273.

Stemmermann, G. N., and Yatani, R., 1973, Diverticulosis and polyps of the large intestine colon, and necropsy study of Hawaii Japanese, *Cancer* **31:**1260–1270.

Thomson, W. H. F., 1975, The nature of hemorrhoids, *Br. J. Surg.* **62:**542–552.

Trowell, H. C. (1975) Ischaemic heart disease, atheroma and fibrinolysis, in: *Refined Carbohydrate Foods and Disease* (D. P. Burkitt and H. C. Trowell, eds.) pp. 195–226, Academic Press, London and New York.

Walker, A. R. P., and Burkitt, D. P., 1976, Colonic cancer—hypotheses of causation, dietary prophylaxis and future research, *Am. J. Dig. Dis.* **21:**910–917.

42

ADVERSE EFFECTS OF SOME FOOD LIPIDS

Lilian Aftergood and Roslyn B. Alfin-Slater

INTRODUCTION

During the last several decades extensive epidemiological studies carried out on a worldwide scale have suggested a strong association between certain dietary components and chronic diseases. A number of these associations have been confirmed by experimental studies with animals. Thus, some naturally occurring toxins, or in some instances, compounds added during food processing, have been shown to be carcinogenic; saturated fats and cholesterol became suspect in atherosclerosis; excessive caloric intake has been positively correlated with obesity; an excessive sugar intake has been linked to increased dental caries in children. Under certain conditions, i.e., kidney disease, an excessive protein intake is undesirable and even an excessive intake of fiber, the latest panacea for all ailments, may be quite harmful in that fiber binds essential trace minerals, making them unavailable. Overdosing with vitamins has also been found to be associated with undesirable side effects. In nutritional science the two basic tenets—namely, moderation and variation—still stand true. The advantage of moderation is obvious and it is not difficult to see that a variation of dietary components tends to insure the provision of a multitude of nutrients, while at the same time, it does not overburden the organism with an excessive amount of potentially harmful agents.

In this chapter, we will be concerned with dietary fat, the potential deleterious effects of some of its constituents, and with cholesterol, stressing some new developments in its purported relation to disease.

FATTY ACIDS

Vegetable vs. Animal Fatty Acids

Several long term epidemiologic studies have implicated an elevated serum cholesterol level as an important risk factor for the development of

Lilian Aftergood and Roslyn B. Alfin-Slater • Nutritional Sciences and Environmental Health Division, School of Public Health, University of California, Los Angeles, California 90024.

myocardial infarction (Heyden *et al.*, 1971; Hill and Wynder, 1976). For men with serum cholesterol levels of 260 mg/dl and above, the risk of developing ischemic heart disease has been found to be twice as high as that for men with cholesterol levels below 220 mg/dl. In general, epidemiological studies indicate that cholesterol levels over 200 mg/dl imply an increased risk of CHD (Coronary Heart Disease) (Kannel *et al.*, 1964).

At the same time, it has been recognized that for humans saturated fat is more cholesterolemic than is unsaturated fat (Ahrens, 1957). The atherogenicity of saturated fat has been shown in animals as well (Kritchevsky *et al.*, 1954). Although most animal fats are saturated and most vegetable fats are unsaturated, there are exceptions. For example, among vegetable oils, coconut oil and peanut oil have been shown to be significantly atherogenic (Kritchevsky, 1979). Coconut oil is 93% saturated and contains predominantly short-chain fatty acids. Peanut oil, even though only 21% saturated and containing approximately 30% of linoleic acid, is characterized by the presence of a significant amount of a 22:0 acid (behenic). Interestingly enough, an interesterification of peanut oil, whereby the structure of triglycerides is modified in that the position of the fatty acids on the glycerol skeleton is changed without affecting their actual composition, results in a reduction of its atherogenicity.

There is considerable evidence that the saturated fatty acids may vary substantially in their hypercholesterolemic effects. It seems that stearic acid ($C_{18:0}$) has less effect on raising serum cholesterol levels than do lauric ($C_{12:0}$), myristic ($C_{14:0}$), or palmitic acids ($C_{16:0}$) (McGandy and Hegsted, 1975).

Milk

An animal fat with somewhat controversial effects is that contained in milk. Milk contains 3.7% fat (predominantly saturated and containing short-chain fatty acids) and 80–90 mg cholesterol/pint; yet it has been shown repeatedly to be hypocholesterolemic (Richardson, 1978), whether it is ingested as yogurt (Mann, 1977) or as a whole milk (Howard and Marks, 1977). The compounds, hydroxymethylglutaric acid and/or orotic acid found in milk, have been suggested as the hypocholesterolemic factors.

Trans Fatty Acids

The utilization of vegetable fat has been greatly promoted and increased in the recent years. Whereas from 1909 to 1913, 26.6% of total calories of the diet were derived from animal fat and 5.5% from vegetable fat, by 1978 these figures have changed to 24.1 and 17.9% respectively, with total calories from fat increasing from 32.1 to 42.0% (Marston and Page, 1979). Processing technology by which vegetable oils form margarine has been significantly improved and today vegetable oils, hydrogenated soybean oil in particular, play an important role in the diet.

At the same time increasingly sophisticated analytical methodology has revealed that partially hydrogenated oils contain large amounts of both geometrical and positional isomers of the naturally occurring unsaturated fatty acids whose nutritional value and biological effects are now being examined.

A small amount of isomeric fat may be found in milk due to the biohydrogenation process taking place in the cow's rumen (Smith *et al.*, 1978), however, the main source of trans isomers of fatty acids consumed by humans is derived from shortenings and margarines which may contain up to 40% of geometric fatty acid isomers (Scholfield *et al.*, 1967; Heckers and Melcher, 1978). It is not surprising, therefore, that trans fatty acids have been found in a variety of human tissues (Heckers *et al.*, 1977).

Since one of the vital roles of fatty acids in the living organism is their influence on the structure and function of cellular membranes, there is always the possibility that a change in their fatty acid composition will affect some of the regulatory properties of such membranes. Thus, the utilization and effects of isomer fatty acids in humans require a careful evaluation (Emken, 1979). The data obtained through many experimental investigations to study the effects of dietary fats containing trans fatty acids on human plasma lipid levels are not always consistent. In general, no short-term toxic effects have been reported for these hydrogenated fats. There is no large increase in plasma lipids above that considered normal. In fact, hydrogenated fats, such as margarines, still maintain at least some of the lipid lowering effect of unsaturated vegetable oils when compared with highly saturated fats (Emken, 1979). Also, the level of trans fatty acids in the aortas of men with advanced atherosclerosis proved to be not significantly different from that in the aortas of men who died from other causes (Heckers *et al.*, 1977).

Recently, however, another disease, cancer, has been found to be possibly correlated with the intake of trans fatty acids-containing dietary fats (Enig *et al.*, 1978).

In animals, it has been shown that essential fatty acid deficiency is accentuated by the ingestion of dietary trans fatty acids (Jensen, 1976; Privett *et al.*, 1977). It has been reported that mitochondria from the heart of animals fed partially hydrogenated fats have a decreased capacity to oxidize substrates (Hsu and Kummerow, 1977). The enzymes involved in various aspects of lipid metabolism also seem to be affected by the presence of trans fatty acids (Mahfouz *et al.*, 1980a), i.e., some microsomal desaturases were found to be inhibited by trans 18:1. Similarly, both body and heart growth rates were found to be decreased in rats fed trans, trans linoleate as 10% of calories. In addition, as a result of this diet, not only the fatty acid composition of heart lipids was affected but less of the phospholipid phosphatidylcholine (lecithin), was present (Yu *et al.*, 1980).

On the other hand, some isomers of trans 18:1 acid were found to be converted to isomers of 18:2 acid which in turn might possibly serve as precursors of unusual polyunsaturated fatty acids (PUFA). It has been suggested (Mahfouz *et al.*, 1980b) that these might lead to unusual prostaglandins, which are normally produced from the naturally occurring cis fatty acids, linolenic and arachidonic. As a result, the metabolism of the common PUFAs, and prostaglandins derived from them, could perhaps be affected through competitive interaction. As a matter of fact, large amounts of dietary trans, trans linoleate have resulted in a decreased biosynthesis of prostaglandins in rats due to decreased concentrations of a precursor fatty acid (Hwang and Kinsella, 1979).

There is no question that the problem of trans fatty acids awaits further clarification. In the meantime, work continues toward the improvement of a technological process leading to "zero trans" margarines (List, *et al.*, 1977).

Potentially Undesirable Effects of PUFA

Since the serum cholesterol-lowering effect appears to be desirable from the standpoint of diminishing risks of heart disease, a decrease in dietary saturated fatty acids and an increase in the proportion of PUFA has been pursued (Anderson *et al.*, 1973; Kummerow, 1975). A positive role of PUFA in human nutrition and metabolism has been claimed for many years, particularly since it has been recognized that essential fatty acid (EFA) deficiency symptoms (i.e., dermatitis) do occur in humans under certain conditions (Söderhjelm *et al.*, 1970; Carroll, 1977). Such symptoms may occur in childhood and adults after prolonged parenteral nutrition with fat-deficient preparations. An intake of EFA as 1–2% of total calories will prevent deficiency.

In the middle-aged human population up to 20% reduction in serum cholesterol can be achieved through a relatively high intake of polyunsaturated fat (up to 20% of total calories as linoleic acid) (Vergroesen, 1972). Nevertheless, clinical trials still have not definitely established the advantage of extremely high dietary PUFA levels in human nutrition (Carroll, 1977; Kaunitz, 1978).

Work with rats has led to the conclusion that PUFA in human diets should not exceed half of the total normal fat calories (Narayan *et al.*, 1974). Recently, it has been advocated that 10% of total fat calories as EFA should be the upper limit (Carroll, 1977). However, the American Heart Association (AHA) recommends a total fat intake of 30% with a P/S ratio of 1:1 (AHA Statement, 1978).

Since two large dietary intervention trials are currently underway, it is hoped that in the near future more definite answers as to the significance of dietary modification in lowering of plasma lipids will be forthcoming (Glueck and Connor, 1979).

There are potentially undesirable effects of dietary polyunsaturated fats. First of all, excessive levels of dietary fats lead not only to obesity but apparently have a positive relationship to certain types of carcinogenesis (Carroll, 1977). Animals on a high fat diet develop mammary tumors more readily than do similarly treated controls on low fat diets (Carroll, 1975). Epidemiological data on human populations reveal a correlation between breast cancer and dietary fat intake, but not exclusively polyunsaturated fat.

Since polyunsaturated fatty acids are easily oxidized to form undesirable lipid oxides and peroxides, some potential adverse effects of the excessive dietary PUFA include an increase in vitamin E requirements associated with the aging phenomenon (Horwitt, 1976). When tissues contain an inadequate amount of antioxidant, an increase in peroxidation of PUFA may occur. Even though most vegetable oils with a high linoleic acid content usually contain considerable amounts of vitamin E, some of it may be lost during storage, processing or cooking. The peroxidation of PUFA can give rise to toxic and possibly carcinogenic products (Artman, 1969; West and Redgrave, 1974).

A higher incidence of gallstones has been observed in animals fed PUFA (Lofland, 1975) as well as in patients on an experimental polyunsaturated fat diet (Sturdevant *et al.*, 1973).

It has been suggested (Hornstra, 1975) that dietary long-chain saturated fatty acids with 14 or more carbon atoms are thrombogenic and their replacement by unsaturated fatty acids decreases thrombogenicity. Linoleic acid has a stronger antithrombotic effect than oleic acid.

EFA are known to be precursors of prostaglandins, prostacyclins, and thromboxanes that are potent vasoactive agents with a variety of other actions depending on the target organ and the type of the compound itself. In acute experiments (Silver *et al.*, 1974), arachidonic acid infused in high concentrations induced sudden death in rabbits due to the formation of pulmonary thrombi. In lower doses, however, arachidonic acid is converted mainly to the prostaglandin PGI_2 (prostacyclin) which inhibits platelet aggregation (Korbut and Moncada, 1978).

A fatty acid peroxide, 15-hydroperoxyarachidonic acid, was found to inhibit the enzyme responsible for the formation of prostacyclin (Moncada and Vane, 1979). Thus, the presence of such lipid peroxides might be of importance in inducing arterial thrombosis by reducing the availability of prostacyclin which has a defensive role in the body.

Heated Oils

Toxic and potentially carcinogenic substances such as polycyclic and aromatic hydrocarbons are produced by heating oils containing high levels of PUFA (Michael *et al.*, 1966). Various symptoms of toxicity, including irritation of the digestive tract, organ enlargement, growth depression, and even death, have been observed when highly abused (oxidized and heated) fats were fed to laboratory animals (Andia and Street, 1975). It was also demonstrated that laboratory-heated oils may act as co-carcinogens (Sugai *et al.*, 1962).

However, it has been shown that commercially used frying fats contain only small amounts of toxic substances and produce no appreciable ill effects in animals consuming them (Nolen *et al.*, 1967). Many biological effects due to changes in fats during heating have been under study in recent years (Alexander, 1978).

Cyclopropenoid Fatty Acids

Cyclopropenoid fatty acids, found in many plants that are sources of food for man and animals, have been known to exert adverse biological effects (Mattson, 1973). Two such acids, sterculic and malvalic, have been identified. They are incorporated into animal fat, but are present in the food chain at levels below the threshold dose for deleterious effects. Commercial cottonseed salad oils may contain 0.04–0.05% of cyclopropenoid acids; cottonseed meal has about 0.01%. Eggs and chicken fat contain them as well. The level of these compounds in food products is being decreased by advances in food technology, and thorough cooking destroys most of the biological activity of cyclopropenoids since they undergo chemical modification with heating (Scarpelli, 1974).

Rapeseed Oil

Rapeseed oil occupies an important place in the world production of edible oils. It is characterized by a high level (usually above 40%) of erucic acid (22:1). The variety of the seed and the environment in which the seed grows can influence its fatty acid composition. Within the last two decades rapeseed plants producing "zero" erucic oil have been developed and grown (Downey *et al.*, 1969). The reason for this preference is that erucic acid has been found to have deleterious effects in animals. The growth of young rats was severely depressed when rapeseed oil at levels of 15% of oil in the diet (Jacquot *et al.*, 1969) was given. The erucic acid fracton of the oil is particularly poorly digested (Rocquelin and Leclerc, 1969). This growth-depressing effect has been confirmed in several other animal species.

The deposition of erucic acid in animal tissues is generally limited and it seldom exceeds 7% of total fatty acids. It has been observed, however, that erucic acid and similar very long chain monoenoic fatty acids are rapidly taken up by heart muscle after intestinal absorption (Abdellatif, 1972). The inability of cardiac tissue to readily metabolize these fatty acids leads to their incorporation into triglycerides which are deposited *in situ.* Degenerative changes may follow. Hydrogenation of erucic acid-containing oils resulted in a lowered cardiopathogenicity (Beare-Rogers, 1979).

Even though it is difficult to extrapolate these results to man, rapeseed oil has been withdrawn from infant formulas (Vles, 1975), as a precautionary measure.

CHOLESTEROL

Occurrence and Functions

Cholesterol is an essential metabolite synthesized in the body at a rate sufficient to meet body needs, i.e., 0.5–1.0 g/day; about one-half of this production takes place in the liver. Cholesterol is a major constituent of cell membranes, essential for their structure and function. As a component of myelin, it is necessary for nerve conduction and brain function. After conversion to bile acids, it participates in fat digestion and absorption. It is an essential component of plasma lipoproteins and, therefore, it is involved in lipid transport. It is a precursor of adrenal and sex hormones and of vitamin D.

Since it is found in all animal tissues, it is ingested daily by all carnivores. The average daily intake has remained constant for many decades, i.e., about 500 mg per capita per day. While cow's milk contains about 34mg/8oz, human milk ranges from 26–52 mg of cholesterol per 8 oz (Jensen *et al.*, 1978).

It has been assumed, but not absolutely confirmed, that the blood and total body cholesterol equilibrate with each other (Sodhi and Mason, 1977).

Dietary fats aid in cholesterol absorption; even though the percent absorption is greater when dietary cholesterol intake is low, the total amount absorbed per day increases as the intake increases. Therefore, the amount absorbed is variable and on the average, probably below 50% (Grundy, 1978). The absorbed cholesterol affects its endogenous synthesis (more so in certain animals than in man) and consequently its distribution and metabolism.

Increased hepatic cholesterogenesis results when the caloric intake is elevated. Conversely, caloric restriction reduces the synthesis. The saturation of dietary fat apparently does not affect the rate of synthesis (Grundy, 1978).

Lipoproteins are actively involved in cholesterol transport. Most of the plasma cholesterol is carried in the low density lipoproteins (LDL) (Thompson and Bortz, 1978). However, the high density lipoproteins (HDL) carry approximately 20% of plasma cholesterol that is relatively insensitive to changes in the diet but is affected by physical activity, moderate alcohol ingestion, and estrogens. HDL has been reported to confer protection against the development of atherosclerosis and might, therefore, be considered an antiatherosclerotic factor (Miller and Miller, 1975).

Factors Affecting Serum Cholesterol Levels

Epidemiological Studies

It has been generally accepted that there is an increased risk of coronary heart disease in those who have hypercholesterolemia. Much evidence suggests that an elevated serum cholesterol is an important risk factor which, together with other risk factors, affect the disease, atherosclerosis (McGill, 1979). Therefore, it is of importance to establish which factors are responsible for affecting serum cholesterol levels.

Over several decades, and in various parts of the world, numerous epidemiological studies have been performed attempting to correlate dietary cholesterol with the morbidity due to atherosclerosis (Eskin, 1971; Armstrong *et al.*, 1975; Connor, 1961). While some of the investigations showed a direct association, the others simply added to the confusion. Keys *et al.* (1950) found no difference in serum cholesterol concentrations between men who consumed diets high in cholesterol and those whose diets were low in cholesterol. The review of many such studies leads to thc conclusion that even though there may be an association between dietary and serum cholesterol concentration, there is no proof of a cause and effect relationship (McGill, 1979).

In studies based on individual values, little or no association of dietary cholesterol with either serum cholesterol, coronary atherosclerosis, or coronary heart disease has been found (Yano *et al.*, 1978).

Experimental Studies

In addition to epidemiological surveys many experimental studies have been performed in humans. Since many variables contribute to and confuse the values (such as age, length of dietary exposure, previous diet, and accompanying fat) such studies are difficult to compare and assess. Some of these investigations suggest a possible relationship between the dietary cholesterol and serum cholesterol, but the relationship is weak and varies greatly among individuals. Most often when dietary factors are considered the correlation between the saturation of the dietary fatty acids and serum cholesterol levels is much stronger.

It has been pointed out (Mahley, 1979) that most animal species are more sensitive to dietary cholesterol than is man, and, on the other hand, that most animals are less sensitive to saturated fat than is man.

Several recent publications show evidence of marked variation in plasma cholesterol response in man to food cholesterol challenges (Slater *et al.*, 1976; Porter *et al.*, 1977; Kummerow *et al.*, 1977; Flynn *et al.*, 1979). In each of these studies increasing dietary cholesterol as eggs or as meat did not significantly increase the plasma cholesterol level in the great majority of the men.

The possibility remains, of course, that dietary cholesterol could affect atherogenesis without changing the serum cholesterol concentration. Recently it was reported (Mahley *et al.*, 1978) that the addition of 4–6 eggs per day to the diets of eleven free-living men and women increased an HDL subfraction regardless of the effect on plasma cholesterol.

In one of the most recent studies from the Netherlands (Bronsgeest-Schoute *et al.*, 1979b) no correlation was found between changes in serum cholesterol and the number of eggs eaten. A very variable response was confirmed in the human population toward dietary cholesterol. In addition, it was concluded (Bronsgeest-Schoute *et al.*, 1979a) that the presence of a high content of linoleic acid in the diet diminished the possible effect of dietary cholesterol on serum cholesterol, if the cholesterol was provided as egg yolk. The investigators emphasize the need for more research to facilitate the identification of "hyperresponders" who are more sensitive to changes in dietary cholesterol. This problem should be solved prior to recommending drastic modifications in the diet.

Early Intake vs. Cholesterolemia

Some years ago a hypothesis was proposed that high cholesterol intake early in life helps to establish a permanent mechanism for maintaining low serum cholesterol concentration in adulthood (Reiser *et al.,* 1979; Reiser and Sidelman, 1972). This presumed stabilization of the anabolic and catabolic processes of cholesterol in adult life by preconditioning young rats with high cholesterol was not confirmed in recent studies (Kris-Etherton *et al.,* 1979). But even though an early exposure to exogenous cholesterol did not protect the rat against the subsequent dietary-induced hypercholesterolemia, cholesterol metabolic systems were affected in early life. There have been attempts to test this hypothesis in humans. In an abnormal (hyperlipoproteinemic) population of infants, it was reported (Glueck and Tsang, 1972) that a commercial formula low in cholesterol but high in PUFA led to lower plasma cholesterol levels than those seen in infants fed regular bovine milk formula. No plasma cholesterol values were followed in the adult. In another study (Friedman and Goldberg, 1975) with breast-fed vs. formula-fed infants, no differences in plasma cholesterol were found in young children. Similarly, a later investigation with slightly older children also proved to be inconclusive (Hodgson *et al.,* 1976). However, here again, the theory was not tested in the adult.

A recent study with rats (Naseem *et al.,* 1980) has indicated that a prenatal exposure to a high cholesterol–high fat diet influences both the synthetic and degradative enzymes of cholesterol metabolism. The inhibition of HMGCoA reductase (an important cholesterol synthesizing enzyme) was in excess of 75% in suckling and mature animals derived from mothers fed the experimental diet (1% cholesterol). On the other hand, cholesterol hydrolase was greatly increased.

Autooxidation Products

Recently a new direction has been pursued in this area of research. It has been noted that some autooxidation products of cholesterol at extremely low dosages have a lethally toxic effect on arterial smooth muscle cells, whereas pure cholesterol has but a minimal angiotoxic effect (Imai *et al.,* 1976; Peng *et al.,* 1978). Dehydrated foods such as powdered milk and powdered egg yolk are highly susceptible to autooxidation of the cholesterol. There is a suggestive evidence that endogenously synthesized cholesterol is protected from autooxidation by antioxidants in the animal organism. On the other hand, traces of oxidation products of cholesterol have been isolated from human atheromata (Smith and Van Lier, 1970). There is an indication that pure cholesterol (either endogenously synthesized or chemically isolated pure cholesterol) is not atherogenic (Taylor *et al.,* 1979). Present methods used in isolation and storage of cholesterol and some cholesterol-containing foods seem to promote the formation of angiotoxic derivatives of cholesterol. It may be advisable to protect cholesterol-containing foods with antioxidants to minimize the formation of these toxic derivatives.

Association with Cancer

Finally, ecologic associations of the occurrence of coronary heart disease and cancer have led to suggestions that cholesterol may play a role in both instances (Rose *et al.,* 1974; Lea, 1966). For instance, initial levels of cholesterol were found to be lower in persons subsequently developing colon cancer; low serum cholesterol levels may pose its own noncardiovascular risks.

Cholesterol, by virtue of its having a regulatory role in biological membranes, may function as a bioregulator in the development and inhibition of leukemia (Inbar and Shinitzky, 1974).

When more than 3000 individuals in Evans County, Georgia were followed for over a decade, it was observed that in cancer cases (127) the mean serum cholesterol level was significantly lower than that of the noncancer population. This finding was consistent for the various cancer sites, with particular significance for males (Kark *et al.,* 1980).

CONCLUSIONS

While the various hypotheses as to the cause of atherosclerosis are being investigated, it is premature to advise drastic dietary changes. A number of risk factors for cardiovascular disease have been identified from epidemiologic studies. These include male sex, genetic trait, hypercholesterolemia, hypertension, obesity, diabetes, smoking, and physical inactivity. In addition, high concentrations of LDL, enhanced platelet aggregation, transformation of smooth muscle cells, altered prostaglandin metabolism, and effects of various steroid hormones on arterial metabolism are implicated.

Some modification of the diet with respect to the level and kind of fat, and amount of dietary cholesterol for sensitive individuals, can result in a more advantageous blood lipid pattern. A major factor appears to be the saturated fat, with dietary cholesterol having the least impact. Other factors in addition to the diet influence serum lipid values of free-living persons in an as yet unpredictable manner. Some studies employing diets containing 35–40% of calories from fat and higher P/S ratios have shown equivocal effects on coronary disease and have been accompanied by a somewhat greater incidence of gastrointestinal disease (Ahrens, 1976).

REFERENCES

Abdellatif, A. M. M., 1972, Cardiopathogenic effects of dietary rapeseed oil, *Nutr. Rev.* **30:**2.

American Heart Association Statement, 1978, Diet modification to control hyperlipidemia, *Circulation* **58:**381A.

Ahrens, E. H., Jr., 1957, Nutritional factors and serum lipid levels, *Am. J. Med.* **23:**928.

Ahrens, E. H., Jr., 1976, The management of hyperlipidemia: Whether, rather than how, *Ann. Intern. Med.* **85:**87.

Alexander, J. C., 1978, Biological effects due to changes in fats during heating, *J. Am. Oil Chem. Soc.* **55:**711.

Andia, A. G., and Street, J. C., 1975, Dietary induction of hepatic microsomal enzymes by thermally oxidized fats, *J. Agr. Food Chem.* **23:**173.

Anderson, J. T., Grande, F., and Keys, A., 1973, Cholesterol-lowering diets, *J. Am. Diet. Assoc.* **62:**133.

Armstrong, B. K., Mann, J. J., Adelstein, A. M., and Eskin, F., 1975, Commodity consumption and ischaemic heart disease mortality with special reference to dietary practices, *J. Chronic Dis.* **28:**455.

Artman, N. R., 1969, The chemical and biological properties of heated and oxidized fats, *Adv. Lipid Res.* **7:**245.

Beare-Rogers, J. L., 1979, Partially hydrogenated rapesed and marine oils, in: *Geometrical and Positional Fatty Acid Isomers* (E. A. Emken and H. J. Dutton, eds.), pp. 131–149, American Oil Chemists' Society Monograph, Champaign, Illinois.

Bronsgeest-Schoute, D. C., Hautvast, J. G. A. J., and Hermus, R. J. J., 1979a, Dependence of the effects of dietary cholesterol and experimental conditions on serum lipids in man I, *Am. J. Clin. Nutr.* **32:**2183.

Bronsgeest-Schoute, D. C., Hermus, R. J. J., Dallinga-Thie, G. M., and Hautvast, J. G. A. J., 1979b, Dependence of the effects of dietary cholesterol and experimental conditions on serum lipids in man III, *Am. J. Clin. Nutr.* **32:**2193.

Carroll, K. K., 1975, Experimental evidence of dietary factors and hormone-dependent cancers, *Cancer Res.* **35:**3374.

Carroll, K. K., 1977, EFA: What level in the diet is most desirable?, *Adv. Exp. Med. Biol.* **83:**535.

Connor, W. E., 1961, Dietary cholesterol and the pathogenesis of atherosclerosis, *Geriatrics* **16:**407.

Downey, R. K., Craig, B. M., and Young, C. G., 1969, Breeding rapeseed for oil and meal quality, *J. Am. Oil Chem. Soc.* **46:**121.

Emken, E. A., 1979, Utilization and effects of isomeric fatty acids in humans, in: *Geometrical and Positional Fatty Acid Isomers* (E. A. Emken, and H. J. Dutton, eds.) pp. 99–129, American Oil Chemists' Society Monograph, Champaign, Illinois.

Enig, M. G., Munn, R. J., and Keeney, M., 1978, Dietary fat and cancer trends—a critique, *Fed. Proc. Fed. Am. Soc. Exp. Biol.* **37:**2215.

Eskin, F., 1971, The role of the egg as a factor in the aetiology of coronary heart disease, *Community Health (Bristol)* **2:**179.

Flynn, M. A., Nolph, G. B., Flynn, T. C., Kahrs, R., and Krause, G., 1979, Effect of dietary egg on human serum cholesterol and triglycerides, *Am. J. Clin. Nutr.* **32:**1051.

Friedman, G., and Goldberg, S. J., 1975, Concurrent and subsequent serum cholesterol of breast and formula fed infants, *Am. J. Clin. Nutr.* **28:**42.

Glueck, C. J., and Connor, W. E., 1979, Diet and atherosclerosis: Past, present and future, *West. J. Med.* **130:**117.

Glueck, C. J., and Tsang, R. C., 1972, Pediatric type II hyperlipoproteinemia effects of diet on plasma cholesterol in the first year of life, *Am. J. Clin. Nutr.* **25:**224.

Grundy, S. M., 1978, Cholesterol metabolism in man, *West. J. Med.* **128:**13.

Heckers, H., and Melcher, F. W., 1978, Trans-isomeric fatty acids present in West German margarines, shortenings, frying and cooking fats, *Am. J. Clin. Nutr.* **31:**1041.

Heckers, H., Körner, M., Tüschen, T. W. L., and Melcher, F. W., 1977, Occurrence of individual trans-isomeric fatty acids in human myocardium, jejunum and aorta in relation to different degrees of atherosclerosis, *Atherosclerosis* **28**:389.

Heyden, S., Walker, L., Hames, C. G., and Tyroler, H. A., 1971, Decrease of serum cholesterol levels and blood pressure in the community, *Arch. Intern. Med.* **129**:982.

Hill, P., and Wynder, E. L., 1976, Dietary regulation of serum lipids in healthy, young adults, *J. Am. Diet. Assoc.* **68**:25.

Hodgson, P. A., Ellefson, R. D., Elvebach, L. R., Harris, L. E., Nelson, R. H., and Weidman, W. H., 1976, Comparison of serum cholesterol in children fed high, moderate or low cholesterol milk diets during neonatal period, *Metabolism* **25**:739.

Hornstra, G., 1975, Specific effects of types of dietary fat on arterial thrombosis, in: *The Role of Fats in Human Nutrition* (A. J. Vergroesen, ed.), pp. 303–330, Academic Press, New York.

Horwitt, M. K., 1976, Vitamin E: A reexamination, *Am. J. Clin. Nutr.* **29**:569.

Howard, A. N., and Marks, J., 1977, Hypocholesterolemic effect of milk, *Lancet* **2**:255.

Hsu, C. M. L., and Kummerow, F. A., 1977, Influence of elaidate and erucate on heart mitochondria, *Lipids* **12**:486.

Hwang, D. H., and Kinsella, J. E., 1979, The effects of tt-methyl linoleate on the concentration of prostaglandins and their precursors in rat, *Prostaglandins* **17**:543.

Imai, H., Werthessen, N. T., Taylor, C. B., and Lee, K. T., 1976, Angiotoxicity and arteriosclerosis due to contaminants of USP grade cholesterol, *Arch. Pathol. Lab. Med.* **100**:565.

Inbar, M., and Shinitzky, M., 1974, Cholesterol as a bioregulator in the development and inhibition of leukemia, *Proc. Nat. Acad. Sci.* **71**:4229.

Jacquot, R., Rocquelin, G., and Potteau, B., 1969, L'huile de colza et son usage alimentaire, *Cah. Nutr. Diet.* **4**:51.

Jensen, B., 1976, Rat testicular lipids and dietary isomeric fatty acids in essential fatty acid deficiency, *Lipids* **11**:179.

Jensen, R. G., Hagerty, M. M., and McMahon, K. E., 1978, Lipids of human milk and infant formulas: A review, *Am. J. Clin. Nutr.* **31**:990.

Kannel, W. B., Dawber, T. R., Friedman, G. D., Glennon, W. E., and McNamara, P. M., 1964, Risk factors in CHD. The Framingham Study, *Ann. Intern. Med.* **61**:888.

Kark, J. D., Smith, A. H., and Hames, C. G., 1980, The relationship of serum cholesterol to the incidence of cancer in Evans County, Georgia, *J. Chronic Dis.* **33**:311.

Kaunitz, H., 1978, Toxic effects of polyunsaturated vegetable oils, in: *Pharmacological Effects of Lipids* (J. J. Kabara, ed.), pp. 203–210, American Oil Chemists' Society Monograph, Champaign, Illinois.

Keys, A., Mickelsen, O., Miller, E. O., and Chapman, C. B., 1950, The relation in man between cholesterol levels in the diet and in the blood, *Science* **112**:79.

Korbut, R., and Moncada, S., 1978, Prostacyclin (PGI_2) and thromboxane A_2 interaction *in vivo*, *Thromb. Res.* **13**:489.

Kris-Etherton, P. M., Layman, D. K., York, P. V., and Frantz, I., Jr., 1979, The influence of early nutrition on the serum cholesterol of the adult rat, *J. Nutr.* **109**:1244.

Kritchevsky, D., 1979, Atherosclerosis and nutrition, *Adv. Nutr. Res.* **2**:181.

Kritchevsky, D., Moyer, A. W., Tesar, W. C., Logan, J. B., Brown, R. A., Davies, M. C., and Cox, H. R., 1954, Effect of cholesterol vehicle in experimental atherosclerosis, *Am. J. Physiol.* **178**:30.

Kummerow, F. A., 1975, Lipids in atherosclerosis, *J. Food Sci.* **40**:12.

Kummerow, F. A., Kim, Y., Hull, J., Pollard, J., Illinov, P., Dorossiev, D. L., and Valek, J., 1977, The influence of egg consumption on the serum cholesterol level in human subjects, *Am. J. Clin. Nutr.* **30**:664.

Lea, A. J., 1966, Dietary factors associated with death rates from certain neoplasms in man, *Lancet* **2**:332.

List, G. R., Emken, E. A., Kwolek, W. F., Simpson, T. D., and Dutton, H. J., 1977, "Zero trans" margarines: Preparation structure and properties of interesterified soybean oil-soy trisaturate blends, *J. Am. Oil Chem. Soc.* **54**:408.

Lofland, H. B., 1975, Cholelithiasis, animal model of human disease, *Am. J. Pathol.* **79:**619.

Mahfouz, M. M., Johnson, S., and Holman, R. T., 1980a, The effect of isomeric *t* 18:1 acids on the desaturation of palmitic, linoleic and eicosa, 8-11-14, trienoic acids by rat liver microsomes, *Lipids* **15:**100.

Mahfouz, M. M., Valicenti, A. J., and Holman, R. T., 1980b, Desaturation of isomeric *t* 18:1 acids by rat liver microsomes, *Biochem. Biophys. Res. Commun.* **618:**1.

Mahley, R. W., 1979, Dietary fat, cholesterol, and accelerated atherosclerosis, in: *Atherosclerosis Reviews* (R. Paoletti and A. Gotto, eds.), Vol. 5, pp. 1–34, Raven Press, New York.

Mahley, R. W., Bersat, T. P., Innerarity, T. L., Lipson, A., and Margolis, S., 1978, Alterations in human high density lipoproteins, with or without increased plasma cholesterol, induced by diets high in cholesterol, *Lancet* **2:**807.

Mann, G. V., 1977, A factor in yogurt which lowers cholesterolemia in man, *Atherosclerosis* **26:**335.

Marston, R., and Page, L., 1979, Dietary sources of fat, *J. Am. Oil Chem. Soc.* **55:**257.

Mattson, F. H., 1973, Potential toxicity of food lipids, in: *Toxicants Occurring Naturally in Food*, 2nd ed., pp. 189–209, National Academy of Sciences, Washington, D.C.

McGandy, R. B., and Hegsted, D. M., 1975, Quantitative effects of dietary fat and cholesterol on serum cholesterol in man, in: *The Role of Fats in Human Nutrition* (A. J. Vergroesen, ed.), pp. 211–230, Academic Press, New York.

McGill, H. C., 1979, The relationship of dietary cholesterol to serum cholesterol concentration and to atherosclerosis in man, *Am. J. Clin. Nutr.* **32:**2664.

Michael, W. R., Alexander, J. C., and Artman, N. R., 1966, Thermal reactions of methyl linoleate, *Lipids* **1:**353.

Miller, G. J., and Miller, N. E., 1975, Plasma HDL concentration and development of ischemic heart disease, *Lancet* **1:**1619.

Moncada, S., and Vane, J. R., 1979, Arachidonic acid metabolites and the interactions between platelets and blood vessel walls, *N. Engl. J. Med.* **300:**1142.

Narayan, K. A., McMullen, J. J., Butler, D. P., Wakefield, T., and Calhoun, W. K., 1974, The influence of a high level of dietary corn oil on rat serum and liver lipids, *Nutr. Rep. Int.* **10:**25.

Naseem, S. M., Khan, M. A., Heald, F. P., and Nair, P. P., 1980, The influence of cholesterol and fat in maternal diet of rats on the development of hepatic cholesterol metabolism in the offspring, *Atherosclerosis* **36:**1.

Nolen, G. A., Alexander, J. C., and Artman, N. R., 1967, Long term rat feeding study with used frying fats, *J. Nutr.* **93:**337.

Peng, S. K., Taylor, C. B., Tham, P., and Mikkelson, B., 1978, Effect of auto-oxidation products from USP grade cholesterol on aortic smooth muscle cells, *Arch. Pathol. Lab. Med.* **102:**57.

Porter, M. W., Yamanaka, W., Carlson, S. D., and Flynn, M. A., 1977, Effect of dietary egg on serum cholesterol and triglyceride of human males, *Am. J. Clin. Nutr.* **30:**490.

Privett, O. S., Phillips, F., Shimasaki, H., Nozawa, T., and Nickell, E. C., 1977, Studies of effects of trans fatty acids in the diet on lipid metabolism in essential fatty acid deficient rats, *Am. J. Clin. Nutr.* **30:**1009.

Reiser, R., and Sidelman, Z., 1972, Control of serum cholesterol homeostasis by cholesterol in the milk of the suckling rat, *J. Nutr.* **102:**1009.

Reiser, R., O'Brian, B. C., Henderson, G. R., and Moore, R. W., 1979, Studies on a possible function for cholesterol in milk, *Nutr. Rep. Int.* **19:**835.

Richardson, T., 1978, The hypocholesterolemic effect of milk—a review, *J. Food Protection* **41:**226.

Rocquelin, G., and Leclerc, J., 1969, L'huile de colza, riche en acide erucique et l'huile de colza sans acide erucique, *Ann. Biol. Anim. Biochim. Biophys.* **9:**413.

Rose, G., Blackburn, H., Keys, A., Kannel, W. B., Paul, O., Reid, D. D., and Stamler, J., 1974, Colon cancer and blood cholesterol, *Lancet* **1:**181.

Scarpelli, D. G., 1974, Mitogenic activity of sterculic acid, a cyclopropenoid fatty acid, *Science* **185:**958.

Scholfield, C. R., Davison, V. L., and Dutton, H. J., 1967, Analysis for geometrical and positional isomers of fatty acids in partially hydrogenated fats, *J. Am. Oil Chem. Soc.* **44:**648.

Silver, M. J., Hoch, W., Kocsis, J. J., Ingerman, C. M., and Smith, J. B., 1974, Arachidonic acid causes sudden death in rabbits, *Science* **194:**1085.

Slater, G., Mead, J., Dhopeshwarkar, G., Robinson, S., and Alfin-Slater, R. B., 1976, Plasma cholesterol and triglycerides in men with added eggs in the diet, *Nutr. Rep. Int.* **14:**249.

Smith, L. L., and Van Lier, J. E., 1970, Sterol metabolism—part 9. 26-hydroxycholesterol levels in human aorta, *Atherosclerosis* **12:**1.

Smith, L. M., Dunkley, W. L., Franke, A., and Dairiki, J., 1978, Measurement of trans and other isomeric unsaturated fatty acids in butter and margarine, *J. Am. Oil Chem. Soc.* **55:**257.

Söderhjelm, L., Wiese, H. F., and Holman, R. T., 1970, The role of polyunsaturated acids in human nutrition and metabolism, *Prog. Chem. Fats Other Lipids* **9:**4.

Sodhi, H., and Mason, D., 1977, New insights into the homeostasis of plasma cholesterol, *Am. J. Med.* **63:**325.

Sturdevant, R. A. L., Pearce, M. L. M. and Dayton, S., 1973, Increased prevalence of cholelithiasis in men ingesting a serum cholesterol-lowering diet: Experience from five clinical trials, *N. Engl. J. Med.* **288:**24.

Sugai, M., Witting, L. A., Tsuchiyama, H., and Kummerow, F. A., 1962, The effect of heated fat on the carcinogenic activity of 2-acetylaminofluorene, *Cancer Res.* **22:**510.

Taylor, C. B., Peng, S. K., Werthessen, N. T., Tham, P., and Lee, K. T., 1979, Spontaneously occurring angiotoxic derivatives of cholesterol, *Am. J. Clin. Nutr.* **32:**40.

Thompson, P., and Bortz, W. M., 1978, Significance of HDL cholesterol, *J. Am. Geriatr. Soc.* **26:**440.

Vergroesen, A. J., 1972, Dietary fat and cardiovascular disease: Possible modes of action of linoleic acid, *Proc. Nutr. Soc.* **31:**323.

Vles, R. O., 1975, Nutritional aspects of rapeseed oil, in: *The Role of Fats in Human Nutrition* (A. J. Vergroesen, ed.), pp. 433–477, Academic Press, New York.

West, C. F., and Redgrave, T. G., 1974, Reservations on the use of polyunsaturated fats in human nutrition, *Search* **5:**90.

Yano, K., Rhoads, G. G., Kagan, A., and Tillotson, J., 1978, Dietary intake and the risk of coronary heart disease in Japanese men living in Hawaii, *Am. J. Clin. Nutr.* **31:**1270.

Yu, P. H., Mai, J., and Kinsella, J. E., 1980, The effects of dietary *tt*-methyl octadecadienoate on composition and fatty acids of rat heart, *Am. J. Clin. Nutr.* **33:**598.

DIET, NUTRITION, AND CANCER

Lois D. McBean and Elwood W. Speckmann

INTRODUCTION

Cancer (i.e., malignant tumors or neoplasms) is the second leading cause of death in the U.S., exceeded only by cardiovascular disease (U.S. Department Health, Education, and Welfare/Public Health Service, 1979). In 1980, cancer is expected to have claimed the lives of 405,000 individuals (U.S. Department Health, Education, and Welfare/Public Health Service, 1979; American Cancer Society, 1979). Although the age-adjusted mortality rate from lung cancer (the most prevalent type) has been increasing, that from colon–rectum and breast cancers (the next most common types) has leveled off (American Cancer Society, 1979). Carcinogenesis (or tumorigenesis) is a multifaceted process, its initiation and promotion influenced by genetics, life-style, cultural patterns, health status, exposure to specific carcinogens and other unknown factors (Petering, 1978). As such, it is no longer viewed as a single disease with a single etiology (Wynder, 1979). Of great interest, and an area of active research, is the relationship between the incidence of cancer at particular sites and certain nutritional imbalances (Weisburger *et al.,* 1980; Gori, 1979a,b; Shils, 1979a; Wynder, 1979; Petering, 1978; Winick, 1977; Wynder and Gori, 1977; Alcantara and Speckmann, 1976; Rivlin, 1973). It is emphasized, however, that a direct cause–effect relationship between diet and cancer in humans has not been firmly established (Food and Nutrition Board, 1980; Upton and Fink, 1980). It is the purpose of this paper to update information presented previously (National Dairy Council, 1975) on the role of nutrition and diet in the etiology of cancer. In addition, nutritional problems induced by cancer *per se* and its treatment modalities, as well as the nutritional management of the cancer patient, will be discussed.

Lois D. McBean and Elwood W. Speckmann • National Dairy Council, Rosemont, Illinois 60018.

NUTRITION, DIET, AND CANCER ETIOLOGY

Types of Evidence

Sources of information implicating diet in specific forms of cancer include epidemiological surveys, experiments using animal models, and, to a lesser extent, case control studies in humans (Shils, 1979a). According to some epidemiologists, 80–90% of cancer incidence in the U.S. can be attributed to specific environmental factors, and diet and nutrition are related to 60% of all cancers in women and over 40% in men (Gori, 1978a, 1979a,b; Wynder, 1977a; Wynder and Gori, 1977; Rivlin, 1973). Correlations between cancer incidence and dietary habits are evident from migrant population studies in which cancer-incidence patterns of migrants change from that of their native country to that common to the population of the host country (e.g., Japanese who migrate to the U.S.); special population groups that live in the same environment but adhere to different dietary habits (e.g., Seventh Day Adventists in California who follow a unique vegetarian diet); geographical or worldwide high–low correlations; and trends in incidence over time (Gori, 1977, 1978a,b, 1979a,b; Wynder, 1977b). Epidemiologic correlations, however, do not establish causation (Gori, 1978a). Furthermore, epidemiological studies do not pinpoint carcinogenic agents, particularly as a long latent period may typify the disease. While laboratory work with animals allows strict control of dietary factors and the findings tend to corroborate epidemiological data, such findings cannot be extrapolated with assurance to man (Shils, 1979a). Different species may respond in unique ways to the same dietary stimuli, and spontaneous tumors may react differently from experimentally induced tumors (Alcantara and Speckmann, 1976). Animal studies do, however, help to elucidate mechanisms by which environmental factors contribute to carcinogenesis.

Speculated Mechanisms of Carcinogenesis

Various hypothetical mechanisms by which diet and nutrition might influence human carcinogenesis have been reviewed (Gori, 1978b, 1979a,b; McMichael, 1979; Shils, 1979a; Weisburger, 1979; Modan, 1977; Wynder, 1976, 1977b; Alcantara and Speckmann, 1976). Current evidence reveals that dietary patterns or certain nutritional deficiencies or excesses, as opposed to ingestion of specific carcinogens or carcinogen precursors (either naturally occurring in food or present as additives), are more important in the diet–cancer relationship (Gori, 1979a; Shils, 1979a). In terms of the latter, however, saccharin and sodium nitrite are of special concern (Upton and Fink, 1980; Levey, 1979).

Saccharin, the only nonnutritive artificial sweetener approved for use in the U.S., has been demonstrated to cause malignant bladder tumors in rats fed high doses. According to findings of a National Academy of Sciences

(NAS) study, saccharin is a carcinogen of low potency relative to other carcinogens in rats and a potential carcinogen in humans but of unknown risk (NAS, 1978). And the National Cancer Institute (NCI) and Food and Drug Administration (FDA) concluded, on the basis of a large-scale epidemiologic study, that there was no increased risk of bladder cancer among users of artificial sweeteners in the study population. The sweetener, however, was hazardous for persons who were both heavy smokers and heavy users of saccharin (NCI, 1980). Some industries, organizations such as the American Diabetes Association (1979), and consumer groups, insist that saccharin is essential for use in diabetes, obesity, and other health problems. As it stands, the moratorium on the saccharin ban has been extended until June 30, 1981* (Levey, 1979) with the need to find a substitute for saccharin emphasized (American Diabetes Association, 1979).

The question of banning sodium nitrite, an additive used to preserve, flavor, and color foods such as cured meats, continues to be debated (Upton and Fink, 1980; Levey, 1979). Nitrosamines, formed under certain conditions from the combination of nitrite (either added to food or produced by the bacterial reduction of nitrates in the body) with naturally occurring secondary amines, are potent carcinogens for laboratory animals (Newberne, 1979). Direct evidence that nitrosamines are carcinogenic for humans is lacking (Fraser *et al.*, 1980). Moreover, the importance of sodium nitrite for preventing botulism is well recognized (Levey, 1979). Nevertheless, the FDA and USDA favor a gradual phasing out of nitrites in certain products, the timing dependent on how soon a chemical substitute can be found (Levey, 1979). Of note is the possible anticarcinogenic potential of antioxidants such as vitamin C (McMichael, 1979; Shils, 1979a). As an active reducing agent, it has been shown to interact with nitrates (the main sources being vegetables and drinking water) or nitrites inhibiting nitrosamine formation. In fact, the steady decline in stomach cancer in the U.S. over the past 50 years has been hypothesized to be due to the year-round availability of vitamin C-rich foods and refrigeration of foods which both inhibits reduction of nitrate to nitrite and lessens the amount of sodium nitrite preservative needed in foods (Upton and Fink, 1980; Shils, 1979a; Weisburger, 1979).

Foods and nutrients appear to modify rather than to initiate the carcinogenic process (Shils, 1979a; Alcantara and Speckmann, 1976; Wynder, 1976). Also, the way in which diet affects carcinogenesis may differ for distinct types of cancer (Weisburger, 1979). The presence or absence of specific dietary constituents has been shown to influence carcinogenic activity (Shils, 1979a; Alcantara and Speckmann, 1976). Restriction of energy intake or underfeeding decreases the development of spontaneous and transplanted tumors in experimental animals (Gori, 1979a; Shils, 1979a; Petering, 1978; Tannenbaum, 1945, 1959). This raises the question of whether obesity, the converse of undernutrition, augments carcinogenesis. Obesity seems to be associated

* In 1981, the moratorium on the saccharin ban was extended until June 30, 1983.

with only two types of human cancer: endometrium and kidney cancer in women (Wynder, 1976). Breast cancer may also be associated with obesity but the evidence is less clear. Other dietary deficiencies (e.g., vitamin A, riboflavin, protein, iodide, pyridoxine, magnesium, and iron) augment specific types of chemically-induced tumors in experimental animals (Shils, 1979a; Rivlin, 1973). And increased amounts of dietary fat have been positively correlated with the development of colonic and mammary tumors on exposure to certain carcinogens (Gori, 1979a; Shils, 1979a; Wynder, 1976).

Dietary and nutritional factors may protect against tumorigenesis by inactivating carcinogens or inhibiting early stages of carcinogenic activity (Shils, 1979a). For example, various indoles present in vegetables of the Brassicaceae family (e.g., Brussels sprouts, cabbage, and broccoli) induce drug metabolizing (microsomal oxidase) enzymes located in the epithelium of the gastrointestinal tract, lungs, and skin, thereby inactivating primary carcinogens (Gori, 1979a; Shils, 1979a; Petering, 1978). On the other hand, specific nutrient deficiencies may depress these enzymes, reducing the body's defense against chemical carcinogens (Calabrese, 1980). Nutrients that have antioxidant properties (e.g., vitamins E and C) may inhibit carcinogens. The value of vitamin E in this respect is unknown (Shils, 1979a). Of special interest is the potential protective role of vitamin A and related compounds in inhibiting tumor growth (Calabrese, 1980; Anonymous, 1979a; Mettlin and Graham, 1979; Shils, 1979a; Sporn and Newton, 1979; Petering, 1978; Sporn, 1977; Sporn *et al.*, 1976). Fat-soluble vitamin A, which plays an important role in epithelial cell differentiation, is toxic at pharmacologic doses; therefore synthetic retinoid analogs have been developed. These are more effective in preventing precancerous lesions, less toxic, and have a different tissue distribution than natural retinoids with greater concentration in epithelial tissues. The efficacy of synthetic retinoids in preventing environmentally induced epithelial cancers in humans is an area of active research.

Another important way in which nutrition and specific dietary constituents may influence cancer incidence is via the immune system (Shils, 1979a; Petering, 1978). Immune processes, which are affected by nutrition, appear to be involved in removal of precancerous and cancerous cells, and in the resistance of concurrent infection (Shils, 1979a; Petering, 1978). Other possible mechanisms are reviewed in the following discussions of colon and breast cancers.

Colon Cancer

Colon and rectal carcinomas, collectively, are the second most frequently diagnosed cancer by site in the U.S., excluding common skin cancers (U.S. Department Health, Education, and Welfare/Public Health Service, 1979). Epidemiological data reveal that environmental factors, in particular a Western-type diet high in fat and protein and low in fiber, is involved in the etiology of colon cancer (as distinct from cancer of the rectum) (Shils, 1979a; Wynder and Reddy, 1977). Worldwide, colon cancer incidence and mortality display wide geographical variations. With the exception of Japan and to a lesser extent Finland, the more economically developed a society (e.g., the U.S., Canada, and Western Europe), the greater the incidence of colon cancer (Wynder and Reddy, 1977). Developing countries such as Africa and Asia show relatively low incidence rates. Migrant studies also support an environmental influence. Colon cancer mortality progressively increases in a population within a relatively short time (i.e., in first-generation migrants) in Japanese born and living in Japan, those migrating to Hawaii, and those of Japanese descent born in Hawaii (McMichael, 1979; Shils, 1979a; Correa and Haenszel, 1978). Also, an increasing trend in colon cancer in Japan is attributed to a general increase in the Westernization of that country's diet (Weisburger, 1979; Wynder and Reddy, 1977). There is also an associated fall in gastric cancer at the present time in Japan (Shils, 1979a). The above correlations, suggesting a strong environmental influence in the etiology of colon cancer, have led to numerous epidemiological and experimental studies designed to identify specific dietary constituents (Shils, 1979a; Wynder and Reddy, 1977). Geographical correlations between increased colon cancer frequency and mortality and high dietary fat consumption have been observed (Armstrong and Doll, 1975; Wynder, 1975). Also, Seventh Day Adventists in California whose diet contains little fat experience a colon cancer incidence 50–70% of that of the general American population (Phillips, 1975). Mormons and non-Mormons in Utah likewise show a lower colon cancer incidence (Lyon *et al.*, 1976), but little difference in fat and fiber consumption is evident between the Utah population and the U.S. as a whole (Lyon and Sorenson, 1978). The relationship of specific dietary components to colon cancer etiology remains unsettled (Enstrom, 1975).

If diet (in particular, fat) has a role in colon cancer etiology, an important question concerns its mode of action. The current working hypothesis is that high dietary fat intake (1) increases bile secretion and hence the occurrence of bile acids and neutral steroids in the large bowel, and (2) influences the composition and metabolic activity of the intestinal microflora, such that the latter contains an increase in anaerobic clostridia capable of converting bile acids and neutral sterols (primary bile acids) by dehydrogenation to carcinogen or cocarcinogen compounds (secondary bile acids) (Wynder and Reddy, 1977). Although a specific carcinogen has not been identified for colon carcinogenesis in humans (Wynder, 1979), experimental studies in animal models and metabolic studies in man have been carried out to examine the above hypothesis (Wynder and Reddy, 1977). Substantial support for dietary fat and for the colon cancer-enhancing effects of secondary bile acids (e.g., deoxycholic acid and lithocholic acid) has come from a variety of animal studies (Wynder and Reddy, 1977).

In humans, the relationship between diet and bile acids and neutral steroids is less clear. On the one hand, significant increases are reported in the fecal concentration of bile acids and neutral sterols and fecal bacterial flora in various populations with high colon cancer mortality; in populations consuming a high-fat, mixed Western-type diet vs. a vegetarian diet; and in volunteers receiving a high-fat diet vs. a low-fat diet. Similarly, in patients with colon cancer, these fecal constituents are elevated. These findings have been reviewed in detail elsewhere (Shils, 1979a; M. J. Hill, 1978; Reddy and Wynder, 1977; Reddy *et al.,* 1977; Wynder and Reddy, 1977). In contrast, other investigators (Moskovitz *et al.,* 1979; Mower *et al.,* 1979; Mastromarino *et al.,* 1978; Mower *et al.,* 1978; Goldberg *et al.,* 1977) have not shown a significant change in the fecal microflora content and composition with diet, or in patients with carcinoma of the colon or in persons at risk of colon cancer compared with controls. Assessment of the total metabolic activity of the intestinal microflora may be a better means of determining relative risk of colon cancer (Mastromarino *et al.,* 1978). Diet and other environmental factors can modify the metabolic activity of fecal flora (Mastromarino *et al.,* 1978). Specific bacterial enzymes (e.g., β-glucuronidase, nitroreductase, and 7-α-dehydroxylase), known to catalyze reactions that result in the formation of proximal carcinogens, can be elevated by a Western-type diet, and decreased by dietary supplements of *Lactobacillus acidophilus* (Goldin *et al.,* 1980; Goldin and Gorbach, 1980; Ayebo *et al.,* 1979; Goldin and Gorbach, 1977; Reddy and Wynder, 1977). It is suggested that the lower colon cancer incidence in Finland (where fat intake is relatively high) may be due to the Finns' higher consumption of dairy products and the greater numbers of lactobacilli harbored in their fecal microflora (Goldin and Gorbach, 1980; Wynder, 1979). Other investigators place emphasis on the high dietary fiber intake by the Finnish population as a factor contributing to their lower risk of colon cancer (Reddy *et al.,* 1978; Wynder, 1979). Several mechanisms by which dietary fiber may exert a protective role in colon cancer have been advanced (Upton and Fink, 1980; Cummings *et al.,* 1979; Modan, 1979; Burkitt, 1978; Cummings, 1978; Walker, 1976). However, study of the effect of dietary fiber in colon cancer etiology is complicated by the complex nature of this dietary component (i.e., not all types of fiber have the same biological properties) (Miller, 1978a; Spiller, 1978; Maclennan and Jensen, 1977).

Breast Cancer

Breast cancer, a hormone-dependent malignant tumor of mammary tissue, is the leading cause of cancer mortality among women in industrialized countries (Japan being an exception) (U.S. Department Health, Education, and Welfare/Public Health Service, 1979). As in the case of colon cancer, indirect evidence is available from epidemiologic and animal studies showing an association between the environment and risk of breast cancer (MacMahon, 1979; McMichael, 1979; Shils, 1979a; Hankin and Rawlings, 1978; Carroll, 1977; Cole and Cramer, 1977; Miller, 1977). Unlike other major cancers, change in incidence of breast cancer upon migration from a low-risk country (e.g., Japan) to a high-risk country (e.g., the U.S.) is not apparent until the second generation as opposed to being evident in the immigrants themselves (Wynder and Hirayama, 1977). This suggests that mammary tissue may be more sensitive to etiologic factors of breast cancer at times of active cell proliferation (i.e., time of puberty or during teenage years) (MacMahon, 1979; Weisburger, 1979; Miller, 1977).

Both the amount and type of dietary fat are of importance in promoting mammary tumors in experimental animals (Hillyard and Abraham, 1979; Hopkins and Carroll, 1979), but conclusions cannot be drawn as to what type of fat (saturated or unsaturated) is a more potent stimulus of tumor genesis (Hankin and Rawlings, 1978; MacMahon, 1979). A certain amount of polyunsaturated fatty acids as well as a high level of dietary fat appears to be necessary to enhance mammary tumorigenesis in rats exposed to a carcinogen (Hopkins and Carroll, 1979). This finding deserves further investigation.

Greater risk of breast cancer has been associated with conditions influencing endocrine status—increased body height and weight, first pregnancy after 30 years of age, early menarche, and late menopause (Gray *et al.* 1979; MacMahon, 1979; Shils, 1979a; Hankin and Rawlings, 1978). Gray *et al.* (1979) propose that some of the effects of diet on breast cancer may be mediated through the influence of diet on these known risk factors. Diet also may promote or inhibit breast cancer by modifying hormones (androgens, prolactin, estrogens, and possibly other hormones), particularly as hormone patterns are determined in part by diet (Gray *et al.*, 1979; MacMahon, 1979; Hankin and Rawlings, 1978; Wynder, 1977b). One theory proposed is that a high-fat diet increases serum prolactin, or more specifically the prolactin : estrogen ratio, which in turn promotes mammary tumors (Chan and Cohen, 1974, 1975). While it has been reported that a high-fat diet elevates plasma prolactin in experimental animals exposed to a carcinogen, and that a vegetable diet decreases the nocturnal release of prolactin in women (Hill and Wynder, 1976, 1979) the relation of diet modification to hormone activity and to the development or promotion of breast cancer remains unclear (MacMahon, 1979). There is little evidence to suggest that hormones, particularly prolactin, are primary carcinogens for human breast cancer (Hankin and Rawlings, 1978). Not unlike colon cancer, evidence relating diet to breast cancer is circumstantial (Hankin and Rawlings, 1978; Miller, 1978b).

Prevention

The suggestion has been made, similar to that for prevention of cardiovascular disease, that modification of our present dietary intake might decrease cancer incidence. Unfortunately, this suggestion has given rise to a number of food fallacies and quackery with extravagant claims regarding cancer preventive, curative, and causative properties of various dietary constituents (Young and Richardson, 1979).

Adoption of a "prudent" diet—one low in energy, total fat (not to exceed 35% of total energy intake), saturated fat (not to exceed 10% of total energy intake), and cholesterol (not to exceed 300 mg daily)—has been recommended to lessen the risk of cancer (McMichael, 1979; Wynder, 1976, 1977b, 1979;

Wynder and Reddy, 1977). An even stricter type of a "prudent" diet—cholesterol intake not to exceed 100 mg daily and total fat intake to constitute no more than 20% of total energy intake—is recommended to prevent those cancers in which dietary fat is implicated (Wynder, 1976, 1977b; Wynder and Reddy, 1977).

Presently, information is incomplete to predict the efficacy of major changes in food habits and dietary composition in terms of cancer prevention (Gori, 1979a). Of more serious consideration is the possibility that such changes may potentiate carcinogenesis. Preliminary findings from four epidemiologic studies have disclosed an association between low blood cholesterol levels (below 180 mg/dl) and increased mortality from various types of cancer (Anonymous, 1980). Confirmation of this cholesterol–cancer connection is being actively pursued by the National Heart, Lung and Blood Institute and the NCI (Anonymous, 1980). Guidance on this issue of diet and cancer prevention is provided by the Food and Nutrition Board of NAS: "Clearly, a nutritious diet providing adequate amounts of all nutrients and the proper energy content to achieve desirable weight is important for general health and for vigorous defense mechanisms against cancer as well as other diseases" (Food and Nutrition Board, 1980).

NUTRITIONAL IMPACT OF CANCER AND ITS TREATMENT

Cancer *per se* and its various treatments can have profound negative effects on the nutritional status of the cancer patient (Costa and Donaldson, 1979; Harvey *et al.*, 1979; Lawrence, 1979; Shils, 1977a, 1978, 1979b; Theologides, 1979a; Van Eys *et al.*, 1979). Malnutrition can be a most disabling aspect of this disease, not only reducing the patient's quality of life, but contributing to morbidity and mortality (Copeland, 1979; Harvey *et al.*, 1979).

The interrelation of cancer and nutrition has been described in terms of both systemic effects of cancer and localized tumor effects (Costa and Donaldson, 1979; Shils, 1978, 1979b). Cancer cachexia is a most frequent accompaniment of late, advanced cancer, affecting one-third to two-thirds of cancer patients (Theologides, 1977a, 1979a; Costa and Donaldson, 1979). This complex metabolic problem of uncertain etiology is characterized clinically by marked anorexia (decrease or lack of appetite), early satiety, weight loss, wasting, and weakness (Shils, 1979b; Theologides, 1979a). In patients with advanced malignant disease it is a major cause of mortality. Cancer itself and its treatment contribute to this syndrome (Theologides, 1979a).

Neoplasms exert a number of localized effects resulting in nutritional problems (Costa and Donaldson, 1979; Lawrence, 1979; Shils, 1977a, 1978, 1979b; Theologides, 1979a). These are tabulated by Shils (1979b). Interference with food intake, due to partial or complete obstruction of one or more sites of the gastrointestinal tract, is viewed as the most common cause of malnutrition in this general category (Shils, 1979b). Maldigestion or malabsorption associated with various conditions (e.g., pancreatic insufficiency) concomitant with neoplastic diseases can lead to deficiencies of a variety of nutrients (Lawrence, 1979; Shils, 1977a, 1979b).

Although a tumor "toxin" has not been unequivocally identified, tumors may secrete a number of potent pharmacologic substances (ectopic hormones) such as hormone peptides, kinins, and prostaglandins (Costa and Donaldson, 1979; Shils, 1978, 1979b). These, in turn, may produce systemic effects that disturb the patient's nutritional status (Shils, 1978, 1979b). Other localized tumor effects evident in advanced cancer include fluid and electrolyte disturbances (generally due to vomiting and diarrhea), hypoalbuminemia and anemia, and depressed serum and tissue levels of vitamins and minerals (Costa and Donaldson, 1979; Shils, 1979b).

Specific treatment modalities (e.g., surgery, radiotherapy, and chemotherapy), used singly or in combination to eradicate the tumor, may predispose, by way of their effects on normal tissues, to various nutritional problems. Effects of surgery are numerous and varied (Shils, 1977b, 1978, 1979b). Ensuing nutritional problems depend on both the site and extent of the surgical intervention. Radical surgical operations on the head and neck region, for example, may contribute to malnutrition via interference with mastication and swallowing (Costa and Donaldson, 1979; Shils, 1977b, 1979b). Radiotherapy likewise can compromise the patient's nutritional status, the nutritional sequelae dependent on dose and region irradiated (Donaldson and Lenon, 1979; Donaldson, 1977). Without specific dietary intervention, 88–92% of patients receiving high-dose radiation experience significant weight loss (Costa and Donaldson, 1979; Shils, 1978; Donaldson, 1977). Mucosal surfaces of the head and neck region are sensitive to radiation, thus radiotherapy to this area leads to loss of taste sensation ("mouth blindness"), xerostomia (dry mouth), loss of teeth and worsening of dental caries, and difficulty in chewing and swallowing (dysphagia) (Shils, 1978, 1979b; Donaldson, 1977).

Most chemotherapeutic agents, while effective against various malignancies, adversely affect dietary intake as a result of anorexia, nausea, vomiting, mucositis, and, to a lesser extent, diarrhea (Costa and Donaldson, 1979; Donaldson and Lenon, 1979; Shils, 1978; Donaldson, 1977). The effect depends on the type of drug or combination of drugs used, duration of treatment, rates of metabolism, and individual susceptibility (Donaldson and Lenon, 1979). Some chemotherapeutic agents are antimetabolites, and some, like 5-fluorouracil, produce a sprue-like syndrome and increase the requirement for certain vitamins such as thiamin (Dickerson, 1979; Donaldson and Lenon, 1979; Shils, 1979b). If chemotherapy is prolonged, as is often the case, weight loss and progressive debility may be severe (Shils, 1979b). Although only limited data are available, nutritional complications may be magnified by a combination of treatments, such as chemotherapy–radiotherapy (Donaldson and Lenon, 1979; Shils, 1978).

NUTRITIONAL MANAGEMENT

Due to effective antitumor treatments, cancer is increasingly becoming a chronic or protracted illness (Shils, 1979b). As such, correction of malnutrition and its consequences is of utmost importance to improve the cancer patient's quality of life (Drasin *et al.*, 1979; Shils, 1979b; Theologides, 1979b). Shils (1979b,c), summarizing principles of nutritional therapy, states foremost that malnutrition need not be a necessary condition for the cancer patient. Nutritional therapy is described as being supportive, adjunctive, or definitive (Shils, 1979b,c). Preliminary evidence indicates some advantages of nutritional support for the cancer patient, including improved sense of well-being and better preservation of tissue function and repair, and improvement in immunocompetence. Also, the well nourished patient may be better able to undergo pharmacological, radiological and surgical therapies when compared to his or her malnourished counterpart (Copeland, 1979; Drasin *et al.*, 1979; Ota *et al.*, 1979; Shils, 1979c; Theologides, 1977b, 1979b; Wollard, 1979; Alcantara and Speckmann, 1976). However, whether nutritional support results in an overall improvement in outcome for the cancer patient is not proved.

A concern that has been expressed is that nutritional support might harm the host by promoting growth of the tumor (Costa and Donaldson, 1979; Shils, 1979b; Munro, 1977). Tumors have been shown to grow slowly in malnourished animals (Costa and Donaldson, 1979). And while patients do not generally exhibit explosive tumor growth during periods of improved nutrition, it is likely that tumor cells, like nontumor cells, depend on good nutrition (Shils, 1979b). Therefore, it is recommended that nutritional support be accompanied by adequate antitumor treatments (Donaldson and Lenon, 1979; Shils, 1979b).

A variety of nutrition intervention techniques, both oral and enteral, are available. Their advantages, contraindications, and methods of delivery are reviewed elsewhere (Johnston, 1979; Soukop and Calman, 1979; Wollard, 1979; Shils, 1977b). Nutritional assessment is seen as the first step in the nutritional management of the patient presenting with cancer (Copeland, 1979; Costa and Donaldson, 1979; Dwyer, 1979; Harvey *et al.*, 1979; Shils, 1979b; Ven Eys *et al.*, 1979; Wollard, 1979). Based on results of the nutrition assessment, a nutritional therapeutic program can be tailored to each individual's needs. Continual follow-up of the patient's nutritional status is emphasized (Shils, 1979b).

The preferred route for nutritional support is the alimentary tract (oral feeding), although this often necessitates manipulation of diets to account for various eating problems of the patient such as anorexia and taste aversions (De Wys, 1980; De Wys and Herbst, 1977). When oral feeding is contraindicated owing to an ineffective alimentary tract or severe anorexia, two alternatives are available: tube feeding (chemically defined diets) and parenteral feeding [intravenous hyperalimentation (IVH)]. Recent attention has centered on the use of IVH as a viable clinical procedure of delivery of adequate nutrition (Anonymous, 1979b; Copeland, 1979; Shils, 1979b; Van Eys *et al.*, 1979; Copeland *et al.*, 1977; Dudrick *et al.*, 1977). There is increasing evidence that for cancer patients receiving IVH as adjunctive nutrition therapy, risk of sepsis is low and nutrition replenishment may be rapid (often within 10–20 days). Furthermore, tumor growth has not been stimulated by nutritional repletion in human beings (Anonymous, 1979b; Copeland, 1979; Costa and Donaldson, 1979; Drasin *et al.*, 1979; Copeland *et al.*, 1977; Dudrick *et al.*, 1977). However, it should be emphasized that patient selection is an important factor in determining the success or failure of IVH (Elliott, 1980), and the cost and requirement for a well-trained hyperalimentation team must be considered (Costa and Donaldson, 1979).

Unfortunately, a number of nonmedical cancer treatments, many based on myths, and most unsupported by scientific evidence, are being practiced by those with cancer. Included in this category are mega-vitamin therapy (e.g., vitamin C), laetrile, pangamic acid, starving the tumor by starving oneself, and use of natural foods (Creagan and Moertel, 1979; Darby, 1979; Herbert, 1979; Young and Richardson, 1979; Basu, 1976). At this time, there are many suggestive leads but still no conclusive evidence to indicate that reduced or excess amounts of any nutrient have a beneficial role in the treatment of cancer in humans (Creagan and Moertel, 1979; Drasin *et al.*, 1979; Basu, 1976).

SUMMARY

The relationship of nutrition, diet, and cancer can be viewed from three perspectives: (1) diet as a factor in cancer causation; (2) the effect of cancer and its treatment on nutritional status; and (3) nutritional management of the cancer patient. Various types of studies (epidemiologic, animal, and case control) have described a number of highly suggestive associations between diet and cancer in humans, but there is as yet no absolute proof of a direct cause–effect relationship. The role that ingestion of food-borne carcinogens or carcinogen precursors has in causing major human cancers remains to be determined. It is likely that diet has an indirect role, modifying carcinogenesis. Several mechanisms are advanced to explain this effect. For example, it is theorized that excess dietary fat may promote carcinogenesis via its influence on altering bile acid production and/or gut microflora development in colon cancer, and secretion of endocrine glands in breast cancer.

Although there is probably no specific "preventive" diet for cancer, it may be advisable to eat a variety of foods, adjust energy intake to energy expenditure, and avoid moldy food, a deficiency of certain nutrients (e.g., vitamin A) and known dietary carcinogens. Cancer *per se* exerts both systemic effects (e.g., cachexia) and localized effects (e.g., malabsorption due to pancreatic insufficiency) that can lead to profound nutritional problems for the cancer patient. In addition, specific treatment modalities (e.g., surgery, radiotherapy, and chemotherapy), used singly or in combination, may compromise the patient's nutritional status. Malnutrition need not be a necessary condition for the cancer patient. Advantages of nutritional intervention via oral, enteral, or intravenous hyperlimentation include improved well-being, enhanced weight gain, improved immunocompetence, and potentially a better response of the tumor to oncologic treatment. The effect of nutritional support on the overall outcome for the cancer patient is unknown. A concern is the possibility that nutritional support may harm the host by promoting tumor growth. Consequently, it is recommended that nutritional intervention be accompanied by adequate antitumor treatment. To date, there is insufficient evidence to support the suggestion that megadoses or reduced amounts of any essential nutrient, or removal of any normal dietary component, prevents cancer or has a useful role in its treatment in human beings.

REFERENCES

Alcantara, E. N., and Speckmann, E. W., 1976, Diet, nutrition, and cancer, *Am. J. Clin. Nutr.* **29:**1035.

American Cancer Society, 1979, *Cancer Facts and Figures, 1980,* Am. Cancer Soc., New York.

American Diabetes Association, 1979, Policy statement. Saccharin, *Diabetes Care* **2:**380.

Anonymous, 1979a, Vitamin A, tumor initiation and tumor promotion, *Nutr. Rev.* **37:**153.

Anonymous, 1979b, Malnutrition and cancer, *Br. Med. J.* **1:**912.

Anonymous, 1980, Cholesterol and noncardiovascular mortality, *J. Am. Med. Assoc.* **244:**25.

Armstrong, B., and Doll, R., 1975, Environmental factors and cancer incidence and mortality in different countries, with special reference to dietary practices, *Int. J. Cancer* **15:**617.

Ayebo, A. D., Angelo, I. A., Shahani, K. M., and Kies, C., 1979, Effect of feeding *Lactobacillus acidophilus* milk upon fecal flora and enzyme activity in humans, *J. Dairy Sci.* **62(suppl. 1):**44.

Basu, T. K., 1976, Significance of vitamins in cancer, *Oncology* **33:**183.

Burkitt, D. P., 1978, Colonic–rectal cancer: Fiber and other dietary factors, *Am. J. Clin. Nutr.* **31:**S58.

Calabrese, E. J., 1980, *The Vitamins,* Vol. 1 of *Nutrition and Environmental Health,* John Wiley & Sons, New York.

Carroll, K. K., 1977, Dietary factors in hormone-dependent cancers, in: *Nutrition and Cancer* (M. Winick, ed.), pp. 25–40, John Wiley and Sons, New York.

Chan, P. C., and Cohen, L. A., 1974, Effect of dietary fat, antiestrogen, and antiprolactin on the development of mammary tumors in rats, *J. Nat. Cancer Inst.* **52:**25.

Chan, P. C., and Cohen, L. A., 1975, Dietary fat and growth promotion of rat mammary tumors, *Cancer Res.* **35:**3384.

Cole, P., and Cramer, D., 1977, Diet and cancer of endocrine target organs, *Cancer* **40:**434.

Copeland, E. M., 3rd, Daly, J. M., and Dudrick, S. J., 1977, Nutrition as an adjunct to cancer treatment in the adult, *Cancer Res.* **37:**2451.

Copeland, E. M., 3rd, 1979, Nutritional concepts in the treatment of cancer, *J. Fla. Med. Assoc.* **66:**373.

Correa, P., and Haenszel, W., 1978, The epidemiology of large-bowel cancer, *Adv. Cancer Res.* **26:**1.

Costa, G., and Donaldson, S. S., 1979, Current concepts in cancer: Effects of cancer and cancer treatment on the nutrition of the host, *N. Engl. J. Med.* **300:**1471.

Creagan, E. T., and Moertel, C., 1979, Vitamin C therapy of advanced cancer (letter), *N. Engl. J. Med.* **301:**1399.

Cummings, J. H., 1978, Dietary factors in the aetiology of gastrointestinal cancer, *J. Hum. Nutr.* **32**:455.

Cummings, J. H., Hill, M. J., Bone, E. S., Branch, W. J., and Jenkins, D. J., 1979, The effect of meat protein and dietary fiber on colonic function and metabolism, II. Bacterial metabolites in feces and urine, *Am. J. Clin. Nutr.* **32**:2094.

Darby, W. J., 1979, Editorial: Etiology of nutritional fads, *Cancer* **43**:2121.

De Wys, W. D., 1980, Nutritional care of the cancer patient, *J. Am. Med. Assoc.* **244**:374.

De Wys, W. D., and Herbst, S. H., 1977, Oral feeding in the nutritional management of the cancer patient, *Cancer Res.* **37**:2429.

Dickerson, J. W., 1979, Nutrition and breast cancer, *J. Hum. Nutr.* **33**:17.

Donaldson, S. S., 1977, Effect of nutrition as related to radiation and chemotherapy, in: *Nutrition and Cancer* (M. Winick, ed.), pp. 137–153, John Wiley & Sons, New York.

Donaldson, S. S., and Lenon, R. A., 1979, Alterations of nutritional status: Impact of chemotherapy and radiation therapy, *Cancer* **43**:2036.

Drasin, H., Rosenbaum, E. H., Stitt, C. A., and Rosenbaum, I. R., 1979, The challenge of nutritional maintenance in cancer patients, *West. J. Med.* **130**:145.

Dudrick, S. J., MacFadyen, B. V., Jr., Souchon, E. A., Englert, D. M., and Copeland, E. M., 3rd, 1977, Parenteral nutrition techniques in cancer patients, *Cancer Res.* **37**:2440.

Dwyer, J. T., 1979, Dietetic assessment of ambulatory cancer patients: With special attention to problems of patients suffering from head–neck cancers undergoing radiation therapy, *Cancer* **43**:2077.

Elliott, J., 1980, Squaring off over total parenteral nutrition (News), *J. Am. Med. Assoc.* **243**:1610.

Enstrom, J. E., 1975, Colorectal cancer and consumption of beef and fat, *Br. J. Cancer* **32**:432.

Food and Nutrition Board, 1980, *Toward Healthful Diets*, National Research Council, NAS, Washington, D.C.

Fraser, P., Chilvers, C., Beral, V., and Hill, M. J., 1980, Nitrate and human cancer: A review of the incidence, *Int. J. Epidemiol.* **9**:3.

Goldberg, M. J., Smith, J. W., and Nichols, R. L., 1977, Comparison of the fecal microflora of Seventh-Day Adventists with individuals consuming a general diet. Implications concerning colonic carcinoma, *Ann. Surg.* **186**:97.

Goldin, B. R., and Gorbach, S. L., 1977, Alterations in fecal microflora enzymes related to diet, age, lactobacillus supplements, and dimethylhydrazine, *Cancer* **40**:2421.

Goldin, B. R., and Gorbach, S. L., 1980, Effect of *Lactobacillus acidophilus* dietary supplements on 1,2-dimethylhydrazine dihydrochloride-induced intestinal cancer in rats, *J. Nat. Cancer Inst.* **64**:263.

Goldin, B. R., Swenson, L., Dwyer, J., Sexton, M., and Gorbach, S. L., 1980, Effect of diet and *Lactobacillus acidophilus* supplements on human fecal bacterial enzymes, *J. Nat. Cancer Inst.* **64**:255.

Gori, G. B., 1977, Diet and cancer, *J. Am. Diet. Assoc.* **71**:375.

Gori, G. B., 1978a, Diet and nutrition in cancer causation, *Nutr. Cancer* **1**:5.

Gori, G. B., 1978b, Role of diet and nutrition in cancer cause, prevention and treatment, *Bull. Cancer (Paris)* **65**:115.

Gori, G. B., 1979a, Food as a factor in the etiology of certain human cancers, *Food Technol.* **33**:48.

Gori, G. B., 1979b, Dietary and nutritional implications in the multifactorial etiology of certain prevalent human cancers, *Cancer* **43**:2151.

Gray, G. E., Pike, M. C., and Henderson, B. E., 1979, Breast-cancer incidence and mortality rates in different countries in relation to known risk factors and dietary practices, *Br. J. Cancer* **39**:1.

Hankin, J. H., and Rawlings, V., 1978, Diet and breast cancer: A review, *Am. J. Clin. Nutr.* **31**:2005.

Harvey, K. B., Bothe, A., Jr., and Blackburn, G. L., 1979, Nutritional assessment and patient outcome during oncological therapy, *Cancer* **43**:2065.

Herbert, V., 1979, The nutritionally and metabolically destructive "nutritional and metabolic antineoplastic diet" of laetrile proponents, *Am. J. Clin. Nutr.* **32**:96.

Hill, M. J., 1978, Some leads to the etiology of cancer of the large bowel, *Surg. Ann.* **10**:135.
Hill, P., and Wynder, E. L., 1976, Diet and prolactin release (letter), *Lancet* **2**:806.
Hill, P. B., and Wynder, E. L., 1979, Effect of a vegetarian diet and dexamethasone on plasma prolactin, testosterone and dehydroepiandrostrone in men and women, *Cancer Lett.* **7**:273.
Hillyard, L. A., and Abraham, S., 1979, Effect of dietary polyunsaturated fatty acids on growth of mammary adenocarcinomas in mice and rats, *Cancer Res.* **39**:4430.
Hopkins, G. J., and Carroll, K. K., 1979, Relationship between amount and type of dietary fat in promotion of mammary carcinogenesis induced by 7,12-dimethylbenz (a) anthracene, *J. Nat. Cancer Inst.* **62**:1009.
Johnston, I. D., 1979, Parenteral nutrition in the cancer patient, *J. Hum. Nutr.* **33**:189.
Lawrence, W., Jr., 1979, Effects of cancer on nutrition: Impaired organ system effects, *Cancer* **43**:2020.
Levey, B., 1979, Legislative update, *Nat. Food Rev.* **8**:44.
Lyon, J. L., and Sorenson, A. W., 1978, Colon cancer in a low-risk population, *Am. J. Clin. Nutr.* **31**:S227.
Lyon, J. L., Klauber, M. R., Gardner, J. W., and Smart, C. R., 1976, Cancer incidence in Mormons and non-Mormons in Utah, 1966–1970, *N. Engl. J. Med.* **294**:129.
Maclennan, R., and Jensen, O. M., 1977, Dietary fibre, transit-time, faecal bacteria, steroids, and colon cancer in two Scandinavian populations. Report from the International Agency for Research on Cancer Intestinal Microecology Group, *Lancet* **2**:207.
MacMahon, B., 1979, Dietary hypotheses concerning the etiology of human breast cancer, *Nutr. Cancer* **1**:38.
Mastromarino, A. J., Reddy, B. S., and Wynder, E. L., 1978, Fecal profiles of anaerobic microflora of large bowel cancer patients and patients with nonhereditary large bowel polyps, *Cancer Res.* **34**:4458.
McMichael, A. J., 1979, Diet in the causation of cancer, *Food Nutr. Notes Rev.* **36**:187.
Mettlin, C., and Graham, S., 1979, Dietary risk factors ih human bladder cancer, *Am. J. Epidemiol.* **110**:255.
Miller, A. B., 1977, Role of nutrition in the etiology of breast cancer, *Cancer* **39**:2704.
Miller, A. B., 1978a, Epidemiology and colorectal cancer, *Can. J. Surg.* **21**:209.
Miller, A. B., 1978b, An overview of hormone-associated cancers, *Cancer Res.* **38**:3985.
Modan, B., 1977, Role of diet in cancer etiology, *Cancer* **40**:1887.
Modan, B., 1979, Patterns of gastrointestinal neoplasms in Israel, *Isr. J. Med. Sci.* **15**:301.
Moskovitz, M., White, C., Barnett, R. N., Stevens, S., Russell, E., Vargo, D., and Floch, M. H., 1979, Diet, fecal bile acids, and neutral sterols in carcinoma of the colon, *Dig. Dis. Sci.* **24**:746.
Mower, H. F., Ray, R. M., Stemmermann, G. N., Nomura, A., and Glober, G. A., 1978, Analysis of fecal bile acids and diet among the Japanese in Hawaii, *J. Nutr.* **108**:1289.
Mower, H. F., Ray, R. M., Shoff, R., Stemmerman, G. N., Nomura, A., Glober, G. A., Kamiyama, S., Shimada, A., and H. Yamakawa, 1979, Fecal bile acids in two Japanese populations with different colon cancer risks, *Cancer Res.* **39**:328.
Munro, H. N., 1977, Tumor–host competition for nutrients in the cancer patient, *J. Am. Diet. Assoc.* **71**:380.
National Academy of Sciences, 1978, *Saccharin: Technical Assessment of Risks and Benefits—Part 1*, Assembly of Life Sciences, National Research Council and the Institute of Medicine, NAS, Washington, D.C.
National Cancer Institute, 1980, Progress Report to the Food and Drug Administration from the National Cancer Institute Concerning the National Bladder Cancer Study, Office of Cancer Communications, NCI, Bethesda, Maryland.
National Dairy Council, 1975, Nutrition, diet, and cancer, *Dairy Counc. Dig.* **46**:25.
Newberne, P., 1979, Nitrite promotes lymphoma incidence in rats, *Science* **204**:1079.
Ota, D. M., Copeland, E. M., 3rd, Corriere, J. N., Jr., and Dudrick, S. J., 1979, The effects of nutrition and treatment of cancer on host immunocompetence, *Surg. Gynecol. Obstet.* **148**:104.
Petering, H. G., 1978, Diet, nutrition and cancer, in: *Inorganic and Nutritional Aspects of Cancer* (G. N. Schrauzer, ed.), pp. 207–223, Plenum Press, New York.

Phillips, R. L., 1975, Role of life-style and dietary habits in risk of cancer among Seventh-Day Adventists, *Cancer Res.* **35:**3513.
Reddy, B. S., and Wynder, E. L., 1977, Metabolic epidemiology of colon cancer. Fecal bile acids and neutral sterols in colon cancer patients and patients with adenomatous polyps, *Cancer* **39:**2533.
Reddy, B. S., Mastromarino, A., and Wynder, E., 1977, Diet and metabolism: large-bowel cancer, *Cancer* **39:**1815.
Reddy, B. S., Hedges, A. R., Laakso, K., and Wynder, E. L., 1978, Metabolic epidemiology of large bowel cancer: Fecal bulk and constituents of high-risk North American and low-risk Finnish population, *Cancer* **42:**2832.
Rivlin, R. S., 1973, Riboflavin and cancer: a review, *Cancer Res.* **33:**1977.
Shils, M. E., 1977a, Nutritional problems associated with gastrointestinal and genitourinary cancer, *Cancer Res.* **37:**2366.
Shils, M. E., 1977b, Nutrition in the treatment of cancer, in: *Nutrition and Cancer* (M. Winick, ed.), pp. 155–167, John Wiley & Sons, New York.
Shils, M. E., 1978, Nutrition and the cancer patient, *Nutr. and Cancer* **1:**9.
Shils, M. E., 1979a, Diet and nutrition as modifying factors in tumor development, *Med. Clin. North Am.* **63:**1027.
Shils, M. E., 1979b, Nutritional problems induced by cancer, *Med. Clin. North Am.* **63:**1009.
Shils, M. E., 1979c, Principles of nutritional therapy, *Cancer* **43:**2093.
Soukop, M., and Calman, K. C., 1979, Nutritional support in patients with malignant disease, *J. Hum. Nutr.* **33:**179.
Spiller, G. A., 1978, Interaction of dietary fiber with other dietary components: A possible factor in certain cancer etiologies, *Am. J. Clin. Nutr.* **31:**S231.
Sporn, M. B., 1977, Vitamin A and its analogs (retinoids) in cancer prevention, in: *Nutrition and Cancer* (M. Winick, ed.), pp. 119–130, John Wiley & Sons, New York.
Sporn, M. B., and Newton, D. L., 1979, Chemoprevention of cancer with retinoids, *Fed. Proc. Fed. Am. Soc. Exp. Biol.* **38:**2528.
Sporn, M. B., Dunlop, N. M., Newton, D. L., and Smith, J. M., 1976, Prevention of chemical carcinogenesis by vitamin A and its synthetic analogs (retinoids), *Fed. Proc. Fed. Am. Soc. Exp. Biol.* **35:**1332.
Tannenbaum, A., 1959, Nutrition and cancer, in: *Physiopathology of Cancer* (F. Homburger, ed.), pp. 517–562, Hoeber-Harper, New York.
Tannenbaum, A., 1945, Dependence of tumor formation on composition of calorie-restricted diet as well as on degree of restriction, *Cancer Res.* **5:**616.
Theologides, A., 1977a, Cancer cachexia, in: *Nutrition and Cancer* (M. Winick, ed.), pp. 75–94, John Wiley & Sons, New York.
Theologides, A., 1977b, Nutritional management of the patient with advanced cancer, *Postgrad. Med.* **61:**97.
Theologides, A., 1979a, Cancer cachexia, *Cancer* **43:**2004.
Theologides, A., 1979b, Nutritional management during cancer therapy, *Minn. Med.* **62:**547.
U.S. Department Health, Education, and Welfare/Public Health Service, 1979, Healthy People. The Surgeon General's Report on Health Promotion and Disease Prevention, *DHEW/PHS Publication No. 79-55071,* U.S. Government Printing Office, Washington, D.C.
Upton, A. G., and Fink, D. J., 1980, Statement on Diet, Nutrition and Cancer. Hearing before the Subcommittee on Nutrition of the Committee on Agriculture, Nutrition and Forestry, United States Senate (Oct. 2, 1979), U.S. Government Printing Office, Washington, D.C.
Van Eys, J., Seelig, M. S., and Nichols, B. L., Jr. (eds.), 1979, *Nutrition and Cancer,* SP Medical and Scientific Books, New York.
Walker, A. R., 1976, Colon cancer and diet, with special reference to intakes of fat and fiber, *Am. J. Clin. Nutr.* **29:**1417.
Weisburger, J. H., 1979, Mechanism of action of diet as a carcinogen, *Cancer* **43:**1987.
Weisburger, J. H., Reddy, B. S., Hill, P., Cohen, L. A., Wynder, E. L., and Spingarn, N. E., 1980, Nutrition and cancer—on the mechanisms bearing on causes of cancer of the colon, breast, prostate, and stomach, *Bull. N.Y. Acad. Med.* **56:**673.

Winick, M. (ed.), 1977, *Nutrition and Cancer,* John Wiley & Sons, New York.
Wollard, J. J. (ed.), 1979, *Nutritional Management of the Cancer Patient,* Raven Press, New York.
Wynder, E. L., 1979, Dietary habits and cancer epidemiology, *Cancer* **43:**1955.
Wynder, E. L., 1977a, Cancer prevention: A question of priorities, *Nature* **268:**284.
Wynder, E. L., 1977b, The dietary environment and cancer, *J. Am. Diet. Assoc.* **71:**385.
Wynder, E. L., 1976, Nutrition and cancer, *Fed. Proc. Fed. Am. Soc. Exp. Biol.* **35:**1309.
Wynder, E. L., 1975, The epidemiology of large bowel cancer, *Cancer Res.* **35:**3388.
Wynder, E. L., and Gori, G. B., 1977, Contribution of the environment to cancer incidence: an epidemiologic exercise, *J. Nat. Cancer Inst.* **58:**825.
Wynder, E. L., and Hirayama, T., 1977, Comparative epidemiology of cancers of the United States and Japan, *Prev. Med.* **6:**567.
Wynder, E. L., and Reddy, B. S., 1977, Diet and cancer of the colon, in: *Nutrition and Cancer* (M. Winick, ed.), pp. 55–71, John Wiley & Sons, New York.
Young, V. R., and Richardson, D. P., 1979, Nutrients, vitamins and minerals in cancer prevention: Facts and fallacies, *Cancer* **43:**2125.

XIII

CONSEQUENCES OF INFANT FEEDING

44

EFFECTS OF INFANT FEEDING

Charlotte G. Neumann and E. F. Patrice Jelliffe

INTRODUCTION

The possibility that the beginnings of adult disease occur silently in infancy and early childhood, and that the potential for primary prevention exists, creates a challenge to all health-care providers dealing with infants and children. Examination of early nutritional experiences, both under- and overnutrition in the antenatal period, the neonatal period, infancy, and early childhood, in relation to later health outcomes may furnish clues for prevention and early intervention. The physical and mental development of a child can be greatly influenced by the way in which a mother eats during pregnancy and the quality and quantity of the food received by the infant and child.

This chapter will review the long-term effects of nutrition in early life on certain aspects of health. Clear cut data exist in some instances but many statements are emotionally charged and speculative. Although animal data are far more abundant than human data, mainly human studies will be referred to and only very relevant animal studies cited. A rapidly proliferating literature is emerging about early nutritional deprivation and long-term consequences, but these will be described only briefly as fuller reviews can be referred to elsewhere. The potentially adverse effects and later consequences of excessive nutrients will be primarily discussed.

ANTENATAL PERIOD: MALNUTRITION

Intra-uterine Growth Retardation (IUGR)

The role of nutritional deficiency during pregnancy as a cause of low birth weight has been under study for the past several decades. Although it is difficult to establish a definite causal relationship, evidence has been amass-

Charlotte G. Neumann and E. F. Patrice Jelliffe • Population, Family and International Health Division, School of Public Health, University of California, Los Angeles, California 90024.

ing that improved maternal nutrition decreases frequency of low birth weight (Table I).

Epidemiological evidence supports the role of maternal malnutrition in fetal growth retardation (Lechtig *et al.*, 1975). Birth weight correlates well with increased maternal weight gain during pregnancy, the more the mother gains, the heavier the newborn. As to optimal total weight gain, about 25–27 lb is recommended (Hytten and Thomson, 1970). However, large-for-gestational-age (LGA) infants with birth weights exceeding 4200 g may have increased perinatal mortality and complications (Udall *et al.*, 1978). In the developing countries maternal height is indicative of previous nutritional experience and directly correlates with birth weight (Stoch and Smythe, 1967; Lechtig *et al.*, 1975).

Evidence from intervention studies point to a correlation between birth weight and nutritional supplementation. In a Guatemalan study (Freeman *et al.*, 1977) a high calorie, high protein supplement of 300 g was given daily providing 20,000 extra calories during pregnancy. There was significant difference in birth weight outcome in that the supplemented group of mothers gave birth to infants weighing 300 g more than the controls. This correlation has not been demonstrated very clearly in affluent populations (Adams *et al.*, 1978). However, in studies of poor women in Canada (Higgins, 1972) and Boston (Burke *et al.*, 1943) decreased infant mortality and increased birth weight were found in the supplemented group of pregnant women. In studies in Taiwan (Roeder and Chow, 1972), supplementation of pregnant women showed increased birth weight in the newborns of the supplemented group. The concensus reached, despite methodological problems, is that adequate nutrition is important in pregnant women, has a positive effect on development and growth of the fetus, and ameliorates the low birth-weight problem. Protein, in some studies, showed a stronger correlation with birth weight than did calories (Burke *et al.*, 1943). However, calorie lack appears to be the more significant of the problems (Lechtig, 1975).

TABLE I
Antenatal Malnutrition: Adverse Effects

Nutrient deficit	Sequelae[a]
Calories, protein, folic acid, zinc (probable)	Intrauterine growth retardation
	↓ Mental development
	subtle learning problems
	attention deficits
	↓ Physical growth and poor catch-up
	↓ Immunocompetence (cell-mediated)
Iodine	Cretinism
	↓ Brain development
	↓ Skeletal development and physical growth

[a] References in text.

Fetal growth retardation has also been associated with specific nutrient deficiencies. Vitamin A deficiency has been linked to IUGR and decreased placental size (Hurley, 1980a,b). Pyridoxine deficiency has been implicated in IUGR, diminished brain growth, and DNA synthesis. Folic acid, zinc, manganese, and magnesium deficiencies have all been associated with decreased cell division and IUGR in either animals or humans, with folic acid particularly important in humans (Hurley 1980a). Gross *et al.* (1975) demonstrated malnutrition is most dangerous in periods of rapid growth, such as the last trimester of fetal development and the first few months of newborn life. Biochemical and anatomical alterations may result from malnutrition, although their functional significance is not fully known (Brasel, 1974). A large part of the maternal diet is needed for the maintenance of oxidative metabolism of maternal tissue. The energy requirement for fetal growth is actually a small part of the total energy cost of pregnancy. The major effect of nutrition in pregnancy may be to provide energy to insure optimal substrates for oxidative metabolism and to support the placenta (Brasel, 1974). In fetal growth retardation, less than 50% of expected placental cell number is seen with a decreased RNA : DNA ratio (Brasel, 1974). Vascular insufficiency and malnutrition both affect cell division. Placental weight and the trophoblastic mass are decreased, as is the growth of peripheral villi which limit the functional surface for fetal exchange of nutrients (Lechtig, 1975).

Attempts at classification of IUGR have been made. Intrinsic IUGR is thought to be due to congenital anomalies, inborn areas of metabolism, and genetic and chromosomal abnormalities. Extrinsic IUGR has been subdivided into asymmetrical growth retardation where the various organs are differentially affected with diminished placenta size but the brain is spared (Brasel, 1974; Sinclair *et al.*, 1974). The liver is decreased in size and glycogenesis is reduced. This is seen with maternal vascular disease with decreased blood flow to the fetus, and more commonly seen in the affluent nations and among toxemic women in developing areas. This condition has been reproduced experimentally in the rat by ligating placental blood vessels (Minkowski *et al.*, 1974). Symmetrical growth retardation results in the equal reduction in cell number and size of the placenta, brain, liver, and all organs in general. This is usually associated with maternal malnutrition and is seen in areas of impoverishment both in the developing countries and in the depressed parts and disadvantaged areas of the developed countries (Naeye *et al.*, 1973; Brasel, 1974).

Sequelae of IUGR

Long-term sequelae of IUGR have been documented in several areas of human function: cognitive development, physical growth (Fancourt, 1976; Fitzhardinge and Steven, 1972), and immuno-competence (Chandra, 1975; Neumann *et al.*, 1980). Short-term sequelae have included increased perinatal mortality, and metabolic derangements such as hypoglycemia and increased infections (Sinclair *et al.*, 1974).

Cognitive Development in IUGR Infants

Few longitudinal studies exist that have examined long-term sequelae in cognitive function of the IUGR infant. Infants whose birth weights are low relative to gestational age, whether born at term or preterm are at risk for developmental problems (Mönckeberg, 1972).

The precise nature of cognitive risk that results is not fully understood. Implicated as causal factors are the impact of insufficient nourishment on the developing brain, neural mechanisms, and other organs (Dobbing, 1970), as well as medical complications evident at birth (Kopp and Parmalee, 1979). Nonetheless, outcomes for IUGR infants, while varied, appear to be more pessimistic than for infants born at appropriate birth weight.

Follow-up studies of IUGR infants often fail to delineate clearly who is represented in the sample (e.g., infants with and without chromosomal anomalies, infants born IUGR at term vs. IUGR preterm). As a result, interpretation of the literature is impeded. Nonetheless, taken together, available studies depict full and preterm IUGR infants as having underlying problems in attention, and in responsiveness to internal and external stimuli (Als *et al.*, 1976; Als *et al.*, 1979). Developmental tests indicate that IUGR infants lag significantly behind those appropriate-weight-for-gestational-age peers through the first 18 months, but by 24 months score similarly (J. Krakow, personal communication, 1980).

At school age, IUGR infants have a disproportionately high rate of academic difficulty, despite intelligence levels well within the normal range. One notable large-scale study, the British Perinatal survey of 1958, followed 15,000 children through age 7 with data from group tests, design copying, tests of visual-motor and physical coordination, draw-a-man, developmental and pregnancy history, and a home visit by a social worker (Butler, 1974). Severe handicaps such as I.Q. scores less than 50, cereberal palsy, congenital anomalies, and defects of the special senses were more common in those with birth weights less than 10th percentile than in controls. Reading, design copy, and overall intellectual ability were all rated worse in the IUGR groups than in controls. When controlling for social class and for birth order, the IUGR group was at higher risk for mental and educational retardation. Risk was found to decrease with increasing birth weight and then increase with disabilities in the LGA infant (Butler, 1974).

Fitzhardinge and Steven (1972) reported that 50% of the boys and 36% of the girls in a clearly defined follow-up sample of full-term IUGR infants had inadequate school performance with special class placement (44%) or consistent failure in regular classes (56%). Of special note is the fact that all of the IUGR children who were failing regular classes had mean I.Q. scores above 90. About half of these children had "specific learning defects." Similar results were obtained by Rubin *et al.*, (1973). At school age, IUGR infants had a mean I.Q. score of 98, but 55% of them had been left back, assigned to special classes, or received special services. Frances-William and Davies (1974) reported that preterm and IUGR infants, despite a mean I.Q. of 92, also had a high prevalence of educational problems.

None of these studies provides in-depth information about the causes of academic difficulty among IUGR children. However, because most of the children had a normal range of intelligence, subtle cognitive deficits, not measured by standardized I.Q. tests, are implicated. Recent evidence has pointed to difficulty in focusing and sustaining attention on salient tasks as one aspect of cognitive difficulty among learning disabled children (Douglas and Peters, 1979; Keogh and Margolis, 1976; Ross, 1977).

In summary, IUGR infants, both pre- and full-term, may be inattentive and unresponsive during the neonatal period. They have development that, although delayed during the first year, approximates that of preterm infants by age 2, and at school age have I.Q. scores primarily in the normal range. Repeated observations of academic problems have been reported; subtle cognitive deficits, including difficulty in focusing and sustaining attention, are implicated in such learning disorders.

Iodine deficiency during pregnancy and in early infancy can have devastating effects on the developing brain and skeletal growth (Pharaoh *et al.*, 1971). The association between endemic goiter, cretinism, and iodine deficiency was made in the mid-1800s (Stanbury and Querido, 1957; Hurley, 1980a). Cretinism due to iodine deficiency is one of several types of cretinism with thyroxine metabolism defects causing a similar picture. The infant born with cretinism, unless immediately diagnosed and vigorously treated is doomed to a life of severe mental, motor, and physical retardation (Smith *et al.*, 1975). Deafness and deaf mutism are also common (Fierro-Benitez *et al.*, 1974). Iodine administration through injections of salt fortified by iodized oil has prevented cretinism in several intervention studies (Pharaoh *et al.*, 1971). Genetic and environmental factors may well interact with iodine lack in the production of cretinism, as there appear to be individual differential responses to exposure of the fetus to iodine lack (Hurley, 1980a).

Physical Growth

Children who have experienced IUGR tend not to show full catch-up growth and remain smaller than their normal birth weight peers on long-term follow-up (Beck and Van den Berg, 1975; Fitzhardinge and Steven, 1972). This is unlike the situation of children who experience postnatal malnutrition and do show considerable catch up growth with nutritional rehabilitation (Tanner, 1978).

An intervention study of protein deficient pregnant women in Taiwan (Roeder and Chow, 1972) found that the IUGR infants became "nitrogen or metabolic wasters" and that despite excellent postnatal nutrition they did not reach normal growth or size. Venkatachalam *et al.* (1967), in studies of malnourished pregnant women in India, found that IUGR infants born to these women—despite supplemental feeding in the early infancy period—never achieved normal size and remained small. A group of Kenyan infants, followed from birth through their first birthday, showed significant stunting as well as diminished head circumference and weight for age (Neumann *et al.*, 1980). There was also a higher incidence of moderate and severe protein-calorie malnutrition (PCM) based on weight for height compared to the control group of infants.

An 8 year follow-up of IUGR infants showed that weight and heights were below those of their siblings who were born with normal birth weights (Fitzhardinge and Steven, 1972). The effects may be even more long lasting; a group of adolescents who were IUGR at birth, followed to adolescence, were found to be shorter than their siblings for expected height, even taking into account their parents' height (Beck and Van den Berg, 1975).

The nutrition of the mother during her own infancy and childhood is of crucial importance. As is known, short women run the risk of themselves producing low birth-weight infants, and poor pregnancy outcome and reproductive performance in general. Aside from obvious problems of a contracted pelvis due to PCM or rickets, prenatal growth retardation may adversely affect her reproductive performance, producing an intergeneraterial effect of nutritional insult (Thomson *et al.*, 1968; Mata, 1974).

Immune Function

IUGR profoundly affects not only growth of general body organ systems, but of lymphoid tissue with particular effect on the thymus, tonsillar tissue, and lymph nodes (Naeye *et al.*, 1973). The adverse effect on the lymphoid system causes decreased cell-mediated immunity (CMI) (Chandra, 1975; Neumann *et al.*, 1980). The various parameters of CMI have been studied in infants judged to be malnourished *in utero*, and uniformly CMI was found to be impaired. Total lymphocyte count, T-cell number and decreased delayed cutaneous hypersensitivity have been found in infants with IUGR. BCG immunization against tuberculosis, whose efficacy is dependent on CMI when administered at birth in these IUGR infants, resulted in a higher percent of nonresponders and smaller delayed cutaneous hypersensitivity reactions on tuberculin testing (PPD) than infants of a normal weight at birth (Manerikaris *et al.*, 1976; Neumann, 1980). These problems were seen whether the mother had malnutrition or whether she had reduced placental blood flow because of a basic degenerative vascular disorder. Of interest is that infants between 2500–2800 g showed depression of CMI midway between that found in the IUGR infant under 2500 g and the normal-weight infant, indicating a spectral problem dependent on birth-weight deficit. The depression of CMI is long-lasting, anywhere from 1 to 5 years (Ferguson, 1978; Chandra, 1975). Increased infection has been seen in these small infants, as well as problems with adequate responses to immunizations, particularly measles, pertussis, and BCG (Neumann *et al.*, 1980).

TABLE II
Antenatal Overnutrition: Adverse Effects

Nutrient excess	Sequelae[a]
Calories	Neonatal obesity
Vitamin A	Anomalies—CNS, genitourinary tract
Vitamin C	Early pregnancy termination, "conditioned scurvy" in offspring
Amino acid excess: phenylalanine	Mental retardation and brain damage, intrauterine growth retardation, congenital anomalies, spontaneous abortion
Galactosemia	? Incipient CNS damage
Iodine	↑ Neonatal mortality, congenital goiter, hypothyroidism
Alcohol	Fetal alcohol syndrome, intrauterine and postnatal growth failure, neurologic abnormalities, anomalies—face, eyes, heart

[a] References in text

ANTENATAL PERIOD: OVERNUTRITION (Table II)

Caloric Excess

Adverse effects of maternal overnutrition during pregnancy may occur in response to calories and specific nutrients. Fatness of the neonate has now been shown by several important studies to be linked to excessive maternal caloric intake and prepregnancy weight (Udall *et al.*, 1978). The relationship between birth weight and neonatal fatness has been evaluated by weight for height and skinfold thickness, and an examination of maternal variables. Simpson *et al.* (1975) were able to show that a positive correlation exists between birth weight and maternal prepregnancy weight, and between birth weight and pregnancy weight gain, both independently and in an additive fashion. Whitelaw (1977) and Frisancho *et al.* (1977), who used fat fold thickness to assess neonatal and postpartum maternal obesity, found that fatter mothers tended to have fatter infants. However, it was not clear if it was the prepregnant maternal obesity or the obesity acquired from the excessive weight gain during pregnancy that correlated with neonatal obesity.

Udall *et al.* (1978) studied the variability of subcutaneous tissue of neonates in the presence or absence of maternal obesity prior to pregnancy and the amount of prenatal weight gain to see if any correlation existed with neonatal fatness. Weight gain in pregnancy (greater than 40 lb) was significantly greater in these fat LGA infants, who were also longer and had a greater mid-arm fat area. Prepregnant obese mothers had infants with significantly thicker fat folds and on multiple regression analysis, weight gain during pregnancy explained the greater portion of the variance. Therefore, it was concluded that maternal obesity was strongly associated with increased subcutaneous fat in the newborn. It was postulated that increased transfer of free fatty acids occurred across the placenta.

Specific Nutrients

An abundance of animal data exists using well-controlled laboratory experiments to examine the effects of excess nutrients in pregnancy on the fetus. Only studies highly relevant to humans will be discussed here.

Vitamin A

Vitamin A has played a special role in teratology and developmental nutrition. Both deficiency and excess, depending on the timing of the insult, have been shown to have adverse effect on the developing fetus (Hurley, 1980). Excessive vitamin A intake in animals has caused anomalies of the central nervous system (CNS), skull, and eyes. These have included anencephaly, exencephaly, spina bifida, cleft palate (Morriss, 1972), and eye defects. The amount of vitamin A administered and the gestational stage of the animal determined the defect produced. The abnormalities occurred from the differential growth rates among varying tissues and lack of synchronous development (Hurley, 1980a).

Excessive vitamin A intake in humans has produced genito-urinary tract anomalies, and malformations of the CNS as well (Bernhardt and Dorsey, 1974). In a study of mothers of infants with malformations, maternal vitamin levels at birth were significantly higher than in control mothers. Vitamin A in the livers of fetuses who had aborted or were premature have also been found to be higher than in those without malformations (Hutchings *et al.*, 1973; Hurley, 1980).

Vitamin D

Excessive intake of vitamin D in rats during pregnancy has produced infants with low birth weight, low total body calcium and phosphorus (including a low content in the fetus), and impaired osteogenesis and skeletal development. Abnormal bone development persisted even though vitamin D metabolism was normal. Multiple fractures with impaired healing and persistent growth retardation persisted throughout life (Ornoy *et al.*, 1972). There are no data on humans.

Vitamin C

Excessive vitamin C during pregnancy can be detrimental to the fetus. Increased estrogen levels have been associated with premature termination of pregnancy and "conditioned scurvy" in the offspring in a variety of experimental animals (Norkus and Rosso, 1975). When guinea pigs, rats, and mice were given excess vitamin C during pregnancy, abortion increased and the high fetal mortality and infertility resulted.

The ingestion of large doses of vitamin C by pregnant women has been associated with early termination of pregnancy and with increased excretion of estrogen in the urine (in 12 of 16 women), as was found in animal experiments. Also in humans, excessive vitamin C caused "conditioned scurvy" in the offspring. This was attributed to the high maternal intake that perhaps induced increased catabolis rate of vitamin C or was due to vitamin C withdrawal at birth. Experimental work in animals supports this idea (Hurley, 1980).

Amino Acid Excess

Animal experiments show that excessive amounts of single amino acids are detrimental if ingested in excess amounts during pregnancy, resulting in amino acid imbalances. For example, large amounts of lysine or leucine in the diet of pregnant rats results in fetal death or retarded fetal growth (Hurley, 1980).

Phenylalanine in excessive amounts can be extremely damaging to the embryo and fetus in animals, and humans as well. Infant monkeys born to mothers with excessive phenylalanine ingested throughout pregnancy had abnormally low birth weights. In terms of learning behavior, there was significant impairment in those whose mothers had the highest levels (Kerr *et al.*, 1968).

In humans, infants born to phenylketonuric mothers may become mentally retarded (Allen and Brown, 1968). Intrauterine growth retardation, microcephaly, skeletal anomalies, cardiac defects, and esophageal and lung abnormalities were noted. Spontaneous abortions and postnatal growth retardation were common (Hurley, 1980a). An inverse relationship between maternal blood level and intelligence in the offspring was found (Allen and Brown, 1968; Ford and Berman, 1977). If maternal levels are high, fetal brain damage may ensue. In one case report an affected mother with an unrestricted diet during pregnancy had two retarded offspring; with a restricted diet she then produced a normal child (Hurley, 1980).

Congenital galactosemia in humans can lead to mental retardation and cataracts. Clinical signs of galactosemia depends on exposure to galactose and perhaps the passage of galactose through the placenta may initiate the process (Haworth *et al.*, 1969). High galactose diets fed to pregnant rats have been associated with cataracts in newborns (Segal and Bernstein, 1963). In humans heterozygous for galactosemia, a diet low in lactose during pregnancy might decrease the possibility of harming the fetus (Hurley, 1980).

Iodine Excess

Excessive iodine in asthma and bronchitis medications taken by pregnant women, as well as iodine deficiency, have adverse effects on the developing fetus. Abnormal infants have been born to mothers ingesting large amounts of iodides during pregnancy (Carswell *et al.*, 1970). The neonatal mortality was high and mental retardation was seen in some of the survivors. Congenital goiter and hypothyroidism were the main findings, presumably due to suppressive effects of the iodine on development of fetal thyroid (Carswell *et al.*, 1970).

Ingestion of radioactive iodine therapy during pregnancy has also been associated with cretinism in the newborn both in humans and experimental animals. The radioactive iodine presumably caused damage to the fetal thyroid gland (Green *et al.*, 1971).

Alcohol and The Fetal Alcohol Syndrome

Alcohol, although not generally thought of as a food, when ingested in pregnancy can produce adverse effects on the development of the fetus which are severe and permanent (Clarren *et al.*, 1978; Hanson *et al.*, 1978). Alcoholic women have a high risk of having abnormal infants. The characteristic abnormalities produced have become known as the fetal alcohol syndrome (FAS) recognized by French and American workers in the late 1960s and early 1970s. The features include small head size, general growth retardation, anomalies of the eyes, face, heart, joints, and external genitalia, and mental retardation. There are numerous other abnormalities involving the jaw, nasal ridge, lips, ears, as well as cleft palate and organ abnormalities (Clarren *et al.*, 1978; Streissguth *et al.*, 1978).

Both intrauterine and postnatal growth failure occur and there appears to be no catch-up growth. Neurologic abnormalities are apparent in the newborn period and include tremulousness and irritability, some due to alcohol withdrawal, but problems with fine motor coordination and retardation may persist. Abnormalities and underdevelopment of the brain have been found on postmortem examination.

The threshold level for alcohol ingestion is not known. Even moderate drinking, 1 oz/day for 1 month, in early pregnancy may result in growth retardation and abnormalities of the fetus. Steady, daily ingestions of 45 ml/day and heavy binge drinking can cause problems (Hanson *et al.*, 1978).

As for pathogenesis, confounding variables are present in alcoholism among pregnant women. Heavy smoking, use of drugs, stress, violence, and poor prenatal care often co-exist with the drinking. Also, maternal nutritional deficiencies secondary to drinking may compound the problem (Hurley, 1980b). Direct transfer of alcohol across the placenta and direct effect of alcohol on the development of the fetus is a most likely explanation (Mann *et al.*, 1975). Thus, an infant with multiple congenital anomalies, growth retardation, brain damage, and mental retardation may result from long-term ingestion of alcohol.

TABLE III
Postnatal Malnutrition: Adverse Effects

Nutrient deficit	Sequelae[a]
Calories, protein	↓ Physical growth and stunting (some degree of catch up), ↓ CNS cellular growth and myelination, ↓ mental development and attention deficits
Iron	Cognitive and behavioral disturbances, attention and memory deficits
"Breast milk deficiency" (formula feeding)	Iron-deficiency anemia, hyperelectrolytemia, hyperaminoacidemia, low initial levels of vitamin E, zinc and calcium, allergy, cognitive deficits (subtle), deciduous dental caries
Fluoride	Dental caries

[a] References in text

POSTNATAL PERIOD: MALNUTRITION (Table III)

Physical Growth

Protein-calorie malnutrition (PCM) in infancy or early childhood delays growth or even causes it to cease, if energy intake falls below a critical level (Tanner, 1978; Graham, 1972). Zinc and iodine deficiency can also have a profound retarding effect on linear growth (Hambridge, 1977; Smith, 1977). The tempo of growth is the first element of growth to be affected. A slowing-down occurs until such a time when energy intake becomes adequate (Tanner, 1978). Children subject to acute episodes of malnutrition recover, provided that the insult is not too severe or of too long duration (Barr *et al.*, 1972). Also the earlier the insult, the more likely that there will be long-term effects (Graham, 1968). With chronic undernutrition, the child does not achieve a normal adult height (Jelliffe and Jelliffe, 1981; Smith, 1977).

The nature of catch-up growth is dramatic following a period of food restriction or undernutrition (Ashworth, 1969). Compensatory growth is rapid and children have an amazing power to return to their predicted growth curve, growing at rates even above that of the expected velocities for age (Ashworth, 1969). Weight gain proceeds rapidly until the child reaches an appropriate weight for height. Growth then proceeds at a slower rate as the height and weight increase together (Davies and Parkin, 1972). The timing, severity, and nature of the insult influence the degree of deficit seen in growth, and in turn these influence the potential catch up. The earlier and more severe the malnutrition, the more stunting is likely to occur (Graham, 1968) and head circumference is likely to be below normal (Stoch and Smythe, 1967). The timing of the insult is important in relation to a critical period, the critical period being the time of greatest sensitivity of an organ to a special or specific action (McCance, 1976). Nutritional deprivation during such a critical period can cause profound growth disturbance.

A number of studies in humans have been helpful. Some children with severe PCM under a year of age never obtain normal linear growth or head circumference (Graham, 1968; Stoch and Smythe, 1967). However, Garrow and Pike (1967) show that Jamaican children who were severely malnourished in infancy did catch up, and grew even better than their siblings when observed at 2–8 years post discharge. Peruvian children hospitalized between 3 and 15 months, and then discharged to foster homes or adopting homes that were of a socioeconomic status higher than their parents, showed remarkable catch-up growth both in height and head circumference (Graham, 1972). Stoch and Smythe (1976) observed adolescents who were malnourished in infancy and noted that they caught up in height by ages 15–18.

In the U.S., studies have been done of children with severe PCM under a year of age, secondary to malformations of the gastrointestinal tract, cystic fibrosis, and coeliac disease (Barr *et al.,* 1972; Chase and Martin, 1970). After surgical correction and appropriate medical treatment, all showed complete catch-up growth, in terms of physical growth. A group of children who had cystic fibrosis and were severely malnourished under a year of age, also showed satisfactory catch up growth by 7–10 years old (Ellis and Hill, 1975). Eid (1971), in following a group of children who had failed to thrive, secondarily to chronic disease and anomalies found significant growth retardation in these children many years later and he assumed that the interaction with an unfavorable environment played a significant role in permanent growth retardation.

INFANCY: MALNUTRITION

Mental Development

Malnutrition early in life has a negative impact on cognitive and social development (Tizard, 1974). Not only is there concern with severe malnutrition but with the more subtle effects of mild–moderate nutrition. Severe PCM affects 3–5% of young children but a much greater percentage, 20–40%, suffer from mild–moderate PCM both in industrialized and developing countries (Behar, 1968). The establishment of a causal relationship between malnutrition and reduced cognitive development has not been straightforward. Several studies have produced presumptive but convincing evidence for a causal relationship in severe PCM, but little is known about lesser degrees of malnutrition (Freeman *et al.,* 1977).

Cellular Growth of the Human Brain

PCM early in life has been associated with reduction of cell number, cortical thickness, brain weight, and neuron size, and defective myelination.

There have been few studies on the effect of malnutrition in cellular growth of the brain in humans (Winick, 1976b). In infants dying with marasmus under 1 year of life, both wet and dry weight of the brain were reduced, as were total protein, RNA, total cholesterol, and phospholipids (Rosso *et al.*, 1970; Mönckeberg, 1968). Total DNA was also proportionately reduced and DNA synthesis and cell division curtailed, with resultant reduced cell number. Thus, one sees reduced cell number and size as in lower animals, especially if the insult occurs during a proliferative phase (Winick and Rosso, 1969).

Myelination is affected by malnutrition and is analogous to events seen in the pig and the rat. The number of glial cells are reduced if malnutrition occurs early, with selective reduction in concentration of certain gangliosides and dendritic branching, as well as reduced cortical thickness. However, electrophysiologic studies reveal that there is a poor correlation between structure and function (Winick, 1976b).

Whether or not there is an association between malnutrition during a critical period of development and permanant changes in brain structure and function is difficult to decide. The human situation is extremely complex and it is difficult to isolate any one factor. A combination of low socioeconomic status and infection compound the problem of malnutrition (Winick, 1976a).

Human Studies of Cognitive Function

There have been numerous prospective and retrospective studies all sharing the problem of trying to isolate malnutrition as the cause of mental deficiency vs. the effects of poverty, illness, and lack of stimulation (Tizard, 1974; World Health Organization, 1974; Winick, 1976). Retrospective studies are difficult to interpret because one cannot be certain of the criteria that were used to establish the diagnosis of malnutrition and the presence of other intervening social factors. In a Serbian study of marasmic children under 1 yr of age, a large discrepancy was found in I.Q. between the study group and the controls at a later age (Cabak and Nejdanvic, 1965). A similar study was carried out in Indonesia with similar results (Liang *et al.*, 1967). In India, children with PCM under 3 yr and controls were followed and then studied between the ages of 8–11 yr. A complete test battery consisting of perceptual, abstraction, memory, and verbal abilities were performed, with significant difference between the two groups (Champakam *et al.*, 1968). Chase and Martin (1970), in the U.S., studying children who were malnourished early in life due to surgically corrected obstructive lesions, also found a discrepancy in I.Q. between the malnourished and control groups.

The best studies in design to date are by Cravioto *et al.* (1966) in Mexico, and Klein *et al.* (1976) and Freeman *et al.* (1977) in Guatemala. The relationship of performance to diet was examined with significantly positive correlations found between birth weight and maternal height. The earlier the PCM, the more profound the abnormality, particularly with malnutrition occurring under 6 months of age.

Attentional deficits, rather than cognitive problems, have also been found in children who have been rehabilitated from malnutrition. Children given a better diet were more active, independent, demanding, appeared larger, and were treated as older children attracting more care and parental contact than malnourished siblings (Birch, 1972; Chavez *et al.*, 1974).

Iron Deficiency

Iron deficiency is recognized as the most common and universal nutritional problem (Dept. Health, Education, and Welfare, 1972). At particular risk for iron deficiency are children 6 months to 3 years of age from low socioeconomic backgrounds and of low birth weight (Owen *et al.*, 1971). Appreciating the impact of iron deficiency on children is particularly important because of possible long-term developmental effects. Nonetheless, the literature on childhood iron deficiency and behavior is small, results are inconsistent, and studies are frequently plagued by methodological inadequacies severe enough to preclude interpretation of data (Pollitt and Leibel, 1976).

Nonhematologic manifestations of iron deficiency include possible changes in CNS chemistry (Leibel *et al.*, 1979; Dallman, 1974; Oski, 1979). Because some important aspects of human brain development (e.g., glial proliferation and dendrite arborization) occur during infancy, a period when iron levels often fall, it has been suggested that infants may be at particular risk for long-term deleterious effects of iron deficiency (Leibel *et al.*, 1979). One prospective study indicates that iron-deficiency anemia during infancy is significantly related to soft neurological signs at school age, a finding that speaks to the importance of monitoring iron nutrition and understanding the effects of iron deficiency during early development (Cantwell, 1974).

Clinically, iron deficiency has been related to listlessness, irritability and lassitude at all ages (Dallman, 1979). More specifically, several studies of infants and young children suggest that iron deficiency, with or without anemia, is associated with subtle cognitive or behavioral disturbances. Taken together, the studies, although heterogeneous with respect to definitions of iron status and selection of outcome measures, appear to support the hypothesis that iron-deficient status is accompanied by problems sustaining attention to tasks. There is, in addition, some evidence that return to iron repletion brings about improved behavioral performance (Pollitt *et al.*, 1978). The literature indicates that iron deficiency or anemia is associated with nonreversible decreased developmental test scores during toddlerhood (9–26 months) (Lozoff *et al.*, 1979) and decreased but rapidly ameliorated developmental test scores at the same age period. In addition, treated toddlers become significantly more alert, responsive, and coordinated within 2 weeks of iron therapy (Oski and Honig, 1977, Honig and Oski, 1978).

In summary, the following disturbances have been documented: impairment of attention and memory control processes (some reversible) (Pollitt *et al.*, 1978); restricted attention, cognitive deficiencies, and perceptual restriction (all reversible) at ages 3–5 years (Howell, 1971); decreased motivation, learning ability, and vocabulary, but no decrease in short-term attention at age 4–5 years (Sulzer *et al.*, 1973); inattentiveness and hyperactivity at ages 6–7 years (Cantwell, 1974); decreased academic achievement, attention and perception, and increased conduct problems at ages 12–14 years (Webb and Oski, 1973). Since iron deficiency is a major and widespread problem throughout the world, these studies take on great significance and more work needs to be done in the area of cognitive and behavioral aspects of iron deficiency.

Breast Milk Deficiency

Formula feeding preparations are being used as the sole source of nutrients for the infant in the first three months of life. These are comprised mainly or wholly of partly defatted cow's milk, lactose or another carbohydrate, whey as the protein base, vegetable fats, and addition of vitamins A, C, D, and iron, with some containing zinc and copper. Use of these formulas for infant feeding has been called "the largest *in vivo* uncontrolled experiment in infant feeding" by Hambraeus, although reputed to be "just like mother's milk" (Hambraeus, 1977). The modified cow's milk formulas have been humanized in some important ways, but dehumanized in others. Complete knowledge about human milk is not at hand, so formula cannot resemble human milk completely. As new ingredients are discovered, these are added by the formula companies.

Human milk is not a static commodity. What model of breast milk should be taken as the standard? Breast milk differs in composition in the mother of the preterm vs. the term infant (Gross *et al.*, 1980). At birth, the milk varies from morning to evening, from the beginning of a feed to the end of a feed, as the fat content rises. Also, the composition of the milk, particularly of the fat, and vitamins reflect the maternal dietary intake. Some of the nutrients are remarkably stable. Formula manufacturers also add extra nutrients to provide a margin of safety. Protein content usually exceeds that of human milk, in which the protein gradually decreases over the first few months of life (Hambraeus, 1977). With the widespread use of soy-based milks, one does not know the long-term effects of using such a product.

Widdowson, in England, analyzed 32 formulas used in Europe for infant feeding and found tremendous differences in nutrient composition. The protein content varied from 10 to 29 g/100 g of powder, the fat from 8.5 to 31 g/100 g, and carbohydrate from 37 to 67 g/100 g to 526 calories/100 g. All of the formulas contained more protein, sodium, and calcium than did breast milk (Widdowson, 1976). Whether an excess of these nutrients will affect the chemical composition of the infant depends not only on the absorption from the gut but on other factors as well. Can the body excrete the excess in the urine or is there another route for metabolism? If the formula-fed infant's body composition differs from that of the breast-fed infant, as it does in fat composition, does this have any functional significance, particularly in CNS function?

Adverse Effects of the Use of Infant Formula

In the wake of the widespread abandonment of breast-feeding in favor of formula-feeding, infantile malnutrition and diarrheal disease have been noted in impoverished communities with poor environmental sanitation as well as increased mobility in affluent communities (Jelliffe, 1977; Cunningham, 1979). In addition, among the protected infants of the well-to-do, a pattern of specific metabolic derangements is now recognized among some formula-fed infants. These conditions may be due to some inherent intolerance of the neonate to cow's milk, a metabolic overload of certain nutrients, an imbalance of nutrients in the formula, the faulty preparation of the food, or to overfeeding. It is now becoming apparent that iatrogenic conditions are occurring among neonates and young infants and are caused by feeding practices (Jelliffe, 1977).

Aminoacidemia

The ability of premature infants to metabolize tyrosine and phenylalanine appears to be slight, and in cow's milk the concentration of these nutrients is 3–4 times higher than in human milk. The low level of the amino acid cystine in cow's milk equally increases metabolic difficulties as the enzyme, cystothianase, required to convert methionine into cystine, is not present in the brain or liver of premature infants (Sturman *et al.*, 1970). Such immature infants to whom high-protein cow's milk formulas are administered may develop an aminoacidemia that manifests itself with high levels of tyrosine in the blood serum that may be 10–20 times higher than in adults. The infants lose weight because of general apathy and poor appetite (Jelliffe and Jelliffe, 1976). Transient neonatal tyrosinemia (TNT) that occurs in up to 80% of bottle-fed neonates is believed by some to diminish the I.Q. and may result in learning disabilities in later years (Wong *et al.*, 1967; Menkes *et al.*, 1972; Mamunes *et al.*, 1976).

Acrodermatitis Enteropathica (Danbolt-Class Syndrome)

This painful syndrome is characterized by diarrhea and a symmetrical vesicopustular rash localized around the body orifices and the extremities. This rare disease that only occurs in formula-fed infants, if untreated, is fatal (Moynahan, 1974). Recent work has suggested that zinc deficiency may precipitate this syndrome (cow's milk has a lower copper/zinc ratio than human milk). In human colostrum the zinc content is higher than it is in later milk, and in mice, zinc deficiency is precipitated if colostrum is withheld from them

(Nishamura, 1953). In human infants, treatment with zinc has proved successful in the treatment of this syndrome (Neldner and Hambridge, 1975).

Allergy to Bovine Protein

Feeding of cow's milk or a cow's milk-based formula in a vulnerable child, one with a strong family history of allergy, can enhance the chances of clinical allergy manifestations to cow's milk. These can range from inconsequential signs and symptoms to shock and even death (Goldman and Heiner, 1977). Allergy to cow's milk protein in infants and young children is among the most common allergic conditions in the U.S. Goldman has estimated the rate to be from 10,000 to 30,000 young children affected each year. Beta-lactalbumin, casein, alpha lactalbumin, and bovine serum albumin appear to be the main antigens (Goldman, 1976). Even the possibility of intrauterine sensitization exists, as in the case of wheat gluten, where antibodies have been demonstrated in the amniotic fluid (Kuroume *et al.*, 1976). During the vulnerable first 6 months of an infant's life the intestinal mucosa is permeable to foreign proteins and cow's milk and cow's milk-based formula can become highly antigenic and should be avoided (Goldman and Heiner, 1977).

Avoidance of cow's milk products through breast-feeding has been attempted to delay, mitigate, and hopefully avoid the emergence of cow's milk allergy that can cause gastrointestinal, respiratory, or skin problems or even catastrophic anaphylactic shock-type reactions (Hamburger and Orgel, 1976). A number of retrospective studies have demonstrated that breast-feeding can serve as a prophylactic measure for cow's milk and other allergies (Wittig *et al.*, 1978). Prospective studies of breast-fed infants have recently been undertaken in Scandinavia, England, and the U.S. and show that fewer of these infants developed allergic diseases. Those that did develop them had milder cases than non-breast fed infants (Jakobsson and Lindberg, 1979; Saarinen *et al.*, 1979; Matthew *et al.*, 1977).

Chandra (1979) followed the newborn siblings of allergic or atopic children who were raised on breast milk only for the first 6 weeks. Eczema, wheezing, IgE, and complement activation were greater in the cow's milk-fed controls. After a milk challenge hemoagglutinating antibody levels to B-lactalbumin were significantly lower compared to formula-fed age-matched controls. Kaufman and Frick (1976) followed infants of allergic parents in prospective fashion (94 infants) and found that a significantly greater number developed allergy and atopic dermatitis in the formula or cow's milk-fed group than in the breast-fed group.

Iron-Deficiency Anemia

A neonate, if full term and of normal birth weight, has optimal iron as circulating hemoglobin and if fully breast-fed, usually requires no supplementation with iron until the introduction of solid foods between the ages of

4–6 months (McMillan *et al.*, 1977; Dallman, 1979). Over the years much controversy has arisen over the needs for supplementation of formulas with minerals.

Iron assimilation appears to be dependent on the amount of copper in the diet, which in turn may depend on the intake of zinc (Picciano and Guthrie, 1976). In maternal milk both zinc and copper concentrations are higher than in cow's milk. The bioavailability of iron is greater in breast-fed infants as iron is more readily absorbed from the gastrointestinal tract (Dallman, 1979, Bullen *et al.*, 1972). Studies have indicated that iron-deficiency anemia is less prevalent in breast-fed than bottle-fed babies (Oski and Perkins, 1975). Lower protein and calcium concentration in breast milk produce less nonsoluble iron complexes that are poorly absorbed. In breast milk higher lactose content promotes iron absorption as does lactoferrin (Woodruff, 1977). It has been suggested that iron supplementation of nursing infants might not only saturate but inactivate the natural antibacterial properties of lactoferrin found in breast milk (Bullen *et al.*, 1972). Microhemorrhages in the intestines secondary to a "foreign protein enteropathy appear to be associated with cow's milk ingestion (Woodruff, 1977). The need for iron-fortified formulae for bottle-fed infants and iron-rich weaning foods, both processed and homemade, for all infants seems necessary, in view of the high incidence of iron-deficiency anemia in preschool children.

Nonhematologic manifestations of iron deficiency include changes in tissue enzyme levels, retarded physical growth, skin and mucous membrane changes, gastrointestinal abnormalities, and changes in muscle function (Dallman, 1979). Long-term effects of iron deficiency also include decreased cell-mediated immunity (Joynson *et al.*, 1972) and decreased leukocyte antibacterial function (Chandra, 1973), in addition to the attentional and learning problems discussed above.

Neonatal Hypocalcemia

Infants fed cow's milk formula, although it contains a much higher calcium content than human milk, may develop late neonatal hypocalcemia, with infantile tetany, convulsions, and hyperphosphatemia, a condition which may occasionally prove fatal (Barltrop and Oppe, 1970). Less affected children have been shown to subsequently develop hypoplasia of the dental enamel (Stimmler *et al.*, 1973). The serum calcium level is found to be depressed because of a high phosphate intake and blood magnesium levels may also be low. This disease is precipitated by poor absorption of calcium due to its loss in the stools as salts of fatty acid, and to the low level of lactose which causes decreased calcium absorption (Condon *et al.*, 1970). In the breast-fed infant, no apparent difficulties occur with calcium absorption, again because of lower protein and higher lactose concentrations than cow's milk. Some workers have suggested that in addition to the high phosphorus level of cow's milk formula, the occurrence of this disease may be influenced by low maternal vitamin D and calcium intakes during pregnancy. The use of "demineralized" milks modified to resemble the human milk content of calcium, phosphorus and sodium do reduce the incidence of hypocalcemia but should be used with caution because of the possible loss of other nutrients as well (Jelliffe and Jelliffe, 1976).

Diseases Due to Overconcentrated Formula Feeding

In infants fed unmodified cow's milk formulas in the early months of life, chronic hypernatremia occurs due to the increased metabolic load that the high solute content of cow's milk (high protein and mineral content) imposed on their immature renal function. Davies and Saunders (1973) have reported levels of 40% mg of urea in healthy bottle-fed babies among 75–88% of healthy infants investigated. In a consecutive series of 40 infants dying unexpectedly at home, analysis of the vitreous humor of the eye indicated the presence of hypernatremia either with or without uremia in a percentage of these (Emery *et al.*, 1974). The effects of continuously raised urea levels in normal infants are not fully known, but the dangers of overload are always present when mothers who have absolute control of the feeding of their infants may easily increase the strength of the feed under the mistaken impression that this will benefit the child's nutritional status (Jelliffe, 1977).

Many workers have stressed the need to educate mothers in the proper use of cow's milk formula, the mineral content of which should be known by the pediatrician. For example, mothers have mistakenly used undiluted evaporated milk or the concentrated formula, or have used more scoops of powdered milk than recommended or insufficient water to mix with the powder. Others believe that the crying child who may actually be thirsty from hypernatremia requires more food because of hunger, and will thus increase the infant's biochemical overload (Taitz and Byers, 1972). Permanent brain damage, gangrene, and disseminated intravascular coagulation have been noted among children suffering from hyperosmolarity and hypernatremic dehydration (Rosenbloom and Sills, 1975; Comay and Karabus, 1975).

Formula Manufacturers' "Lacunae" Disease

Manufacturers of cow's milk formulas have been mainly concerned with attempting to market a product resembling, as closely as possible, the proximate principles of breast milk. Not unexpectedly, manifestations of disease have occurred among infants fed formulas deficient in certain other nutrients (Jelliffe and Jelliffe, 1976).

Pyridoxine Deficiency

In the 1950s convulsions were noted in infants fed a formula in which the B_6 content was below normal requirements for the normal metabolism of the CNS. Permanent damage due to the convulsions is not well documented (Coursin, 1954), although there have been lawsuits to this effect.

Linoleic Acid Deficiency

Skin lesions and poor growth (Hansen *et al.*, 1958) have been noted in infants fed milk-based formulas in which this fatty acid was markedly deficient. Disappearance of lesions occurs with a diet which provides 1–2% of the calories in the form of linoleic acid (Cuthbertson, 1976).

Vitamin E Deficiency

In premature infants levels of alpha tocopherol are low and a formula containing high levels of unsaturated fatty acids will increase the need for this vitamin in these deficient infants. Human colostrum has high concentrations of vitamin E. Subsequently, both hemolytic anemia and edema may develop if adequate amounts of vitamin E are not supplied to the premature (Horwitt, 1976).

Hypochloremia

Within the past few years a syndrome occurring in young infants, consisting of vomiting, hypochloremic alkalosis, dehydration, collapse, shock, and death, was attributed to feeding with a certain soy protein hydrolysate. The manufacturers, well meaning, heeded the recommendations of American Academy of Pediatrics to lower the sodium chloride content of the formula and lowered the sodium chloride too drastically, to the point of causing the above described syndrome with disastrous results (Garin *et al.*, 1979).

Other Deficiencies

Other deficiency syndromes may appear in the future as the process of further "humanizing" milk will promote manipulation of certain nutrients at the probable expense of micronutrients, the presence and importance of which may as yet not be appreciated. As research into the composition of breast milk continues, nutrients are constantly being discovered and some of these have been incorporated into existing formula. One such nutrient is taurine, an amino acid which promotes optimal infant growth and is involved in brain development, particularly in the newborn period (Sturman *et al.*, 1977; Rassin *et al.*, 1979). Clinical trials with taurine-supplemented formula are currently being conducted. Zinc supplementation of commercial formula was stimulated by the fact that retardation and suboptimal growth result from zinc deficiency (Hambridge, 1977).

Nursing Bottle Syndrome

In this syndrome rampant caries of the deciduous teeth, especially the upper incisors, occurs in some bottle-fed infants (Kroll and Stone, 1967). This type of dental decay is attributed to prolonged nocturnal or "nap" feeding. The main fluids promoting caries in infants and young children are whole milk (which contains 4% lactose), as well as the sweetening agents in formula-feeding or acid-containing liquids (e.g. fruit juices) or any sweetened drinks given to infants and toddlers through the medium of the nursing bottle. Rubber nipples or pacifiers covered with sweetening agents are also implicated. Stagnation of milky and sweetening fluids that promote an acid medium are the causal agents of this disease that is difficult to treat; extraction of teeth is often the only solution (Nizel, 1977).

Learning and Cognitive Function

Two large epidemiologic studies in the U.S. and England have shown that use of high protein formula in infancy has been linked to deficits in cognitive development (Pollitt and Lewis, 1980). Menkes *et al.* (1972) examined the early feeding histories of children with learning disorders in a private practice setting. Thirteen percent with learning disorders had been breast-fed compared to 47.2% of controls who were breast-fed. A study in Great Britain found small but significant differences in nonverbal, reading, and mathematic skills in 8–15-year-old children, favoring those who were breast-fed as infants. Socioeconomic status was adequately controlled for, and I.Q. scores of breast-fed children were slightly higher than in never breast-fed infants (Rodgers, 1978). A U.S. study based on a nationally representative sample showed small but significant differences in I.Q. and school achievement scores (Edwards and Grossman, 1980), also favoring breast-fed children.

It is postulated that the accumulation of certain amino acids that may be toxic to the brain may result from excessive protein intake. An alternate explanation to postulating damaging effects of high protein may be the positive effects on cognitive development, promoted by the psychologic closeness and attachment that may be more intense in a nursing mother and her infant (Klaus and Kennell, 1970). Numerous workers have found more favorable patterns of adaptation of the mother–child relationship in regard to crying differences and development in bottle-fed vs. breast-fed infants (Bernal and Richards, 1970; Thomas *et al.*, 1972).

The above studies suggest that breast-fed infants compared to bottle-fed infants show a small advantage in cognitive tests and educational attainment indicators. Whether the difference is due to biochemical composition of the milks or because of associated child-rearing and child-nurturing variables due to breast-feeding is not clear.

Cretinism and Breast Milk

Mitigation of cretinism by breast-feeding for 10 months was reported by Bode *et al.* (1978). Physical growth proceeded normally and the damaging effects on the neurologic system and mental development were minimized but not prevented as long as the infant was breast-fed. This observation prompted the analysis of several breast milk samples of commonly used infant milks for thyroid hormones and revealed that commercial formulas and pasteurized cow's milk contained negligible amounts of T3 and T4, but that human breast milk contained sufficient thyroid hormone to deliver daily 1–3 μg of T3 and 3–9 μg of T4 to nursing infants. However, the thyroid hormone levels are suboptimal but sufficient to ameliorate some of the harmful somatic growth and CNS effects of cretinism.

POSTNATAL: OVERNUTRITION (Table IV)

Infant–Child Nutrition and Cardiovascular Health

Adult obesity, hypertension, and atherosclerotic coronary heart disease (CHD) are believed to have their origins early in life (Kannel and Dawber, 1972). Risk factors can be identified and acted upon early (Strasser, 1981; Stamler, 1978). Polemics and emotion abound, particularly in regard to the nutritional aspects of cardiovascular disease. Abundant data exist, but are not necessarily demonstrative of causal relationships, but rather of associative

TABLE IV
Postnatal Overnutrition

Nutrient excess	Sequelae[a]
Calories	Obesity, hypertension
Sodium	Hyperosmolarity, hypertension[b]
Cholesterol, saturated fats	Hyperlipidemia[b]
Refined sugar	Dental caries
Iron overload	Liver damage and cirrhosis
Hypervitaminosis	
A	↑ Intracranial pressure skin, bone, joint lesions
D	Metastatic calcification with renal damage
E	? Hepatic damage
C	↑ Renal stones
Fluoride	Fluorosis and dental mottling

[a] References in text.
[b] In susceptible persons.

relationships, linking early diet to adult cardiovascular disease. Long-term longitudinal studies are needed to ascertain the adult outcomes of infant- and child-feeding practices and other risk-factor reduction in childhood (Marmot, 1979).

One school of thought, particularly endorsed by the American Heart Association (1978) and the U.S. Senate Select Committee on Nutrition and Human Needs (1977), is recommending that action be taken at present, that there is enough knowledge to recommend a "prudent diet" and reduction of risk factors early in life (Stamler, 1980a). Another school of thought says that there should not be any intervention at this point, that there is inadequate knowledge and that more harm than good may be done, particularly with recent reports of cancer in groups of people on long-term polyunsaturated fatty acid diets (Epstein, 1977; Rose, 1980). The American public have taken matters into their own hands in regard to diet and exercise, and a sizeable reduction in deaths from CHD has occurred in the past decade in the U.S. (Cooper *et al.,* 1979). Generally good agreement exists as to the necessity to control blood pressure and obesity, but there is much more disagreement concerning dietary modifications (Strasser, 1981).

Obesity

By overfeeding infants with formula containing an overconcentration of protein, calories, carbohydrates, and early feeding of solids, infants are easily receiving nutrients far in excess of what is needed for growth, maintenance, and activity levels (Taitz, 1977a). Is the excessive intake occurring during a period considered vulnerable for hyperplasia of adipocytes and resultant increased fat-cell number and size (Knittle, 1972; Brook, 1972)? The interval from the last months of fetal growth through the second year of life has been considered a sensitive period (Hahn, 1972; Knittle and Hirsch, 1968). Are these infants being programmed for a life of obesity? Is this hyperplasia of adipocytes genetically predetermined in familial obesity? If not under genetic control, are infants being programmed to habitually overeat by being overfed? Infants under 2–3 months of age are very defenseless against overfeeding. Fomon (1974) has shown that it takes consistent overfeeding with several ounces before an infant will refuse feeds or vomit. Also infants less than 2 months of age, if given excessively concentrated formulas do not appear to reduce their intake volumes to compensate for this overconcentrated, excessive amount of formula (Fomon, 1974). Thirsting mechanisms are not available for correction of the overconcentrated feed. Instead of being given needed water in response to their cries, infants are often given more formula (Taitz, 1977b).

What is the effect of a *laissez-faire* weight-gain policy for pregnant women? The pendulum has swung away from severe restriction of weight gain in the vain hope of avoiding toxemia. It is recognized that at least a 24 lb weight gain is physiologic and the storage of 9 kg of fat is needed for fetal energy and lactational energy needs (Hytten and Thomson, 1970). As mentioned earlier, excessive maternal weight gain, over 40 lb, during pregnancy represents a risk to the neonate who is apt to be fatter and large for gestational age (Udall *et al.,* 1978; Whitelaw, 1977; Simpson *et al.,* 1975).

The next question to be answered is whether infants who are obese at birth have an increased likelihood of remaining obese at subsequent ages. Gampel (1965) followed 331 English mothers and their newborn for 1 yr and found no correlation between skin folds thickness of the mothers and their infants or between any other variables by 1 year. However, Frisch *et al.* (1975) examined the weight and length of 1786 infants, using existing growth data from the Minnesota portion of the NIH Collaborative Perinatal Project. Ninety-six infants were classified as obese at birth (weight-to-height ratio greater than 95th percentile). At 4 years of age, 74 of the 96 obese infants were again measured and 26% had a weight-to-height ratio greater than the 90th percentile and 58% had ratios greater than the 70th percentile. When the total number of children judged to be obese at 4 years of age were remeasured at 7 years of age, 78% were found to be fatter than the 90th percentile. The data suggest that a high percent of neonates, judged to be obese at birth, retain their heavy physique at least until age 7 years.

There is a high correlation between obesity in childhood and in adolescence (Heald and Hollander, 1965) and later life. Some studies support this relationship, others refute this (Charney *et al.,* 1975; Lloyd and Wolff, 1976; Weil, 1977).

Obesity is a very important risk factor for the development of hypertension (Chiang *et al.,* 1969). In adults and children hypertension is more frequent among obese than nonobese individuals (Dustan, 1980). Longitudinal studies have shown that weight gain in early life predisposes to hypertension (Lauer *et al.,* 1975). A cogent argument for preventing obesity in infancy and early childhood is the recognition that it can produce hypertension.

Hypertension is the leading risk factor in CHD and other degenerative diseases of the cardiovascular system (Stamler, 1980a; Holland *et al.,* 1980). Children have been studied to much less extent than adults. What is the degree of risk that childhood obesity holds for later development of hypertension? In the Muscatine studies, 4000 school children were examined. For those whose body weights were in the upper deciles, 28.6% had systolic and 28.4% had diastolic pressures greater than the 90th percentile (Lauer *et al.,* 1975). Levine *et al.* (1976) studied 2058 high school students, 5.9% of whom were hypertensive. None of the traditional causes were found except that 64% were obese, with 75% of these with a positive family history of obesity. Barta and Rosta (1961) examined 100 obese 3–14-year-olds. Seventy percent had arterial pressures higher than expected for their age and moderate elevations were found in 32% of the cases. Court (1974) found that 114 out of 209 obese children had diastolic blood pressures greater than 90 mm of mercury and systolic blood pressures greater than 135 mm of mercury. Londe *et al.* (1971) studied 74 hypertensive children and in 69 found no cause except that 53%

were overweight in contrast to the group of 74 age-matched normotensive normal-weight children.

The mechanism of how obesity causes hypertension is not very well understood, with very little work done in animals or humans. Possible mechanisms have implicated increased cardiac output that is ejected into an arterial tree of normal size, increased blood volume, and excessive sodium that is an accompaniment of excessive caloric intake and alteration in pressure receptors. Weight and blood pressure correlations in children are in need of further study, particularly observation of interrelationships of weight, diet, hypertension, and emergence of atherosclerotic disease in adulthood (Dustan, 1980).

A tracking effect, whereby blood pressure at a given age is correlated with blood pressure at a subsequent age, has been found to be present in children as well as in adults (Holland and Beresford, 1975). Several workers have found a correlation between weight and a child's blood pressure and it is reasonable to assume that obese and overweight children will have raised blood pressure levels that may continue to adulthood and predispose to CHD (Holland *et al.,* 1980). The need for a 30 year study is apparent, but hardly feasible. However, proxy or intermediate outcomes can be sought with programs of weight reduction in children to see if their blood pressure actually is reduced upon weight loss. One study carried out in England (Leeder *et al.,* 1976) followed 568 infants born in 1963. The families were studied as well for blood pressure and it was found that the weight percentiles and parental blood pressure levels influenced the blood pressure levels in their children.

Another set of risk factors of importance are serum lipid levels. Cholesterol and triglyceride levels, if elevated, may be reduced by weight loss. This is particularly true for triglycerides (Stamler, 1980a; Court *et al.,* 1974).

Obesity can be prevented in infancy and childhood (Pisacano *et al.,* 1978), but the question is whether or not this will prevent obesity in adulthood. The outcome of treating obesity is very disconcerting and discouraging and investigators feel that any attempt at prevention is worthwhile, the earlier in life the better (Neumann, 1980).

Sodium and Hypertension

The relationship between dietary sodium intake and human hypertension has long been suspected (Dahl, 1972; Page, 1980). The question of whether or not excessive sodium intakes in infancy and early childhood can predispose the individual to hypertension in adolescence and adulthood has been raised by many. The limitation of sodium intake in infants and children has been advocated in the prevention of hypertension in adulthood.

Population studies have demonstrated familial clusters of blood pressure levels (Johnson *et al.,* 1965; Heller *et al.,* 1980). Also, young children, if tracked throughout childhood, tend to adhere to the same blood pressure percentiles over time (Feinleib *et al.,* 1980). A strong genetic influence appears to interact with a shared familial environment of risk factors (Lauer, 1978). Extensive work with genetically-obese rats has shown that in the absence of sodium intake they remain normotensive, but upon exposure to dietary salt they become hypertensive (Dahl and Tassinari, 1963). These rats also have sensitive periods during which sodium intake results in hypertension. In dramatic fashion, this strain of rats was fed commercial infant food in the 1960s and because of the high sodium levels, the rats became hypertensive (Dahl, 1968). Extrapolation by Dahl of the above experiments to humans was in part responsible for the lowering of sodium content of infant foods.

Whether or not there is a critical level of sodium intake at which an individual develops hypertension is not known (Freis, 1976). There have been some 20 population studies where a striking absence of hypertension was associated with a traditional lifestyle and high potassium and low sodium intakes. With acculturation and diet change, hypertension began to develop (Page, 1980).

The relationship of sodium intake (as measured by urinary salt excretion) to hypertension has been examined in a number of studies with a variety of results (Gleibermann, 1973). In a study of 6-month-old infants, Schachter (1980) found no relationship on a population-wide basis. However, for infants whose mothers had blood pressure above the median there was a positive relationship found between blood pressure and sodium intake in the infants. There is probably a marked variability in the genetic susceptibility of individuals to sodium in regard to hypertension (Page, 1980).

Taitz (1977b) also examined sodium intake in children, finding elevated levels associated with elevated blood pressure percentiles. Others point out that the sodium intake of American infants is 3 times greater than the established requirements. The disquieting question is raised as to whether or not the American infant diet with its high sodium content predisposes the child to hypertension in adult life, particularly in families with known hypertension.

Estimated requirements for sodium in the 0–4-month-old infant is 2.5 meq/day and 2.1 meq/day for the 4–12-month-old infant. Breast milk has 0.9 meq/100 kcal, cow's milk 3.3 meq/100 kcal and most infant formula concentrations have sodium concentrations somewhere in between (Fomon, 1974). Infants whose major intake is formula or cow's milk ingest generous amounts of sodium. Many infant foods marketed in the U.S. have over 10 meq sodium/100 kcal and many mothers add salt to improve what they perceive as palatability for the infant (Fomon *et al.,* 1970).

In 1970 the infant food companies moved to eliminate added salt in processed infant foods in response to the Food Protection Committee of the Food and Nutrition Board. The National Academy of Science Ad Hoc Committee on Sodium and Hypertension recommended that infant food manufacturers be required to restrict addition of salt to baby food to an upper limit of 0.25% and stop adding salt to cereals altogether (Filer, 1980). Elimination of salt caused no change in acceptability of the foods by the infants as tested by Fomon in 4–7-month-old infants (Fomon *et al.,* 1970). In 1977, the addition of any salt to processed infant foods was discontinued by all infant food companies (Filer, 1980). In the home preparation of infant food, Kerr *et al.* (1978) demonstrated that homemade infant food may have more salt than commercially processed food, up to 0.4% added salt. Salt had been added in previous preparation of commercially canned or frozen food originally intended for adult consumption.

In light of the above information it would appear prudent to limit sodium intake, with no added salt and avoidance of obviously salted foods in the diets of children who come from families with a positive history for hypertension or who themselves have blood pressures over the 95th percentile. Should severe salt restriction become necessary they would not be so "addicted" to salted food.

Atherosclerotic Coronary Heart Disease

Increasing concern has been expressed that atherosclerotic CHD has its origins in early childhood (Kannel and Dawber, 1972). What role does the diet of a young child play in the emergence of this disease? Identification of risk factors and intervention as early as possible is being recommended in some pediatric circles (Nora, 1980). There has been a definite decrease of CHD in the past decade, perhaps due to the fact that Americans have reduced their smoking, changed their diets to that of lower saturated fat and cholesterol, and have been exercising more (Cooper *et al.*, 1979). Even more dramatic decreases have been seen in Finland due to energetic reversal and removal of risk factors (Puska *et al.*, 1980). An emotional controversy exists as to the role of diet in the prevention of CHD, but genetic factors, hyperlipidemia, smoking, hypertension, and obesity have all been readily accepted as known risk factors (Stamler, 1980a). A firm epidemiologic basis for the prevention of CHD exists, although causation has not been firmly established (Marmot, 1979). The role of major risk factors is appreciated, and these factors are detectable and preventable. Over half of the incidence of CHD is now being explained by hyperlipidemia, hypertension, and smoking (Pooling Project Research Group, 1978).

Prevention of CHD has now become regarded as a pediatric concern (Kannel and Dawber, 1972). The most compelling evidence that the process starts early in life is based on a study that showed that 45% of young men under 30, victims of the Korean War, had already advanced atherosclerotic plaques and changes in their coronary arteries (Enos *et al.*, 1953). This process had obviously started earlier in life with no clinical manifestations or they would not have been inducted into the army. It cannot be certain exactly when the process started. The universal finding of fatty streaking in the aortas of 3-month-olds is probably not the forerunner of the atherosclerotic plaque (McGill, 1968; McMillan, 1973). Some congenitally acquired disorders such as familial hypercholesterolemia (Type II), with abnormal low density lipoproteins (LDL) show premature atherosclerosis (Glueck *et al.*, 1971; Kwiterovich *et al.*, 1973). In cystic fibrosis with fat malabsorption, aorta and coronary blood vessels are remarkably free of atherosclerotic changes at postmortem (Holman *et al.*, 1959).

In prospective studies, individuals with elevated cholesterol levels are more prone to develop CHD and mortality appears to vary in proportion to the average serum cholesterol concentrations obtained (Norum, 1978; Kannel and Gordon, 1970; Stamler, 1978). Further evidence linking saturated fat and cholesterol in the diet to serum cholesterol levels is supported by cross-cultural population studies (Kannel and Gordon, 1970; Keys, 1970; Stamler, 1980; Connor, 1980). Strong circumstantial support exists for the importance of diet in cholesterol, lipids, beta-lipoproteins, and LDL blood levels and their pathogenesis of human disease.

In children, although there are marked individual variations in lipid profiles, there is significant tracking noted for serum and total cholesterol and beta-lipoprotein. Those children who initially had low levels seem to remain in a lower percentile, and those who initially had high levels remain in a higher percentile for a long period of time (Frerichs *et al.*, 1976; Nora, 1980). Determination of these levels, then, can be predictive of the future. A dramatic rise of lipids and lipoprotein occur at year one and again at puberty.

In the Bogalusa studies, diets and snacks were quantified for lipid and cholesterol content (Frerichs *et al.*, 1976). Although there was an overall low correlation as in the Framingham studies (Kannel and Gordon, 1968), by selecting children with high and low cholesterol levels, a significant relationship was found between diet and serum lipids. Diets of infants at 1 yr showed a significant association with saturated fat, calories, sucrose, beta-lipoprotein, and total serum cholesterol.

Tsang *et al.* (1974) studied the effect of diet on neonates with normal cord cholesterols and normal parents, neonates with hypercholesterolemia and hypercholesterolemic parents, and those infants with hypercholesterolemia of normal parents. There were no changes in the infants with high cholesterol intake over a year's time, but in the infants with hypercholesterolemia and normal parents, the levels were intermediate, and in those hypercholesterolemic infants with hypercholesterolemic parents levels were highest at 1 yr Strong genetic influence appears to be present.

In a study of 103 affluent white children, divided into low, mid, and high deciles for cholesterol levels, no correlation was found between the level of serum cholesterol and the mean daily cholesterol, saturated fat, and salt intake (Weidman *et al.*, 1978). Frank *et al.* (1978) also showed a variable response to dietary cholesterol. Whether on a high- or low-fat diet, no differences in cholesterol level were noted. However, if one grouped the children into high and low serum cholesterol groups, correlations were noted between serum levels and dietary intake. McGandy *et al.* (1972) found they could lower cholesterol levels by 16% in students, by dietary means. It appeared that individuals with the highest levels were the most responsive. Hitchcock and Gracey (1977), in a study of mothers and children, found that saturated fat and to a lesser degree dietary cholesterol, affected serum cholesterol, beta-lipoprotein and LDL levels.

In intrapopulation studies, there are individual differences in response to diet that can be lost in the "mean response", and over- and under-responders are missed (Weidman, 1980). Many studies show that when there is a mean decrease in serum cholesterol, with the introduction of a low cholesterol and saturated fat diet, the individuals with high initial serum cholesterols show the greatest fall, and people with normal or near normal levels respond least or not at all. The response of serum cholesterol level to the amount of saturated fat and cholesterol in the diet is the result of a combination of factors which include the contribution of cholesterol and fat in the diet to total body cholesterol. The diet plays less of a role in the individual who is hyporesponsive (Weidman, 1980).

In summary, from the above evidence, it appears that among large population studies or even smaller groups, there is an association between dietary cholesterol intake, serum cholesterol, and CHD. However, when individuals are examined only one time, as in a cross-sectional study, there is poor or little association between current diet, cholesterol and serum levels, and the incidence of CHD. If individuals are examined under rigorously controlled conditions, when eating different diets with varying cholesterol intake, most will show a modest change in cholesterol related to dietary cholesterol.

"Cholesterol Challenge Theory" and CHD

In infancy, the cholesterol level is related to the type of feeding. Breast milk and cow's milk infants have the highest cholesterol levels and those fed with commercial formula that have no cholesterol, the lowest levels (Friedman and Goldberg, 1975; Fomon *et al.,* 1978; Darmady *et al.,* 1972). There is no justification for tampering in the first year of life with feedings except in type II familial hypercholesterolemia, which is basically an inherited genetic disorder. It is exactly in the areas of the world where prolonged breast-feeding is carried on that the lowest rates of CHD are seen. These are mainly in developing areas where less meat, saturated fat, cholesterol, and salt are eaten, and people are physically active and there is little obesity.

Reiser's "cholesterol challenge" hypothesis supports breast-feeding in the first year of life (Reiser and Sidelman, 1972). Although well documented in rats as well as in swine, the evidence is far from clear-cut in humans and perhaps the right studies have not been carried out to date. Reiser and Sidelman (1972) found increased survival of animals fed cholesterol in the newborn period and in early life. He felt that the animals were better able to cope with dietary cholesterol as adult animals because of the induction of enzymes that could handle cholesterol. This capacity was decreased in cholesterol-deprived newborns.

In humans the evidence is inconclusive. Glueck *et al.* (1972) found no difference in serum cholesterol levels in 14-month-old infants fed cholesterol formula or breast milk. Infants studied by Friedman and Goldberg (1975), fed breast milk or formula for the first 4–6 months of life (one group receiving 4–5 mg cholesterol/8 oz in the formula-fed, vs. those receiving 25–52 mg cholesterol/8 oz in the breast milk-fed) showed that at a year of age the latter group had cholesterol levels of 130 mg% and those fed breast milk had levels of 147 mg%. Criticisms of this study are that an inadequate challenge of cholesterol was given at 1 yr and that all the infants after their initial test feeding period for 4–6 months were placed on skim milk and the "prudent" American diet, and that the subjects were highly selected. It was felt that a year was not long enough to follow these children. Hodgson *et al.* (1976) looked at 97 school children 7–12 years old, having known their feeding histories for the first 3 months of life. The formula-fed group had the lowest level of cholesterol, the breast-fed group had the highest level of cholesterol, but the individual intakes showed a wide variation.

On the positive side, Ziegler and Fomon (1980) followed 326 children from birth and were able at 8 years of age to obtain cholesterol levels on 293 of these children as well as complete feeding histories. Their complete lifetime dict intakes were known exactly, as these children were in a longitudinal growth study. For the males who were breast-fed a barely significant but positive correlation was found between serum cholesterol in infancy and at 8 years of age with a correlation coefficient of $r = 0.47$. One could see a correlation of serum cholesterols at infancy and at age 8, such as the tracking that was seen in the Bogalusa studies. Furthermore, in an unpublished study by Valadian in 1979 (Cunningham, 1981) on a 30 year follow-up of adults who were followed for early growth studies and whose exact newborn and early infancy feeding studies were known, the breast-fed group appears to have lower serum cholesterols levels in adulthood.

Should the Reiser hypothesis be laid to rest? One would think not. The studies done to date are perhaps not adequate and one should really carry out a functional test requiring that the handling of a cholesterol load be studied upon an adequate challenge. It is again of interest that in parts of the world such as developing countries where breast-feeding is carried on for 1–2 years, atherosclerotic CHD is strikingly lower than in the U.S., but as mentioned, these are countries of low meat consumption, hard physical work, absence of widespread obesity, and lower salt intakes (Keys, 1970), as well as early mortality in childhood from other causes.

There is general agreement among pediatricians not to interfere with eating habits of infants in the first year of life and especially to encourage breast-feeding (Nutrition Committee, 1972). Should the postweaning child have a modified diet in the hope of warding off risk factors? Can these prudent diets be safely introduced? There is no evidence that slight modifications in the diet as suggested by the American Heart Association prevent difficulties in children (Friedman and Goldberg, 1978). For those children in whom there is a marked hyperlipidemia and whose families have a positive family history of early CHD, there should be some modification of the diet since the athersclerotic disease process appears to start early. Although there is discordance in studies where there are high cholesterol levels and no CHD, multiple factors operate with interaction of genetic and environmental responses in the causation of this disease (Kwiterovich, 1980).

There are questions about the safety of the prudent diet in that cereal grains, plants, and nuts may furnish insufficient high-quality protein, and that the diets are low in energy and of great bulk. It is felt by some that the dietary recommendations may offer a false sense of security (Mann, 1977; Olsen, 1978). Will an increase in polyunsaturated fatty acids (PUFA) and phytosterols not be deleterious in causing gall stones, cancer, and excessive calcium? In adult studies an increased emergence of cancer among senior citizens on PUFA diets has been noted (Rose, 1974). A diet sizeably reduced in saturated fat to 7–8% of total calories, cholesterol under 300 mg, fat reduced to 30% of the total calories, a moderate intake of PUFA (8–10% of the calories), more use of complex carbohydrates, and foods of lower caloric density with more calories coming from grains, legumes, fruits, and vegetables and more protein from fish and fowl, and a deemphasis on baked goods and visible fatty products would "do some good" (U.S. Senate Select Committee on Nutrition and Human Needs, 1977). Some of the adult intervention studies such as the Multiple Risk Factor Intervention Trial (MR. FIT) study (1977) have shown the efficacy of lowering dietary fat and increasing exercise. The overwhelming feeling is that whatever is done to children should be done very moderately and conservatively. The real question is whether or not the current American diet for young children is a safe diet in terms of cardiovascular health.

In summary, the risks and benefits of modifying the diet conservatively after the first year in children show that there is no harm (Friedman and Goldberg, 1978). However, the direct evidence that reducing the cholesterol and fat intake early in life will decrease CHD in adult life is lacking. Such a study would take some 30 years. The dilemma now is whether to act in the light of epidemiologic knowledge or wait for definitive evidence. Since serum cholesterol has a strong positive and independent association with the probability of developing CHD at least among adults, the inference is to reduce cholesterol intake. Presumably the greatest effect would be if these measures were started in childhood, particularly in high-risk families, although other factors no doubt affect cholesterol and the risk of CHD. A *modest reduction* is suggested. The position is somewhat akin to the problems of linking salt intake to hypertension. If the child comes from a high-risk family, modifications in the diet should be taken. The use of a modified prudent diet for all children after 1 year that includes modest dietary changes probably is wise (Stamler, 1980b). It appears there are those individuals that can handle cholesterol and take any amount in the diet with no effect on serum cholesterol levels and there are those that cannot (Kwiterovich, 1980). Until one has a definitive test to pick out hyper- and hyporesponders, perhaps the most prudent thing is to modify the diet.

The implementation of modified behavioral patterns leading to avoidance of overeating, reducing salt intake, modifying the fat intake to 35% of total energy calories, routine physical activity, and no smoking, is the stumbling block (Stamler, 1980a). The roles of fiber, vitamin C, yogurt plant sterols, vitamin B_6, and vegetable proteins have been seen to lower cholesterol and one must keep an open mind for further evidence in this regard.

Dental Caries

Dental caries is one of the most prevalent diseases among children, affecting close to 100% of the population. By age 10, more than 80% of the children have caries of their permanent teeth and by the time the permanent teeth are erupted, the average American child has about 6 decayed, missing, or filled teeth. Half of 2-year-olds and two-thirds of 3-year-olds already have dental caries (Nizel, 1977).

Carbohydrates

The cariogenicity of various sugars has been rated as following in decreasing order: Sucrose, glucose, maltose, fructose, sorbitol and xylitol. Sucrose is the worst cariogenic agent of all for the human being (Nizel, 1977). Attempts to replace sucrose by glucose or fructose in foods have been made, hoping that the microorganisms might adapt to the new sugar. Sorbitol has also been used in chewing gum (Marthaier, 1967).

Epidemiologic studies of the relation of sucrose consumption and dental caries in humans have been carried out. It is of interest that during the Second World War when sugar was not available in Norway, caries decreased (Toverud *et al.,* 1961). The sticky form of sucrose that adheres to the teeth is the worst form. Cariogenic streptococcus in the mouth allows an anerobic metabolism of sucrose to fructose, from which long-chain polymers of glucose (dextrans) or fructose (levans) form, contributing to formation of dental plaque. Gelatinous sticky plaque protects the acids produced by cariogenic microbes from the buffering effect of saliva. Fructose enters the cell and glycogen is stored intracellularly. Demineralization of enamel is caused by the bacterial acids (Nizel, 1977). Casein-containing products such as cheese and milk that also contain calcium and phosphorus may be protective. Casein may reduce enamel solubility. Fibrous foods such as apple, celery, and carrots have a detergent and mechanical cleansing action (Nizel, 1977).

Fluoride

Incorporation of adequate amounts of fluoride in the outer layers of the tooth enamel increases resistance to caries (Newbrun, 1975). Amounts of 0.7–1.2 mg/liter of fluoride throughout life reduces caries by 50–60% in the permanent teeth and slightly less in the primary teeth, probably due to the relatively low intake of preschoolers where water is actually fluoridated (Dean, 1942). Less than half of the United States receives desirable intakes of fluoride from any source. Infants may benefit from added fluoride (Aasenden and Peebles, 1974). Fluoride in Europe is added to milk, salt, flour, and water (Miller and Barmes, 1981). The mechanism of fluoride is that it effects conversion of enamel, a mineral hydroxy appetite, to fluoro-appetite with reduction in the solubility of the enamel in acid. Fluoride promotes the rate of remineralization of enamel and it may depress the metabolism of bacteria and interfere with plaque formation (Nizel, 1977). Fluoride in human milk is low, 0.05 mg/liter and it does not seem to increase in areas of water fluoridation. Cow's milk has 0.03–0.1 mg/liter, and milk formulas vary; in some areas it is high and some it is low (Dirk, 1974).

There is a considerable margin of safety before one sees fluorosis, particularly if the dose of fluoride in infancy is restricted to 0.5 mg/day. Ingestion of 4.4 ppm will cause moderate fluorosis in 46% and severe fluorosis in 18% of individuals (Newbrun, 1975). Where water is not fluoridated, the daily intake should not exceed 1.6 ppm. If the water is not fluoridated, fluoride can be given once a day or as a single dose of topical fluoride occasionally. Preparations come in a variety of concentrations and must be adjusted to the fluoride content of the water supply. Therefore, attention must be paid to the form and concentration of fluoride before prescribing it (Nutrition Committee, 1979). The effectiveness ratio in caries prevention by fluoridation, elimination of sweetened beverages, particularly via infant bottles, and avoidance of sucrose-laden snacks is very substantial.

Vitamin and Iron Excess

Hypervitaminosis

Hypervitaminosis may occur from purposive self-medication, or accidental overdosage due to confusion as to concentrations delivered by various vitamin preparations. Toxicity is most commonly seen with the fat-soluble vitamins, A and D. Even megadoses of most vitamins have not led to long-term sequelae, but rather short-term toxicity that is rapidly reversible upon cessation of ingestion of the given vitamins. Vitamin C, the B-complex vitamins, and folic acid, mainly the water soluble vitamins, fall into this latter category (Anderson and Fomon, 1974). However, megadoses of vitamin C in some individuals can precipitate renal and bladder stones with the potential for renal damage (McLaren, 1979; Stein *et al.*, 1976).

Vitamin A toxicity, although alarming and dramatic in its clinical manisfestations, with CNS symptoms of increased intracranial pressure such as papillaedema, diplopia, headache, and vomiting, is completely reversible upon cessation of vitamin A ingestion. Hair, skin, and joint changes also occur. Chronic overdosage may also lead to osteoporosis and premature epiphyseal closure (Persson *et al.*, 1965; Furman, 1973; Nutrition Committee, 1971).

Unfortunately, hypervitaminosis D can lead to long-term and severe sequelae. Metastatic calcifications occur, involving the soft tissues, including the kidney. Failure to thrive, irritability, hypercalcemia, severe renal damage, and chronic renal failure can ensue, which is not reversible (McLaren, 1979; Anderson and Fomon, 1974; Nutrition Committee, 1973, 1976).

Not much is known about the long-term sequelae of ingesting vitamin E in excessive amounts. There have been some reports of liver toxicity and damage. Large doses of vitamin E (menadione) in the neonate have caused anemia, hyperbilirubinemia, and even kernicterus with brain damage (Gross and Melborn, 1972).

In infancy, mild overdosage for as little as 1 month can lead to vitamin A and D toxicity and careful instructions must be given in the administration of vitamin preparations, particularly in infancy and childhood (McLaren, 1979). The ease of obtaining high potency vitamins over the counter, particularly in health food stores, aggravates the problem.

Iron Overload

Iron overload, through dietary ingestion or therapeutic overload, is relatively rare in children, but adult manifestations of this condition may begin silently in children. The excess iron is deposited predominately in the reticuloendothelial system and to a lesser extent in parenchymal cells such as hepatocytes.

In areas of the world such as Ethiopia the eating of a staple grain called tef provides 20–100 mg iron/100 g of grain eaten. The iron comes from contaminations of the grain with iron-rich soil. Older children may ingest up to 750 mg/day (Hofvander, 1968). In South Africa the excess iron comes from cook pots which are made of iron. When these pots are used also for grain fermentation in beer-making, even more iron is absorbed by the food. An insidious process of liver parenchymal damage starts early in life culminating in liver cirrhosis by mid-adulthood (Lynch *et al.*, 1974). In therapeutic overdosage ingestion of massive doses or repeat blood transfusions can lead to parenchymal hemosiderosis and liver cirrhosis (Choudhry and Verma, 1971).

CONCLUDING REMARKS

A wide range of nutrients, but mainly under- and overingestion of calories and protein, have been explored from the point of view of adverse effects in early life and possible long-term sequelae. Some nutritional stresses may cause no apparent short-term abnormalities but express themselves later in life—some cause immediate as well as long-term sequelae. The former situation is the more difficult to identify and evaluate unless some biochemical, physical, or functional marker is available for determining the group at risk. Cohort studies are expensive but effective in this regard, but case-control studies are more feasible and practical.

Innumerable factors and unfavorable events affect an individual and it is not always easy to demonstrate clear-cut cause and effect of the adverse effects of certain nutrients. Unanswered questions are numerous:

To what extent does a nutritionally deprived fetus or child "make good" their deficit and realize their full potential?

How does one identify the child who is at "silent risk" for overt sequelae at a later age? For example, what markers exist for the hypertension- or CHD-prone child?

What are the criteria for intervention from a nutritional point of view in light of incomplete knowledge of cause and effect of certain factors and poor adult outcome? Where does lack of harm of a given measure justify diet modification for an uncertain benefit?

What are the limits and definitions of truly critical sensitive periods in growth and development in the preconception, prenatal, and postnatal periods of life in terms of retarded or defective growth and function or irreversible damage? Insults occurring during critical or vulnerable times often have more serious, less reversible, and more severe sequelae both in situations of nutrient excess or deficiency.

How can genetic endowment be evaluated in determining different levels of risk for a condition or problem in a given individual in determining outcome? What markers or screening tools can be developed to identify those at risk? To date, aside from some specific screening tests for relatively rare conditions, the most inexpensive and potent instrument is a thoughtful and thorough family history of selected morbidity and mortality. It may not be practical to subject all individuals to the same preventive programs or measures for certain serious multifactorial conditions or diseases such as hypertension, obesity, and atherosclerosis.

What is the role of socioeconomic cultural circumstances contributing to recovery from nutritional insult? Exciting research has pointed to the positive role of infant and young child stimulation in recovery from the adverse effects of malnutrition, cognitive function, and social behavior and learning (Winick, 1976a; Cravioto *et al.*, 1966).

In conclusion, long-term sequelae and adult health outcomes as they relate from early adverse nutritional experiences in the preconceptual, prenatal, postnatal, and early childhood periods are determined by the nature and the timing of the insult, the genetic endowment of the individual, and the sociocultural economic situation during the period of insult and recovery. These factors, in interactive and synergistic fashion, shape the outcome of adult health status (Wharton, 1976). Prevention and intervention must also then be multifactorial in approach.

REFERENCES

Aasenden, R., and Peebles, T. C., 1974, Effects of fluoride supplementation from birth on human deciduous and permanent teeth, *Arch. Oral Biol.* **19:**321n326.

Adams, S. O., Barr, G. D., and Huenemann, R. L., 1978, Effect of nutritional supplementation in pregnancy: Effect on outcome in pregnancy, *J. Am. Diet. Assoc.* **72:**144–147.

Allen, J. D., and Brown, J. K., 1968, Maternal phenylketonuria and foetal brain damage. An attempt at prevention by dietary control, in: *Some Recent Advances in Inborn Errors of Metabolism* (K. S. Holt and V. P. Coffee, eds.), E. & E. Livingstone, Edinburgh.

Als, H., Lester, B., and Brazelton, B., 1979, Dynamics of the behavioral organization of the premature infant, a theoretical perspective, in: *Infants Born at Risk: Behavior and Development* (T. M. Field, A. Miller, A. M. Sostek, S. Goldberg and H. H. Shuman, eds.), pp. 173–192, SP. Medical and Scientific Books, New York.

Als, H., Tronick, E., Admason, L., and Brazelton, B., 1976, The behavior of the full-term but underweight newborn infant, *Dev. Med. Child Neurol.* **18:**590–602.

American Heart Association Committee on Nutrition, 1978, *Diet and coronary disease,* New York American Heart Association.

Anderson, T. A., and Fomon, S., 1974, Vitamins, in: *Infant Nutrition* (S. Foman, ed.), pp. 209–244, W. B. Saunders, Philadelphia.

Ashworth, A., 1969, Growth rates in children recovering from protein–calorie malnutrition, *Br. J. Nutr.* **23:**835–845.

Barltrop, D., and Oppe, T. E., 1970, Dietary factors in neonatal homeostasis, *Lancet* **2:**1333–1335.

Barr, D. G. D., Schmerling, D. H., and Prader, A., 1972, Catch-up growth in malnutrition studied in celiac disease after institution of gluten-free diet, *Pediatr. Res.* **6:**521.

Barta, L., and Rosta, J., 1961, Juvenile obesity: Investigations of carbohydrate and fat metabolism, *Ann. Pediatr.* **196:**189–203.

Beck, G. J., and Van den Berg, B. J., 1975, The relationship of the rate of intrauterine growth of low-birth weight infants to later growth, *J. Pediatr.* **86:**504–511.

Behar, M. 1968, Prevalence of malnutrition among pre-school children in developing countries, in: *Malnutrition, Learning and Behavior* (N. S. Scrimshaw and J. Gordon, eds.), p. 30, M.I.T. Press, Cambridge.

Bernal, J. and Richards, M. P. M., 1970, The effects of bottle and breast feeding on infant development, *J. Psychosom. Res.* **14:**247–252.

Bernhardt, I. B., and Dorsey, D. J., 1974, Hypervitaminosis A and congenital renal anomalies in a human infant, *Obstet. Gynecol.* **43:**750–755.

Birch, H. G., 1972, Malnutrition, learning and intelligence, *Am. J. Public Health* **62:**773–784.

Bode, H. H., Vanjonack, W. J., and Crawford, J. D., 1978, Mitigation of cretinism by breast-feeding, *Pediatrics* **62:**13–16.

Brasel, J. A., 1974, Cellular changes in intrauterine malnutrition in: *Nutrition and Fetal Development* (M. Winick, ed.), pp. 13–26, John Wiley, New York.

Brook, C. G. D., 1972, Evidence for a sensitive period in adipose cell replication in man, *Lancet* **11:**624–627.

Bullen, J. J., Rogers, H. J., and Leigh, L., 1972, Iron-binding proteins in milk and resistance to *E. coli* infection in infants, *Br. Med. J.* **1:**69–75.

Burke, B. S., Beal, V. A., Kirkwood, S. B., and Stuart, H. C., 1943, The influence of nutrition during pregnancy upon the condition of the infant at birth, *J. Nutr.* **26:**569–583.

Butler, N., 1974, Early postnatal consequences of fetal malnutrition, in: *Nutrition and Fetal Development* (M. Winick, ed.) pp. 173–178, John Wiley, New York.

Cabak, V., and Najdanvic, R., 1965, Effect of under-nutrition in early life on physical and mental development, *Arch. Dis. Child.,* **40:**532–534.

Cantwell, R. J., 1974, The long term neurological sequelae of anemia in infancy, *Pediatr. Res.* **8:**342.

Carswell, F., Kerr, M. J., and Hutchison, J. H., 1970, Cngenital goitre and hypothyroidism produced by maternal ingestion of iodides, *Lancet* **1:**1241–1243.

Champankan, S., Srikantia, S. G., and Gopalan, C., 1968, Kwashiorkor and mental development, *Am. J. Clin. Nutr.* **21:**844–852.

Chandra, R. K., 1973, Reduced bactericidal capacity of polymorphs in iron deficiency, *Arch. Dis. Child.* **48:**864–866.

Chandra, R. K., 1975, Fetal malnutrition and post-natal immuno-competence, *Am. J. Dis. Child.* **129**:450–454.

Chandra, R. K., 1979, Prospective studies of the effect of breast feeding on incidence of infection and allergy, *Acta. Paediatr. Scand.* **68**:691–694.

Charney, E., Goodman, H. C., McBride, M. et al., 1975, Childhood antecedents of adult obesity: do chubby infants become obese adults? *N. Engl. J. Med.* **295**:6–9.

Chase, H. P., and Martin, H. P., 1970, Undernutrition and child development, *N. Engl. J. Med.* **282**:933–938.

Chavez, A., Martinez, C., and Yaschine, T., 1974, The importance of nutrition and stimuli on child mental and social development, in: *Symposia of the Swedish Nutrition Foundation XII, Early Malnutrition and Mental Development* (J. Cravioto, L. Hambraeus and B. Vahlquist, eds.) p. 211, Almquist & Wiksell, Uppsala.

Chiang, B. N., Perlman, L. V., and Epstein, F. H., 1969, Overweight and hypertension, *Circulation* **39**:403–421.

Choudhry, V. P., and Verma, I. C., 1971, Hemosiderosis due to excessive intake of iron, *Indian J. Pediatr.* **38**:306.

Clarren, S. K., Alvord, E. C., Jr., Sumi, S. M., Streissguth, A. P., and Smith, D. W., 1978, Brain malformation related to prenatal exposure to alcohol, *J. Pediatr.* **92**:64–67.

Comay, S. C., and Karabus, C. D., 1975, Peripheral gangrene in hypernatremic dehydration of infancy, *Arch. Dis. Child.* **50**:616–619.

Condon, J. R., Nassim, J. R., and Millard, F. J. C., 1970, Calcium and phosphorus metabolism in relation to lactose tolerance, *Lancet* **1**:1027–1029.

Connor, W. E., 1980, Cross-cultural studies of diet and plasma lipids and lipoproteins, in: *Childhood Prevention of Atherosclerosis and Hypertension* (R. M. Lauer and R. B. Shakelle, eds.), pp. 99–111, Raven Press, New York.

Cooper, R., Stamler, J., Dyer, A., and Garside, D., 1979, The decline in mortality from coronary heart disease, U.S.A., *J. Chronic Dis.* **31**:709–720.

Coursin, D. B., 1954, Convulsive seizures in infants with pyridoxine deficient diets, *J. Am. Med. Assoc.* **154**:406–408.

Court, M. M., Dunlop, M., and Leanard, R. F., 1974, Plasma lipid values in childhood obesity, *Aust. Paediatr. J.* **10**:10–14.

Cravioto, J., DeLicordie, E. R., and Birch, H. G., 1966, Nutrition, growth and neurointe grative development: An experimental and ecologic study, *Pediatrics* **38**:319–372.

Cunningham, A. S., 1979, Morbidity in breast-fed and artificially fed infants, II, *J. Pediatr.* **95**:685–689.

Cunningham, A. S., 1981, Breast-feeding and morbidity in industrialized countries: An update, in: *Advances in International Maternal and Child Health* (D. B. Jelliffe and E. F. P. Jelliffe, eds:) Vol. 1, pp. 128–168, Oxford Press, London.

Cuthbertson, W. F. J., 1976, Essential fatty acid requirements in infancy, *Am. J. Clin. Nutr.* **29**:559–568.

Dept. Health, Education, and Welfare, 1972,: Ten-State Nutritional Survey, *DHEW* No. (HSM) 72-8132, Washington, D.C.

Dahl, K. K., and Tassinari, L., 1963, High salt content of western infant's diet: Possible relationship to hypertension in the adult, *Nature,* **198**:1204–1205.

Dahl, L. K., 1968, Salt in processed baby foods, *Am. J. Clin. Nutr.* **21**:787–792.

Dahl, L. K., 1972, Salt and hypertension, *Am. J. Clin. Nutr.* **25**:231–244.

Dallman, P. R., 1974, Tissue effects of iron deficiency, in: *Iron in Biochemistry and Medicine* (A. Jacobs and M. Worwood, eds.), pp. 437–475, Academic Press, New York.

Dallman, P. R., 1979, Nutritional anemias, in: *Pediatric Nutrition Handbook,* pp. 263–274, Amer. Acad. Pediatrics, Evanston, Illinois.

Darmady, J. M., Fosbrooke, A. S., and Lloyd, J. K., 1972, Prospective study of serum cholesterol levels during the first year of life, *Br. Med. J.* **2**:685–688.

Davies, D. P., and Saunders, R., 1973, Blood urea: Normal values in early infancy relating to feeding practices, *Arch. Dis. Child.* **48**:563–567.

Davies, T. F., and Parkin, J. M., 1972, Catch-up growth following childhood malnutrition, *East Afr. Med. J.* **49:**672–686.
Dean, H. T., 1942, Domestic waters and dental caries, *Public Health Reports,* **57:**1155–1179.
Dirk, O. B., 1974, Total and free ionic fluoride in human and cow's milk as determined by gas-liquid chromatography and the fluoride electrode, *Caries Res.* **8:**181–186.
Dobbing, J., 1970, Undernutrition and the developing brains, *Am. J. Dis. Child.* **120:**411–415.
Douglas, V., and Peters, K., 1979, in: Attention and Cognitive Development, Plenum, New York.
Dustan, H. P., 1980, Obesity and hypertension, in: *Childhood Prevention of Atherosclerosis and Hypertension* (R. M. Lauer and R. B. Shekelle, eds.), pp. 305–312, Raven Press, New York.
Edwards, L. N., and Grossman, M., 1980, The relationship between children's health and intellectual development, in: *Health: What is it Worth* (S. Mushkin, ed.), Pergamon Press, New York.
Eid, E. E., 1971, A follow-up study of physical growth following failure to thrive with special reference to a critical period in the first year of life, *Acta Paediatr. Scand.* **60:**39–48.
Ellis, C. E., and Hill, D. E., 1975, Growth, intelligence and school performance in children with cystic fibrosis who have had an episode of malnutrition during infancy, *J. Pediatr.* **87:**565–568.
Emery, J. L., Swift, P. G. T., and Worthy, E., 1974, Hypernatremia and uraemia in infancy, *Arch. Dis. Child.* **49:**686–692.
Enos, W. F., Holmes, R. H., and Beyer, J., 1953, Coronary disease among U.S. soldiers killed in action in Korea, *J. Am. Med. Assoc.* **152:**1090–1093.
Epstein, F. H., 1977, Preventive trials and the "diet-heart" question: Wait for results or act now, *Atherosclerosis* **26:**515–523.
Fancourt, R., 1976, Follow-up study of small-for-date babies, *Br. Med. J.* **1:**1435–1437.
Feinleib, M., Garrison, R. J., and Havlik, R. J., 1980, Environmental and genetic factors affecting the distribution of blood pressure in children, in: *Childhood Prevention of Atherosclerosis and Hypertension* (R. M. Lauer and R. B. Shekelle, eds.) pp. 271–280, Raven Press, New York.
Ferguson, A. C., 1978, Prolonged impairment of cellular immunity in children with intrauterine growth retardation, *J. Pediatr.* **93:**52–56.
Fierro-Benitez, R., Ramirez, I., Garces, J., Jaramillo, C., Moncayo, F., and Stanbury, J. B., 1974, The clinical pattern of cretinism as seen in highland Ecuador, *Am. J. Clin. Nutr.* **27:**531–543.
Filer, L. J., 1980, Salt in infant foods, *Nutr. Rev.* **29:**27–30.
Fitzhardinge, P. M., and Steven, E. M., 1972, The small-for-date infant. I. Later growth patterns, *Pediatrics* **49:**671–681.
Fomon, S. J., 1974, Voluntary food intake and its regulation, in: *Infant Nutrition* (2nd ed., pp. 20–33, W. B. Saunders, Philadelphia.
Fomon, S. J., Ziegler, E. E., Filer, L. J., Anderson, T. A., Edwards, B. B., and Nelson, S. E., 1978, Growth and serum chemistry of normal breastfed infants, *Acta Paediatr. Scand.* **Suppl. 273.**
Fomon, S. J., Thomas, L. N., and Filer, L. J., 1970, Acceptance of unsalted strained foods by normal infants, *J. Pediatr.* **76:**242–246.
Ford, R. C., and Berman, J. L., 1977, Phenylalanine metabolism and intellectual functioning among carriers of phenylketonuria and hyperphenyl-alaninaemia, *Lancet* **1:**767–771.
Frances-William, J., and Davies, P. A., 1974, Very low birth weight and later intelligence, *Dev. Med. Child Neurol.* **16:**709–728.
Frank, G. C., Berenson, G. S., and Webber, L. S., 1978, Dietary studies and the relationship of diet to cardiovascular disease risk factor variables in 10 year old children—The Bogalusa heart study. *Am. J. Clin. Nutr.* **31:**328–340.
Freeman, H. E., Klein, N. E., Kagan, J., and Yarbrough, C., 1977, Relations between nutrition and cognition in rural Guatemala, *Am J. Public Health* **67:**233–239.
Freis, E. D., 1976, Salt, volume and the prevention of hypertension, *Circulation,* **53:**589–595.
Frerichs, R. R., Srinivasan, S. R., Webber, L. S., and Berenson, G. S., 1976, Serum cholesterol and triglyceride levels in 3,446 children from a bi-racial community—The Bogalusa heart study, *Circulation* **54:**302–309.

Friedman, G., and Goldberg, S. J., 1975, Concurrent and subsequent serum cholesterols of breast-fed and formula-fed infants, *Am. J. Clin. Nutr.* **28:**42–45.

Friedman, G., and Goldberg, S. J., 1978, An evaluation of the safety of a low-saturated fat, low cholesterol diet beginning in infancy, *Pediatrics,* **58:**655–657.

Frisancho, A. R., Klayman, J. E., and Matos, J., 1977, Influence of maternal nutritional status on prenatal growth in a Peruvian urban population, *Am. J. Phys. Anthropol.* **46:**265–274.

Frisch, R. O., Bilek, M. S., and Ulstrom, R., 1975, Obesity and leaness at birth and their relationship to body habitus in later childhood, *Pediatrics,* **56:**521–527.

Furman, K. I., 1973, Acute hypervitaminosis A in an adult, *Am. J. Clin. Nutr.* **26:**575–577.

Gampel, B., 1965, The relation of skinfold thickness in the neonate to sex, length of gestation, size at birth and maternal skinfold, *Hum. Biol.* **37:**29–37.

Garin, E., Geary, D., and Richard, G., 1979, Soybean formula (NeoMullsoy): Metabolic alkalosis in infancy, *J. Pediatr.* **95:**985–987.

Garrow, J. S., and Pike, M. C., 1967, The long-term prognosis of severe infantile malnutrition, *Lancet* **1:**1–3.

Gleibermann, L., 1973, Blood pressure and dietary salt in human populations, *Ecol. Food Nutr.* **2:**143–156.

Glueck, C. J., Heckman, F., Schoenfeld, M., Steiner, P., and Pearce, W., 1971, Neonatal familial type II hyperlipoproteinemia: Cord blood cholesterol in 1800 births, *Metabolism* **20:**597–608.

Glueck, C. J., Tsang, R., and Balistreri, W., 1972, Plasma and dietary cholesterol in infancy: Effects of early low or moderate dietary cholesterol on subsequent response to increased dietary cholesterol, *Metabolism* **21:**1181–1192.

Goldman, A. S., and Heiner, D. C., 1977, Clinical aspects of food sensitivity: Diagnosis and management, *Pediatr. Clin. North Am.* **24:**133–140.

Graham, G. G., 1968, The later growth of malnourished infants: Effects of age, severity and subsequent diet, in: *Calorie Deficiencies and Protein Deficiencies* (R. A. McCance and E. M. Widdowson, eds.), Churchill, London.

Graham, G. G., 1972, Environmental factors affecting the growth of children, *Am. J. Clin. Nutr.* **25:**1184–1188.

Graham, G. G., and Adrianzen, T., 1972, Late "catch-up" growth after severe infantile malnutrition, *Johns Hopkins Med. J.* **131:**204.

Green, H. G., Gareis, F. J., Shepard, T. H., and Kelley, V. C., 1971, Cretinism associated with maternal iodide I 131 therapy during pregnancy, *Am. J. Dis. Child.* **122:**247–249.

Gross, R. L., Reid, J. V. O., Newberne, P. M., Burgess, B., Marston, R., and Hift, W., 1975, Depressed cell-mediated immunity in megalo-blastic anemia due to folic acid deficiency, *Am. J. Clin. Nutr* **28:**225–232.

Gross, S., and Melborn, D. K., 1972, Vitamin E, red cell lipids and red cell stability in prematurity, *Ann. N.Y. Acad. Sci.* **203:**141.

Gros, S. J., David, R. J., Bauman, L., and Tomarelli, R. M., 1980, Nutritional composition of milk produced by mothers delivered pre-term, *J. Pediatr.* **96:**641–644.

Hahn, P. V., 1972, Lipid metabolism and nutrition in the prenatal and postnatal period, in: *Nutrition and Development* (M. Winick, ed.), pp. 99–134, John Wiley, New York.

Hambraeus, Leif, 1977, Proprietary milk versus human breast milk in infant feeding, *Pediatr. Clin. North Am.* **24:**17–36.

Hambridge, K. M., 1977, The role of zinc and other trace metals in pediatric nutrition and health, *Pediatr. Clin. North Am.* **24:**95–106.

Hamburger, R. N., and Orgel, H. A., 1976, The prophylaxis of allergy in infants debate, *Pediatr. Res.* **10:**387.

Hansen, A. E., Haggard, M. E., Boelshe, A. N., Adams, D. J. D., and Weise, H. F., 1958, Essential fatty acids in infant nutrition III—Clinical manifestations of linoleic acid deficiency, *J. Nutr.* **66:**565–576.

Hanson, J. W., Streissguth, A. P., and Smith, D. W., 1978, The effects of moderate alcohol consumption during pregnancy on fetal growth and morphogenesis, *J. Pediatr.* **92:**457–460.

Haworth, J. C., Ford, J. D., and Younoszai, M. K., 1969, Effect of galactose toxicity on growth of the rat fetus and brain, *Pediatr. Res.* **3**:441–447.

Heald, F. P., and Hollander, R. J., 1965, The relationship between obesity in adolescence and early growth, *J. Pediatr.* **67**:35–38.

Heller, R. F., Robinson, N., and Peart, W. S., 1980, Value of blood pressure measurement in relatives of hypertensive patients, *Lancet* **1**:1206–1208.

Higgins, A., 1972, Workshop: Nutritional supplementation and the outcome of pregnancy, National Academy of Sciences/National Research Council Washington D.C.

Hitchcock, N. E., and Gracey, M., 1977, Diet and serum cholesterol, *Arch. Dis. Child.* **52**:790–793.

Hodgson, P. A., Ellefson, R. D., Elveback, L. R., Harris, L. E., Nelson, R. A., and Weidman, W., 1976, Comparison of serum cholesterol in children fed high, moderate, or low cholesterol milk diets during neonatal period, *Metabolism* **25**:739–746.

Hofvander, Y., 1968, Hematological investigations in Ethiopia with special reference to high iron intake, *Acta Med. Scand. Suppl.* **494**:7–46.

Holland, W. W., and Beresford, S. A. A., 1975, in: *Epidemiology and Control of Hypertension* (O. Paul, ed.), pp. 375–386, Stratton Medical Book Corp., Miami.

Holland, W. W., Chinn, S., and Wainwright, A., 1980, Weight and blood pressure, in: *Childhood Prevention of Atherosclerosis and Hypertension* (R. M. Lauer and R. B. Shekelle, eds.), pp. 331–341, Raven Press, New York.

Holman, R. L., Blanc, W. A., and Anderson, D., 1959, Decreased aortic atherosclerosis in cystic fibrosis of the pancreas, *Pediatrics* **24**:34–39.

Honig, A. S., and Oski, F. A., 1978, Developmental scores of iron deficient infants and the effects of therapy, *Inf. Behav. Dev.* **1**:168–178.

Horwitt, M. K., 1976, Vitamin E: A re-examination, *Am. J. Clin. Nutr.* **29**:569–578.

Howell, D., 1971, Summary Proceedings of workshop of Food and Nutrition Board, National Academy of Science, Washington, D.C.

Hurley, L. S., 1980a, Nutritional influences on embryonic and fetal development, in: *Developmental Nutrition* pp. 65–74, Prentice Hall, Englewood Cliffs, New Jersey.

Hurley, L. S., 1980b, Alcohol, in *Developmental Nutrition,* pp. 228–233, Prentice Hall, Englewood Cliffs, New Jersey.

Hutchings, D. E., Gobbon, J., and Kaufman, M. A., 1973, Maternal vitamin A excess during the early fetal period: Effects on learning and development in the offspring, *Dev. Psychobiol.* **6**:445–459.

Hytten, F. E., and Thomson, A. M., 1970, Maternal physiological adjustments, in: *Committee on Maternal Nutrition,* Food and Nutrition Board, NRC, NAS, U.S. Government Printing Office, Washington, D.C..

Jakobsson, I., and Lindberg, T., 1979, A prospective study of cow's milk protein intolerance in Swedish infants, *Acta Paediatr. Scand.* **68**:853–859.

Jelliffe, D. B., and Jelliffe, E. F. P., 1976, Biochemical considerations, in: *Human Milk in the Modern World* pp.26–58, Oxford Univ. Press, London.

Jelliffe, D. B., and Jelliffe, E. F. P., 1981, *Assessment of Nutritional Status of the Community* 2nd ed., Oxford Press, London (In Press).

Jelliffe, E. F. P., 1977, Infant feeding practices: associated isotrogenic diseases, *Pediatr. Clin. North Am.* **24**:49–62.

Johnson, B. C. et al, 1965, Distributions and familial studies of blood pressure and serum cholestorol levels in a total community, *J. Chronic Dis.,* **18**:147–160.

Joynson, D. H. M., Jacobs, A., Walker, D. M., and Dolby, A. E., 1972, Defect of cell-mediated immunity inpatients with iron-deficiency anemia, *Lancet,* **2**:1058–1059.

Kannel, W. B., and Dawber, R., 1972, Atherosclerosis as a pediatric problem, *J. Pediatr.* **80**:544–555.

Kannel, W. B., and Gordon, T. (eds.) 1970, *The Framingham Study: An Epidemiological Investigation of Cardiovascular Disease,* Sect. 24 and 25, Government Printing Office, Washington D.C.

Kaufman, H. S., and Frick, O. L., 1976, The development of allergy in infants of allergic parents: A prospective study concerning the rule of heredity, *Ann. Allergy* **37**:410–415.

Keogh, B., and Margolis, J., 1976, Learn to labor and learn to wait: Attentional problems of children with learning disabilities, *J. Learn. Disabil.* **9:**276–286.

Kerr, C. M., Reisinger, K. S., and Plankey, F. W., 1978, Sodium concentration of homemade baby foods, *Pediatrics* **62:**331–335.

Kerr, G. R., Chamove, A. S., Harlow, H. F., and Waisman, H. A., 1968, Fetal PkU: The effect of maternal hyperphenylalaninemia during pregnancy in the rhesus monkey, *Pediatrics* **42:**27–36.

Keys, A. (ed.), 1970, Coronary heart disease in seven countries, *Circulation,* **41(Suppl. (1):**1–211.

Klaus, M. H., and Kennell, J. H., 1970, Mothers separated from their newborn infants, *Pediatr. Clin. North Am.* **17:**4–1015-1037.

Klein, R. E., Arenales, P., Delgado, H., Engle, P. L., Guzman, G. et al, 1976, Effect of maternal nutrition on fetal growth and infant development. *PAHO Bull.* **10:**301–306.

Knittle, J. L., 1972, Obesity in childhood: A problem in adipose tissue cellular development, *J. Pediatr.* **81:**1048–1059.

Knittle, J. L., and Hirsch, J., 1968, Effect of early nutrition on the development of rat epididymal fat pads to cellularity and metabolism, *J. Clin. Invest.* **47:**2091–2098.

Kopp, C. B., and Parmalee, A. H., 1979, in: *Handbook of Infant Development,* (J. Ofosley, ed.), Wiley, New York.

Kroll, R. G., and Stone, J. H., 1967, Nocturnal bottle-feeding as a contributory cause of rampant dental caries in the infant and young child, *J. Dent. Child.* **34:**454–459.

Kuroume, T., Oguri, M., Matsumura, T., Iwasaki, I., and Yuzuru, K., 1976, Milk sensitivity and soy bean sensitivity in the production of eczematous manifestations in breast-fed infants with particular reference to intrauterine sensitization, *Ann. Allergy* **76:**41–46.

Kwiterovich, P. O., 1980, Can an effective fat-modified diet be safely recommended after weaning for infants and children in general? in: *Childhood Prevention of Atherosclerosis and Hypertension* (R. M. Lauer and R. B. Shakelle, eds.), pp. 375–382, Raven Press, New York.

Kwiterovich, P. O., Levy, R. I., and Fredrickson, D. S., 1973, Neonatal diagnosis of familial type II hyperlipoproteinemia, *Lancet* **1:**118–122.

Lauer, R. M., 1978, Blood pressure and its significance in childhood, *Postgrad. Med. J.* **54:**206–210.

Lauer, R. M., Connor, W. E., Leaverton, P. E., Reiter, M. A., and Clarke, W. R., 1975, Coronary heart disease risk factors in school children: The Muscatine study, *J. Pediatr.* **86:**697–706.

Lechtig, A., Habicht, J. P., Delgado, H., Klein, R. E., Yarbrough, C., and Matorell, R., 1975, Influence of maternal nutrition on birth weight, *Am. J. Clin. Nutr.* **28:**1223–1233.

Leeder, S. R., Corkhill, R., Irwig, L. M., and Holland, W. W., 1976, Influence of family factors on the incidence of lower respiratory illness during the first year of life, *Br. J. Prev. Soc. Med.* **30:**203–212.

Leibel, R., Greenfield, D., and Pollitt, E., 1979, Biochemical and behavioral aspects of sideropenia, *Br. J. Haematol.* **41:**145–150.

Levine, L. S., Lewy, J. E., and New, M. I., 1976, Hypertension in high school students: Evaluation in New York City, *N.Y. State J. Med.* **76:**4–44.

Liang, P., Hie, T. T., Jan, O. H., and Glok, L. T., 1967, Evaluation of mental development in relation to early malnutrition, *Am. J. Clin. Nutr.* **20:**1290–1294.

Lloyd, J. K., and Wolff, O. H., 1976, Obesity, *Recent Adv. Pediatr.* **5:**305–332.

Londe, S., Bourgoignie, J. J., Robson, A. M., and Goldring, D., 1971, Hypertension in apparently normal children, *J. Pediatr.* **78:**569–577.

Lozoff, B., Brittenham, C., Viteri, F. E., Urrutia, J. J., and Oski, F. A., 1979, Developmental test deficit in infants with iron deficiency anemia, *Pediatr. Res.* **13:**334.

Lynch, S. R., Lipschitz, D. A., Bothwell, T. H., and Charlton, R. W., 1974, Iron and the reticulo-endothelial system, in: *Iron in Biochemistry and Medicine* (A. Jacobs and M. Wormwood, eds.), Academic Press, London.

Mamunes, P., Prince, P. E., Thornton, N. H., Hunt, P. A., and Hitchcock, E. S., 1976, Intellectual deficits after transient tyrosinemia in the term neonate, *Pediatrics* **57:**675–680.

Manerikaris, S., Malaviya, A. N., and Singh, M. B., 1976, BCG vaccination in newborns with intrauterine growth retardation, *Clin. Exp. Immunol.* **26:**173–175.

Mann, G. V., 1977, Diet–heart: End of an era. *N. Engl. J. Med.* **297:**644–650.
Mann, L. I., Bhakthavathsalan, A., and Liu, M., 1975, Placental transport of alcohol and its effect on maternal and fetal acid base balance, *Am. J. Obstet. Gynecol.* **122:**837–844.
Marmot, M. G., 1979, Epidemiological basis for the prevention of coronary heart disease, *Bull W.H.O.* **57:**331–349.
Marthaier, T. M., 1967, Epidemiological and clinical dental findings in relation to intake of carbohydrates, *Caries Res.* **1:**222–238.
Mata, L. J., 1974, Maternal factors and fetal growth, in: *The Children of Santa Maria Cauque': A Prospective Field Study* pp. 137–151, M.I.T. Press, Cambridge,
Matthew, D. J., Taylor, B., Norman, A. P., and Turner, M. W., 1977, Prevention of eczema, *Lancet* **12:**321–324.
McCance, R. A., 1976, Critical periods of growth, *Proc. Nutr. Soc.* **35:**309.
McGandy, R. B., Hall, B., Ford, C., and Stare, F. J., 1972, Dietary regulation of blood cholesterol in adolescent males: A pilot study, *Am. J. Clin. Nutr.* **25:**61–66.
McGill, H. C., Jr., 1968, Fatty streaks in the coronary arteries and aorta, *Lab. Invest.* **18:**560–564.
McLaren, D. S., 1979, Vitamin deficiency, toxicity and dependancy, in: *Textbook of Pediatric Nutrition* (D. S. McLaren and D. Burman, eds.), pp. 147–172, Churchill Livingstone, Edinburgh.
McMillan, G. C., 1973, Development of arteriosclerosis, *Am. J. Cardiol.* **31:**542.
McMillan, J. A., Landaw, S. A., and Oski, F. A., 1977, Iron sufficiency in breast-fed infants and the availability of iron from milk, *Pediatrics* **58:**686–691.
Menkes, J. H., 1977, Early feeding history of children with learning disorders, *Dev. Med. Child. Neurol.* **19:**169–171.
Menkes, J. K., Welcher, D. W., Levi, J. S., Dallas, J., and Gratsby, N. W., 1972, Relation of elevated blood tyrosine to the ultimate intellectual performance of premature infants, *Pediatrics* **49:**218–224.
Miller, J., and Barmes, D. E., 1981, Oral health, in *Childhood Prevention of Adult Health Problems* (F. Falkner, ed.), pp. 107–120, World Health Organization, Geneva.
Minkowski, A., Roux, J. M., and Fordet-Caridroit, C., 1974, Pathophysiologic changes in intrauterine malnutrition, in: *Nutrition and Fetal Development* (M. Winick, ed.), pp. 45–79, John Wiley, New York.
Mönckeberg, F., 1968, Effects of early marasmic malnutrition on subsequent physical and psychological development, in: *Malnutrition, Learning and Behavior* (N. S. Scrimshaw and J. E. Gordon, eds.), pp. 269–278, M.I.T. Press, Cambridge.
Mönckeberg, F., 1972, PAG statement on relationship of pre-and postnatal malnutrition on children and mental development, *Learning and Behavior,* PAG Bulletin No. 18 (U.N.) Vol. 2, p.23.
Morriss, G. M., 1972, Morphogenesis of the malformations induced in rat embryos by maternal hypervitaminosis A, *J. Anat.* **113:**241–250.
Moynahan, E. M., 1974, Acrodermatitis enteropatorica: A lethal inherited zinc deficiency disorder, *Lancet* **2:**399–400.
Multiple Risk Factor Intervention Trial Research Group, 1977, *Circulation,* **56(Suppl 3):**113.
Naeye, R. L., Blanc, W., and Paul, C., 1973, Effects of maternal nutrition on the human fetus, *Pediatrics* **52:**494–503.
Neldner, K. H., and Hambridge, K. M., 1975, Zinc therapy of acrodermatitis enteropathica *N. Engl. J. Med.* **292:**879–882.
Neumann, C. G., 1980, Prevention of obesity in infancy and childhood in: *Childhood Prevention of Atherosclerosis and Hypertension* (R. M. Lauer and R. B. Shekelle, eds.), pp. 367–374, Raven Press, New York.
Neumann, C. G., Stiehm, E. R., and Swendseid, M., 1980, Longitudinal study of immune function in intrauterine growth retarded infants, *Fed. Am. Soc. Exp. Biol.* **39:**888.
Newbrun, E., 1975, Water fluoridation and dietary fluorides, in: *Fluorides and Dental Caries* (E. Newbrun, ed.), Charles C. Thomas, Springfield.

Nishamura, H., 1953, Zinc deficiency in suckling mice deprived of colostrum, *J. Nutr.* **49:**79–97.

Nizel, A. E., 1977, Preventing dental caries: The nutritional factors, *Pediatr. Clin. North Am.* **24:**141–155.

Nora, J., 1980, Identifying the child at risk for coronary disease as an adult, *J. Pediatr.* **97:**706–714.

Norkus, E. P., and Rosso, P., 1975, Changes in ascorbic acid metabolism of the offspring following high maternal intake of this vitamin in the pregnant guinea pig, *Ann. N.Y. Acad. Sci.* **258:**401–409.

Norum, K. R., 1978, Some present concepts concerning diet and prevention of coronary heart disease, *Nutr. Metab.* **22:**1–7.

Nutrition Committee, Am. Acad. Pediatrics, 1973, The prophylactic requirement and the toxicity of vitamin D, *Pediatrics* **31:**512–525.

Nutrition Committee, Am. Acad. Pediatrics, 1976, Relation of infantile hypercalcemia and vitamin D; Public Health implications in No. America, *Pediatrics* **40:**1050–1061.

Nutrition Committee, American Academy of Pediatrics, 1971, The use and abuse of vitamin A, *Pediatrics* **48:**655–656.

Nutrition Committee, American Academy of Pediatrics, 1972, Childhood diet and coronary heart disease, *Pediatrics* **49:**305–307.

Nutrition Committee of the Canadian Paediatric Soc. and the Committee on Nutrition of the Am. Acad. Peds., 1978, Breast feeding, *Pediatrics,* **62:**591–601.

Nutrition Committee, American Academy of Pediatrics, 1979, Fluoride supplementation: Revised dosage schedule, *Pediatrics* **63:**150–152.

Olsen, R. E., 1978, Clinical nutrition, an interface between human ecology and internal medicine, *Nutr. Rev.* **36:**161–178.

Ornoy, A., Kaspi, T., and Nebel, L., 1972, Persistant defects of bone formation in young rats following maternal hypervitaminosis D2, *Isr. J. Med. Sci.* **8:**943–949.

Oski, F. A., 1979, The non-hematologic manifestations of iron deficiency, *Am. J. Dis. Child.* **133:**315–322.

Oski, F. A., and Honig, A., 1977, The effects of therapy on the developmental scores of iron deficient infants, *Pediatr. Res.* **11:**380–384.

Oski, F. A., and Perkins, K. C., 1975, Lack of iron deficiency in infants exclusively breast fed for long periods of time, *Pediatr. Res.* **9:**326–328.

Owen, G. M., Lubin, A. H., and Garry, P. J., 1971, Preschool children in the United States: Who has iron deficiency? *J. Pediatr.* **79:**563–568.

Page, L. B., 1980, Dietary sodium and blood pressure: Evidence from human studies, in: *Childhood Prevention of Atherosclerosis and Hypertension* (R. M. Lauer and R. B. Shekelle, eds.), pp. 291–303, Raven Press, New York.

Persson, B., Tunell, R., and Ekengren, K., 1965, Chronic vitamin A intoxication during the first half year of life, *Acta Paediatr. Scand.* **54:**49–60.

Pharaoh, P. O., Buttfield, I. H., and Hetzel, B. S., 1971, Neurological damage to the fetus resulting form severe iodine deficiency during pregnancy, *Lancet* **1:**308–310.

Picciano, M. F., and Guthrie, H. A., 1976, Copper, iron and zinc contents of mature human milks, *Am. J. Clin. Nutr.* **29:**242–254.

Pisacano, J. C., Lichter, H., Ritter, B. S., and Siegel, A. P., 1978, *Pediatrics* **61:**360–364.

Pollitt, E., and Leibel, R. I., 1976, Iron deficiency and behavior, *J. Pediat.* **88:**372–381.

Pollitt, E., and Lewis, N., 1980, Nutritional and educational achievement, *Food Nutr. Bull.* **2:**33–37.

Pollitt, E., Greenfield, D., and Leibel, R., 1978, Behavioral effects of iron deficiency among preschool children, *Fed. Proc. Fed. Am. Soc. Exp. Biol.* **37:**487.

Pooling Project Research Group, 1978, Relationship of blood pressure, serum cholestorol, smoking habit, relative weight and ECG abnormalities to incidence of major coronary events: Final report, *J. Chron. Dis.,* **31:**201–306.

Puska, P., Tuomilehto, J., Nissinen, A., Salonen, J., Mäki, J., and Pa Uonen, U., 1980, Changing the cardiovascular risk in an entire community: The North Karelia project, in: *Childhood*

Prevention of Atherosclerosis and Hypertension (R. M. Lauer and R. B. Shekelle, eds.), pp. 441–451, Raven Press, New York.

Rassin, D. K., Järvenpää, A. L., Niels, C. R., Rdihä, and Gaull, G. E., 1979, Breast feeding vs. formula feeding in full-term infants: Effects on taurine and cholesterol, *Pediatr. Res.* **13:**406.

Reiser, R., and Sidelman, Z., 1972, Control of serum cholesterol homeostasis by cholesterol in the milk of the suckling rat, *J. Nutr.* **102:**1009–1016.

Rodgers, B., 1978, Feeding in infancy and later ability and attainment: A longitudinal study, *Dev. Med. Child. Neurol.* **20:**421–426.

Roeder, L. M., and Chow, B. F., 1972, Maternal nutrition and its long-term effects on the offspring, *Am. J. Clin. Nutr.* **25:**812–821.

Rose, G., 1980, Relative merits of intervening on whole populations vs. high-risk individuals only, in: *Childhood Prevention of Atherosclerosis and Hypertension* (R. M. Lauer and R. B. Shekelle, eds.), pp. 351–356, Raven Press, New York.

Rosenbloom, L., and Sills, J. A., 1975, Hypernatraemic dehydration and infant mortality, *Arch. Dis. Child.* **50:**750.

Ross, A. O., 1977, *Learning Disability: The Unrealized Potential,* McGraw-Hill, New York.

Rosso, P., Hormazabal, J., and Winick, M., 1970, Changes in brain weight, cholesterol, phospholipid and DNA content in marasmic children, *Am. J. Clin. Nutr.* **23:**1275–1279.

Rubin, A., Rosenblatt, C., and Balow, B., 1973, Psychological and educational sequelae of prematurity, 1973, *Pediatrics* **52:**352–363.

Saarinen, U. M., Kajosaari, M., Backman, A. and Siimes, M. A., 1979, Prolonged breast-feeding as prophylaxis for atopic disease, *Lancet* **28:**163–166.

Schachter, J., 1980, discussant in: Dietary sodium and blood pressure, in: *Childhood Prevention of Atherosclerosis and Hypertension* (R. M. Lauer and R. B. Shekelle, eds.), pp. 305–312, Raven Press, New York.

Segal, S., and Bernstein, H., 1963, Observations on cataract formation in the newborn offspring rats fed a high galactose diet, *J. Pediatr.* **62:**363–369.

Simpson, J. W., Lawless, R. W., and Mitchell, A. C., 1975, Responsibility of the obstetrician to the fetus: Influence of pre-pregnancy weight and pregnancy weight gain in birthweight, *J. Obstetr. Gynecol.* **15:**481–487.

Sinclair, J. C., Saigal, S., and Yeung, C. Y., 1974, Early post-natal consequences of fetal malnutrition, in: *Nutrition and Fetal Development,* pp. 147–172, John Wiley, New York.

Smith, D. W., 1977, Growth deficiency disorders—postnatal onset, in: *Growth and its Disorders,* pp. 85–106, Saunders, Philadelphia.

Smith, D. W., Klein, A. M., and Henderson, J. R., 1975, Congenital hypothyroidism sign said symptoms in the newborn period, *J. Pediatr.* **87:**958–962.

Stamler, J., 1978, Lifestyles, major risk factors, proof and public policy, *Circulation,* **58:**3–19.

Stamler, J., 1980a, Improved life styles: Their potential for the primary prevention of atherosclerosis and hypertension in childhood, in: *Childhood Prevention of Atherosclerosis and Hypertension* (R. M. Lauer and R. B. Shekelle, eds.), pp. 3–30, Raven Press, New York.

Stamler, J., 1980b, Can an effective fat-modified diet be sagely recommended after weaning for infants and children in general? in: *Childhood Prevention of Atherosclerosis and Hypertension* (R. M. Lauer and R. B. Shekelle, eds.), pp. 387–404, Raven Press, New York.

Stanbury, J. B., and Querido, A., 1957, On the nature of endemic cretinism, *J. Clin. Endocrinol. Metab.* **17:**803–804.

Stein, H. G., Hasan, A., and Fox, I. H., 1976, Ascorbic acid-induced uricosuria. A consequence of megavitamin therapy, *Ann. Intern. Med.* **84:**385–388.

Stimmler, L., Snodgrass, G. J. A., and Jaffe, E. A., 1973, Dental defects associated with neonatal symptomatic hypocalcemia, *Arch. Dis. Child.* **48:**217–220.

Stoch, M. B., and Smythe, P. M., 1967, The effect of undernutrition during infancy on subsequent brain growth and intellectual development, *S. Afr. Med. J.* **41:**1027–1030.

Stoch, M. B., and Smythe, P. M., 1976, 15 year developmental study on effects of severe malnutrition on subsequent physical growth and intellectual functioning, *Arch. Dis. Child.* **51:**327–336.

Strasser, T., 1981, Prevention in childhood of major cardiovascular diseases of adults, in: *Prevention in Childhood of Health Problems in Adult Life* (F. Falkner, ed.), pp. 71–85, World Health Organization, Geneva.

Streissguth, A. P., Herman, C. S., and Smith, D. W., 1978, Intelligence, behavior, and dysmorphogenesis in the fetal alcohol syndrome: A report on 20 patients, *J. Pediatr.* **92:**363–367.

Sturman, J. A., Gaull, G., and Raika, N. C. R., 1970, Absence of cystathianase in human fetal liver; Is cystine essential? *Science* **69:**74–75.

Sturman, J. A., Rassin, D. K., and Gaull, G., 1977, Taurine in developing rat brain: Transfer of (35S) taurine to pups via the milk, *Pediatr. Res.* **11:**28–33.

Sulzer, J. L., Honsche, W. J., and Koenig, F., 1973, in: *Nutrition, Development and Social Behavior* (D. J. Kallen, ed.), U.S. Dept. HEW (DHEW Pub. No. NIH) 73-242, U.S. Govt. Printing Office, Washington, D.C.

Taitz, L. S., 1977a, Obesity in pediatric practice: Infantile obesity, *Pediatr. Clin. North Am.* **24:**107–115.

Taitz, L. S., 1977b, Sodium intake and health in infancy, *J. Hum. Nutr.* **31:**325–328.

Taitz, L. S., and Byers, H. D., 1972, High calorie/osmolar feeding and hypertonic dehydration, *Arch. Dis. Child.* **47:**257–260.

Tanner, J. M., 1978, Cells and the growth of tissues, in: *Fetus into Man: Physical Growth from Conception to Maturity,* pp. 6–23, Harvard University Press, Cambridge.

Thoman, E. B. M., Lieberman, P. H., and Olson, J. P., 1972, Neonate–mother interaction during breast-feeding, *Dev. Psychol.* **6:**110–118.

Thomson, A. M., Billewicz, W. Z. and Hytten, 1968, The assessment of fetal growth, *J. Obstet. Gynaec. Brit. Cwlth.* **75:**903–916.

Tizard, J., 1974, Early malnutrition, growth and mental development in man, *Br. Med. Bull.* **30:**169–174.

Toverud, G., 1961, Dental caries in Norwegian children during and after the second World War, *J. Am. Diet. Assc.* **26:**673–680.

Tsang, R. C., Fallat, R. W., and Glueck, C. J., 1974, Cholesterol at birth and age 1: Comparison of normal and hypercholesterolemic neonates, *Pediatrics* **53:**458–470.

Udall, J. N., Harrison, G. G., Vaucher, Y., Watson, P. D., and Morrow G., III, 1978, Interaction of maternal and neonatal obesity, *Pediatrics* **62:**17–21.

U.S. Senate Select Committee on Nutrition and Human Needs, 1977, *Dietary Goals for the United States,* 2nd ed., No. 052-270-03913-2, U.S. Government Printing Office, Washington, D.C.

Vander Berg, B. J. and Yerushalmy, J., 1966, The relationship of the rate of intrauterine growth of infants of low birth weight to mortality and morbidity and congenital anomalies, *J. Pediat.* **69:**531–545.

Venkatachalam, P. S., Susheela, T. P., and Rau, P., 1967, Effect of nutritional supplementation during early infancy in growth of infants, *J. Trop. Pediatr.* **13:**70–76.

Webb, T. E., and Oski, F. A., 1973, Iron deficiency anemia and scholastic achievement in young adolescents, *J. Pediatrics* **82:**827–830.

Weidman, W. H., 1980, Effect of change in diet on level of serum cholesterol, in: *Childhood Prevention of Atherosclerosis and Hypertension* (R. M. Lauer and R. B. Shekelle, eds.), pp. 137–143, Raven Press, New York.

Weidman, W. H., Elveback, L. R., Nelson, R. A., Hodgson, P. A., and Ellefson, R. D., 1978, Nutrient intake and serum cholesterol level in normal children 6 to 16 years of age, *Pediatrics* **61:**354–359.

Weil, W., 1977, Current controversies in childhood obesity, *J. Pediatr.* **9:**175–187.

Wharton, B. A., 1976, Genes, clocks and circumstances—The effects of over and undernutrition, in: *Early Nutrition and Later Development* (A. W. Wilkinson, ed.), pp. 112–123, Pitman Medical Pub., Bath.

Whitelaw, A. G. L., 1977, Infant feeding and subcutaneous fat at birth and at one year, *Lancet* **2:**1098–1099.

World Health Organization, 1974, Malnutrition and development, *WHO Chron.* **28:**95–102.

Widdowson, E. M., 196, Artificial milks and their effects on the composition of the infant, in: *Early Nutrition and Later Development* (A. W. Wilkinson, ed.), pp. 71–78, Yearbook Medical, Chicago.

Winick, M., 1976a, Malnutrition and mental development, in: *Malnutrition and Brain Development,* pp. 128–155, Oxford University Press, London.

Winick, M., 1976b, Nutrition and cellular growth of the brain, in: *Malnutrition and Brain Development,* pp. 63–97, Oxford University Press, 1976.

Winick, M., and Rosso, P., 1969, The effect of severe early malnutrition on cellular growth of human brain, *Pediatr. Res.* **3:**181.

Wittig, H. J., McLaughlin, E. T., Leifer, K. I., and Belloit, J. D., 1978, Risk factors for the development of allergic disease: Analysis of 2190 patient records, *Am. Allergy* **41:**84–88.

Woodruff, C. W., 1977, Iron deficiency in infancy and childhood, *Pediatr. Clin. North Am.* **24:**85–94.

Wong, P. W. K., Lambert, A. M., and Komrowe, G. M., 1967, Tyrosinaemia and tyrosinuria in infancy, *Dev. Med. Child. Neurol.* **9:**551.

Ziegler, E. E., and Fomon, S. J., 1980, Infant feeding and blood lipid levels during childhood, in: *Childhood Prevention of Atherosclerosis and Hypertension* (R. M. Lauer and R. B. Shekelle, eds.), pp. 121–125, Raven Press, New York.

45

METABOLIC ALKALOSIS FROM CHLORIDE-DEFICIENT INFANT FORMULA

Shane Roy, III

INTRODUCTION

The metabolic consequences of chloride deficiency have recently been reviewed by Simopoulos and Bartter (1980). Chloride-depletion syndromes include cystic fibrosis with excessive chloride loss through sweating, gastrointestinal chloride loss resulting from congenital hypertrophic pyloric stenosis, surreptitous vomiting, laxative abuse or congenital chloride diarrhea, and urinary chloride loss secondary to diuretic use or abuse or in Bartter's syndrome. Salt deprivation in man can result in metabolic alkalosis but the effect of chloride deprivation alone, though studied in rats (Holliday *et al.*, 1961; Levine *et al.*, 1974), has not been reported in man. Hypochloremic metabolic alkalosis with failure to thrive resulting from the feeding of chloride deficient infant formulas (*Neo-Mull-Soy* and *CHO-Free*) (Roy and Arant, 1979) is a new syndrome of chloride deficiency which further emphasizes the importance of the chloride ion in acid–base balance and in nutrition.

DESCRIPTION OF INDEX PATIENTS

Six unrelated infants were evaluated at Le Bonheur Children's Medical Center in Memphis during the same month in 1979 because of failure to thrive (Roy and Arant, 1980). Each infant had hypochloremia, metabolic alkalosis, hyponatremia, hypokalemia, hyperaldosteronism, elevated plasma renin activity, normal blood pressure, and microscopic hematuria. Muscular

---

Shane Roy, III • Department of Pediatrics, University of Tennessee Center for the Health Sciences, Memphis, Tennessee 38163.

weakness, delayed motor development, constipation, and anorexia for solid foods were also observed. The infant formula *Neo-Mull-Soy* (*NMS*) (Syntex Labs, Palo Alto, California) had been the primary source of nutrition for each infant for 2–5 months prior to their evaluation.

Since chloride loss in vomitus, urine, stool, or sweat could not be documented in any of these infants, an analysis of their *NMS* formula revealed that, contrary to product information, the chloride concentration was less than 2 meq/liter and provided each infant less than 0.3 meq/kg per day of chloride as opposed to the recommended 3–5 meq/kg per day (Winters, 1973; Ziegler and Fomon, 1974). The sodium and potassium contents of the formula were 17 and 25 meq/liter, respectively. The citrate content (240 mg/dl) provided an equivalent bicarbonate load of 2 meq/kg per day in these infants.

LABORATORY DATA AND COURSE OF INDEX PATIENTS

In Fig. 1 plasma electrolytes, blood urea nitrogen, and weight for two affected infants are plotted over four periods of observation. The first period of observation for both infants illustrates hypochloremia, hyponatremia, hypokalemia, elevated plasma bicarbonate, and elevated blood urea nitrogen during *NMS* feedings alone. When potassium chloride (3–4 meq/kg per day) was added to *NMS* in the second period of observation, plasma potassium, chloride, and bicarbonate returned to normal in a few days. Hypokalemia, hypochloremia, and alkalosis recurred promptly during the third period of observation, when potassium chloride was discontinued, and *NMS* alone was fed. In the fourth period of observation, the electrolyte abnormalities were corrected again by substituting *Nutramigen* (Chloride = 23 meq/liter) and *Similac* (Chloride = 18 meq/liter) for the chloride-deficient *NMS*.

Renal biopsies were performed in these two infants because of unexplained hematuria, proteinuria, and elevated serum creatinine in association with metabolic alkalosis. Juxtaglomerular hyperplasia was not observed but tubular calcification was observed in one and interstitial fibrosis in both.

Electrolyte and urinary abnormalities similar to those in the first two index infants were observed in four other infants (Table I) during *NMS* feedings and were corrected within 6–13 days after changing to a different formula with a chloride concentration of at least 13 meq/liter. Elevated plasma renin activity (PRA) of 81, 62, and 56 ng/ml per hour decreased to 16, 13.6, and 38.6 ng/ml per hr in patients A, B, and C respectively one week after the formula was changed to one sufficient in chloride. PRA in patients B and C was 0.5 and 3.3 ng/ml per hr 6 months later. Initial plasma aldosterone (PA) values in patients A, B, C, and E were 12, 35, 41, and 119 ng/dl, respectively. One week after changing to a chloride-sufficient formula PA was 28, 34, and 34 ng/dl in patients A, B, and C respectively. Hematuria disappeared while plasma creatinine and electrolyte concentrations remain normal in each infant after 6 months of follow-up.

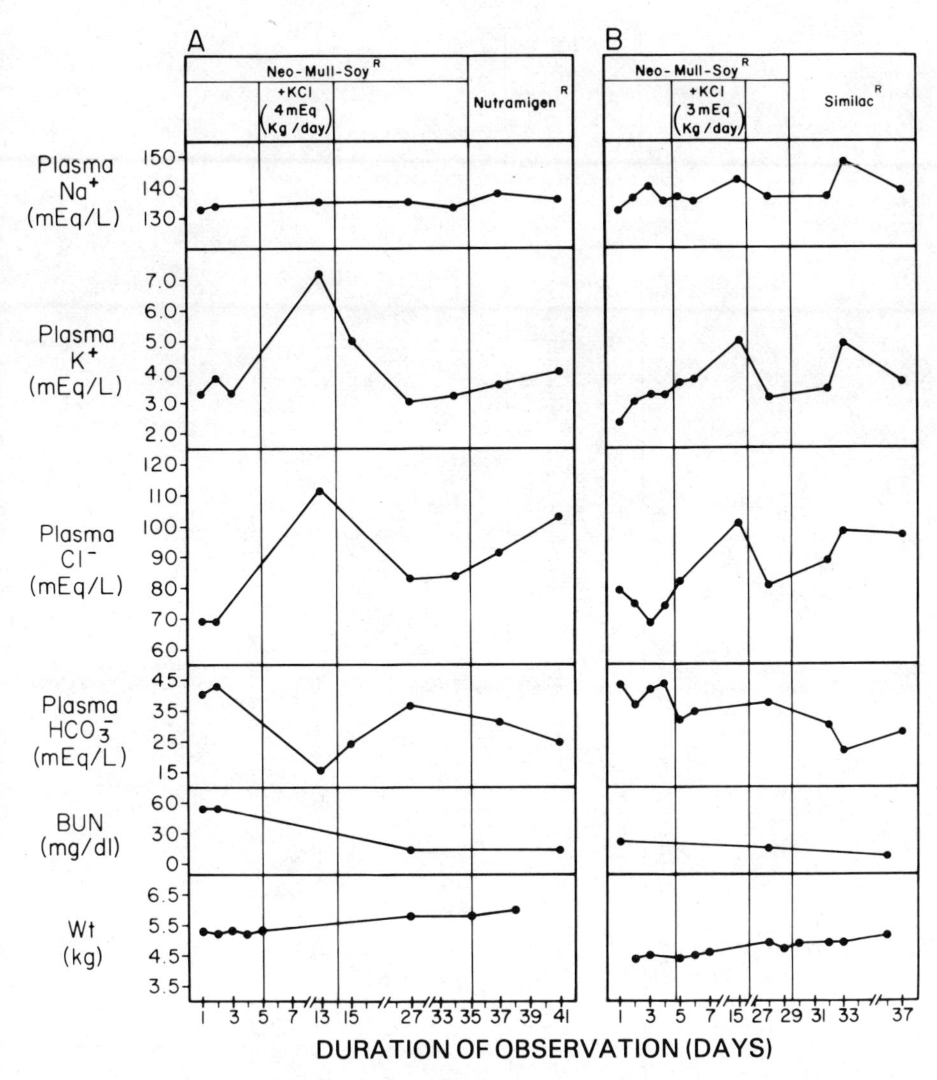

FIGURE 1. Course of infants A and B on *Neo-Mull-Soy* without and with supplemental KCl and on a chloride-adequate formula.

Dietary intakes of sodium, potassium, and chloride at the time of the initial evaluation are compared to daily urinary excretion in Table II. Chloride intake was < 0.3 meq/kg per 24 hr in each of the six infants while sodium intake ranged from 1.4 to 2.8 meq/kg per 24 hr and potassium intake ranged from 2.1 to 4.2 meq/kg per 24 hr during feeding of the chloride deficient formula. Twenty-four hour urinary chloride excretion was < 1.0 meq/kg in each infant.

TABLE I

Summary of Initial Laboratory Data in the Infants

Patient	Plasma										Urine		
	(meq/L)				(mg/dl)			Hct	Plasma[a] renin activity	Plasma aldosterone			
	Na^+	K^+	Cl^-	HCO_3^-	BUN	Cr	Blood pH	(%)	(ng/ml per hr)	(ng/dl)	pH	SG	RBC/ hpf
A	133	3.3	69	40	53	0.8	7.61	33	81	12	8.0	1.010	40–60
B	132	2.3	74	43	21	0.4	7.55	36	62	35	7.5	1.009	20–30
C	130	2.6	61	46	22	0.7	7.73	32	56	41	7.5	1.008	8–15
D	135	2.9	88	28	22	—	7.52	32	—	—	5.0	1.005	0
E	134	3.3	76	36	26	0.8	—	43	>30	119	7.0	1.005	0
F	127	3.9	75	23[c]	19	—	—	—	—	—	7.0	1.006	[d]

[a] Normal values for infants 3 to 12 months of age are 6.25 ± 1.23 (mean ± SD) ng/ml/hour.
[b] Normal range of values for infants 4 to 12 months of age are 3–10 ng/dl.
[c] NMS discontinued 24 hours before blood value was obtained.
[d] Small amount of blood by *Multistix* (Ames Co., Elkhart, IN); (–) laboratory test not done during initial evaluation.

TABLE II
Dietary Intake and Urinary Excretion of Electrolytes at the Time of Initial Evaluation of the Infants

Patient	Dietary intake			Urinary excretion		
	Na^+	K^+ (meq/kg per 24 hr)	Cl^-	Na^+	K^+ (meq/kg per 24 hr)	Cl^-
A	2.3	3.4	0.3	1.0	1.4	<0.1
B	2.8	4.2	0.3	0.5	1.8	<0.1
C	2.5	3.6	0.3	1.4	1.9	<0.1
D	1.4	2.1	0.2	2.1	2.2	1.0
E	1.7	2.6	0.2	—	—	<0.1
F	2.1	3.1	0.2	—	—	<0.1

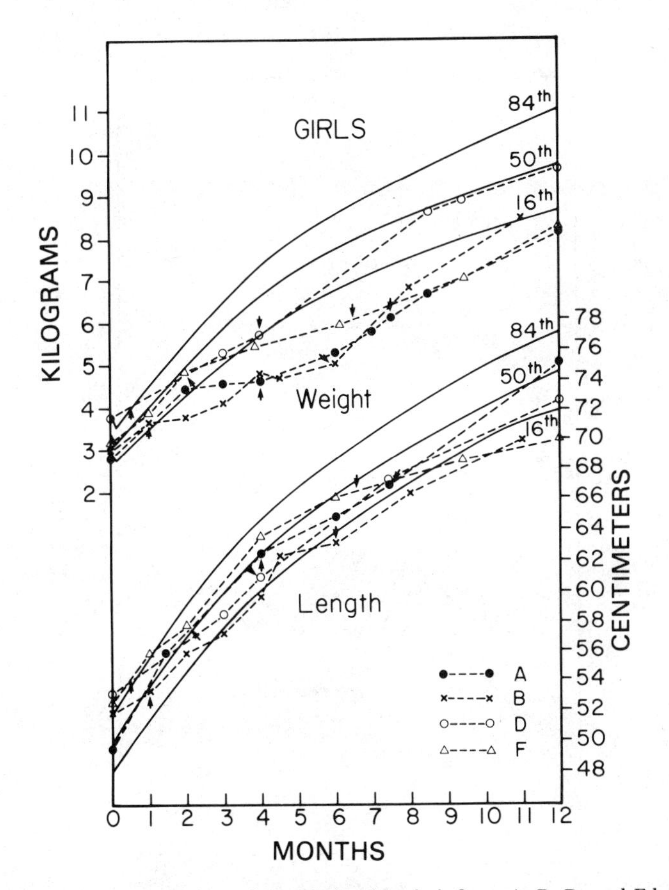

FIGURE 2. Comparison of weight and length for female infants A, B, D, and F before, during, and after *Neo-Mull-Soy*. Upward arrows indicate beginning and downward arrows indicate discontinuing *Neo-Mull-Soy*.

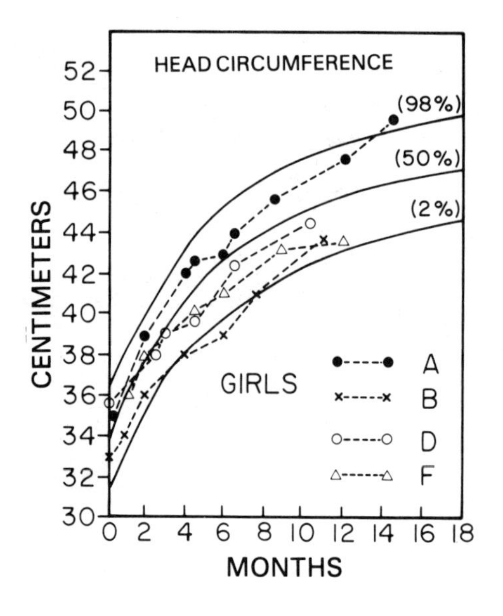

FIGURE 3. Head circumference percentiles for female infants A, B, D, and F before, during, and after *Neo-Mull-Soy*.

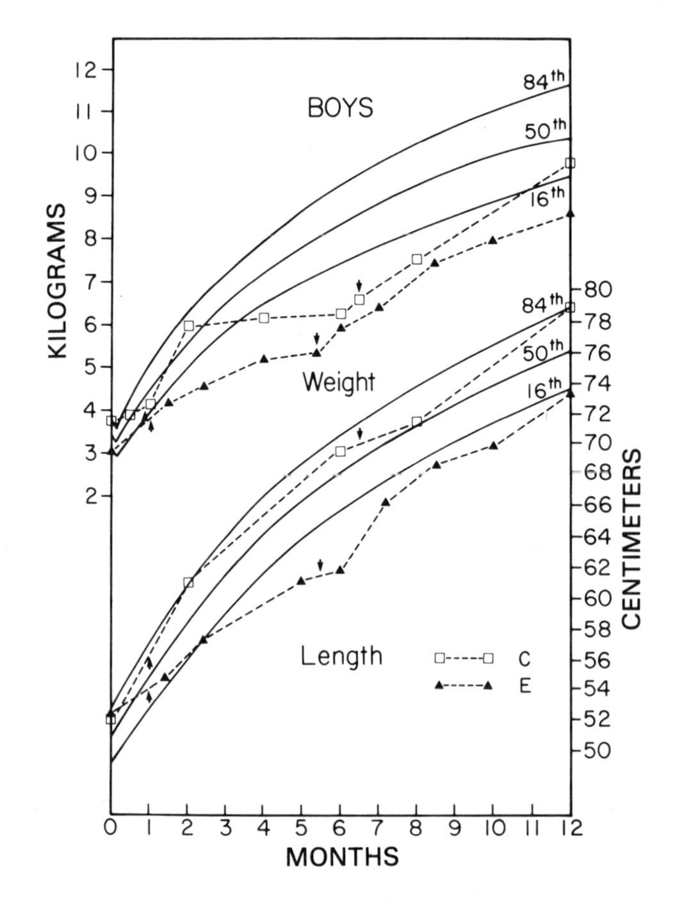

FIGURE 4. Comparison of weight and length for male infants C and E before, during, and after *Neo-Mull-Soy*. Arrows as in Fig. 2.

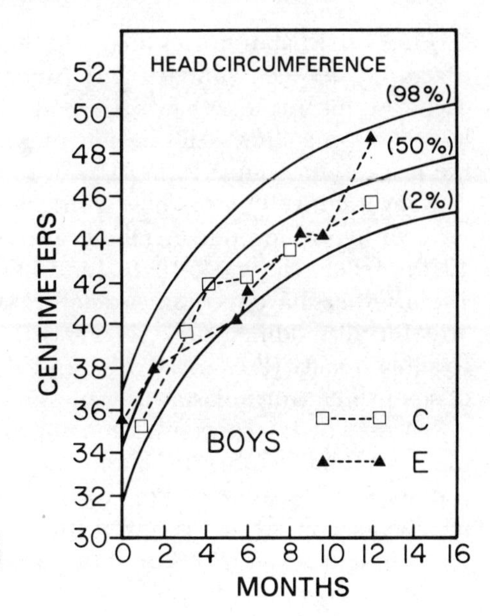

FIGURE 5. Head circumference percentiles for male infants C and E before, during, and after *Neo-Mull-Soy*.

GROWTH DATA OF INDEX PATIENTS

Longitudinal growth patterns for length, body weight, and head circumference for each infant from birth until 4–8 months after *NMS* was discontinued are compared to National Center for Health Statistics Growth Charts, 1976, in Figs. 2, 3, 4, and 5. Growth data for each infant were also compared as the ratio of actual increment in length, weight and head circumference to the expected monthly increment for an infant of the same sex growing at the 50th percentile. Mean ratios ± SE of 0.65 ± 0.16, 0.96 ± 0.11 and 0.85 ± 0.12 for length (cm/month), weight (g/month), and head circumference (cm/month), respectively, during modified milk formula feeding changed to 0.77 ± 0.04 (paired t-test, p = NS) 0.51 ± 0.08 ($p < 0.01$) and 0.60 ± 0.16 ($p = 0.1$) during the period *NMS* was fed the infants. Four months after discontinuing *NMS,* ratios of actual : expected growth rates at the 50th percentile increased to 0.97 ± 0.15 (p = NS), 1.29 ± 0.12 ($p < 0.001$) and 1.55 ± 0.19 ($p < 0.001$) for length, weight, and head circumference; however, head circumferences in four infants were still < 10th percentile for chronological age 6 months after *NMS* was discontinued.

DESCRIPTIONS OF SIMILAR PATIENTS

After our findings were reported to Syntex Laboratories and the Center for Disease Control (CDC), a survey by the CDC identified 31 infants who had developed unexplained metabolic alkalosis during 1979 (CDC, 1979). Of 27 of these infants, for whom a feeding history was available, 26 were receiving

NMS. Syntex Laboratories analyzed 99 case lots of *NMS* and *CHO-Free* manufactured between January and June 1979. The chloride concentration in each case lot was less than 5.1 meq/liter, but 66% of the case lots contained less than 2 meq/liter. All case lots of *NMS* and *CHO-Free* were recalled from the market on August 2, 1979.

Several investigators have subsequently described other infants who developed hypochloremic metabolic alkalosis while receiving *NMS* (Garin *et al.*, 1979; Hellerstein *et al.*, 1979; Lindshaw *et al.*, 1980; Wolfsdorf and Senior, 1980). Others have recommended the adoption of minimum electrolyte standards for all modified milk formulas (Holliday, 1980), regulations to ensure formula quality (Barness, 1980), and short-term rather than continuous use of soy-protein formulas as the sole food source for infants (Finberg, 1979).

A total of 120 cases of formula-associated infant metabolic alkalosis identified between October 1978 and August 1979 were reported in March 1980 (Center for Disease Control, 1980). Although most infants who developed metabolic alkalosis had received the deficient formula as their sole source of nourishment for at least 2 months, alkalosis did not develop in a majority of infants receiving the formula. Presumably they were receiving enough other foods to prevent chloride deficiency.

DISCUSSION OF PROBABLE MECHANISM OF METABOLIC ALKALOSIS

Suggested chloride requirements for the growing infant range from 2–3 meq/day (Ziegler and Fomon, 1974) to 5 meq/100 cal metabolized per day (Winters, 1973), but chloride in the infant's diet is usually provided in amounts equivalent to sodium. The chloride concentration in breast milk, however, is 1.7 times greater than that of sodium (Macy and Kelly, 1961). Chloride intake in the infants described above was 0.2–0.3 meq/kg per day (Table II).

In man metabolic alkalosis is generated by a net loss of acid, with or without a concomitant depletion of extracellular fluid (ECF) and is maintained by chloride deficiency, alkali administration, reduction in ECF volume, hypokalemia and/or hyperaldosteronism (Seldin and Rector, 1972). It has been clearly shown, however, that chloride is essential for correcting hypokalemic metabolic alkalosis in both man and dogs (Atkins and Schwartz, 1962; Kassirer *et al.*, 1965; Schwartz *et al.*, 1955).

Urine chloride excretion was < 1.0 meq/kg per 24 hr and loss of chloride in stool, although not measured, was minimal since constipation was a presenting complaint in each infant. Chloride deprivation was, therefore, the most likely cause of the hypochloremia. The decrease in filtered chloride resulted in diminished reabsorption of chloride in the loop of Henle and an increased distal tubular delivery of sodium for reabsorption, leading to increased potassium and hydrogen secretion (Bank and Schwartz, 1960; Schwartz *et al.*, 1955). Citrate, the major anion in *NMS* (25 meq/liter of

$K_3C_6H_5O_7H_2O$), is metabolized in the body to bicarbonate. It has proven difficult to produce metabolic alkalosis in adults by giving large quantities of bicarbonate (25 meq/kg) (Van Goidsenhoven *et al.*, 1954), but the equivalent of 1–2 meq/kg per day of bicarbonate provided these chloride-deficient infants likely augmented the generation of the alkalosis. A primary role for citrate alone in producing the alkalosis is doubtful, however, because citrate intake continued during the times that the addition of chloride to the NMS corrected the alkalosis in two index infants. No infant had evidence of compromised pulmonary function. Thus, the net acid loss required to generate the metabolic alkalosis in these infants was produced by the initial loss of hydrogen ion into the urine and perhaps augmented by exogenous alkali administration.

Continued chloride deprivation, the ingestion of exogenous alkali, anorexia for solid foods, with decreased intake of all nutrients, and gradual dehydration maintained the metabolic alkalosis in the infants. The resultant ECF volume depletion stimulated renin release and aldosterone secretion (Bull *et al.*, 1970) that, along with bicarbonaturia (Malnic *et al.*, 1971), would facilitate potassium and hydrogen ion excretion.

The element of growth in these chloride-deprived infants adds a new dimension to the proposed mechanism of metabolic alkalosis (Simopoulos and Bartter, 1980). During periods of rapid growth the organism produces new extracellular fluid at a rapid rate. Without chloride such fluid must be low in volume, contain an abnormally high ratio of bicarbonate to chloride, or both. Therefore, hypochloremic alkalosis can be produced without loss of chloride; the ECF produced simply lacks chloride from the onset due to deficient intake of chloride.

In preliminary studies in three mongrel dogs fed chloride-deficient *NMS* the syndrome of hypochloremia and hypokalemia with elevated plasma bicarbonate, plasma renin activity, and plasma aldosterone was produced. Urinary chloride excretion ceased after 3–4 weeks of chloride deficiency. Extracellular fluid volume decreased progressively, venous hematocrit increased, and weight loss or poor weight gain was observed. The syndrome was corrected similarly in two dogs fed either newly constituted, chloride-adequate *NMS* (Chloride = 13–14 meq/liter) or *Isomil* (Chloride = 13 meq/liter). Since the syndrome was corrected as rapidly by chloride-adequate *NMS*, with 240 mg/dl of citrate, as by *Isomil*, with a much lower citrate content, it appears that chloride deficiency was the primary, if not the sole, etiologic factor operative in producing the metabolic alkalosis.

FOLLOW-UP OBSERVATIONS IN INFANTS

Delayed onset of expressive language has been noted in 6 of 31 infants who received *NMS* and have been followed for a mean of 6.6 months (Roy *et al.*, 1980). Speech evaluations in four infants between 18 and 27 months of age have demonstrated receptive language skills which were equal to or above those expected for their chronological age. Their expressive language, however, is comparable with an age range 6–18 months younger than their chronological age.

Because of learning disabilities described by Klein *et al.* (1975) and Berglund *et al.* (1973) in children following malnutrition and metabolic alkalosis secondary to pyloric stenosis during early infancy, it is our recommendation that infants who received the chloride-deficient *NMS* formula be followed prospectively with special attention to language skills and the possible future development of preceptual problems by school age.

CONCLUSION

The temporal relationship of this syndrome to the removal of all added salt to infant foods is interesting. In 1977, the baby food industry removed all added salt from infant foods. In 1976, the Committee on Nutrition of the American Academy of Pediatrics set minimum standards for the content of electrolytes and other ingredients in prepared infant formulas (1976). A minimum of 11 meq/liter of chloride was recommended for all infant formulas. The unavailability of prepared infant foods with added salt after the first 6–9 months of 1978, coupled with the recent recommendation to delay feeding of solid foods until 5–6 months of age (Fomon *et al.*, 1979), may have contributed to the appearance of this chloride-deficient syndrome. Closer cooperation and understanding among the formula and infant food industries, the medical community, and governmental regulatory agencies, in addition to strict formula quality-assurance programs, should decrease the chances of another technical catastrophe in infant formula manufacture. A detailed dietary and nutritional history must be obtained in each infant with subnormal weight gain or failure to thrive. Furthermore, one must consider proceeding beyond simply changing the infant's formula and must question both the adequacy and the composition of the formula before eliminating inadequate nutrition as the cause of an infant's failure to thrive.

REFERENCES

Atkins, E. L., and Schwartz, W. B., 1962, Factors governing correction of the alkalosis associated with potassium deficiency: The critical role of chloride in the recovery process, *J. Clin. Invest.* **41:**218.

Bank, N., and Schwartz, W. B., 1960, The influence of anion penetrating ability on urinary acidification and the excretion of titratable acid, *J. Clin. Invest.* **39:**1516.

Barness, L. A., 1980, Formula manufacture and infant feeding, *J. Am. Med. Assoc.* **243:**1075.

Berglund, G., and Rabo, E., 1973, A long-term follow-up investigation of patients with hypertrophic pyloric stenosis—with special reference to the physical and mental development, *Acta Paediatr. Scand.* **62:**125.

Bull, M. B., Hillman, R. S., Cannon, P. J., and Laragh, J. H., 1970, Renin and aldosterone secretion in man as influenced by changes in electrolyte balance and blood volume, *Circ. Res.* **27:**953.

Center for Disease Control, 1979, Infant metabolic alkalosis and soy-based formula—United States, *Morbidity Mortality Weekly Report* **28:**358.

Center for Disease Control, 1980, Follow-up on formula-associated illness in children, *Morbidity Mortality Weekly Report* **29:**124.

Committee on Nutrition, American Academy of Pediatrics, 1976, Commentary on breast feeding and infant formulas, including proposed standards for formulas, *Pediatrics* **57:**278.

Finberg, L., 1979, One milk for all—not ever likely and certainly not yet, *J. Pediatr.* **96:**240.

Fomon, S. J., Filer, L. J., Anderson, T. A., and Ziegler, E. E., 1979, Recommendations for feeding normal infants, *Pediatrics* **63:**52.

Garin, E. H., Geary, D., and Richard, G. A., 1979, Soybean formula (*Neo-Mull-Soy)* metabolic alkalosis in infancy, *J. Pediatr.* **95:**985.

Hellerstein, G., Guggan, E., Grossman, H. M., McCamman, S., Sharma, P., and Welchert, E., 1979, Metabolic alkalosis and *Neo-Mull-Soy, J. Pediatr.* **95:**1083.

Holliday, M. A., 1980, Alkalosis in infancy and commercial formulas, *Pediatrics* **65:**639.

Holliday, M. A., Bright, N. H., Schultz, D., and Oliver, J., 1961, The renal lesions of electrolyte imbalance. III. The effect of acute chloride depletion and alkalosis on the renal cortex, *J. Exp. Med.* **113:**971.

Kassirer, J. P., Berkman, P.M., Lawrenz, D. R., and Schwartz, W. B., 1965, The critical role of chloride in the correction of hypokalemic alkalosis in man, *Am. J. Med.* **38:**172.

Klein, P. S., Forbes, G. B., and Hader, P. R., 1975, Effects of starvation in infancy (pyloric stenosis) on subsequent learning abilities, *J. Pediatr.* **87:**8.

Levine, D. Z., Roy, D., Tolnai, G., Nash, L., and Shah, B. G., 1974, Chloride depletion and nephrocalcinosis, *Am. J. Physiol.* **227:**878.

Lindshaw, M. A., Harrison, H. L., Gruskin, A. B., Prebis, J., Harris, J., Stein, R., Jayaram, M. R., Preston, D., DiLiberti, J., Baluarte, H. J., Elzouki, A., and Carroll, N., 1980, Hypochloremic alkalosis in infants associated with soy-protein formula, *J. Pediatr.* **96:**635.

Macy, I. G., and Kelly, H. J., 1961, Human milk and cow's milk in infant nutrition, in *Milk: The Mammary Gland and its Secretion* (S. K. Kon and A. T. Cowie, eds.), pp. 359, Academic Press, New York.

Malnic, G., Mello-Aires, M., and Giebisch, G., 1971, Potassium transport across renal distal tubule during acid–base disturbances, *Am. J. Physiol.* **221:**1192.

Roy, S., III, and Arant, B. S., Jr., 1979, Alkalosis from chloride-deficient *Neo-Mull-Soy, N. Engl. J. Med.* **301:**615.

Roy, S., III, and Arant, B. S., Jr.,1981, Hypokalemic metabolic alkalosis in normotensive infants with elevated plasma renin activity and hyperaldosteronism: The role of dietary chloride deficiency, *Pediatrics* **67:**423.

Roy, S., III, Stapleton, F. B., and Arant, B. S., Jr., 1980, Hypochloremic metabolic alkalosis in infants fed a chloride-deficient formula, *Pediatr. Res.* **14:**509.

Schwartz, W. B., Jenson, R. L., and Relman, A. S., 1955, Acidification of the urine and increased ammonium excretion without change in acid-base equilibrium. Sodium reabsorption as a stimulus to the acidifying process, *J. Clin. Invest.* **34:**673.

Seldin, D. W., and Rector, F. C., Jr., 1972, The generation and maintenance of metabolic alkalosis, *Kidney Int.* **1:306.**

Simopoulos, A. P., and Bartter, F. C., 1980, The metabolic consequences of chloride deficiency, *Nutr. Rev.* **38:**201.

Van Goidsenhoven, G. M. T., Gray, O. V., Price, A. V., and Sanderson, P. H., 1954, The effect of prolonged administration of large doses of sodium bicarbonate in man, *Clin. Sci.* **13:**383.

Winters, R. W., 1973, Maintenance fluid therapy, in: *The Body Fluids in Pediatrics,* (R. W. Winters, ed.), p. 113, Little, Brown, Boston.

Wolfsdorf, J. E., and Senior, B., 1980, Failure to thrive and metabolic alkalosis. Adverse effects of a chloride-deficient formula in two infants, *J. Am. Med. Assoc.* **243:**1068.

Ziegler, E. E., and Fomon, S. J., 1974, Major minerals, in: *Infant Nutrition* 2nd ed. (S. J. Fomon, ed.), p. 267, W. B. Saunders, Philadelphia.

XIV

AVOIDANCE OF ADVERSE DIETS

46

SELECTION OF AN OPTIMAL DIET

Modern Criteria

J. Michael Gurney

Man, being a social animal living in a wide range of habitats and consuming a highly varied diet, challenges any simple definition of his food requirements. For example a traditional daily Kikuyu diet contains approximately 22 g of fat and 390 g of carbohydrate. This contrasts with 162 g of fat, an enormous intake, and 59 g of carbohydrate in Eskimo diets (Weiner, 1964). With this and many other examples of cultural variability, an optimal diet for the promotion of long life and well-being can be defined and classified only in the broadest terms.

SOCIOCULTURAL AND PSYCHOLOGICAL NEEDS FOR FOOD

Foodways are deeply woven into the backcloth of all cultures. Man's contemporary eating habits reflect origins stretching right back to when he was a fruit-picking primate who turned hunter (Morris, 1977). Cultures have differentiated, and food habits changed, so much so that we now find a variation in human food and nutrition behavior that cannot be explained on a purely nutritional basis (Montgomery, 1978).

Certain common sociocultural uses for food can however be distinguished even though these vary in strength and manifestation between cultures. De Garine (1970) has categorized foodways into sociocultural phenomena, signifiers of cultural identity, enhancers of communication, economic forces, and supernatural and religious symbols. He associates food in traditional societies with the collective subconscious and contrasts this with the more individualistic nature of food cults and food fads in industrial societies. He considered such cults and fads as attempts to restore the symbolic and emotional role of food.

J. Michael Gurney • Caribbean Food and Nutrition Institute, P. O. Box 140, Kingston, 7, Jamaica.

While food selection in traditional societies retains a strong cultural basis, it appears that in the U.S. such selection sometimes becomes individualized in the face of an apparent width of choice (Jerome, 1975).

The hierarchy of needs defined by Maslow (1970) provides a useful cross-cultural analytical tool particularly valuable in measuring changing motivations that influence food choice. The basic need is *physiological*—to relieve hunger; the next most basic need is for *security* to ensure a safe food supply. Only if these two needs are satisfied can man use food to satisfy his need for *belongingness* in his cultural group. The fourth need in the hierarchy is for *esteem* or status, and the last is *self-actualization* in which, the other needs being satisfied, the individual feels able to experiment.

Maslow's theory of the hierarchy of needs helps to explain the diversity of dietary responses of individuals and groups to culture change such as emigration. Some rapidly adopt the diet patterns of their hosts, while others retain their native food habits even though they adopt the patterns of their hosts in other respects (Burgess and Dean, 1962).

A limitation of Maslow's hierarchy of needs may be its bias toward personal needs. This may well apply fully in young and individualistic cultures, but should be taken in the context of the very real and deep sociological motivations with regard to food that occur in many societies worldwide.

To avoid unconscious bias entering into an evaluation, it is useful to categorize food habits under four headings: *beneficial*—needing promotion and support; *neutral*—to be let alone; *unclassifiable*—needing further investigation; and *harmful*—ill effects needing to be demonstrated to encourage willing change (Jelliffe and Bennett, 1961).

In the modern world, a new set of considerations has emerged: the widespread use of chemicals in food processing and in agriculture. For example, food additives, pesticides, and the spillover of industrial pollutants into the food chain are features of contemporary life. The selection of an optimal diet clearly includes minimizing by all means possible the intake of any such substances, especially for vulnerable pregnant and lactating mothers.

PHYSIOLOGICAL FUNCTIONS OF FOODS

There are three basic physiological functions of foods: energy release and utilization; growth, maintenance and repair; and regulation of reactions. McNutt and McNutt (1978) emphasize the following crucial interrelationships between these basic functions:

1. Energy release can occur only in the presence of regulatory enzymes and cofactors.
2. Growth, maintenance, and repair of the body are dependent upon an adequate supply of energy and of regulatory enzymes and cofactors.
3. The reactions that are regulated by enzymes and cofactors are also dependent upon an adequate supply of energy; the ultimate purpose

of these regulated reactions is the growth, maintenance, and repair of the body.

NUTRIENT REQUIREMENTS

This century has seen a great increase in scientific understanding of the elements in the diet that contribute to the three basic physiological functions of foods. This knowledge has led the World Health Organization (WHO) and the Food and Agriculture Organization (FAO) (Passmore *et al.*, 1974) and a number of countries to formulate recommended dietary intakes. In general, these guidelines specify requirements for dietary energy, proteins (taking into account the amounts of amino acids likely to be present in the proteins), vitamins, and certain minerals (the number varies). The guidelines separate recommended "intakes" or "allowances" by age, sex, and sometimes activity levels.

Such tables perform three main functions: they act as reference points against which the diets of communities and individuals can be assessed; they provide a basis for planning diets; and they enable national food policies to be developed in relation to nutritional needs. They are essential tools for anyone working in these areas. However, tables of recommended intakes have three major limitations:

1. They are limited by knowledge about nutritional requirements at the time of compilation. However, knowledge is increasing. While many needed nutrients not included in current tables are unlikely to be missing or deficient in normal diets, some, particularly minerals, may be. Consequently, we can expect to see a proliferation of various minerals in future tables of nutrient requirements.
2. The method of estimating recommended intakes is controversial. For example, the FAO/WHO recommendations represent an average requirement augmented by a factor that takes into account interindividual variability. They are thus the amounts considered sufficient for the maintenance of health in nearly all people (Passmore *et al.*, 1974). The one exception is dietary energy, for which an average requirement is specified on the grounds that excess is harmful. As human requirements for nutrients are quite variable this generous approach means that most individuals need considerably less than the recommended intake. The excess does no harm to the consumer; it merely makes a fair distribution of resources more difficult. Application of the recommendations in national planning results in estimates of dietary needs well above requirements (although shortfalls, wastage, and maldistribution must also be taken into account when planning food supplies).
3. If diets are to be assessed or devised in terms of their nutritional content, nutrient requirement tables need to be used in association

with food composition tables. These are extremely useful but by their nature can only provide averages that may differ quite widely from the amounts of nutrients in the actual food eaten. Davidson *et al.* (1975) list food composition tables for different countries and regions of the world.

NUTRIENT DENSITIES

The requirements of all nutrients, with the exception of energy, are expressed in terms of the weight needed each day. It would be useful to be able to categorize all the nutrient components of diets in one unit so that nutrient satisfactions can be compared and limiting nutrients identified. For the macronutrients that provide energy this is not difficult. For example, the British Department of Health and Social Security (1969) has calculated recommended intakes of dietary protein in terms of g/day on the assumption that 10% of the total energy in the diet should be provided by protein. A combined index of protein quantity and quality in the total diet is the net dietary protein : energy ratio that similarly expresses protein in units of energy (Platt *et al.,* 1961). Carbohydrates, fats, and alcohol can similarly be measured in energy units.

Recently the suggestion has been made to express other nutrients in terms of energy. This is done by calculating the ratio of the nutrient in the diet (or food) to that recommended per 1000 kcal (Hansen *et al.,* 1978). The resulting ratio is called the *nutrient density* or *nutritional quality index* of foods (Hansen *et al.,* 1979). The densities of each nutrient can be compared and limiting nutrients identified.

This concept of nutrient density should not be confused with that of *energy density,* which provides an index of bulk by expressing the amount of energy provided in a unit volume of cooked food, and is especially important in infant foods.

KNOWLEDGE ABOUT NUTRIENTS—THE REDUCTIONIST SLANT

Much of the current knowledge of protein, fat, carbohydrates, minerals, and energy balance was accumulated during the chemico–analytical era of nutrition, 1750–1900 (Schneider, 1958). The understanding that certain clinical syndromes were caused by a lack of specific micronutrients (soon called vitamins) came later during the biological era.

Scientific advances in understanding nutritional requirements have been mainly limited to discovery of dietary components. Only recently have the complex interrelationships between dietary elements themselves and with other factors such as infections, infestations, and endogenous metabolic processes—in phenylketonuria and diabetes mellitus for example—begun to be elucidated.

In the last decade the importance of fiber has been investigated more scientifically. Fiber is a dietary element but it plays no known active role in body metabolism. It was previously thought unnecessary, or even harmful. Now it is thought to protect against a large number of diseases that are common among the middle-aged and elderly in industrialized countries (Burkitt, 1975; Tudge, 1979). This has led to a renewed interest in traditional food patterns that include good amounts of fiber.

Hall (1977) expresses frustration regarding the limitations of recommended dietary allowances in public policy in the following statement:

> Nourishment is a very complex biological phenomenon in which one life form gives up its life and is converted into another life form. People eat food that at one time was either a living plant or animal and all the biological complexities resident in that living (ex) flesh are transformed into the unique complexities of the human consumer. Science has not had much success in defining precisely the nature of this phenomenon and except for an understanding of the principles, the details remain hazy. How then can we go about assigning numbers to something as nebulous as nourishment and should we even try? Might as well try to measure precisely the dimensions of a cumulous cloud.

Scientific knowledge of the nutrient basis of an optimal diet is a prime example of classical reductionist thinking in which understanding of the system is built up from knowledge of its isolated parts. However, folk understanding of food requirements usually is part of a romantic holistic view of the world and man's place in it. To quote Pirsig (1974) in a related context:

> Persons tend to think and feel exclusively in one mode or the other and in doing so tend to misunderstand and underestimate what the other mode is all about. But no one is willing to give up the truth as he sees it, and as far as I know, no one now living has any real reconciliation of these truths or modes. There is no point at which these visions of reality are unified . . . In terms of ultimate truth a dichotomy of this sort has little meaning but it is quite legitimate when one is operating within the classic mode used to discover or create a world of underlying form.

Clearly if we are to translate knowledge of nutrition into "an optimal diet" this dichotomy must be resolved—scientific knowledge and folk understanding of food and diet must be integrated.

FOOD REQUIREMENTS IN A HOLISTIC CONTEXT

The great adaptibility of mankind has enabled our species to remain at the top of the food chain in all ecological niches we inhabit. A risk of this "king-of-the-castle" position is that, like the peregrine falcon, we can become the repository of toxins that enter and pass up the food chain. Poisons present in the environment and ingested through the diet provide the major theme of this book. They will not be discussed specifically in this chapter except to draw attention to the need to avoid them in choosing an adequate diet.

Man's diet has evolved in relation to his environment over many thousands of years. Scientifically based knowledge of nutritional and dietary needs remains limited. Therefore it would seem wise for scientists to develop a concept of optimal nutrition that is based on man's sociocultural tradition. Metabolic, psychological, sociological, or economic understanding can thus be translated into advice through tradition rather than in conflict with it. Traditional dietary behavior is based on foods rather than nutrients.

Jelliffe (1967, 1968a) has developed a conceptual basis of diets based upon this approach and incorporating modern nutrition knowledge. He has pointed out that most, perhaps all, traditional cultures have a "cultural superfood" (Jelliffe, 1968b). This takes up much of the agricultural resources of the society, is present as a daily, or almost daily, component of the diet, and is associated with myths, rituals, and religious observances. Such cultural superfoods contain the largest proportion of the dietary energy and often of protein and some micronutrients as well. Heterogenous societies such as are found in the Caribbean or in large cities may not have one single cultural superfood but rely on a number of staples.

In all but hunter–gatherer societies, the cultural superfood falls into one of two nutritional categories: the cereals, and the starchy fruit, roots, and tubers (SFRT) group. Both groups supply most of their dietary energy through carbohydrates, particularly starch, and both provide around 1 kcal/g of cooked product. This estimate fluctuates somewhat between products, and quite largely with the method of cooking. Cereals provide around 10% of their energy as protein (the figure varies) and the SFRT group provides less.

In the cereal-based cultures worldwide, legumes form a regular part of the diet. This cereal–legume mix has considerable nutritional benefits, particularly as regards the well-known complementarity of amino acids between cereals and legumes (Aykroyd and Doughty, 1964).

Legumes are not of such benefit when taken with the SFRT group. However, this latter group is often eaten with food of animal origin—e.g., the relationship between cassava, where it is well established, and fish. SFRT-based habitats are less ecologically stable than are those dependent upon a cereal, because of the low protein content of the staple unless complemented with food from animals and because of the greater landspace required to produce an energy yield equivalent to cereals (Gurney, 1975).

THE MULTIMIX PRINCIPLE

The *multimix principle* developed by Jelliffe (1967, 1968a) is based on the observations outlined above, and represents an intercultural approach both to analyze diets and to establish a culture-specific basis for nutrition education.

In the multimix principle the staple (cultural superfood) is considered as the basis for the diet, with three additional add-on categories of legumes, foods from animals, and dark green leafy or yellow vegetables. Two-, three- or four-mix meals can be identified and quantities of each group specified.

This principle applies to almost all traditional societies. However, when food choices are available from foods that do not conform nutritionally to what might be expected from the name, appearance, taste, or color, this and all other food-grouping systems fall down.

The multimix principle is of equal value in planning family diets and those for infants and young children (Cameron and Hofvander, 1977). It is only the youngest baby, for whom the optimal diet is unsupplemented breast-milk, who does not need a mixture of foods (Jelliffe *et al.*, 1975; Jellife, 1977; Joint WHO/UNICEF Meeting on Infant and Young Child Feeding, 1979).

Douglas (1975) has developed a somewhat similar concept to Jelliffe's multimix principle, that is derived from an anthropological approach to British dietary patterns. This incorporates a center food (protein rich), a staple food (such as potatoes or bread), a trimming (such as vegetables), and a dressing (such as gravy). Hollingsworth (1977) has discussed the incorporation of this concept into public education.

There are three areas where the original multimix principle was incomplete. It did not take account of fruits and did not include sugar or fats and oils.

In many traditional societies fruits are readily eaten as snacks and particular reference to them may be unnecessary. However, the ascorbic acid (vitamin C) content of sharp-tasting fruits is important both in its own right as an antiscorbutic and in its role as an enhancer of iron absorption (Björn-Rasmussen and Hallberg, 1974). Orange-colored fruits, such as the mango, are important sources of carotene, the precurser of vitamin A. As urbanized societies worldwide do not have ready access to free fruits, it is important that they are included in the food budget.

Most settled traditional societies consume little fat or oil. Most industrialized societies consume more than is good for them. Yet fats or oils are important ingredients of weaning foods that would otherwise be too dilute in energy. A reduction in fat and oil (and sugar) consumption is important in the treatment of obesity. There is thus a value in considering fats and oils as a separate food group. Sugar need only be included so as to warn against it. These problems are dealt with in the English-speaking Caribbean (where undernutrition coexists with diabetes mellitus and obesity) by basing most nutrition education and diet and hospital-menu planning on the four food groups of the multimix principle (staples, legumes, food from animals, and dark green and yellow vegetables) with the addition of two extra groups comprising fruits, and fats and oils (Caribbean Food and Nutrition Institute, 1980).

FOOD SCIENCE AND INDUSTRIAL PROCESSING

The great advances that have taken place in food science this century have been accompanied by an enormous growth in the food-processing industry. This means that in industrialized countries there is now a wide variety of processed foods available on supermarket shelves. This has not been accompanied by an equivalent increase in the food components from which these products are constructed; in fact, they are rather limited. There are at least 60,000 different brand-name, processed foods in the U.S. alone. However, 75% of the total amount of food in the U.S. comes from only 100 generic items (Molitor, 1980). The food variety has been introduced by processing methods and by various additives and subtractives rather than from new foods. Consequently, the processed foods available in supermarkets in the industrialized and industrializing world, while they may or may not contain considerable amounts of the nutrients specified in recommended dietary allowances, do not fit well into any of the classifications commonly used by nutritionists.

Industrially processed foods may appear to be made up of ingredients that can be categorized into food groups while in reality they are not. For example, there are frozen pizzas sold in North America that contain artificial tomato paste, artificial cheese and simulated Italian sausage containing no meat. The food ingredients are derived from wheat, maize, soybean, and sugarcane sugar. Emulsifiers, acidulants, coal-tar dyes, artificial flavors, and various conditioning agents are added (Hall, 1978). It is not easy to categorize this pizza into any conventional classification system based on food groups.

Clearly, the statement that "even young children know that. . . . pizza represents cereal, dairy, and meat groups and often vegetables as well" (Harper, 1978) is no longer a valid defense of the basic four food classifications widely used in the U.S. Something else is needed.

DIETARY GUIDELINES IN RESPONSE TO CHANGE

Périssé *et al.* (1969) have shown how, with increasing affluence (or food availability), intakes of starch (and presumably fiber) tend to fall and those of fat and sugar to rise (and *vice versa*) (Fig. 1). Concern over these trends has stimulated considerable debate on food policy issues involving, for instance, legislation, crop and livestock goals, and pricing structure.

Industrialized and newly urbanized people in rich and poor countries are consuming diets and adopting lifestyles often very different from those of their grandparents. Consequently the constraints imposed by tradition become inappropriate and the information provided in tables of nutrient requirements is no longer sufficient. There are particular concerns, based on a growing body of knowledge, about the optimal intakes of cholesterol, fat, sucrose, alcohol, salt, and fiber. Bray (1980), writing for the American Society for Clinical Nutrition (1979), makes the following five recommendations, which are broadly in line with dietary guidelines adopted by other industrialized countries (Molitor, 1980):

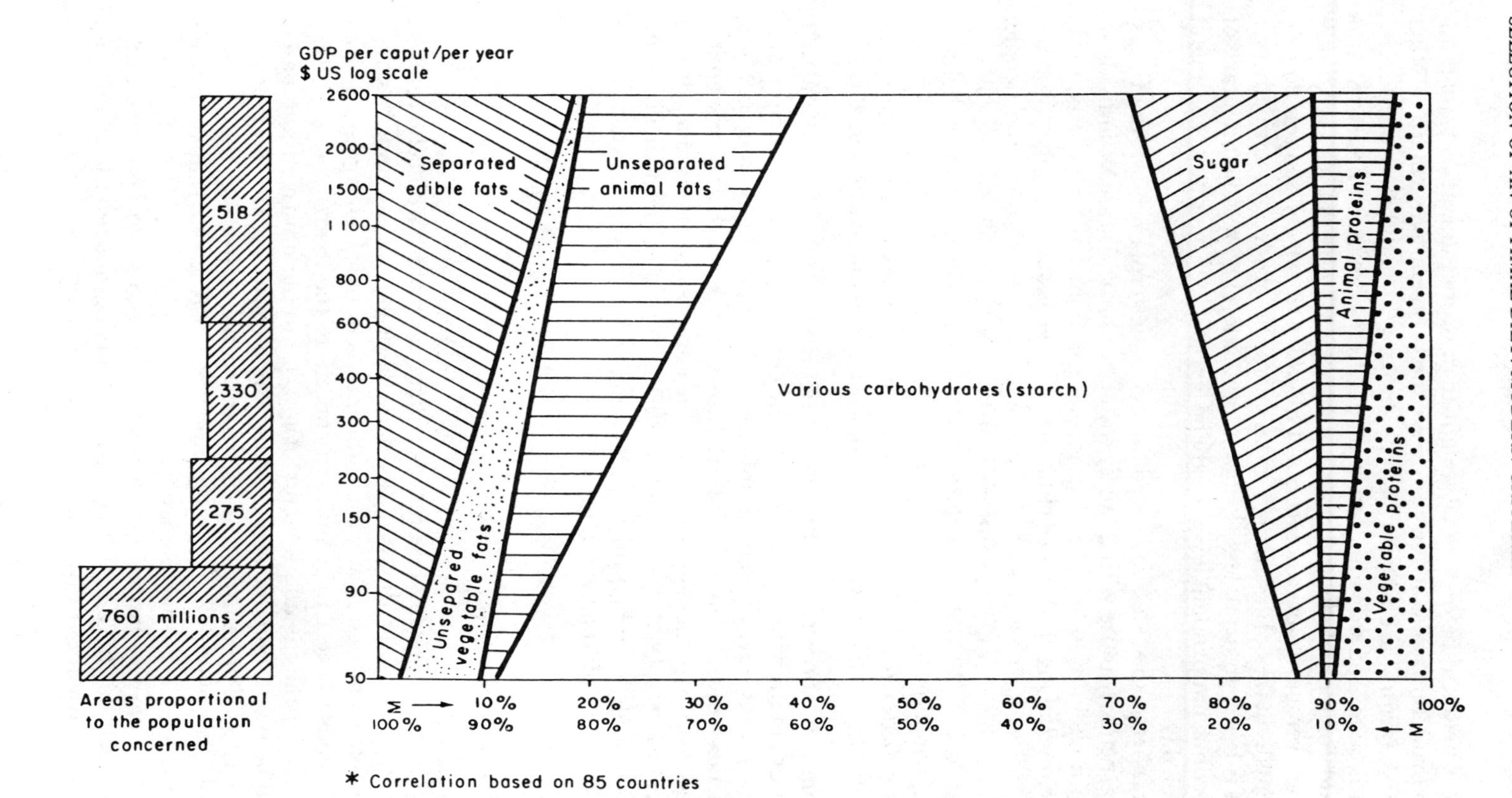

FIGURE 1. Calories derived from fats, carbohydrates, and proteins as percentage of total calories according to the income of the countries. From Périssé *et al.* (1969).

1. It is better to be of normal weight than overweight. If overweight, it is prudent to take action to correct the problem, particularly if other "risk factors" are present such as smoking or high blood pressure.
2. It is better for individuals in families where high blood pressure exists and for individuals with existing hypertension to eat less salt. Since the average American eats 5.8–11.6 g/day of salt (or sodium equivalents), intakes of less than 5 g/day might be prudent dietary advice.
3. It is better to drink less alcohol. Optimal levels would be no intake of alcohol, but prudent levels might be less than one beer or its equivalent per day.
4. It is better to eat less saturated fat and cholesterol for most Americans, particularly those with a family history of heart disease or individuals with cholesterol levels above normal (prudent levels would be those below 220 mg/day). Since cholesterol intake is 300–350 mg/day on the average, lower levels of intake would be prudent for some people. Since the average consumption of fat represents 35–40% of calories, intakes of 30–35% of calories as fat might be prudent in light of information now available.
5. It is better to eat less sugar, and more complex carbohydrates, particularly for those who are at risk for high frequency of dental caries.

Recommendations such as the above are helpful for general use and take some account of individual phenotypes. The genetic makeup of an individual can react with his environment to influence his susceptibility to certain diseases. Dietary control can probably reduce the risk for highly susceptible individuals with a certain genotype. An example is cardiovascular disease (Fredrickson *et al.*, 1973; Anonymous, 1980). The same individualized approach may well prove helpful in preventing, for example, cancers, hypertension, diabetes, and obesity (Hegsted, 1980).

NUTRIENT LABELING

Food-grouping systems become inadequate for making a diet choice when a high proportion of the diet is industrially fabricated. An attempt has been made in some countries to overcome this problem by requiring labels of all processed foods to specify the nutrient content of the food.

Nutrient labeling is undoubtedly of interest to the informed purchaser. It can also be used, sometimes misleadingly, in advertising. However, as anyone who has tried to construct diets based on the nutritional content of a number of available foods can attest, nutrient labels are extremely difficult to use in practice as a principle criteria for menu planning. The number of nutrients that have to be reconciled is simply too great to be manageable. Various suggestions have been made for simplifying nutrition labeling, e.g., by using the nutrient density approach described earlier.

A possible way to avoid risk might be to ensure that every processed food contained a full complement of nutrients balanced one to the other and with energy. It would only remain to "count the calories." This approach has three drawbacks: first, it would divorce food from tradition and culture; second, as all nutrient requirements are not known, relying on such a diet may result in unexpected deficiencies; and third, individual foods complement each other and need not each be complete meals in themselves.

"NATURAL" FOODS

An alternative approach to diet choice, which is gaining popularity in societies where processed foods are widely available, is to base the optimal diet on "natural" foods, and to avoid what is considered as excessive processing. For example, a representative study of food and nutritional concerns of the British people revealed four major issues: the importance of a "natural diet," belief in a "balanced diet," belief in the role of individual foods in causing or preventing illness, and belief in the concept of "an ideal diet for optimum health" (McKenzie, 1979). Bearing in mind the known and possible unknown risks associated with all stages in the production of many processed foods, there is some validity to these views.

An argument for "natural" foods that, while teleological in nature, is convincing, runs as follows. During his evolution as a hunter–gatherer, man and his precursers ate minimally-processed food, and even when he settled as a cultivator, processing was limited to roasting, baking, smoking, and later, when he had developed utensils, boiling. Man remains physiologically, if no longer culturally, adapted to a hunter–gatherer diet. Nourishment is a very complex biological phenomenon and modern science is only just beginning to explain and quantify the implications of this complexity in terms of optimum diets. Consequently reliance on tradition, tempered by scientific knowledge, is prudent. Some processing is necessary but some may have undesirable consequences, such as the addition of possibly harmful additives, as well as yet undiscovered ill-effects.

Foodways being so deeply imbedded in culture and personality, it is not surprising that the natural-food movement has associated with it religious and philosophical beliefs of varying rationality. The En-Trophy Institute has devised a logical conceptualization of food types entitled "Levels of Eating" (En-Trophy Institute, 1979). This takes the form of a two-way matrix. The vertical columns represent food types. These are fruits, vegetables, grain, meat, fish, eggs, dairy, oils and fats, legumes and seeds, nuts, drink, and sweeteners. This list could be simplified—into the multimix categories, for example.

The horizontal rows categorize four "quality levels" with quality dropping from raw (unprocessed) to the highly processed: from no additives to items rich in additives, from the natural to the imitation and from biological diversity to limited diversity.

The cells of the matrix contain illustrative examples of the food type at the quality level indicated. Users are expected to categorize the various food types in their habitual diet into the appropriate qualitative level and then, if necessary, to aim for a gradual upgrading.

Conceptually, the idea is useful for people in industrial cultures who wish to optimize their diets in the direction of "natural foods." It is unfortunate that nutritional science is not yet developed enough to facilitate a definitive scientific assessment of the approach. This should not disconcert us; nutrition has advanced rapidly this century and we can expect the trend to continue.

REFERENCES

American Society for Clinical Nutrition, 1979, The evidence relating six dietary factors to the nation's health, *Am. J. Clin. Nutr.* **32**:2620.

Anonymous, 1980, Dietary fiber, exercise and selected blood lipid constituents, *Nutr. Rev.* **38**:207.

Aykroyd, W. R., and Doughty, J., 1964, Legumes in human nutrition, *Nutrition Studies Series No. 19,* Food and Agriculture Organization, Rome.

Björn-Rasmussen, E., and Hallberg, L., 1974, Iron absorption from maize. Effect of ascorbic acid on iron absorption from maize supplemented with ferrous sulphate, *Nutr. Metab.* **16**:94.

Bray, G. A., 1980, Dietary guidelines: The shape of things to come, *J. Nutr. Educ.* **12**:97.

British Department of Health and Social Security, 1969, Recommended daily intakes of energy and nutrients for the UK.

Burgess, A., and Dean, R. F. A., (eds.), 1962, *Malnutrition and Food Habits: Report of an International and Interprofessional Conference,* Tavistock, London.

Burkitt, D. P., 1975, *Refined Carbohydrate Food and Disease: Some Implications of Dietary Fibre,* Academic Press, London.

Cameron, M., and Hofvander, Y., 1977, *Manual on Feeding Infants and Young Children,* 2nd ed., United Nations, New York.

Caribbean Food and Nutrition Institute, 1980, *Nutrition Education Handbook for Supervisors of Day Care Centres and Nursery Schools* (CFNI-NE4-80-1K), CFNI, Kingston, Jamaica.

Davidson, S., Passmore, R., Brock, J. F., and Truswell, A. S., 1979, *Human Nutrition and Dietetics,* 7th ed., pp. 163–165, Churchill Livingstone, Edinburgh.

De Garine, I., 1970, The social and cultural background of food habits in developing countries (traditional societies), in: *Food Cultism and Nutrition Quackery* (G. Blix, ed.), pp. 34–46, Almqvist and Wiksells, Uppsala.

Douglas, M., 1975, The sociology of bread, in: *Bread: Social, Nutritional and Agricultural Aspects of Wheaten Bread* (A. Spicer, ed.), Applied Science, London.

En-Trophy Institute, 1979, Strategy for Wellness, *En-Trophy* **2(6)**.

Fredrickson, D. S., Levy, R. I., Bonnell, M., and Ernst, N., 1973, *Dietary Management of Hyperlipoproteinemia. A Handbook for Physicians and Dietitians,* Pub. No. (NIH) 37, National Heart and Lung Institute, Department of Health Education and Welfare, Bethesda.

Gurney, J. M., 1975, Nutritional considerations concerning the staple foods of the English-speaking Caribbean, *Ecol. Food Nutr.* **4**:171.

Hall, R. H., 1977, The RDA's and Public Policy, *En-Trophy* **1(1)**.

Hall, R. H., 1978, Thirty years of *laissez-faire* in human nourishment, *En-Trophy* **1(3)**.

Hansen, R. G., Wyse, B. W., and Brown, G., 1978, Nutrient needs and their expression, *Food Technol.* **32(2)**:44.

Hansen, R. G., Wyse, B. W., and Brown, G., 1979, *Nutritional Quality Index of Foods,* Avi, Westport, Connecticut.

Harper, A. E., 1978, What are appropriate dietary guidelines? *Food Technol.* **32(9)**:48.

Hegsted, D. M., 1980, Dietary guidelines: Where do we go from here? *J. Nutr. Educ.* **12**:100.
Hollingsworth, D. F., 1977, Translating nutrition into diet, *Food Technol.* **31(2)**:38.
Jelliffe, D. B., 1967, Approaches to village-level infant feeding (1) Multimixes as weaning foods, *J. Trop. Pediat.* **13**:46–48.
Jelliffe, D. B., 1968a, *Infant Nutrition in the Subtropics and Tropics,* 2nd ed. (Monograph Series No. 29), World Health Organization, Geneva.
Jelliffe, D. B., 1968b, *Child Nutrition in Developing Countries,* U.S. Government Printing Office, Washington, D.C.
Jelliffe, D. B., 1977, Breast is best: Modern meanings, *N. Engl. J. Med.* **297**:912.
Jelliffe, D. B., and Bennett, F. J., 1961, Cultural and anthropological factors in infant and maternal nutrition, *Fed. Proc. Fed. Am. Soc. Exp. Biol.* **18**:185.
Jelliffe, D. B., Gurney, M., and Jelliffe, E. F. P., 1975, Unsupplemented human milk and the nutrition of the exterogestate fetus, in: *Proceedings of the 9th International Congress of Nutrition,* Mexico, 1972, Vol. 2, (A. Chavez, H. Bourges, and S. Basta, eds.), pp. 77–85, Karger, Basel.
Jerome, N. W., 1975, On determining food patterns of urban dwellers in contemporary United States society, in: *Gastronomy: The Anthropology of Food Habits* (M. L. Arnott, ed.), pp. 91–111, Mouton, The Hague.
Joint WHO/UNICEF Meeting on Infant and Young Child Feeding, 1979, World Health Organization, Geneva.
Maslow, A. H., 1970, *Motivation and Personality,* Harper and Row, New York.
McKenzie, J. C., 1979, The consumers' view of problems and priorities in nutrition, *Proc. Nutr. Soc.* **38**:219.
McNutt, K. W., and McNutt, D. R., 1978, *Nutrition and Food Choices,* Science Research Associates, Chicago.
Molitor, G. T. T., 1980, The food system in the 1980s, *J. Nutr. Educ.* **12**:103.
Montgomery, E., 1978, Anthropological contributions to the study of food-related cultural variability, in: *Progress in Human Nutrition,* Vol. 2 (S. Margen, ed.),pp. 42–56.
Morris, D., 1977, *Manwatching: A Field Guide to Human Behaviour,* Johnathan Cape, London.
Passmore, R., Nicol, B. M., and Narayana Rao, M., 1974, *Handbook of Human Nutritional Requirements* (Monograph Series No. 61), World Health Organization, Geneva.
Périssé, J., Sizaret, F., and François, P., 1969, The Effect of Income on the Structure of the Diet, *Nutr. Newsletter* **7**:1.
Pirsig, R. M., 1974, *Zen and the Art of Motorcycle Maintenance,* pp. 66,67, Corgi, Ealing.
Platt, B. S., Miller, D. S., and Payne, P. R., 1961, Protein values of human foods, in: *Recent Advances in Human Nutrition* (J. F. Brock, ed.), p. 360, Churchill, London.
Schneider, H. A., 1958, What has happened to nutrition? *Perspect. Biol. Med.* **1**:278.
Tudge, C., 1979, The ins and outs of roughage, *New Sci.* **82, No. 1160**:988–990.
Weiner, J. S., 1964, Nutritional ecology, in: *Human Biology, An Introduction to Human Evolution, Variation and Growth* (G. A. Harrison, J. S. Weiner, J. M. Tanner, and N. A. Barnicot, eds.), pp. 413–440, OUP, London.

INDEX